Hydroelasticity in Marine Technology, Faltinsen et al. (eds) © 1994 Balkema, Rotterdam, ISBN 90 5410 387 6

Table of contents

Hydroelasticity in Marine Technology, Faltinsen et al. (eds) © 1994 Balkema, Rotterdam, ISBN 90 5410 387 6

Preface

Hydroelasticity is important to many areas of marine technology, such as offshore and deep water systems, conventional and high-speed ships, floating airports, flexible containers of coated fabrics and fish farms. Analysis for design of such structures requires integration of hydrodynamics and structural mechanics and innovative use of theoretical and experimental techniques. Often there are experts on either hydrodynamics or structural mechanics while hydroelastic problems need to be solved through a complete understanding of both disciplines. Also when dealing with cases where the environmental oscillatory loads act quasi-statically on an elastic structure, increased interdisciplinary expertise in hydrodynamics and structural mechanics should be encouraged.

The purpose of the present conference has been to bring together experts on both hydrodynamics and structural dynamics from various fields of application and hence encourage exchange of experience from different types of hydroelastic phenomena.

One important application dealt with in these proceedings is the behaviour of rigid and flexible risers on floating production platforms. The effect of both oscillatory fluid motion and current are studied. Vortex-induced vibrations is one area of concern. Similar problems arise for cables and pipelines. Impact loads due to waves (slamming) are considered both for conventional and high-speed vessels. Both global and local hydroelastic effects are studied. The global effects are called whipping. The local hydroelastic effects due to slamming are most pronounced for wetdeck slamming. The global steady state hydroelastic oscillations of ships and Tension Leg Platforms (TLPs) are called springing and are dealt with in the proceedings. The importance of springing for ships increases with increasing speed and/or increasing ship length. Similar problems are of concern for very large floating structures like floating airports. Hydroelastic behaviour of membrane type of structures like the seal bag system of a Surface Effect Ship (SES) or flexible containers is another important topic in the proceedings.

NTH, March 1994

Odd Faltinsen

Risers

Hydroelasticity in Marine Technology, Faltinsen et al. (eds) © 1994 Balkema, Rotterdam, ISBN 90 5410 387 6

Prediction of vortex-induced oscillation of cylinders in oscillatory flow

P.W. Bearman, X.W. Lin & P.R. Mackwood

Department of Aeronautics, Imperial College, London, UK

ABSTRACT: The paper describes a method to predict the transverse response of a circular cylinder in oscillatory flow using model equations to represent the transverse force due to vortex shedding. A Van der Pol type non-linear oscillator equation was tried but although this seems to work well for unidirectional flows it predicts unrealistic responses in oscillatory flow. An alternative model equation due to Bearman et al (1984) has also been used. The equation is modified for a flexible cylinder by the introduction of hydrodynamic damping and inertia terms. The predictions show the correct form of response with the appropriate frequency composition. Two formulations are used: one where the vortex shedding pattern repeats itself every half cycle and one where a mirror image pattern occurs. Experimental evidence indicates that the resulting two types of response occur in practice.

1 INTRODUCTION

Compliant and floating structures and components such as risers respond under the action of hydrodynamic forces. Over the years a valuable database of hydrodynamic force coefficients for fixed structures has been built up through both experiments and the development of a number of computer prediction methods. However, as a result of response, interactions may occur between the motion and the loading and it is not always appropriate to design structures using loading coefficients derived for fixed structures. Techniques based on ideal flow diffraction theory can be used to predict the response of large volume structures. In some cases, such as cross-flow response of cylinders in waves, where the forcing is a result of vortex shedding there is little guidance available as to how to predict response.

The aim of research being carried out at Imperial College is to develop and test prediction procedures for the response of cylinders in flow regimes where viscous effects are known to be of prime importance. This work is greatly assisted by recent improved understanding of vortex shedding in oscillatory flows and the availability of carefully measured response data for circular cylinders, such as that by Bearman et al (1992). Under the majority of conditions, the response in the in-line direction can be fairly well described by using the relative motion form of Morison's Equation. Response in

the transverse direction is a direct result of vortex shedding and exhibits a number of complex characteristics. An interesting behaviour occurs whereby the response frequency of the cylinder synchronises with an integer multiple of the flow oscillation frequency. The cylinder oscillation frequency can be shifted to the nearest integer multiple frequency of the flow, above or below its natural frequency measured in still water. The higher the ratio of the cylinder natural frequency to the flow frequency the higher the Keulegan Carpenter number range for maximum transverse response. For circular cylinders free to move in two-dimensional motion and having low values of the ratio of cylinder frequency to flow frequency, experiments indicate that as Keulegan Carpenter number is increased from zero, there is likely to be a range of KC where the transverse response is larger than the in-line one. Naturally, at sufficiently high KC the in-line response was always found to be larger than the response in the transverse direction.

The objective of the work described in this paper was to develop a reliable prediction method for estimating response in the transverse direction, using an appropriate model equation for the transverse force. Selecting a suitable equation for the transverse force represents a major difficulty. A number of researchers have had considerable success in predicting the transverse response of a cylinder in a steady current flow by applying a nonlinear oscillator model coupled to the motion of the cylinder. A model of this type with a non-linear fluid damping term, similar to the non-linear damping term in Van der Pol's equation, was first applied to predict the response of a flexible cylinder by Hartlen and Currie (1970). They found that it works well in current flows and correctly predicts the build up of vortex shedding to a steady limit cycle and the locking-in of the vortex shedding frequency to the body frequency. In our investigation this model was applied to the case of a cylinder in oscillatory flow.

An alternative model equation to predict the time history of the transverse force on a cylinder in oscillatory flow has been proposed by Bearman et al (1984). This equation, termed the Quasi-Steady model, assumes that at any instant the transverse force is similar to that on a cylinder in a steady flow with the same velocity. As the velocity in an oscillatory flow increases or decreases, the vortex shedding frequency is predicted to change so as to keep the Strouhal number based on the instantaneous velocity constant. Bearman et al show that the model fits experimental data surprisingly well and reproduces the amplitude and frequency modulation seen in time histories of the transverse force. A further interesting observation is that Fourier analysis of the predicted force shows components at integer multiples of the flow oscillation frequency. Since the experiments of Bearman et al (1992) indicated responses at integer multiples of the flow frequency it was thought that this model showed considerable promise for the prediction of transverse response. Predictions will be compared with those obtained by using the non-linear oscillator equation. In addition, by comparing predictions with experiment, it was hoped that some further light might be shed on the physical processes that control the development of transverse forces.

2. PREDICTION OF TRANSVERSE RESPONSE

The equation of motion for the transverse direction is written as:

$$M_y \ddot{y} + C_y \dot{y} + K_y y = F_y(t) ,$$

where M_y, C_y and K_y are the structural mass, damping and stiffness. It is normal to express the structural damping as a coefficient, ξ_y, where $\xi_y = C_y / (2 M_y \omega_{ny})$. $F_y(t)$ is the total fluid force acting on the cylinder and contains contributions to hydrodynamic damping and added mass. Although the analysis can be extended to motion in two

dimensions all the results presented in this paper are for a cylinder free to move in only the transverse direction.

The cylinder natural frequency in air, or more strictly in vacuo, ω_{ny}, is given by $\omega_{ny} = 2\pi f_{ny} = \sqrt{K_y / M_y}$. The cylinder oscillation frequency in oscillating water, f_y, can be deduced from the motion equation and is given by:

$$f_y = \frac{1}{2\pi} \left(\frac{K_y}{M_y + \rho \forall C_A} \right)^{\frac{1}{2}} .$$

In this expression C_A is the added mass coefficient for a cylinder accelerating in a fluid and is related to the inertia coefficient as follows; $C_A = C_M - 1$. Its value is very difficult to predict because it is influenced by the formation and shedding of vortices.

In experiments it is common to measure the cylinder oscillation frequency in still water. If it is assumed that the cylinder is displaced by a small amount and released, then the added mass coefficient can be expected to be close to unity. Hence the cylinder oscillation frequency is given by:

$$f_{nw} = \frac{1}{2\pi} \left(\frac{K_y}{M_y + \rho \forall} \right)^{\frac{1}{2}} .$$

For response in the in-line direction, the response frequency is expected to lie between the still water frequency, f_{nw}, ($C_A = 1$), and the cylinder natural frequency ($C_A = 0$). It is not clear if the same bounds apply to the transverse response frequency. If the mass ratio, $M_y / \rho \forall$, is very much greater than 1 then the three frequencies defined above will be the same. However, in practical applications it is expected that this ratio will be in the range 1-10 and hence response frequency is likely to be affected by the value of the mass ratio.

3. NON-LINEAR OSCILLATOR MODEL

There is no model equation for the transverse force that is as universally accepted as Morison's equation is for the in-line force. However, following the successful prediction by Hartlen and Currie (1970) of the transverse response of a cylinder in a steady current flow using a non-linear oscillator model of the Van der Pol type, this approach is first investigated here. The model can be coupled to the motion of a cylinder by adding a forcing term to the right-hand-side of the oscillator equation. In steady flows, the equation has the form:

$$C_L - \alpha\omega_0 C_L + \frac{\gamma}{\omega_0}(C_L)^3 + \omega_0^2 C_L = bf(y,\dot{y},\ddot{y}),$$

where α, γ and b are constants; f is some function of the response, say y, $\dot{y}$ or $\ddot{y}$ or perhaps some combination of two or all three; and

$$\omega_0 = \frac{SU}{f_{ny}D},$$

where S is the Strouhal number, and f_{ny} the cylinder natural frequency. When the right hand side of the oscillator equation is made equal to zero it represents an equation for the transverse force on a fixed cylinder in a current. The combination of the linear and non-linear damping terms generate an oscillatory force with a self-limiting amplitude. By careful selection of the constants in the equation the amplitude and frequency of the predicted transverse force can be made similar to measured values. For a flexible cylinder Hartlen and Currie selected $b\dot{y}$ for the forcing term on the right-hand-side, and the transverse force appearing in the equation of motion is expressed as:

$$F_y(t) = \frac{1}{2}\rho AU^2 C_L.$$

They found that a coupling term of this form gave good agreement with experiment, including locking in of the vortex shedding frequency to the cylinder frequency.

In oscillatory flows, the water velocity varies in magnitude and direction. For a sinusoidally oscillating flow, the water velocity is taken as:

$$U = U_0 Sin(2\pi f_w t).$$

The equation for ω_0 is maintained the same as for steady flows. However, the original oscillator model equation has to be modified since ω_0 will be zero when flow reverses (i.e. when $U = 0$) and the non-linear damping term would then be infinite. A simple modification is introduced such that the

characteristics of the oscillator model are maintained and a solution can be obtained. The revised oscillator model equation is given by:

$$C_L - \alpha\omega_0^2 C_L + \gamma(C_L)^3 + \omega_0^2 C_L = bf(y,\dot{y},\ddot{y}),$$

where constants α, γ, b and the function, f, play a similar role to those appearing in the equation for steady flows.

In the original and modified equations, α and γ can be related to, C_{L0}, the magnitude of the transverse force on a fixed cylinder by (provided the value of α used is small):

$$C_{L0} = \left(\frac{4\alpha}{3\gamma}\right)^{\frac{1}{2}}.$$

For a fixed cylinder b can be put to zero and C_{L0} can be taken from experimental measurements.

A calibration process had to be carried out to obtain the variation of C_{L0} with KC such that the oscillator model equation would produce the same rms transverse force for a fixed cylinder in oscillating flows as that measured in experiments. Following Hartlen and Currie (1970), a value of α of 0.02 and a Strouhal number of 0.2 are used. Figure 1 shows C_{L0} versus KC number obtained from this calibration process together with the predicted and measured rms transverse force coefficients. The measured values are due to Obasaju et al (1988). The agreement between measured and predicted rms transverse force coefficients is excellent, but it should be noted that prediction is sensitive to the value of C_{L0} and that the variation of C_{L0} with KC behaves very oddly.

Time histories of transverse forces produced by the oscillator model equation at KC values equal to 10, 17.5, 26.2, 35 and 43.4 are displayed in figure 2. These time histories are very similar to those produced by the Quasi Steady model of Bearman et al (1984), particularly at the higher KCs, and show similar characteristics to those seen in experiments.

The coefficients obtained from the above calibration are then inserted into the oscillator equation and this equation is used, together with the equation of motion for the cylinder, to predict the transverse response. In the first computations, the oscillator model equation was uncoupled to the response, i.e. the right hand side of the equation was put equal to zero. These predictions were made

5

for a value of the ratio of the cylinder frequency in-still-water to the water oscillation frequency of 1.81. Comparing the predicted response with measurements, the computations were found to greatly over-predict displacement with large peaks occurring around KC values equal to 16, 32 and 48.

It seemed from these initial computations that some form of additional damping has to be introduced to limit the response. Hence, it was decided to include a Morison type term as a fluid reaction force to oppose the transverse force developed by the oscillator equation. In this revised approach, the transverse force is assumed to comprise three terms: a force generated by vortex shedding and modelled by $05\rho AU^2C_L$, a drag term, $-05\rho AC_D y|y|$, and an added mass term, $-\rho\forall C_A y$, i.e.

$$F_y(t) = 05\rho AU^2C_L - 05\rho AC_D y|y| - \rho\forall C_A y.$$

It should be noted that the drag and added mass coefficients (C_D, C_A) may not necessarily be the same as those appropriate for in-line response.

When the new force formulation is used with values of C_D of 1.8 and C_A of 1.0 the predictions still over-estimate the response over most of the KC range, as shown in figure 3. Large predicted responses can still be seen around KC equal to 16, 32 and 48. Figure 3 also shows that varying the added mass coeffcient with KC, according to values expected for planar oscillatory flow, produced little change. The addition of a coupling term on the right-hand-side of the oscillator model equation did not improve the predictions. The coupling terms studied included y, y, yy, $y|y|$, $y|y|$, combined with a range of positive and negative values of the constant b.

For all the predictions made, the response is under-predicted for KCs smaller than about 10 and greatly over-predicted at KCs around 16, 32, 48. The predicted response frequency in the KC range from 4 to 15 is usually equal to the flow frequency, while the response frequency observed in experiments is twice the flow frequency. The reasons for this discrepancy are not clearly understood. It was thought that, perhaps, the chosen Strouhal number of 0.2 is not appropriate for oscillatory flow at low KC. However, using other values of the Strouhal number did not improve the response frequency prediction. Since this approach did not seem to model adequately the

physics in oscillatory flow, further development was abandoned.

4. QUASI STEADY MODEL

In a second approach to modelling the transverse force, the Quasi Steady model proposed by Bearman et al (1984) was studied. The Morison type reaction force discussed above is retained and the model has been used in the following two forms:

$$F_y(t) = \frac{1}{2}\rho AU^2\dot{C}_L Cos(KC\cdot S[1 - Cos(2\pi t/T_w)] + \psi)$$
$$-\frac{1}{2}\rho AC_D y|y| - \rho\forall C_A y,$$

and

$$F_y(t) = \frac{1}{2}\rho AU|U|\dot{C}_L Cos(KC\cdot S[1 - Cos(2\pi t/T_w)] + \psi)$$
$$-\frac{1}{2}\rho AC_D y|y| - \rho\forall C_A y,$$

where $\dot{C}_L$ is a type of transverse force coefficient (appropriate to the Quasi Steady model) and ψ a phase angle. The values of these quantities were taken from Figure 1 of Bearman et al (1984). In the above two equations T_w is the period of water oscillation. The two models differ in the form of the first term: one uses U^2 and the other $U|U|$ for the velocity term in the vortex shedding force. When the flow reverses it is assumed that there are two possible vortex shedding patterns: either the previous vortex shedding pattern is repeated in the next half cycle or is 180° out of phase, i.e. a mirror image. Hence response will either be excited at predominately even or odd multiples of the flow oscillation frequency. The two model equations reproduce this behaviour but there is no guide as to when to use one equation in favour of the other, apart from recourse to experimental observations.

Predictions of response obtained using the two forms of the model equation have been compared with measurements made by Mackwood (1992). 12 cases were studied and they are listed in Table 1. For each value of KC, the mean value of $\dot{C}_L$ given in Figure 1 of Bearman et al (1984) was used together with ψ of 20° and a Strouhal number of 0.2. In the case of drag and inertia coefficients (C_D, C_A), the values for a fixed cylinder at β of 750

Table 1 Mass, Damping Coefficient and Frequencies

Case	f_{nw}/f_w	f_{ny}/f_w	M_y (Kg)	ξ_y	γ	D (m)	L (m)
1	1.70	1.94	4.205	7.0×10^{-4}	3.32	0.05	0.626
2	1.76	2.01	4.109	7.0×10^{-4}	3.32	0.05	0.626
3	1.79	2.04	4.135	7.0×10^{-4}	3.32	0.05	0.628
4	1.81	2.06	4.135	7.57×10^{-4}	3.32	0.05	0.628
5	2.01	2.29	4.135	7.0×10^{-4}	3.32	0.05	0.628
6	2.13	2.43	4.135	7.0×10^{-4}	3.32	0.05	0.628
7	2.47	2.82	4.135	7.0×10^{-4}	3.32	0.05	0.628
8	2.72	3.10	4.135	7.0×10^{-4}	3.32	0.05	0.628
9	3.00	3.42	4.135	7.0×10^{-4}	3.32	0.05	0.628
10	3.07	3.50	4.135	7.0×10^{-4}	3.32	0.05	0.628
11	3.50	3.99	4.135	7.0×10^{-4}	3.32	0.05	0.628
12	2.72	2.89	10.265	7.0×10^{-4}	8.32	0.05	0.628

Note: γ is the mass ratio defined by $M_y / \rho \forall$.

are used in the hope that they are close to those that are appropriate for transverse response. Unfortunately, there is no other guidance available from experiments on what values to use.

Figures 4 and 5 show the predicted rms responses of the first 11 cases in Table 1 using the Quasi Steady model with the U^2 flow velocity term. Interestingly ,the response at f_{nw}/f_w of 2.01 is not the largest and the responses at f_{nw}/f_w of 1.70, 1.76, 1.79 and 1.81 are very close to each other. For f_{nw}/f_w below 2.13, large responses occur in the upper range of KC, as can be seen in the figure. For f_{nw}/f_w higher than 2.13, generally smaller responses are predicted, and apart from the case of f_{nw}/f_w equal to 3.5, the peak responses at high KC are suppressed.

Using this form of the model it is observed that the response frequency is at even multiples of the flow oscillation frequency (in the experiments the flow frequency was 0.3003 Hz). A typical plot of predicted rms transverse response, divided by cylinder diameter, is shown in Figure 6 for f_{nw}/f_w of 1.81. There are three predominant peaks close to KC values of 19, 38 and 56. The variation of response frequency (in Hz) with KC is also shown and it is either at twice the flow frequency (0.6006Hz) or four times the flow frequency (1.2012Hz). It was found that the region of higher frequency expands with increase of f_{nw}/f_w . The full rms transverse force coefficient (defined as transverse force on the responding cylinder divided by $05_\rho A U_0^2$) and the force coefficient due only to

the Quasi Steady term (ie the force predicted for a fixed cylinder) are also plotted in figure 6 against KC. There is a substantial difference between the two force coefficients at low KC values.

Plotted in Figure 7, from top to bottom, are time histories for KC=12 and f_{nw}/f_w of 1.81 of the oscillatory flow velocity and predicted transverse response divided by the cylinder diameter and the transverse force coefficient. The responses are quite similar to those obtained from experiments at a similar KC and the same frequency ratio as shown in figure 8.

Root mean square transverse responses for the first 11 cases using the Quasi Steady model with the $U|U|$ term are shown in figures 9 and 10. For f_{nw}/f_w below 2.13, the responses are greatly reduced compared with those produced by the Quasi Steady model with the U^2 term, and the largest response is at f_{nw}/f_w of 2.13. The response peaks at high KC are not as pronounced as those obtained using the other form of the Quasi Steady equation. Figure 10 shows, for f_{nw}/f_w greater than 2.13, large responses are predicted in a band of KC around 44. This is in contrast to the results using the Quasi Steady model with the U^2 term.

For this form of the Quasi Steady model, figure 11 shows rms transverse response divided by the cylinder diameter, response frequency, the full rms transverse force coefficient and the transverse force coefficient due only to the Quasi Steady term for f_{nw}/f_w of 1.81. The response frequency is seen to be now at odd multiples of the flow oscillation

water frequency (0.3003 Hz). Again the regions of higher frequency were observed to expand with increasing f_{nw}/f_w. Figure 12 displays the predicted time histories of transverse response divided by the cylinder diameter and transverse force coefficient together with flow velocity for KC=12 and f_{nw}/f_w of 1.81. However, unlike for the previous form of the model, the response at this KC and frequency ratio is significantly different from that seen in experimental results plotted in figure 8.

It is interesting to see comparisons between the predicted and measured responses. Figures 13 to 14 are typical and they show the predicted rms responses, obtained by using the two forms of the Quasi Steady model, and the measured responses. It seems from comparisons such as these that the U^2 model is better at predicting the response at small KCs while the $U|U|$ model is better at predicting the response at high KCs.

The comparison between predicted and measured responses shown in figure 14 is particularly interesting. For this frequency ratio the measurements show a sudden large increase in response at KC around 13. Below 13 the measurements are closest to the prediction using the model with the $U|U|$ term and above 13 the form U^2 is closest. This suggests that at KC=13 the vortex shedding switches from a mirror image form to a repeating pattern.

To sum up, it seems that the two forms of the Quasi Steady model are complementary and both are needed to predict correctly the transverse response. However, we have no means of knowing, apart from experimental observations, when to apply the U^2 model and when to use the $U|U|$ model.

5. CONCLUSIONS

Predictions of transverse force for a fixed cylinder obtained with the Van der Pol oscillator model appear to be in reasonably good agreement with measurements. However, predictions of transverse response are generally in poor agreement with experiment. There are spurious response peaks occurring in the high KC ranges around 32 and 48 and even at low KC the responses tend to be much higher than those measured. Closer agreement with experiment was obtained if a Morison type force, based on cylinder velocity and acceleration, was assumed to act in opposition to the vortex shedding transverse force. Various forms were used for a coupling term that links the non-linear oscillator equation to the response of the cylinder. While some performed better than others, no satisfactory term was found. On the other hand, the transverse response predictions made using the Quasi Steady transverse force model, with the inclusion of the transverse Morison force, appear much more realistic. The predictions show the correct form of response and the appropriate frequency composition. To fit the data, it is found that two formulations of the Quasi Steady model are required, one with a U^2 term and the other with a $U|U|$ term. This takes account of different patterns of shedding when the flow reverses, either a repeated pattern or a mirror image one. Unfortunately there is no way to predict from the model, apart from comparing with experimental data, when it is appropriate to use one form rather than the other. Nevertheless, by using both forms it is possible to assess the range of possible values of response.

6. ACKNOWLEDGEMENTS

The wprk described in this paper was part of a programme of research entitled Behaviour of Fixed and Compliant Offshore Structures. This research was promoted by MTD Ltd and sponsored by SERC, government and the offshore industry.

7. REFERENCES

BEARMAN, P.W., GRAHAM, J.M.R. and OBASAJU, E.D. 1984. A Model Equation for the Transverse Forces on Cylinders in Oscillatory Flows *Applied Ocean Research* Vol. 6 No 3 pp 166-172

BEARMAN, P.W., LIN, X.W. and MACKWOOD, P.R. 1992. Measurement and Prediction of Response of Circular Cylinders in Oscillating Flow *Proceedings of the Sixth International BOSS Conference,* London, UK. May 1992

HARTLEN, R.T. and CURRIE, I.G. 1970. Lift-Oscillator Model of Vortex-Induced Vibration *J. of Engineering Mechanics Division ASCE,* pp 577-591

MACKWOOD, P.R. (1992) Private communication

OBASAJU, E.D., BEARMAN, P.W. and GRAHAM, J.M.R. 1988. A Study of Forces, Circulation and Vortex Patterns around a Circular Cylinder in Oscillating Flow *J. Fluid Mechanics, Vol 196, pp 467-494*

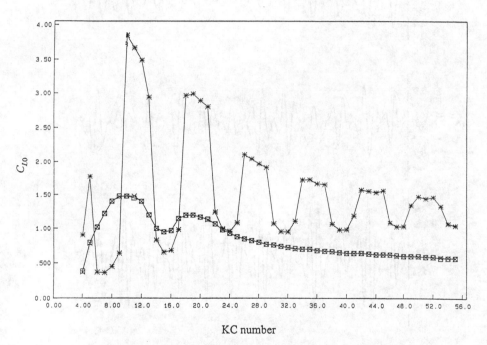

KC number

Figure 1   Calibrated C_{L0} for Van der Pol's Equation versus KC

—*— C_{L0},   —□— Predicted rms Lift Force,   —×— Measured rms Lift force

9

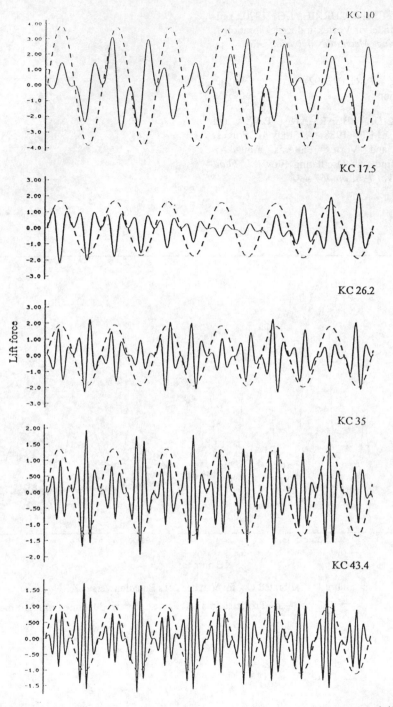

Figure 2   The Time Histories of Lift Force at KC 10.0, 17.5, 26.2, 35.0 and 43.4
————   Water velocity,   ———— Lift force

10

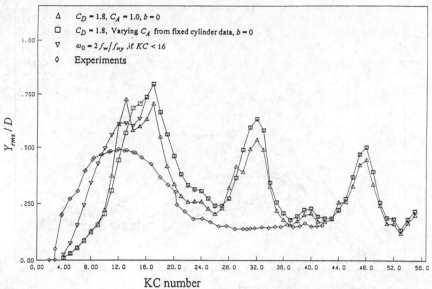

Figure 3　Transverse Responses predicted using the Oscillator Model of Van der Pol type
Compared with Experiments for f_{nw} / f_w of 1.81

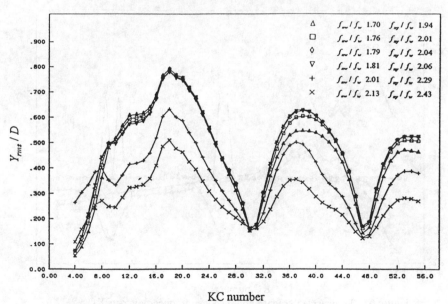

Figure 4　Transverse Response predicted by the U^2 Quasi-steady Model

11

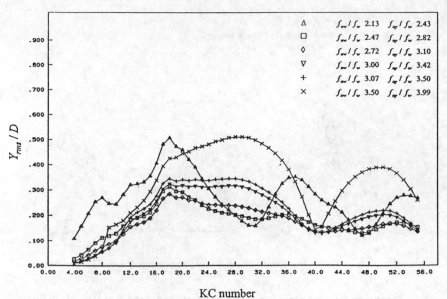

Figure 5　Transverse Response predicted by the U^2 Quasi-steady Model

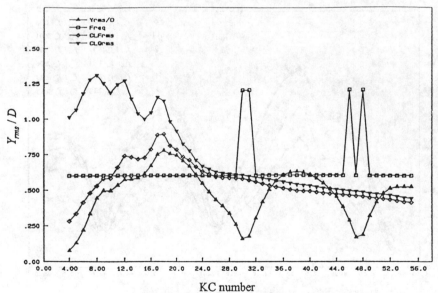

Figure 6　Transverse Response and response frequency versus KC
at f_{nw} / f_w of 1.81, U^2 Quasi-steady Model

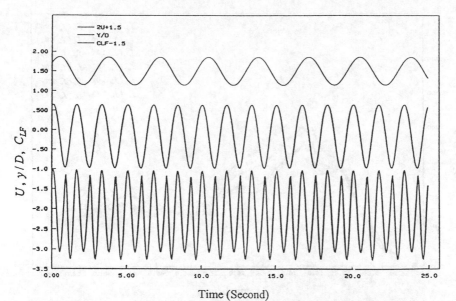

Figure 7   Time histories of Transverse force coefficient predicted by
the U^2 Quasi-steady Model at KC 12 and f_{nw} / f_w 1.81.
From top to bottom Water velocity, transverse response and lift force coefficient

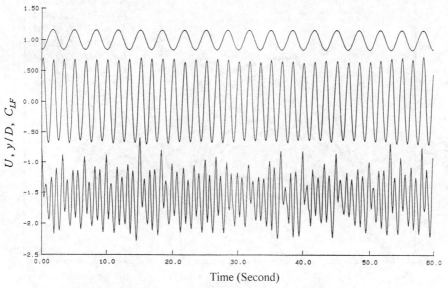

Figure 8   The reconstructed Transverse force from measured
transverse response at KC 10.62 and $f_{nw} / f_w = 1.81$;
From top to bottom, Water Velocity, Transverse Response, Transverse Force

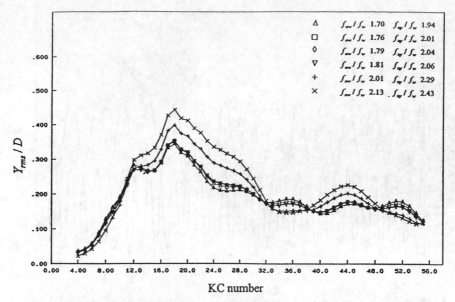

Figure 9   Transverse Response predicted by the $U|U|$ Quasi-steady Model

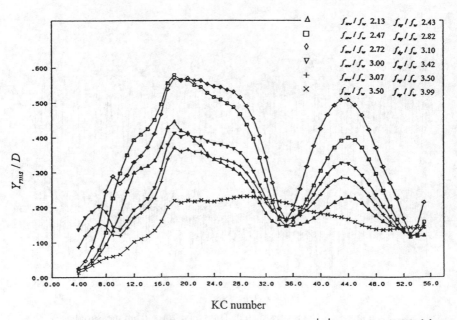

Figure 10   Transverse Response predicted by the $U|U|$ Quasi-steady Model

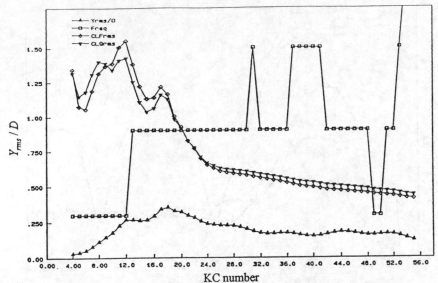

Figure 11 Transverse Response and response frequency versus KC
at f_{nw} / f_w of 1.81, $U|U|$ Quasi-steady Model

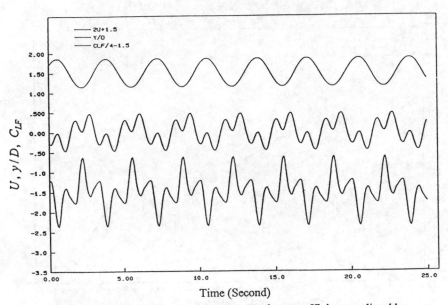

Figure 12 Time Histories of Transverse force coefficient predicted by
the $U|U|$ Quasi-steady model at KC 12 and f_{nw} / f_w 1.81.

From top to bottom Water velocity, transverse response and transverse force coefficient

15

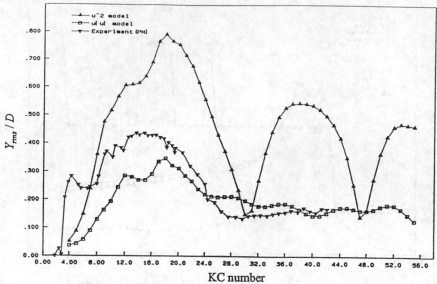

Figure 13 Comparison of Response predicted by the two forms of the Quasi-steady Model and measured in experiments at f_{nw} / f_w 1.70 with mass ratio 3.32

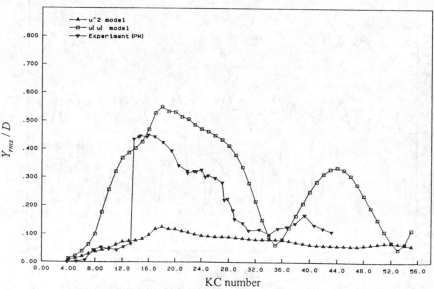

Figure 14   Comparison of Response predicted by the two forms of the Quasi-steady Model and measured in experiments at f_{nw} / f_w 272 with mass ratio 8.32

16

Hydroelasticity in Marine Technology, Faltinsen et al. (eds) © 1994 Balkema, Rotterdam, ISBN 90 5410 387 6

Hydroelasticity of tensioned buoyant platform tethers at low and negative tension

M. H. Patel, G. J. Lyons & T. Wilne
Department of Mechanical Engineering, University College London, UK

ABSTRACT

Analytical and numerical investigations into the tethers of tensioned buoyant platforms (TBPs) at very low to negative tensions indicate that TBPs with relatively low mean tether tension compared to contemporary designs may be feasible.

This paper presents the preliminary results of an experimental investigation aimed at verifying some of the predictions obtained from these theoretical investigations. This has been done by carrying out tests at model scale into the physical phenomena encountered with submerged TBP tethers cycling through low to negative tensions and back again. Measurements are presented of tether tension and of transient transverse vibrations at points along the tether through the reducing and increasing tension parts of the cycle.

The paper also gives a brief review of the analysis used to predict tether behaviour under low and negative tensions.

1. INTRODUCTION

A conventional TBP is a structure floating at the ocean surface and connected to a sea bed foundation by vertical tethers which are kept in tension by excess buoyancy over weight of the surface platform. The use of TBPs as alternatives to oil production from fixed or freely floating offshore structures is now well established because a TBP possesses a combination of desirable features - Figure 1 and Table 1 gives details of three applications. The absence of a fixed rigid structure from sea surface to sea bed reduces cost while still giving the platform excellent station keeping characteristics. The tensioned tethers radically change platform sea keeping characteristics so as to yield very much reduced heave, roll and pitch motions. The vertical tethers are, however, compliant to horizontal wind, current and wave forces permitting the platform to respond in surge, sway and yaw. This compliance is desirable since it ensures that these horizontal environmental loads are not fully reacted by the platform structure.

The merits of TBPs outlined above have to be set against a number of drawbacks which can limit cost effective operational applications. Firstly, the surface platform of a TBP must have sufficient excess buoyancy to keep the tethers at tensions that do not reduce to zero at low tidal and high wave conditions. These high tensions, in their turn, impose structural requirements to the TBP and tether/hull connections which requires an increase in structural buoyancy. The consequently bigger volume of the structure also induces high dynamic forces in the tethers themselves and require foundations capable of withstanding large vertical forces. The need to maintain excess buoyancy also makes a TBP more sensitive to payload growth and platform mass distribution. This leads to a vicious design circle which limits platform payload, performance and cost and is impossible to avoid with conventional TBP design criteria.

A substantial amount of research work and the results of design engineering have been carried out on the hydrodynamics and structural response of TBPs. These started with the earliest work of Horton (1975) through to more recent studies - by Rainey (1977), Faltinsen et al (1982), Jefferys and Patel (1982) and Lyons and Patel (1984) among many others. Patel and Witz (1991) give an overview of these. The result of typical design engineering studies are reported by Mercier (1982) and there have been several patents granted for tethers and tether tensioning systems - see Vickers Ltd (1981) and Conoco Inc (1982).

Consideration of TBPs operating at low mean tensions requires consideration of two operating configurations - the first is associated with the platform operating at significantly lower mean tensions than is usual in present day designs. However, the corollary to this is that under extreme conditions of tide, waves and platform loading, the tethers may well be subjected to transient periods of compression.

Consider the case of tethers operating at reduced tension first. Such tethers are simultaneously subjected to two sources of dynamic excitation to which they become increasingly susceptible as their mean tension reduces. The first is due to motion at the tether top induced by horizontal platform oscillations. This is a forcing excitation whereas the second source of dynamic excitation is due to changes in tether axial force due to variations in vertical force on the platform; called parametric excitation. Much research work has been carried out on each of these excitation sources applied separately to TBP tethers. The horizontal top end forcing excitation problem has been extensively researched - Patel and Witz (1991) give a review and Jefferys and Patel (1982) presents one typical investigation. The parametric excitation problem - that of determining lateral vibration of the tether induced by time-varying axial force - has been investigated by Hsu (1975) and Strickland and Mason (1981) among others.

However, recent studies (Park and Patel, 1992) have shown that as the mean tension is reduced, time varying axial forces play an important role in increasing lateral motions of the tethers but that this lateral motion is still limited by quadratic damping in water. In these cases of reduced tension, acceptable platform and tether motions are achievable in normal and moderately severe operating conditions.

However, during combined occurrences of extreme wave, tide and variable weight,

individual tether tensions will reduce to zero and become compressive for part of a wave cycle. This transient tension loss can also be defined as a dynamic pulse buckling phenomenon - the application of an axial compressive force larger than the static Euler buckling value for a short period of time.

In dynamic pulse buckling, it is known that a slender column can survive (with acceptable stress levels), a sudden compressive axial load much greater than the static Euler load as long as the load duration is short enough. Pulse buckling has been extensively studied in other branches of engineering - for example in the design of aircraft landing struts, ballistic missiles and shock or blast resistant structures. An early researcher in this field, Meier (1945), investigated time deflection relations when an axial force was very rapidly applied to a nearly straight bar. Following Meier's work, the dynamic pulse buckling problem for slender bar type structures was extensively studied by many researchers including Gerard and Becker (1952), Sevin (1960), Lindberg (1965) and Holzer and Eubanks (1969). Lindberg and Florence, in their book of 1987, present an integrated treatment of dynamic pulse buckling including much of the research carried out in the last two decades. However, all of this work has been carried out for structures in air with pure axial forcing and no lateral added mass or quadratic damping effects. Brekke and Gardner (1988) report one of the few numerical investigations into this tension loss case.

This report gives a brief review of the analytical techniques and results used for consideration of the reduced tension and tether pulse buckling configuration. This is followed by the results of an experimental study in a tether tank to study the behaviour of tethers in air and water when subjected to a transient tension loss.

2. TETHER BEHAVIOUR AT LOW TENSION

This section of the paper gives a brief overview of tether analysis for the cases of a tether operating at reduced tension and one which is subjected to transient compression.

2.1 The Reduced Tension Case

The response of a tether at reduced tension is considered when the tether is subjected to combined lateral and axial vibrations - the former from platform horizontal motions in waves and the latter from changes in vertical wave force acting on the platform.

A generalised TBP tether is idealised as a straight, simply supported column of uniform cross section. Figure 2 shows the idealised configuration under combined excitation and gives the notation being used. The governing equation of lateral motion for the tether is written as:

$$m\frac{\partial^2 y}{\partial t^2} + EI\frac{\partial^4 y}{\partial x^4} - (T_0 - S\cos\omega t)\frac{\partial^2 y}{\partial x^2} + B_v\left|\frac{\partial y}{\partial t}\right|\frac{\partial y}{\partial t} = 0$$

(1)

where m is the total physical plus hydrodynamic added mass per unit length of the tether, EI is the structure flexural rigidity, T_0 is constant axial tension, S is the time-varying axial force amplitude, ω is the angular frequency of the time-varying axial force (parametric excitation) and $B_v = 0.5\ C_D\ \rho_w\ d_o$, where C_D is a drag coefficient, d_o is the outer diameter of the tether and ρ_w is sea water density. The tension T_0 is taken to be constant here so as to develop analytic results demonstrating the effects of combined forcing on tethers with tensions that are high compared to their self weight. Following, the same argument, the time-varying axial force is assumed to be sinusoidal. However the formulation of the equation does include the nonlinear drag induced damping force.

The partial differential equation (1) is reduced to an ordinary non-linear differential equation and solved numerically by using a combination of Romberg's method and a fourth order Runge-Kutta method. Patel and Park (1992) give further details.

The effects of this combined excitation on a TBP with reduced tensions are illustrated here by using the tether of the Snorre platform as an example. Table 1 gives a summary of the physical data used whereas Figure 2 presents the lateral vibration of the tethers for various tensions from a normal operation tension of 8 MN to a reduced tension of 6 MN.

Figure 2(a) displays tether response at the operating tension condition and with a time-varying axial force whose amplitude is taken to be the same as this tension. As can be seen in Figure 2(a), even at the operating pre-tension conditions, there exists some small amplitude, high frequency response due to the axial force excitation. It is interesting to speculate that this response looks remarkably like that observed in model tests and characterised as tether ringing. However, the link, if any, between these combined excitation responses and tether ringing is still to be proven or disproven. As the pretension of the Snorre TBP tethers reduces, its high frequency response increases. When the pretension is reduced to 88% of operating tension (Figure 2(c)), lateral displacements are somewhat large and from $T_{Top} = 6.5$ MN downwards the Snorre tethers are close to failure.

It should be noted that this is only an example case for illustration with the axial force amplitude for Figures 2(a) to (e) constant and very large - that is equal to the pre-tension at the design operating condition.

2.2 The Transient Tension Loss Case

The second approach to tethers operating at low tension is to consider the transient tension loss or pulse buckling case expected to occur at extremes of tidal, wave and platform variable weight conditions.

Using the idealisation of Figure 1, the governing equation for lateral motion can be written as:

$$m\frac{\partial^2 y}{\partial t^2} + EI\frac{\partial^4 y}{\partial x^4} + P_c\frac{\partial^2(y+y_i)}{\partial x^2} + B_v\left|\frac{\partial y}{\partial t}\right|\frac{\partial y}{\partial t} = 0 \quad (2)$$

where EI is the flexural rigidity, P_c is a compressive axial force assumed to be constant here, m_s is the physical mass per unit length, m_a is the added mass per unit length, B_v is the damping coefficient and y_i is the initial deflection of the tether.

Equation (2) is non-dimensionalised using axial and lateral characteristic lengths and solved analytically for the case of zero damping and numerically using the fourth order Runge-Kutta for the square law damping case. Park (1992) gives further details of the solution.

It is shown below that tether buckling due to transient tension loss can occur at higher bending modes with an amplification function arising out of the analysis giving the preferred mode of buckling.

Taking

$$m = m_s + m_a$$

where m_s and m_a are the physical and hydrodynamic masses per unit length respectively and using the parameters

$$k^2 = \frac{P_c}{EI} \quad r^2 = \frac{I}{A_s} \quad c^2 = \frac{A_s E}{m_s} \quad e^2 = 1 + \frac{m_a}{m_s}$$

equation (2) can be reduced to

$$\frac{e^2}{r^2 c^2}\frac{\partial^2 y}{\partial t^2} + \frac{\partial^4 y}{\partial x^4} + k^2\frac{\partial^2 y}{\partial x^2} + \frac{B_v}{EI}\left|\frac{\partial y}{\partial t}\right|\frac{\partial y}{\partial t} = -k^2\frac{\partial^2 y_i}{\partial x^2} \tag{3}$$

Since the wavelengths of interest in pulse buckling are short, in equation (3), the tether length is non-dimensionalised with respect to characteristic length ξ and lateral deflections are non-dimensionalised with respect to r - the radius of gyration of the cross-sectional second moment of area about a diameter.

Then using the variables,

$$w = y/r \qquad \xi = kx$$

$$\tau = (k^2 r c\, t)/e \quad \beta = (B_v\, r)/(m_s + m_a) \tag{4}$$

where $B_v = 0.5\,\rho_w\, C_d\, d_0$ and $r = \sqrt{I/A_s}$.

Equation (3) becomes

$$\frac{\partial^2 w}{\partial \tau^2} + \frac{\partial^4 w}{\partial \xi^4} + \frac{\partial^2 w}{\partial \xi^2} + \beta\left|\frac{\partial w}{\partial \tau}\right|\frac{\partial w}{\partial \tau} = -\frac{\partial^2 w_i}{\partial \xi^2} \tag{5}$$

The boundary conditions are

$$w = \frac{\partial^2 w}{\partial \xi^2} = 0 \quad \text{at} \ \xi = 0 \ \text{and} \ \xi = l\,(= kL)$$

For the above boundary conditions, the solution of Equation (5) can be put in the form

$$w(\xi,\tau) = \sum_{n=1}^{\infty} g_n(\tau)\sin\eta\xi \tag{6}$$

$$w_i(\xi) = \sum_{n=1}^{\infty} a_n \sin\eta\xi \tag{7}$$

where an axial wave number η is introduced by

$$\eta = n\pi/l \tag{8}$$

Substituting Equations (6) and (7) into Equation (5) gives

$$\frac{d^2 g_n}{d\tau^2} + \varepsilon\left|\frac{dg_n}{d\tau}\right|\frac{dg_n}{d\tau} + \eta^2(1-\eta^2)g_n = \eta^2 a_n \tag{9}$$

where $|\sin\eta\xi|\sin\eta\xi = \dfrac{8}{3\pi}\sin\eta\xi$ is used and thus

$$\varepsilon = 8\beta/3\pi \tag{10}$$

The solution of Equation (9) becomes hyperbolic for $\eta < 1$ and trigonometric for $\eta > 1$. The trigonometric form gives stable motion, that is, the compressive axial force is less than the Euler buckling load. Thus only the hyperbolic form is considered here. For such a condition, Equation (9) becomes

$$\frac{d^2 g_n}{d\tau^2} + \varepsilon\left|\frac{dg_n}{d\tau}\right|\frac{dg_n}{d\tau} - \eta^2(1-\eta^2)g_n = \eta^2 a_n \tag{11}$$

The non-linear damping term makes it difficult to obtain a closed form exact solution of Equation (11). The closed form approximate solution is obtained in the following form (see Park 1992)

For $\quad 0 < \tau < \dfrac{1}{\sqrt{\eta^2(1-\eta^2)}}l_n\dfrac{2(1-\eta^2)}{\varepsilon a_n}$

$$g_n(\eta,\tau) = \frac{a_n}{(1-\eta^2-2\varepsilon a_n)}$$

$$\left\{\cosh\sqrt{1-\frac{2\varepsilon a_n}{1-\eta^2}}\cdot\sqrt{\eta^2(1-\eta^2)}\ \tau - 1\right\} \tag{12}$$

For $\quad \tau > \dfrac{1}{\sqrt{\eta^2(1-\eta^2)}}l_n\dfrac{2(1-\eta^2)}{\varepsilon a_n}$

$$g_n(\eta,\tau) = \frac{1}{2\varepsilon} - \frac{a_n}{1-\eta^2} + \frac{1}{4\varepsilon}$$

$$\left\{ \sqrt{\eta^2(1-\eta^2)}\,\tau - l_n \frac{2(1-\eta^2)}{\varepsilon a_n} + 2\sqrt{0.5 + \frac{\varepsilon a_n}{(1-\eta^2)}} \right\}^2 \quad (13)$$

Equation (11) can also be solved numerically by using the fourth-order Runge-Kutta method.

When the hydrodynamic damping is not considered, i.e., $\varepsilon = 0$, the solution can be given in the following form by putting $\varepsilon = 0$ in Equation (12)

$$g_n(\eta,\tau) = \frac{a_n}{1-\eta^2}(\cosh \eta\sqrt{1-\eta^2}\,\tau - 1) \quad (14)$$

The ratio between the Fourier coefficient, a_n, of the initial deflection and the coefficient $g_n(\tau)$ of the response deflection is called the amplification function and is expressed as follows :

$$G_n(\tau) = \frac{g_n(\tau)}{a_n} =$$

$$\frac{1}{1-\eta^2}\left[\cosh(\eta\sqrt{1-\eta^2}\tau - 1)\right] \quad (15)$$

The preferred mode, i.e., the most amplified mode can be obtained by differentiating the amplification function with regard to wave number and setting the result to zero. It can be assumed that hydrodynamic damping does not affect the preferred mode of buckling since the preferred mode is decided before the hydrodynamic damping forces come into effect. In other words, the preferred mode of buckling can be obtained by finding η which satisfies the following condition

$$\frac{dG}{d(\eta^2)} = 0 \quad (16)$$

Then the resulting preferred mode is taken as

$$\eta_{cr} = 1/\sqrt{2} \quad (17)$$

or $\eta_{cr} = \dfrac{1}{\sqrt{2}}\sqrt{\dfrac{\tau}{\tau-2}}$ for a better estimate

The corresponding wavelength is found from

$$\eta_p\,\xi_p = 2\pi \quad \text{or} \quad \xi_p \equiv 2\pi\sqrt{2} \quad (18)$$

The corresponding wave length in dimensional units is obtained from Equation (4) as

$$x_p = 8.88/k \quad (19)$$

Now for a transient tension loss, the recovery of tension will straighten the tether back again without significant lateral deflections and bending moment. Nevertheless, this excitation of high modes will permit allowable compressive tension for significant durations.

This feature is quantified by using the date of Table 1 for Hutton and the additional data of tether inner diameter d_i=0.076, yield stress=795MPa and hydrodynamic drag coefficient C_d=1.1 to derive the pulse buckling envelope of Figure 4. This plots maximum compressive load on the vertical axis with the maximum duration for which the load can persist without exceeding allowable stresses in the tether. For example, if the tension loss duration is less than 1.50 seconds, the maximum allowable axial compressible load can be up to 0.80 MN.

However, if the compressive part of the transient tension loss were to persist for some time, the high mode tether deformations are expected to reform in to lower modes with the buckling phenomenon tending towards the static case.

The experiments described in this paper are an attempt to quantify this transient loss mechanism and to obtain experimental data to compare with the above analysis.

3. EXPERIMENTAL WORK

A specially built tether and riser testing facility at University College London was used for the tests.

This is a 4.5 m deep still water tank with a 1.5 m square base (see Figure 5). A hydraulically driven carriage is mounted on the top with tensioning equipment for the tether model. The tether mounting position on the carriage is capable of being moved in a plane through any pre-defined horizontal and vertical motions to simulate the movement of a tether mounting point at the platform base. Glass viewing panels down one side of the tank perpendicular to the direction of carriage motion permit measurement of in-line displacement of the tether model by using a television based

21

monitoring system. Additionally, there is a video camera which may be positioned at any level within the tank to record transverse vibrations of the tether. The tether model can also, of course, be instrumented with strain gauges and tension load cells at the upper and lower mounting points.

Three test models were made from CAB (Cellulose Acetyl Bulyrate) material with outer diameters of 9.5 mm, 12.7 mm and 19.1 mm to model three tethers with different flexural stiffness values. No attempt was made at this stage to model the stiffnesses and mass per unit length of specific tethers - nor was an attempt made to model the tether according to Froude or Reynolds number scaling. It should also be noted that the pinned connectors used at either end of the tether models limited their movement in one vertical plane only.

Table 2 gives physical data for the tethers whereas Figure 6 shows the monitoring points used for the tests. These consisted of three stations A, B and D along the tether at which transverse displacements were measured simultaneously with the tether bottom tension. A further measurement station C was also used by moving the lateral displacement sensing camera from position A to C.

The tether tank carriage was set up to generate vertical motions only along the vertical centre line of the tether and moved through a square wave cycle to simulate a sudden tension loss followed by a constant compressive loading and a rapid recovery of tension. It should be noted that the compressive loading was maintained for some period of time so as to observe resultant tether behaviour before the tension recovery.

The subsequent lateral displacements of the tether were recorded by utilising the television monitoring system, which followed a point source of reflected light on the tethers as they moved thus producing a voltage output proportional to the movement. The point light source was achieved by fixing a ball bearing to the tether and illuminating it from above, which resulted in a sufficiently defined point reflection. The voltage output for each monitoring position was fed into a data acquisition package in a PC based system, from which the necessary calibration and scaling of the displacements could be achieved using MATLAB.

It was also necessary to measure the compressive force exerted on the tether. This was recorded using a load cell fixed at the bottom of the tether which sent a signal to the data acquisition package via a signal conditioning unit.

Initially a relatively slow compression of the tether was tested to make sure that any residual out-of-straightness in the tether models did not induce unusual behaviour. These tests showed for all three tethers that they buckled in the conventional way with increasing deflection at their first mode. Following these preliminary tests, a series of measurements were made for the 3 tethers at 3 levels of compression in the square wave and for 2 different displacement camera measuring positions. Each of these tests were repeated 10 times to give a total of 180 measurements.

4.   RESULTS AND DISCUSSION

The experimental results presented here are the first to emerge from a work programme being carried out to compare measured tether response under transient loss conditions with predictions from the theory presented in section 2.

Each of the three tethers were subjected to a severe rapid tension loss, this being maintained for a relatively long period before tension was restored equally severely. This "square wave" application of tension is shown in Figure 7(a) through the output of a tension cell at the tether lower end. Table 2 presents the tether and test data showing how a transient compressive load of 11.25 N (measured at the tether base) was applied to each of the three tethers. This corresponded to a ratio R of transient compressive load to static Euler buckling load of from 2.5 to 33.1.

All of the experiments described in this paper are in air with comparative measurements in water being part of a follow on study.

Figures 7(b), (c) and (d) present typical time histories of lateral response for each of three tethers of 9.5, 12.7 and 19.1 mm diameter respectively. The responses of each of the tethers following the tension loss are characterised by low frequency vibrations followed by higher frequency 'springing' during the tension recovery phase. The latter is an expected phenomenon during the straightening and retensioning of the tethers. During the period just after tension loss the three tethers exhibit low frequency vibrations at frequencies of 0.90 Hz, 1.22 Hz and 2.47 Hz for the 9.5 mm, 12.7 mm and 19.1 mm diameters respectively.

Tether behaviour following tension loss is more effectively presented by Figure 8, 9 and 10 for the 9.5, 12.7 and 19.1 mm diameter tethers. In each of these figures, following tension loss, the tether instantaneous behaviour is presented. Straight lines are used between the four measurement camera positions to give the tether vibration shapes being measured.

Taking Figure 8 for illustration, the left hand part of the figure starts from the straight taut tether prior to tension loss - denoted by 1. The lines labelled 2 show successive tether shapes for the first 30 sampling points ie. from 0 up to 0.15s of the motion. In the same way, the lines labelled 3 denote the first 60 sampled points (0 up to 0.30s) and 4 denotes the first 90 points (0 up to 0.40s). Looking from left to right then gives a view of tether shapes developing in time immediately after the tension loss. Figures 9 and 10 use a similar presentation.

Comparisons of tether behaviour after tension loss in Figures 8, 9 and 10 demonstrates a higher mode response. The 9.5 mm tether presents a response between first and second mode but closer to the first mode whereas the 12.7 mm tether exhibits a distinct second mode response. Turning finally to the 19.1 mm stiffest tether, it shows a clear third mode response. This feature demonstrates the presence of specific preferred modes which are amplified during the transient tension loss phenomenon. A detailed comparison of experimental results with the theory of section 2 remains to be done. However, these preliminary results show that the amplification of preferred modes does occur with the tethers of high stiffness giving the higher mode response. A part of this comparison is the extension of the theory of section 2 to account for variation of axial compression with length for a tether with significant self-weight compared to the applied load.

The comparison of experiment and theory will also investigate further the potential amplitude reducing effects of viscous drag. Further work is also planned to investigate the nature and severity of the tension recovery "springing" phenomenon and the way in which its magnitude is influenced by viscous forces and axial tether, platform and foundation stiffnesses.

REFERENCES

Brekke J N and Gardner T N, 1988, 'Analysis of brief tension loss in TLP tethers', Journal of Offshore Mechanics and Arctic Engineering, Vol 110, pp 43-47.

Conoco Inc Patent Application, 1982, 'Controlling the tension in platform supporting tension legs', European Patent Office, Patent Application, EP 0 072 692 A2, Filed 16/8/82.

Faltinsen O I, Fylling I J, van Hooft R, Teigin P S, 1982, 'Thoeretical and experimental investigations of tension leg platform behaviour, BOSS82, Vol 1, p411-423.

Gerard G and Becker H, 1952, 'Column behaviour under conditions of impact', Journal of Aeronautical Sciences, Vol 19, pp 58-62 and 65.

Holzer S M and Eubanks R A, 'Stability of columns subjected to impulsive loading', Journal of the Engineering Mechanics Division, ASCE, Vol 95, No EM4.

Horton E, 1975, 'Tension leg platform prototype completes Pacific coast test', Ocean Industry, pp245-247, September.

Hsu C S, 1975, 'The response of a parametrically excited handing string in fluid', Journal of Sound and Vibration, 39, p305-316.

Jefferys E R and Patel M H, 1982, 'On the dynamics of taut mooring systems', J of Engineering Structures, Vol 4, p37-43.

Lindberg H E and Florence A L, 1987, 'Dynamic Pulse Buckling', Martinus Nijhoff Publishers.

Lindberg H E, 1965, 'Impact buckling of a thin bar', Journal of Applied Mechanics, Transactions ASME, pp 315-22.

Lyons G J and Patel M H, l984, 'Comparisons of theory with model test data for tensioned buoyant platforms, Transactions of the ASME, J of Energy Resources Technology, Vol 106, December, pp426-436.

Meier J H, 1945, 'On the dynamics of elastic buckling', Journal of the Aeronautical Sciences, Vol 12, pp 433-40.

Mercier J A, 1982, 'Evolution of tension leg platform technology', Proc of 3rd International Conference on the Behaviour of Offshore Structures, Massachusetts Institute of Technology.

Park H I and Patel M H, 1992. 'Dynamics of tension leg platform tethers at low tension, part 2 - combined excitation', Proceedings of the 1992 Offshore Mechanics and Arctic Engineering Conference, Calgary.

Park H I, 1992, 'Dynamic stability and vibrations of slender marine structures at low tension', PhD Thesis, University of London, England.

Patel M H and Witz J A, 1991, 'Compliant offshore structures', Ch 6 Tensioned Buoyant Platforms, Butterworth-Heinemann, pp137-188.

Rainey R C T, 1977, 'The dynamics of tethered platforms', Paper No 6, Spring Meeting, Royal Institution of Naval Architects.

Sevin E, 1960, 'On the elastic bending of columns due to dynamic axial forces including effects of axial inertia', Journal of Applied Mechanics, Transactions ASME, pp 125-31.

Strickland G E and Mason A B, 1981, 'Parametric response of TLP tendons - theoretical and numerical analysis', Proceedings of Offshore Technology Conference 3, p45-54.

Vickers Ltd Patent Application, 1981, 'Method of forming a vertical stressed mooring tether in a floating oil platform', UK Patent Application No: GB 2068 321 A, Filed 23/1/81.

Table 1: Nominal Data for Hutton, Jolliet and Snorre TBP Tethers.

TLP Tether System Dimensions	Hutton	Snorre	Jolliet
Length (m)	114.0	310.0	510.0
Top tension (N)	8.0×10^6	15.7×10^6	3.75×10^6
Flexural rigidity ($N\ m^2$)	5.29×10^7	14.57×10^8	3.22×10^8
Outer diameter (m)	0.26	0.812	0.6
Dry mass (kg/m^3)	472	726.3	276

Table 2: Model Tether Parameters

Outer Diameter mm	Inner Diameter mm	EI Nm^2	Static Euler Buckling Load N	Ratio R
9.5	2.9	0.57	0.34	33.1
12.7	6.0	1.73	1.02	11.0
19.1	12.3	7.71	4.53	2.5

Tether length = 4.1m
Tether material density = 1440 kg/m3
Tether material Youngs Modulus = 1.43 GPa
Applied transient compressive load (at tether base) = 11.25 N
Ratio R = Transient compressive load/Static Euler buckling load

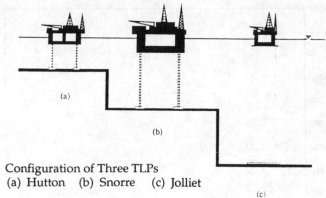

Figure 1: Configuration of Three TLPs
(a) Hutton   (b) Snorre   (c) Jolliet

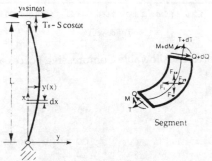

Figure 2: Tether Model under Combined Excitation Condition

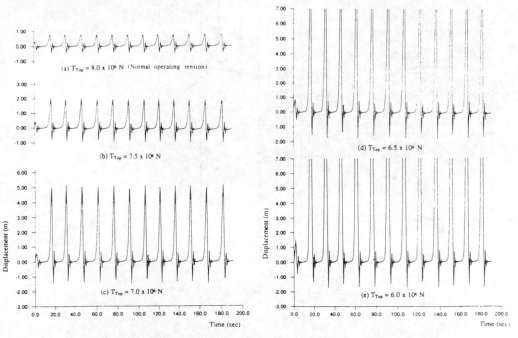

Figure 3: Time Histories of Lateral Displacement at the Mid-Point of Snorre Tether
Subjected to Combined Excitation

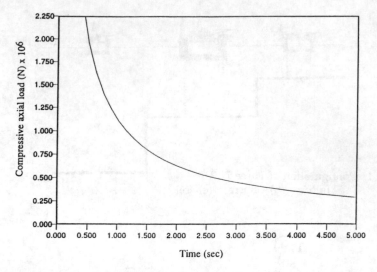

Figure 4: Relationship between Allowable Compressive Forces and Duration Time for Hutton Tether

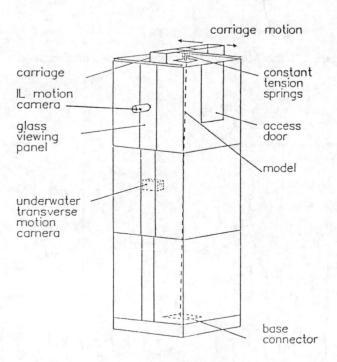

Figure 5: UCL Tether Testing Tank - Schematic

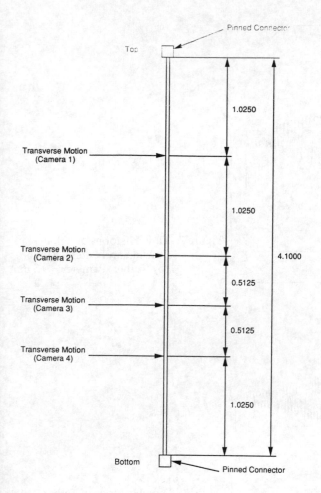

Figure 6: Monitoring Points on Model

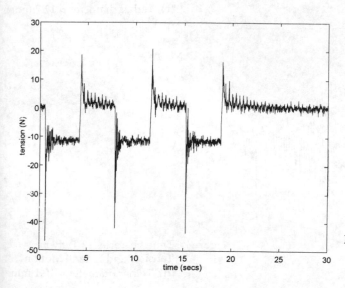

Figure 7: Time Histories of Tether
Tension and Lateral Motion
(a) Tension Response

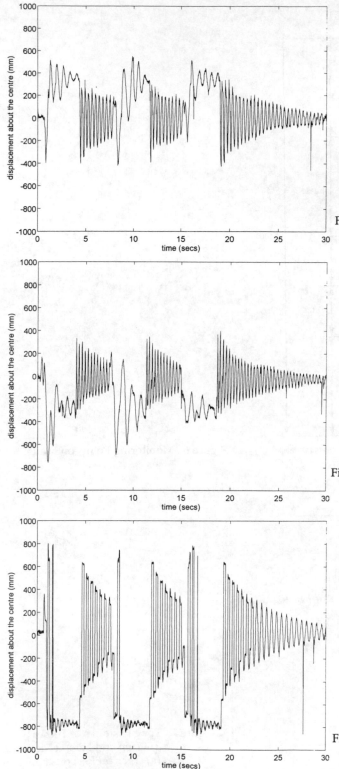

Figure 7: Time Histories of Tether
Tension and Lateral Motion
(b) Tether diameter = 9.5 mm

Figure 7: Time Histories of Tether
Tension and Lateral Motion
(c) Tether diameter = 12.7 mm

Figure 7: Time Histories of Tether
Tension and Lateral Motion
(d) Tether diameter = 19.1 mm

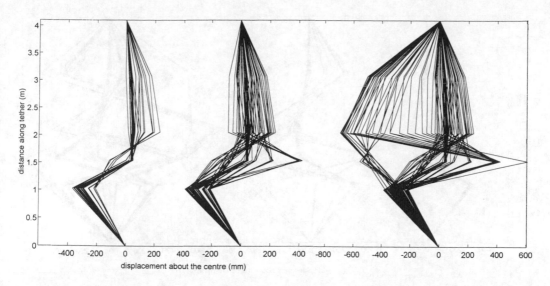

Figure 8: Tether Tension Loss Response - 9.5 mm Tether

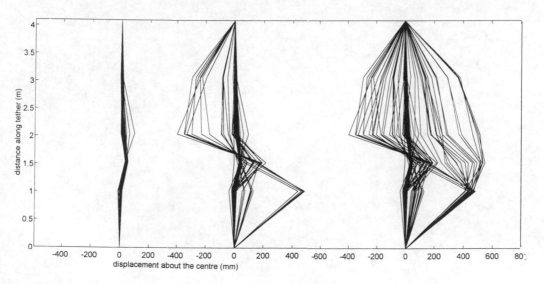

Figure 9: Tether Tension Loss Response - 12.7 mm Tether

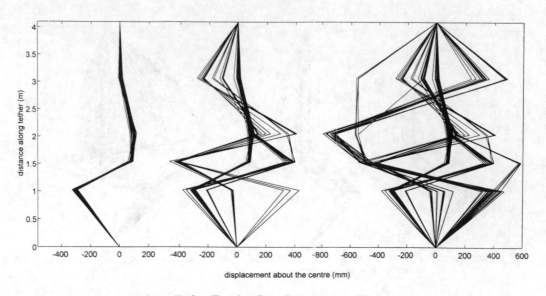

Figure 10: Tether Tension Loss Response - 19.1 mm Tether

Vortex-induced vibrations in a sheared flow: A new predictive method

M.S.Triantafyllou & R.Gopalkrishnan
Department of Ocean Engineering, Massachusetts Institute of Technology, Cambridge, Mass., USA

M.A.Grosenbaugh
Woods Hole Oceanographic Institution, Mass., USA

ABSTRACT: It is shown that a linear hydrodynamic damping term is an intrinsic feature of the vortex-induced vibrations of slender cylinders in the lock-in regime. The damping coefficient can be directly evaluated from experimental measurements of the force acting on a section of the cylinder forced to move in a uniform flow.

1 INTRODUCTION

Elastically mounted cylinders and long, flexible cylinders undergo vortex-induced vibrations when placed normal to a flow. The amplitude of this process is self-limiting with a maximum value approximately equal to one to two cylinder diameters. Laboratory experiments have been conducted to measure forces on rigid cylinders that are oscillated at a specific amplitude and frequency transversely to a uniform flow. These tests confirm that there is power input into the cylinder vibrations at small amplitudes of motion, for frequencies close to the Strouhal frequency of natural vortex formation, whereas there is dissipation for larger amplitudes (King 1977, Staubli 1983, Bearman 1984). When a cylinder oscillates with a frequency that is within a narrow range about the Strouhal frequency, the vortex formation process synchronizes with the motion of the cylinder in what is called a condition of *lock-in*. Under lock-in conditions, a vibrating cylinder is subject to a significantly increased drag force, up to three or more times higher than that of a stationary cylinder.

Hartlen and Currie (1970) and several other authors (Bearman 1984) used the van der Pol oscillator to represent qualitatively the self-limiting nature of the excitation lift force. Alternatively, the concept of energy balance has been incorporated in models to predict the response of long, flexible cylinders (Vandiver 1988). In these models, the direction of energy transfer is dependent on whether or not the motion of the cylinder at a particular point is correlated with the vortex formation process. Energy is assumed to be transferred from the fluid to the cylinder at points where the motion is synchronized with vortex shedding (lock-in condition), while it is assumed that the cylinder loses energy to the fluid at points where the motion is not correlated with vortex shedding. At these points, the loss of energy is modelled by an "equivalent" hydrodynamic damping term, calculated by linearizing the quadratic drag force acting on the cylinder.

In this paper we show, on the basis of experimental results, that the vortex-induced lift force depends on the amplitude of the cylinder vibration in a manner which is characteristic of a process containing a purely linear damping term. This provides a direct way of evaluating the damping coefficient using laboratory measurements.

In §2, we derive a model of hydrodynamic damping for a simple harmonic response. The model is extended to the more general case of a narrow-band response in §3. In §4, we incorporate the model into a simple scheme that predicts the vortex-induced response of flexibly mounted, rigid cylinders and long, flexible cylinders. We compare the predictions to previously published experimental results.

2 HARMONIC RESPONSE

The force acting on a section of a slender circular cylinder of diameter d, vibrating harmonically in the transverse direction relative to an oncoming flow of velocity V, is a nonlinear function of the

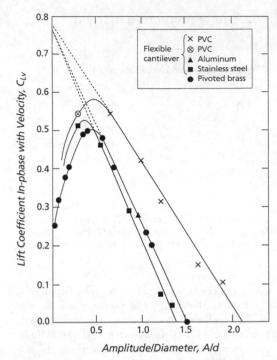

Figure 1. The lift coefficient in-phase with velocity as a function of the amplitude-to-diameter ratio (King 1977).

motion. We denote the lift force per unit span that is in-phase with the velocity by $L_v(t)$ and its amplitude by L_o and proceed to nondimensionalize it to obtain the lift coefficient that is in-phase with velocity, C_{L_v}:

$$C_{L_v} = \frac{L_o}{\frac{1}{2}\rho dV^2} \qquad (1)$$

where ρ denotes the fluid density.

Figure 1 shows a plot of the measured coefficient, C_{L_v}, of rigid pivoted cylinders versus the vibration amplitude for a nondimensional oscillation frequency close to the Strouhal number (King 1977). Power input occurs when C_{L_v} is positive and dissipation occurs when C_{L_v} is negative. Except for small amplitudes, when the vortex formation process is not well correlated along the span of the cylinder, there is clearly a linear relation between the lift coefficient and the amplitude of motion. Over a range of practical interest, typically for amplitude to diameter ratios higher than 0.4, the curve can be approximated by a straight line with negative slope. This is a distinct feature of

a nonlinear process that contains a term that can be modelled through a linear damping coefficient. The damping coefficient can be directly obtained from the slope of the line. A simple representation of the lift force curve is

$$C_{L_v} = C_o - \lambda\frac{A}{d} \qquad (2)$$

where $\frac{A}{d}$ is the amplitude-to-diameter ratio and C_o and λ are curve-fitting constants. Equation (2) is accurate if the cylinder vibrates with an amplitude that is larger than the threshold amplitude.

It should be noted that the methodology to replace an amplitude-dependent excitation by equivalent motion-dependent terms has been applied before in other fields to analyze nonlinear phenomena, such as the value of wave-drift damping estimated from second order wave forces (Faltinsen 1990).

Experiments have been conducted in the MIT Testing Tank Facility on rigid circular cylinders of diameter 2.54 cm and span 30 cm, forced to move in a prescribed motion transversely to a flow with constant velocity V (Gopalkrishnan 1992). Figure 2 shows several plots of the coefficient, C_{L_v}, for harmonic motion versus the amplitude-to-diameter ratio for various imposed frequencies, which are near the frequency of the maximum in-phase lift coefficient. It is interesting to note that the slope of the various curves varies little over a range of nondimensional frequencies, $\frac{fd}{V}$, where f is the oscillation frequency in Hertz.

For a purely sinusoidal force at circular frequency $\omega = 2\pi f$, equation 2 provides the component of the lift force in phase with velocity:

$$L_v(t) = \frac{1}{2}\rho dV^2 \left(C_o - \lambda\frac{A}{d}\right)\sin\omega t \qquad (3)$$

If we define

$$L_e(t) = \left(\frac{1}{2}\rho dV^2\right)C_o\sin\omega t \qquad (4)$$

$$b_h = \left(\frac{1}{2}\rho dV^2\right)\frac{\lambda}{\omega d} \qquad (5)$$

$$v(t) = \omega A\sin\omega t \qquad (6)$$

where $v(t)$ is the cylinder velocity, we can write equation 3 more simply as

$$L_v(t) = L_e(t) - b_h v(t) \qquad (7)$$

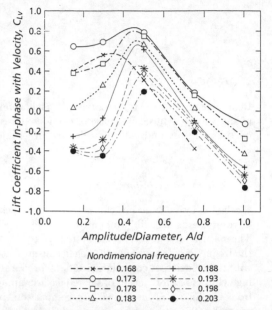

Figure 2. Experimental measurements (Gopalkrishnan 1992) of the lift coefficient in-phase with velocity as a function of the amplitude-to-diameter ratio for various values of the nondimensional frequency parameter $\frac{fd}{V}$.

In equation 7, the lift force that is in-phase with the cylinder velocity is decomposed into two parts, one of which is a pure excitation force $L_e(t)$ and a second term which is a linear damping force $b_h v(t)$. Equation 5 can be used to obtain a direct estimate of the hydrodynamic damping coefficient b_h, once the curve-fitting constant λ is determined from experimental data. We define the hydrodynamic damping ratio, ζ_h, as:

$$\zeta_h = \frac{b_h}{2m\omega} = \frac{\rho}{\rho_c} \frac{1}{4\pi^3} \frac{\lambda}{St^2} \qquad (8)$$

where m is the mass per unit length of the cylinder and ρ_c is the cylinder density. We have assumed that the harmonic motion is at the Strouhal frequency $\omega_v = 2\pi f_v$ and write the Strouhal number as $St = \frac{f_v d}{V}$. Then, taking a metallic cylinder with specific density equal to 5.0, specifying a Strouhal number of $St = 0.17$, and calculating $\lambda = 1.36$ from the experimental data in figure 2, we find that $\zeta_h = 0.076$ (7.6% of critical).

Both the damping coefficient, b_h, and the pure excitation force in the direction of the velocity, L_e, are independent of amplitude. The excitation force, however, is phase-correlated with the velocity. This is important in any numerical calcula-

tion, as shown in the sequel. For most applications in water, the structural damping is small in comparison to hydrodynamic damping and may be neglected. In air, the structural damping is significant and may be added directly to b_h.

3 NARROW-BAND RESPONSE

Lock-in of a flexibly mounted cylinder, or a flexible structure is usually characterized by a narrow-band response with characteristic beating oscillations. We can extend heuristically the derivation of §2 to apply to these cases when the response is not harmonic. For example, a three-dimensional plot of the lift coefficient in phase with velocity as function of the amplitude-to-diameter ratio and the frequency of oscillation can be constructed from figure 2. Such plots have been provided by Staubli (1983) and Gopalkrishnan (1992). The lift force that is in-phase with the velocity can be then represented then by the following, more general equation

$$C_{L_v}(\omega) = H(\omega)C_o - \Lambda(\omega)\lambda\frac{A}{d} \qquad (9)$$

which is similar to equation 2, but includes the frequency dependence ω in the curve-fitting parameters H and Λ. The functional form of the curve-fitting parameters is determined from experimental data, and C_o and λ are as defined before. Because of the similarity in the the shapes of the curves in figure 2, we conclude that $\Lambda(\omega)$ is very nearly constant over a narrow frequency range and is equal to one. This results in considerable simplification for use in numerical calculations.

The accuracy of equation 9 is subject to the same amplitude-threshold considerations as those related to equation 2. In addition, we note that in order for equation 9 to apply to a multi-frequency response, linearity must be assumed. This is not correct for other parameters relevant to vortex-induced oscillations. For example, the excitation force for monochromatic excitation at nonlock-in conditions contains an additional component at the Strouhal frequency. Thus, in order to employ equation 9, we must assume that the dominant force component has a frequency content that is within a narrow band around a specific frequency ω, and that the response is still within the lock-in regime. Triantafyllou and Karniadakis (1989) have shown numerically and Gopalkrishnan (1992) and Gopalkrishnan *et al.* (1992) have shown experimentally that, in the case of a beating oscillation, i.e., an oscillation consisting of two (or three equidistant) sinusoidal components, the harmonic results can be used to predict the lift force in a multi-frequency response, provided that the frequencies are sufficiently close together and within the lock-in regime. However, the drag force in a multi-frequency response can not be calculated on the basis of harmonic results.

Hence, assuming that harmonic data can be used to calculate the lift force in a narrow-band response, we can write H and Λ as integro-differential operators in the time domain. The damping, as expressed by the term containing Λ, is still linear and resembles, in form, the well-known, frequency-dependent damping of floating bodies in the presence of a free surface (Faltinsen 1990).

If the cylinder motion has a narrow-band spectrum about $\omega = \omega_v$, then we can exploit the fact that the slope of the lift-force coefficient in-phase with the velocity for a given imposed amplitude appears to be nearly frequency-independent (figure 2), and we can write the time-dependent lift force approximately as

$$L_v(t) = \left[\frac{v(t)}{\hat{v}(t)}\right] L_{eo} - b_h v(t) \qquad (10)$$

where $\hat{v}(t)$ is the slowly varying envelope of $v(t)$. The damping coefficient, b_h, is given by equation 5 with $\omega = \omega_v$, and L_{eo} is approximately given as

$$L_{eo} = H(\omega_v) C_o \left(\frac{1}{2}\rho dV^2\right) \qquad (11)$$

As with the case of the purely sinusoidal response, the expression for b_h is simple and can be determined directly from experimental data (figure 2). The difficulty in this case consists of ensuring that the excitation is indeed properly correlated with the velocity. This is straightforward in time-domain simulations, since one must calculate the envelope of the velocity at each time step before using equation 10. Often, however, frequency domain techniques are employed, resulting in considerable savings in computational expense; an additional requirement must then be imposed, to ensure that the excitation is properly correlated, viz.

$$\left[\lim_{T\to\infty}\frac{1}{T}\int_0^T L_e(t)v(t)dt\right]^2 = \qquad (12)$$

$$\frac{1}{2}L_o^2\left[\lim_{T\to\infty}\frac{1}{T}\int_0^T v(t)v(t)dt\right]$$

where

$$L_o = \left(\frac{1}{2}\rho dV^2\right) C_o \qquad (13)$$

4 APPLICATIONS

Below, we provide the results of simple calculations of the vortex-induced response of cylinders based on the concepts and equations derived in §2 and §3.

4.1 *Cylinder in a Uniform Current*

We begin by considering the narrow-band, lock-in response of a flexibly-mounted, rigid cylinder. The natural frequency of the system is equal to the frequency of maximum lift coefficient in-phase with the velocity. A compilation of data for vibrating cylinders as a function of the reduced damping from Griffin (1981) is shown, for comparison, in figure 3. The reduced damping is defined as the ratio of the structural damping ratio ζ_s and the quantity μ, where

$$\zeta_s = \frac{b_s}{2m\omega} \qquad (14)$$

$$\mu = \frac{\frac{1}{2}\rho dV^2}{m\omega^2 d} \qquad (15)$$

The term b_s is the structural damping coefficient per unit span.

Superimposed on the figure are calculations by the present method. Here, we have modelled the transverse motion, $y(t)$, of a rigid cylinder of unit span, having a mass m, mounted on a spring of constant k and a linear dashpot of constant b_s, and placed transversely to a constant flow of velocity V. The following is the equation of motion that is used for the calculations

$$m\frac{d^2y(t)}{dt^2} + b_s\frac{dy(t)}{dt} + ky(t) = f(t) \qquad (16)$$

The right hand side of the equation, $f(t)$, is the fluid force, which is written as the sum of an added mass term and the lift force in-phase with the velocity, which is further decomposed in accordance with equation 9. The method of harmonic balance together with equation 12 provides the solution plotted in figure 3. The calculations show good agreement with experimental data, even for large values of structural damping when the response is smaller than the threshold value.

34

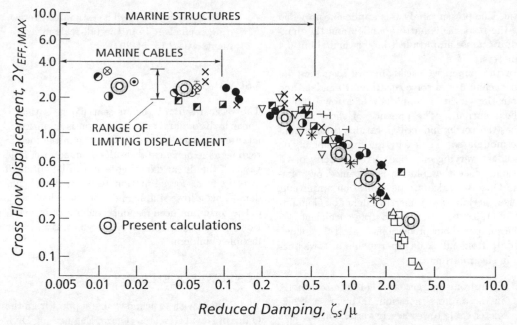

Figure 3. Comparison between the measurements of the maximum double-amplitude motion of a circular cylinder as a function of the reduced damping $\frac{\zeta}{\mu}$ (Griffin 1981) and calculations using the method outlined in this paper.

4.2 *Taut String in a Shear Current*

For the next application, we consider a taut string of length L placed normal to a spatially-varying current with nominal velocity V. The transverse response of the string, $y(t,s)$, is assumed to be described accurately by the following linear structural model and a hydrodynamic force, $f(s,t)$,

$$m\frac{\partial^2 y}{\partial t^2} + b_s\frac{\partial y}{\partial t} = \frac{\partial}{\partial s}\left[T(s)\frac{\partial y}{\partial s}\right] + f(s,t) \quad (17)$$

where s is the Lagrangian coordinate along the string, $T(s)$ is the static tension, m is the mass per-unit-length and b_s is the structural damping per-unit-length. The force, $f(s,t)$, can be decomposed into the approximate form of equation 10 with a hydrodynamic damping force and an excitation force that is in-phase with the velocity. The decomposition also yields a term that represents the added mass force. It is further assumed that the characteristic wavelength of the string oscillations is much smaller than the length of the string, hence the response is effectively that of an infinitely long string.

An important consideration in studying the response of long structures is the length over which the vortex formation process can be assumed to be correlated. It is assumed herein that vortex shedding is fully correlated over half of a wavelength of a travelling wave. This is based on experimental measurements by Ramberg and Griffin (1976), who evaluated the cross-correlation between velocities measured at two locations in the wake of a vibrating cable, separated by a distance s along the axis of the cable: They found nearly perfect correlation for all points between two successive nodes of the vibrating cable for vibrational amplitudes above a threshold value. Also, Gharib (1989) showed through visualization of the response of a flexible cylinder that there is full correlation in the vortex formation process between two successive nodes, while, at the nodes, longitudinal vortical structures destroy any vortex interconnection. The frequency of excitation within a half wavelength is assumed to be equal to the frequency at the anti-node, where the maximum amplitude occurs.

Equation 17 can be solved together with equation 10 to provide the time-domain response of a cable, even when the response is not monochromatic. The present analysis is applicable provided that the response is narrow-banded and the maximum response amplitude is larger than about 0.4 diameters.

For this paper, we used standard frequency domain techniques to solve equations 10 and 17 and obtain the vibration amplitude of a tow cable in a shear current. The presence of shear current causes the vortex-induced vibrations to be amplitude modulated. The excitation force depends on the slowly varying envelope of the velocity of vibration, hence the solution is obtained by iteration. Once we calculate the vibration amplitude, we use the laboratory measurements of Gopalkrishnan (1992) to estimate the drag coefficient.

We compared our predictions against the following data from full-scale experiments of towed cables in shear currents:

1. Data from Yoerger et al. (1991) for run 1A in the authors' notation, involving a 1,200-meter cable towed nearly vertical at 0.5 m/s in the presence of a measured shear current. The configuration of the cable was recorded using acoustic transponders, and from these measurements the drag coefficient was estimated to be equal to 2.47 ±0.24. By using the measured shear current and the procedure outlined above, we obtained a spatially varying drag coefficient along the cable length, between the values of 2.0 and 2.7, with an average value of 2.21.

2. Data from Yoerger et al. (1991) for run 2A involving an 800-meter cable towed at 0.5 m/s in the presence of a shear current. The calculations provided an average drag coefficient of 2.05. The measured full-scale drag coefficient was 2.24 ±0.24.

3. Data from figure 14 in Grosenbaugh (1991) for a cable 1,200-meters long towed nearly vertically in a transient condition. Our calculations gave a spatially varying drag coefficient in the range of 1.7 to 2.7, with an average value of 2.08. The average drag coefficient from the full-scale measurements was 1.95 ±0.20.

4. Data from figure 3 in Grosenbaugh (1991) corresponding to a 1,200-meter tow cable that had reached steady-state conditions. The calculation provided a spatially varying drag coefficient in the range of 1.6 to 2.4, with an

average value of 1.95. The estimated average drag coefficient from the full-scale measurements was 2.15 ±0.20.

5 SUMMARY

The basic result of the present paper is that a linear hydrodynamic damping term is an intrinsic feature of vortex-induced vibrations in the lock-in regime, as experimental results demonstrate. The value of the linear damping term can be obtained directly from forced-motion tests on rigid cylinders. This allows simple and efficient calculations of the vortex-induced response under lock-in conditions of flexibly-mounted, rigid cylinders and long flexible cylinders.

ACKNOWLEDGEMENTS

The authors wish to acknowledge support from the Office of Naval Research (Ocean Engineering Division) under grant numbers N00014-89-J-3061 and N00014-92-J-1269.

REFERENCES

Bearman, P. 1984. Vortex shedding from oscillating bluff bodies. *Annual Review of Fluid Mechanics* **16**, 195-222.

Gharib, Morteza 1991. Ordered and chaotic vortex streets behind vibrating circular cylinders. *Marine Industry Collegium Symposium*, April 23-24, MIT, Cambridge, Massachusetts.

Gopalkrishnan, R. 1992. Vortex-induced forces on oscillating bluff cylinders. *Ph.D. Dissertation, Joint Program of Massachusetts Institute of Technology and Woods Hole Oceanographic Institution.*

Gopalkrishnan, R., Grosenbaugh, M.A., and Triantafyllou, M.S. 1992. Amplitude modulated cylinders in constant flow: Fundamental experiments to predict response in shear flow. *Proc. Third International Symposium on Flow Induced Vibrations and Noise* ASME, Anaheim, California.

Griffin, O.E. 1981. OTEC cold water pipe design for problems caused by vortex-excited oscillations. *Ocean Engng.* **8**, 129-209.

Grosenbaugh, M.A. 1991. The effect of unsteady motion on the drag forces and flow-induced vibrations of a long vertical tow cable. *Int. J. Offshore and Polar Engng.* **1**, 18-26.

Hartlen, R.T. and Currie, I.G. 1970. Lift oscillator model of vortex-induced vibrations, *J. Eng. Mech. Div. ASCE* **96**, 577-591.

Faltinsen, O. 1990. *Sea Loads on Ships and Offshore Structures.* Cambridge University Press.

King, R. 1977. A review of vortex shedding research and its applications. *Ocean Engng* **4**, 141-171.

Ramberg S.E. and Griffin O.M. 1976. Velocity correlation and vortex spacing in the wake of a vibrating cable. *Journal of Fluids Engineering* **98**, 10-18.

Staubli, T. 1983. Calculation of the vibration of an elastically mounted cylinder using experimental data from forced oscillation. *Journal of Fluids Engineering* **105**, 225-229.

Triantafyllou, G.S. and Karniadakis, G.E. 1989. Forces on a vibrating cylinder in steady cross-flow. *Proc. Eighth International Conference on Offshore Mechanics and Arctic Engineering, The Hague, The Netherlands,* **2**, 247-252.

Vandiver, J.K. 1988. Predicting the response characteristics of long, flexible cylinders in ocean currents. *Ocean Structural Dynamics Symposium '88, Corvallis, Oregon.*

Yoerger, D.R., Grosenbaugh, M.A., Triantafyllou, M.S. and Burgess, J. J. 1988. Drag forces and flow-induced vibrations of a long vertical tow cable. Part I: Steady-state towing conditions. *J. Offshore Mech. Arctic Engng. ASME* **113**, 117-127.

Hydroelasticity in Marine Technology, Faltinsen et al. (eds) © 1994 Balkema, Rotterdam, ISBN 90 5410 387 6

Riser response to vertical current profiles and regular waves

J. M. Niedzwecki
Department of Civil Engineering, Texas A&M University, College Station, Tex., USA

G. Thoresen & S. Remseth
The Norwegian Institute of Technology, Trondheim, Norway

ABSTRACT: The objective of this study was to examine the dynamic response of a large scale flexible cylinder under controlled current and wave conditions in order to better examine and quantify the observed dynamic behavior. Optical and video techniques combined with more standard techniques were used to obtain measurements at selected elevations along the riser model. The results illustrate the complex multi-modal behavior of the response behavior.

1 INTRODUCTION

The response of deep water risers and tendons excited by ocean current and ocean surface waves is of great interest to a large segment of the offshore community. These slender flexible structures exhibit complex dynamic behavior which is not completely understood. Design methods are available but they yield results which engineers recognize as being imprecise and leave considerable room for interpretation. Significant field studies have been conducted and reported by Vandiver (1993) and Huse (1993). However, to gain a better understanding of the fluid/structure interaction phenomena large scale laboratory studies, in which the current and wave conditions could be more adequately controlled need to be performed. The results of one such study are presented in this article.

This study had three major objectives. The first was to generate and characterize a scaled version of the American Petroleum Institute (API) design current for deep water in the Gulf of Mexico (GOM). The second objective was to gain experience in the simultaneous use of optical and underwater video techniques in the direct measurement of riser displacement behavior. The final objective was to evaluate the robustness of present thinking regarding the response of long flexible cylinders in current and wave-current conditions.

2 EXPERIMENTAL DESIGN

The experimental design consisted of three parts, the specification of the environment, the measurement strategy and implementation, and the model design and construction. Each of these tasks must be carefully planned so that an integrated test plan to accurately measure the variables of interest can be achieved. Further, rigorous testing and evaluation of the environment and the model proceed any experiments in the wave basin in order to insure the quality of the final experiments.

2.1 *Environment*

Hurricane driven current profile guidelines for platforms designed for the GOM were reported by Petrauskas, Heideman, and Berek (1993), along with a procedure for their use in design calculations. Of particular interest to this study was the deep water current profile.

The deep water current profile propagated to a depth of 182.8 m (600 ft) beneath a flat ocean surface. Between water depths of 182.8 m (600 ft) and 91.4 m (300 ft) a uniform current profile with a speed of 0.2 kts was specified. From a water depth of 91.4 m (300 ft) to 60.9 m (200 ft) the current profile varied linearly from 0.2 kts to 2.1 kts. Above 60.9 m (200 ft) the current profile was assumed to be uniform with a current speed of 2.1 kts. The challenge to reproduce this type of sheared vertical current profile at reasonable model scales was undertaken in this study.

2.2 *Setup and Instrumentation*

The experiments were performed in the wave basin operated by the Offshore Technology

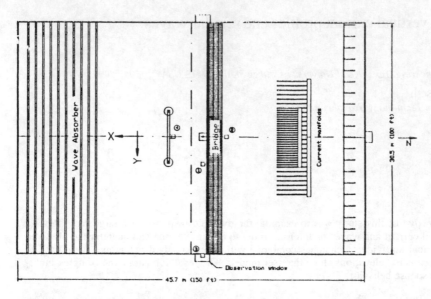

Figure 1. Plan view of OTRC wave basin.

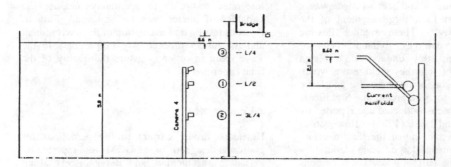

Figure 2. Side view of experimental setup in wave basin.

Research Center (OTRC). The wave basin is 45.7 m (150 ft) long and 30.5 m (100 ft) wide. The water depth in the basin is 5.8 m (19 ft), and there is a deep pit in the basin which provides a total water depth in the pit of 16.76 (55 ft). The current generation system consisted of three 6.1 m (20 ft) long manifolds with 25 jet nozzles per manifold. The nozzles had a diameter of 5 cm (2 in) and were spaced 23 cm (9 in) apart. Flow through each manifold could be adjusted as could the flow though each nozzle. For this series of experiments two manifolds were placed next to each producing a near surface current and the third was placed at a second elevation to adjust the profile.

A schematic of the wave basin and the experimental setup is shown in Figure 1. Figure 2 provides a corresponding schematic drawing which illustrates the positioning of the manifolds used to generate the currents, the placement of the riser model and the camera setup.

The measurements included optical and video tracking at selected elevations along the riser model above and below the free surface. Inline and transverse displacements and reactions were measured as well as the free surface elevation and current velocities. Additional details on the optical and video systems can be found in the thesis by Guerandel (1994).

40

Figure 3 presents the scaled (1:20) deep water current profile and that obtained in the wave basin. Two other profiles not reported in this article were obtained and are discussed in the thesis reports by Thoresen (1994) and Guerandel (1994). The first graph in Figure 3 was included to show the shape of the scaled API current. The second graph shows the actual measurements obtained using electromagnetic current meters. As indicated by this figure, the current does fluctuate with time. Evaluation of the time series of the current meter data indicates that the mean oscillates with a low frequency. Further, high peaks do not last long which is important when considering the model behavior. Spectral analyses of the current velocity records indicated that indeed there was significant energy at the lower frequencies. Above, about 6 Hz the current velocity spectra in the inline and transverse directions was nearly constant.

The turbulence level of the current which is defined as the ratio of the standard deviation of the velocity to the maximum velocity was computed along the current profile. The turbulence level was less than 0.2 for the upper part of the current profile. This is in the region where the current is directly coming from the nozzles. In the transition zone where the generated current and the still water mix, the turbulence intensity increased and mean velocity was lower. The standard deviation in the upper layers was on the order of 21 mm/s. In the transition zone it was about 48 mm/s and in the lower water depths it was approximately 21 mm/s.

2.3 *Riser model*

The model was constructed such that it was characteristic of deep water risers. This meant that the mass distribution, stiffness and the natural frequencies were modeled very carefully. Care was taken to insure that the riser model represented a generic riser.

Engineering numbers selected for the prototype riser, the computer model are presented in Table 1. Also shown are the actual numbers from the laboratory model of the riser. The riser model was built around a central wire which could be tensioned. Lead weights were attached to the wire to assist in obtaining the desired frequency behavior of the model. The exterior shell of the riser model was constructed from ABS-plastic. As can be seen in Table 1 the mass ratio was approximately 3.55 for the riser model.

3 PREDICTIVE MODELS

3.1 *Natural frequency*

The first eight natural frequencies of the model were 0.40, 0.83, 1.3, 1.9, 2.6, 3.3, 4.2 and 5.2 Hz. The experimental values obtained from free vibration tests were 0.38, 0.8, 1.4, 2.0, 2.6, 3.4, 4.2 and 5.4 Hz. Thus, the agreement was excellent.

3.2 *Damping*

Using the time series from a free vibration test, the critical damping ratio was estimated to be 0.01. This is also referred to as the relative structural damping. The total relative damping in still water was estimated to be 0.03.

3.3 *Lock-in*

Figure 4 presents a graphical view of the riser frequencies excited by vortex shedding. The vertical lines represent the natural frequencies of the riser. The vertical lines which have diamonds or squares indicate the frequencies which were excited. The interior symbols represent elevations when the response was measured. The dashed line on the graph represents the vortex shedding frequency estimated using a Strouhal number of 0.21. The dominant frequency of excitation was the fifth natural frequency. By changing the current profile it was observed that different frequencies could be excited and other frequencies would dominate (Thoresen 1994). It was found that the dominant frequency of excitation could be estimated by using the mean current velocity over the depth. This tendency was observed for all the current profiles which were studied (Thoresen 1994).

Comparison of the shear fraction, excitation bandwidth, vortex shedding bandwidth and reduced velocity are presented in Table 2. The shear fraction for these experiments was estimated to be in the range of .64 - 1.0 with lock-in. The excitation bandwidth estimated with a 200 mm/s velocity difference was 3.3 Hz. It was predicted that seven modes would be predicted and experimentally five were observed to be excited.

3.4 *Response behavior*

The mean and maximum inline displacement profiles of the riser model in current flow is shown in Figure 5. The maximum and standard deviation profiles of the transverse displacement

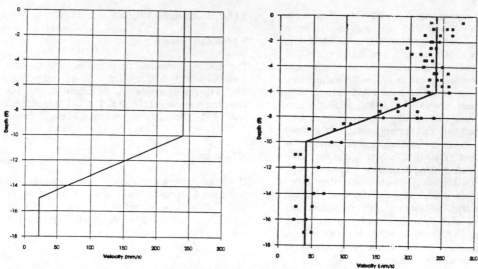

Figure 3. API deep water current profile and laboratory profile.

Table 1. Riser prototype, computer model and laboratory data.

	PROTOTYPE	MODEL	LAB
RISER			
Scale, L_H		1:20	1:20
Scale, L_V		1:91	1:88
Depth (m)	500	5.495	5.671
Outer diameter (N/m²)	0.2445	0.0127	0.0127
Lateral stiffness (N/m²)	1.19E+07	0.0395[1]	2.18
Mass/length (kg/m)	170	0.4587[2]	0.4464
Added mass/length (kg/m)	46.95	0.1267	0.1267
Weigth/length (N/m)	1667.7	3.2568	5.6221
Bouyancy/length (N/m)	460.6	1.2421	1.2427
Effective weigth/length (N/m)	1207	3.2568	4.3794
Total effective weigth (N/m)	686935[3]	23.31[3]	23.31
Top tension (N)	8.55E+05	29.0[4]	28.0
Top tension/tot.eff.w. (N/m)	1.24	1.24	1.22
Mass ratio	3.63	3.63	3.55
CURRENT			
Scale, L_H		1:20	1:20
Scale, L_V		1:20	1:33
Velocity (mm/s)	1801	240	240
Depth (m)	61	3.05	1.83

[1]   Scaled with Froude scaling.
[2]   Scaled with constant mass ratio.
[3]   Prototype rising 50 m above water surface.
      Model rising 1.21 m above water surface.
[4]   Scaled with constant relation of top tension/total effective weight.

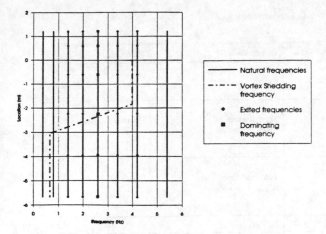

Figure 4. Frequencies excited by the vortex shedding.

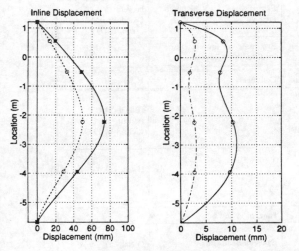

Figure 5. Inline and transverse displacements.

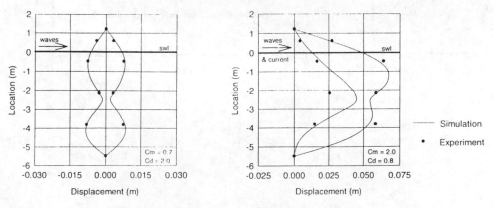

Figure 6. Regular wave-current interaction data and simulation results.

43

Table 2. Comparison of experimental results with earlier studies.

Parameter	Predicted Value	Experimental Value	Comments	Reference
U [mm/s]		< 40 - 240 >	Current velocity	
$\dfrac{\Delta U}{U_{max}}$		0.83	Shear fraction (ΔU = 200mm/s)	(Vandiver , 1988)
$\dfrac{U_{st.dev}}{U_{max}}$		0.15	Turbulence level ($U_{st.dev}$ = 35mm/s)	(Vandiver, 1988)
Re		< 4.5e02 - 2.7e03 >	Subcritical	
S_t	0.21			(Lienhard, 1966)
C_d	1.0	0.9	Exp: mean reaction - inertia	
	2.7 / 1.5 1.6 2.6 / 1.7	1.4	Exp: max reaction - inertia Pred: $A_{y(max)}$ / $A_{y(rms)}$ $f_n = f_s$	curvefit (Vandiver, 1983) (Skop, 1977)
$C_{L(max)}$ $C_{L(rms)}$	(1.4) (0.4)	1.4 0.4	Exp: reaction - low freq. Pred: for other Reynolds #	

Parameter	Predicted Value	Experimental Value	Comments	Reference
$\dfrac{A_{y\ max}}{D}$	0.7 1.1 0.5	0.8 - 1.1		(Blevins, 1977) (Griffin, 1982) (Sarpkaya, 1979)
$\gamma = \dfrac{A_{y_{max}}}{A_{y_{mean}}}$	1.155			(Iwan, 1975)
δ_r		0.35 - 1.34	Func. of m and ξ	
$\Delta f = \dfrac{S \cdot \Delta U}{D}$	3.3	2.8	Exitation Bandwith	(Vandiver)
$N_s = \dfrac{\Delta f}{f_1}$	7	5	Bandwith of vortex shedding	(Vandiver)

are also shown in Figure 5. It is evident from this comparison that nonlinear modes control the response in the transverse direction. The dominant mode was identified in Figure 4. Upon examination of the corresponding time series for the transverse displacement, a low frequency drifting of the cylinder to the sides was observed. This was easily filtered out and is believed to be a result of the flow in the basin.

Estimates of maximum response were obtained using well known formulas and these were compared with the experimental results. Generally, the predicted results compared favorably with the laboratory data. Different prediction models for estimating the transverse displacement are given by Blevins (1990). They are all a function of the reduced damping. Blevins and Griffins formulas are empirical and are based upon the numbers for reduced damping which the authors found to correspond to their experiments. Sarpkaya's formula was analytically based. In applying these expressions to this experiment, the same value for reduced damping was used. The mode shaped factor in Iwan's expression was assumed to have a value of 1.155. Later studies have shown that this factor can be improved (Moe 1991). Comparisons between the predicted and experimental displacements are shown in Table 2.

The comparison of simulation and experimental results for the combined wave-current condition is presented in Figure 6 (Guerandel 1994). This figure illustrates that for regular waves the envelope predictions are quite good. However, with the introduction of combined wave and current flow some research still remains.

4 CONCLUSION

The objective of this study was to examine the dynamic response of a large scale flexible cylinder under controlled current and wave conditions in order to better examine and quantify the observed dynamic behavior.

The use of jet nozzle arrays to generate the vertically sheared current profile has been characterized and some new information regarding this type of current generating system has been found which will help improve the system for future studies. It should be noted that this was the first real test of current profile generation using the recently completed system. Further, the combined use of optical tracking and underwater video tracking has been demonstrated for the direct and accurate measurement of displacement. An accelerometer was located on the riser and initial comparison of the measurements did not yield any apparent discrepancies with the optical tracking systems.

The new results presented included the comparison of key parameters and response envelopes. Some discrepancies were noted and investigation of these aspects is underway. The interaction of currents with random seas is the subject of the thesis by Guerandel (1994).

ACKNOWLEDGMENTS

The writers gratefully acknowledge the support of the OTRC in conducting the experimental program with special thanks to Dr. A.S. Duggal and Mr. V. Guerandel. The first writer was supported in part by NSF Engineering Research Centers Grant No. CDR-8721512 and ONR Grant No. N00014-93-1-0620 Further the writers gratefully acknowledge the financial support of the Norwegian Institute of Technology for Mr. Thoresen and the opportunity for him to pursue research at the OTRC. A special thanks is also due to Professor G. Moe for providing technical information to Mr. Thoresen to help him in his studies.

REFERENCES

Blevins, R.D. 1990. *Flow-Induced Vibration*, 2nd ed. Van Nostrand Reinhold: New York.
Guerandel, V. 1994. *Experimental study of a single riser in waves and currents*, TAMU Thesis: Department of Civil Engineering.
Huse, E. 1993. Interaction in deep-sea riser arrays, *Perch. 25th Offshore Technology Conference*: 7237, 313-322. Houston: Texas.
Moe, G. 1991. An experimentally based model for the prediciton of lock-in in riser motions, *OMAE*.
Petrauskas, C., J.C. Heideman & E.P. Berek 1993. Extreme wave-force calculation procedures for 20th edition of API RP-2A, *Perch. 25th Offshore Technology Conference*: 7153, 201-211. Houston: Texas.
Thoresen, G. 1994. *Current interaction with a TLP riser*, NTH Thesis: Department of Civil Engineering.
Vandiver, J.K. 1988. Predicting the response characteristics of long, flexible cylinders in ocean currents, *Ocean Structural Dynamics Symposium*: 1-41. Corvallis: Oregon.
Vandiver, J.K. 1993. Dimensionless parameters important to the prediction of vortex-induced vibration of long flexible cylinders in ocean currents, *Journal of Fluids and Structures: 7*, 423-425.

Hydroelasticity in Marine Technology, Faltinsen et al. (eds) © 1994 Balkema, Rotterdam, ISBN 90 5410 387 6

Response analysis of slender drilling conductors

Henrik Nedergaard & Erik Bendiksen
LICengineering A/S, Esbjerg, Denmark

Kristian Kudsk Andreasen
Mærsk Olie og Gas AS, Esbjerg, Denmark

ABSTRACT: A method for calculating the response of a drilling conductor is presented and compared with full scale measurements from the North Sea.

The response model is based upon the impulse-response method, in which any non-linear loading term can be included, eg. quadratic drag and vortex shedding load. The loading is arising from irregular 3-dimensional wave time-series combined with sheared current.

Results from the model are presented and compared with measurements from an instrumented drilling conductor in the North Sea. The significant dynamic behaviour in the measured sea-states is found to be modelled with good agreement, both in-line and cross flow. The different sources of the cross-flow response (3-D wave effects and vortex shedding resonance) are isolated and quantified.

1. INTRODUCTION

This paper describes the analysis of a drilling conductor and the results are compared with measured data. Fig. 1 depicts the jack-up and drilling conductor in focus. The drilling conductor is installed from the jack-up Santa Fe situated in the British sector of the North Sea. The water depth is 93.3 m. Measurements have been carried out in three sea states. The first and most severe of these - on the 8th of March 1992 - has been analysed in the present paper. On this day the casing programme consisted of an outer conductor and two inner casings of 20" and 14". Up to elevation -79.7 m the conductor is 36" and from there and up it is 30". A Blow Out Preventer (BOP) is installed approximately at elevation +30 m. In this elevation top tension is applied and the conductor is supported horizontally just below the BOP. Below seabed the weight of the casings is transferred to the conductor by a mudline suspension system.

The bending moment in the conductor is measured by strain gauges at three elevations - just above mudline, at midspan and just below the BOP. Comparison have been made for bending moments above mudline.

The measuring system and its setup are in detail described by Noble Denton 1993 and further by Hamel-Derouich *et al.* 1994.

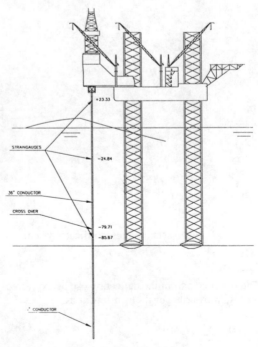

Fig. 1   Arrangement at jack-up

2. METHOD OF ANALYSIS

The response model is a general 3 dimensional model for calculation of the response of the conductor exposed to wave and current forces. Cross-flow vortex shedding induced vibrations are included in the time domain by use of a force coefficient method, see Ottesen Hansen 1982 and Nedergaard *et al.* 1994. Here the analysis method is only summarised.

2.1 Response model

The equation of motion in the x-direction can for a beam section be expressed by (Fig. 2):

$$EI(z)\frac{\partial^4 x}{\partial z^4} - N(z)\frac{\partial^2 x}{\partial z^2} + c(z)\frac{\partial x}{\partial t} + m(z)\frac{\partial^2 x}{\partial t^2} = F_x(z,t) \quad (1)$$

Similar expressions for the y direction. Assuming that the motion can be divided into a forced and a dynamic motion the following relation can be set up (see Brouwers 1982):

$$x(t,z) = x_f(t,z) + x_d(t,z) \quad (2)$$

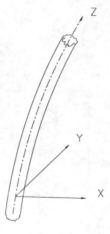

Fig. 2 Beam section and coordinate system

The dynamic part of the deflection can be expressed by a sum of orthogonal eigen functions:

$$x_d(t,z) = \sum_{i=1}^{N} X_i(t)\cdot\psi(z) \quad (3)$$

The undamped eigenvalue solution is derived from:

$$\omega_0^2\int_0^L m(z)\psi_i(z)^2 dz - $$
$$\int_0^L (EI(z)\psi_i(z)\psi_i^{IV}(z) - N(z)\psi_i(z)\psi_i^{II}(z))dz \quad (4)$$

By inserting Eq. 2 into 1 and integrate along the pipe and further utilize Eqs. 3 and 4 the following expression can be obtained:

$$\ddot{X}_i + 2\beta\omega_{oi}\dot{X}_i + \omega_{oi}^2 X_i = $$
$$\frac{1}{\bar{m}_i}\int_0^L [F_x(z,t) - m(z)\frac{\partial^2 x_f}{\partial t^2}]\psi_i(z)dz \quad (5)$$

The solution to Eq. 5 can be obtained by use of the impulse-response method:

$$X_i(t) = \frac{1}{\bar{m}_i\omega_i}\int_0^t exp^{-\beta\omega_{oi}(t-\tau)}\sin(\omega_i(t-\tau))\,\bar{F}_{ix}(\tau)d\tau$$

in which: (6)

$$\omega_i^2 = \omega_{oi}(1-\beta_i)^2$$
$$\bar{F}_{ix}(t) = \int_0^L [F_x(z,t) - m(z)\frac{\partial^2 x_f}{\partial t^2}]\psi_i(z)dz$$

The above solution procedure is simultaneously calculated for both the x- and y-direction. The response calculation of the direct response and vortex shedding is in each time step calculated separately, i.e. Eq. 6 is evaluated four times for each time step.

2.2 Load model

The in-line hydrodynamic loading is included by the Morison equation with inclusion of the terms arising from the movements of the pipe itself. The force intensity per unit length is calculated by:

$$F = \rho A\dot{U} + \rho C_m A(\dot{U}-\dot{U}_p) + \frac{1}{2}\rho C_D(U-U_p)|U-U_p|D \quad (7)$$

This equation is calculated for the relative fluid direction (direction of $U-U_p$) and then projected onto the x- and y-direction.

The cross-flow (relative) response of the pipe is the motion exerted by the vortices. In each vortex cell the following force per unit length is present:

$$F_L = \frac{1}{2}\rho\, D\, C_L\, U(z,t)^2 \sin(\omega_{st}t) \qquad (8)$$

This forcing is present in all cells, however, only when there is resonance with a structural eigen mode response will develop, $\omega_{st} = \omega_i$. This phenomenon is called lock-in, it occurs under the following conditions:

a) An acceleration criterion - the change in relative velocity during the shedding of one vortex must be limited, within 10%.

b) A resonance criterion - reduced velocity for the eigen mode shall be within certain limits. For cross flow induced vibrations the criteria are:

$$4.8 < U_{R,i} < 8$$

in which: $\qquad (9)$

$$U_{R,i} = \frac{U(z,t)}{f_i\, D(z)}$$

When resonance - lock-in - the specific length of the pipe which satisfies the above criteria will be forced. The corresponding generalized force is:

$$F_i =$$

$$\sqrt{\int_{L_{i,1}(t)}^{L_{i,2}(t)} \int_{L_{i,1}(t)}^{L_{i,2}(t)} F_L(z_1,t)\, F_L(z_2,t)\, \psi_i(z_1)\, \psi_i(z_2) \exp^{\frac{-2|z_1-z_2|}{l_c(z_1,t)}}\, dz_1\, dz_2}$$

$$(10)$$

where $[L_{i,1}(t); L_{i,2}(t)]$ is the region with lock-in for mode i at time t.

l_c - correlation length - is an important factor in the above formulae for the vortex shedding generalized force. It is an instantaneous function of the actual vortex shedding response amplitude, see Ottesen Hansen 1982 and 1984. This concept models the phenomenon that with growing vibrational amplitude the vortices are gradually tuned into equal phase.

The damping for the cross flow response is defined by:

$$\beta_i = \beta_s + \beta_{h,i} - \beta_{v,i} \qquad (11)$$

where β_s is the structural damping, $\beta_{h,i}$ is the hydrodynamic damping based upon the instantaneous velocity squared, and finally $\beta_{v,i}$ is a term called the "negative" damping calculated over the lock-in region by:

$$\beta_{v,i} =$$

$$\frac{\rho D C_v}{\omega_{o,i}\, m_i}\sqrt{\int_{L_{i,1}(t)}^{L_{i,2}(t)} \int_{L_{i,1}(t)}^{L_{i,2}(t)} u(z_1,t)\, u(z_2,t)\, \psi_i(z_1)\, \psi_i(z_2) \exp^{\frac{-2|z_1-z_2|}{l_c(z_1,t)}}\, dz_1\, dz_2}$$

$$(12)$$

The negative damping is caused by alteration of the flow by the presence of the vortex shedding and is in fact a driving force which cause that the damping within the lock-in region is very limited. C_v is based upon tests set to 0.75, corresponding to a vortex shedding coefficient $C_L = 0.9$ which produces the most accurate results for high KC-numbers.

3. COMPARISON WITH FULL-SCALE MEASUREMENTS

3.1 Eigenfrequency analysis

The conductor is modelled by a finite element system using 58 beam-column elements. Because a linear model is needed for the eigen value analysis, gaps and other non-linearities are neglected. Therefore the conductor/seabed interaction is modelled by linear springs. The correct axial force in the beam elements is essential for the bending stiffness. The axial force in each beam element is found from the top tension and the weight of the conductor/casings and BOP system. Due to the mud line suspension system the axial force is quite well determined. During the actual storm the inner 14" casing was still being run i.e. it was supported vertically only by the derrick and not by the mudline suspension system. By assuming that the inner casings follows the conductor they contributes to the total sectional properties and axial force. The length of the inner casing has in Noble Denton, 1993, been estimated to 7000 ft, equivalent tension is added. Including this effect the lowest six eigen frequencies are found.

The two lowest calculated eigen frequencies marked on Fig. 3 are seen to compare well with the peaks on the measured transfer function. The correct modelling of the axial force was found to be very important, without the tension of the inner casing the lowest frequency was determined to 0.19 Hz instead of 0.24 Hz.

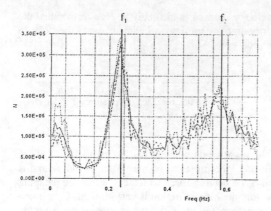

Fig. 3   Measured transfer function and calculated eigen frequencies

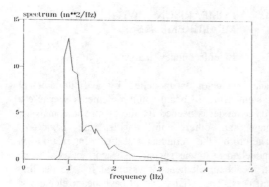

Fig. 4   Measured wave spectra

3.2   Environmental condition and force coefficients

The wave kinematics was measured and is represented by a spectrum, Fig 4. Directional spreading in the sea is modelled by use of a squared cosine function with preserved energy. The spectrum is transferred into time-domain by discrete wavelets with random phases, both in frequency and direction. The higher frequency wavelets are assumed to "ride on the back" of the lower in order to model the wave kinematics in the wave crest realistically.

Current measurements was unfortunately not carried out, but a wind driven current is included in the analysis. The wind speed was measured (Noble Denton 1993) to 25 knots in an elevation of 113 m with nearly the same direction as the wave propagation. This wind speed is transferred to

elevation 10 m by assuming a logarithmic profile. The induced surface current is then calculated by use of the assumption that it is 2% of the wind speed in elev. +10 m. The wind driven current profile is assumed to be linear and zero in 30 m of water depth.

Based upon Noble Denton, 1993, an uniform current (tidal) of 0.3 m is added to the wind driven current. The water velocity time-series derived upon the above assumptions is analysed for determination of the hydrodynamic force coefficients. The statistic is carried out for elevation 0 m. The root mean square (RMS) value of the horizontal wave velocity is calculated and then scaled to a local maximum wave velocity based upon assumption of sinusoidal velocity variation. The Reynolds number (RE) is calculated based upon the sum of the wave and current velocity, whereas the Keuligan Carpenter number is adjusted due to current according to the formula:

$$KC = \frac{U_{wave} T_z}{D}(1 + \frac{U_{cur}}{U_{wave}})^2 \qquad (13)$$

The statistical values is summarized in table 1.

Table 1   Statistical values for the synthetic wave and current velocity time-series

U_{wave} (m/s)	0.56
U_{cur} (m/s)	0.50
U_{tot} (m/s)	1.06
RE (-)	$5.4 \ 10^5$
KC (-)	15.1

Based upon table 1 the drag and inertia coefficient is selected to $C_D = 1.2$ and $C_M = 2.0$. The rather high drag coefficient is selected due to the instrument umbilical (piggy backed to the conductor) and increased diameter of connectors. The process is a little more tedious for the lift coefficient. The very low KC-number reveal that in the oscillatory flow the previous shed vortices will be swept back over the conductor and thus increase the lift. This phenomenon is measured by Bearman et al. 1981, an extract is seen Fig. 5. Here it can be seen that the lift coefficient (vortex shedding force coefficient) is around a factor of 3 higher in the range of KC = 10-20 compared with KC > 40. Based on these findings $C_L = 2.5$ was selected, to be compared with $C_L = 0.9$ for high KC-numbers for present model.

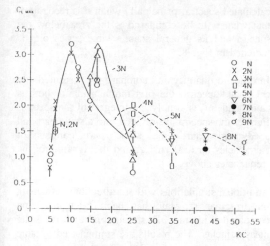

Fig. 5 Lift coefficient as function of KC-number, N is the number of vortices shed per half wave period, Bearmann *et al.* 1981.

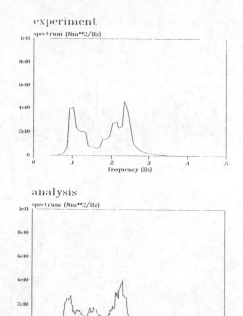

Fig. 6 In-line spectra, measured and calculated

3.3 Time domain simulations

By use of the method described in section 2, the eigen frequencies from 3.1 and wave kinematics and coefficients from 3.2, the response of the conductor was calculated. A structural damping coefficient of 3 % was selected due to the gab between the inner casings and conductor (non-linearities causes energy to dissipate - damping). This value is high compared with standard values for steel structures 0.2 - 0.5 %.

The response analysis has been carried out for 13.5 min and compared with a measurement period of 20 min. Both the calculated and measured results are FFT (Fast Fourier Transformation) analysed in order to determine the spectrum of the response - bending stress at mudline.

In Fig. 6 the measured spectrum for the bending stress at mudline for the fibre in the wave direction compared with the calculated. Generally, a two peaked spectrum can be seen, excitation with the wave frequency (0.12 Hz) and resonance with the first eigen frequency (0.24 Hz). Concerning the cross-flow response, Fig. 7 presents the spectra. A much smaller peak is found near the wave frequency (directionality in the waves) whereas a much higher peak is found near the first eigen frequency, this is due to both resonance with the waves and especially vortex shedding induced vibrations. A quite good agrement can be seen between the measured and

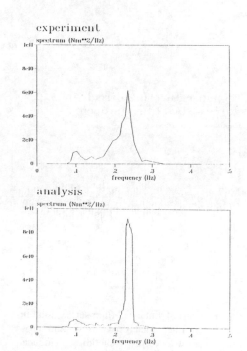

Fig. 7 Cross flow spectra, measured and calculated

51

response in main direction
total

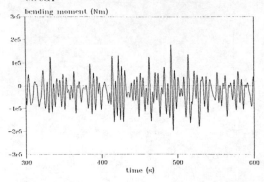

response in cross direction
total

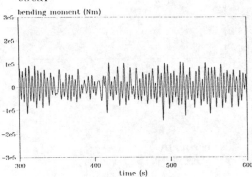

response in cross direction
vortex shedding induced

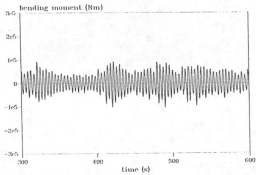

Fig. 8   Part of simulated time-series, total in-in-line, total cross flow and cross flow vortex shedding induced vibrations response

calculated spectra, principally when it is recognized that what is to be compared is the area below the spectra and its general shape. Time series plot is included in Fig. 8.

In order to quantify and compare the measured and calculated spectra, the first moment m_0 for both is calculated, m_0 is simply the integral of the spectra. Further the spectra are divided into a static part and a dynamic part, the static part consists of frequencies up to 0.16 Hz, and the dynamic of the frequencies above.

Additional simulations without directional effects in the irregular sea-state has been calculated, and compared with the 3D simulations and full-scale measurements. The results are summarized in table 2.

Table 2   Comparison between measured response and 3D and 2D analysis.

	m_0 10^9 $(Nm)^2$		
Item	static	dynamic	total
Experiment			
In-line	1.30	2.61	3.90
Cross flow	0.39	2.54	2.92
3D analysis			
In-line	1.02	2.18	3.20
Cross flow	0.25	2.28	2.53
2D analysis			
In-line	1.62	2.84	4.45
Cross flow	0.07	1.22	1.29

From table 2 it can be seen that the 3D analysis gives results close to the measured, both considering the static, dynamic and the total. Further it can be seen that for the 2D analysis the in-line response is over estimated, whereas the cross flow will be under estimated. Finally it can be seen that a pure static "in-line" analysis will give results which underestimate the measurements severe.

Another interpretation of the results in table 2 is the "significant bending moment" defined by:

$$M_s = 2\sqrt{m_0} \qquad (14)$$

Table 3 M_s significant bending moment. Total refers to combined in-line and cross flow response.

Item	M_s 10^6(Nm)
Experiment	
In-line	0.125
Total	0.140
3D analysis	
In-line	0.113
Total	0.128
2D analysis	
In-line	0.133
Total	0.136
Pure Static	
In-line	0.063
Total	0.063

From table 3 it can be seen that the 3-D analysis yields results for the bending moment very close to the experimental values, the combined in-line and cross flow response is underestimated with only 8 %. The results from the 2-D analysis yields in-line results at the "high side" as expected. These results is very satisfactory bearing in mind that the input to the analysis was a measured wave spectra with no specific data concerning directional spread and current. Further the relative short measuring and analysis period (20-13 minutes) does not necessarily gives stable result due to the stochastic process.

It can be seen that a pure static analysis (with all the wave frequencies) results in under estimating the response with 55 %. With respect to fatigue damage the static analysis under estimates the experimental results with a factor in the range of 7 - 15.

4.   CONCLUSION

From section 3 - the results of the analysis and comparison with full-scale measurements - it can be concluded:

-      The presented model describes the response

very satisfactory, both in-line and cross flow including vortex shedding induced vibrations

-      Vortex shedding induced vibrations plays a significant role for the cross flow response

-      Concerning fatigue analysis, both the effect of dynamics and cross flow response is very important

The above findings is in line with the results found by use of the model, see Nedergaard *et al.* 1992. Herein it is demonstrated that the fatigue damage of drilling conductors is heavily governed by resonance with the direct wave load. The results of the full scale experiments referred herein support these conclusions. For both the extreme wave response analysis and the total fatigue analysis the following conclusions can be drawn:

-      In the extreme wave response analysis "static" response can be representative but the lock-in to vortex shedding induced vibrations shall be considered, the magnitudes may be of same order.

-      Fatigue response of drilling conductors is governed by resonance with direct wave loading. The dynamic amplification in an irregular sea must be included.

5.   ACKNOWLEDGEMENTS

Mærsk Olie & Gas AS is greatly acknowledged for releasing data and the given support for the present re-analysis of the full-scale experiment. Also their willingness to introduce new and more accurate methods, especially in the area of vortex shedding induced vibrational response, shall be recognised.

6.   REFERENCES

Bearman, P.W., J.M.R. Graham, P. Naylor and E.D Obasaju, 1981. The Role of Vortices in Oscillatory Flow about bluff Cylinders. Proceedings of the International Symposium on Hydrodynamics in Ocean Engineering, The Norwegian Institute of Technology, 1981.

Brouwers, J.J.H., 1982. Analytical Methods for Predicting the response of Marine Risers, Publication 614, Koninklijke/Shell Exploratie en Laboratorium. Rijswijk, The Netherlands

Hamil-Derouich D., Robinson R., Stonor R., 1994. Assessment of an Analysis of an Instrumented Drilling Jack-up Conductor OTC 1994 (paper no. 7459)

Nedergaard, H., N.-E. Ottesen Hansen and C, Eilersen. 1992. Dynamic Design of Well Conductors. Proceedings of the Second (1992) International Offshore and Polar Engineering Conference, San Francisco, USA, 14-19 June 1992.

Nedergaard, H., N.-E. Ottesen Hansen, Fines S. 1994. Response of Free Hanging Tethers. Proc. BOSS'94, Boston july 1994.

Noble Denton, 1993. Instrumentation of the conductor tube on the Santa Fe Galaxy - 1 at Ranger block 29/4B. Main Report and Appendix A-D, 28th May 1993, .

Ottesen Hansen, N.-E., 1982. Vibrations to Pipe Arrays in Waves. Proceedings of BOSS'82, Boston Aug. 1982.

Ottesen Hansen, N.-E., 1984. Vortex Shedding in Marine Risers in Directional Seas. Symposium on Description and Modelling of Directional Seas, June 18-20, 1984, Copenhagen, Denmark.

C_M	inertia coefficient
C_m	added mass coefficient
KC	Keulegan-Carpenter number
U_R	reduced velocity
ω	cyclic frequency
β	relative damping
Ψ	mode shape
N	number of modes

indices

i	mode number
f	forced
d	dynamic
o	undamped
p	pipe (motion)
s	structural
h	hydrodynamic
v	"negative damping"
st	Strouhal

7.   NOMENCLATURE

x,y	lateral coordinates
z	axial coordinate
X,Y	amplitude functions
EI	bending stiffness
N	axial force
A	cross sectional area
D	diameter of pipe
L	Length of structure
l_c	correlation length
η	vortex shedding amplitude
c	damping coefficient
f	frequency
F	force
t,τ	time
m	mass pr. unity
$\overline{m}$	generalised mass
ρ	density of water
U	water velocity
C_D	drag coefficient
C_L	vortex shedding force coefficient
C_v	"negative" damping coefficient

Hydroelasticity in Marine Technology, Faltinsen et al. (eds) © 1994 Balkema, Rotterdam, ISBN 90 5410 387 6

Collision criteria for deep-sea TLP riser arrays

Erling Huse
MARINTEK, Trondheim, Norway

SUMMARY: A method for calculating current force on individual elements in arrays of rigid cylinders in fixed positions has been developed and published previously. The method is based on the general theory of turbulent wakes and momentum balance.

The present paper describes how these principles can be extended to calculate the static deflection of each riser of a deep sea tension leg platform. This in turn makes it possible to predict whether or not collisions between individual risers will take place. Such collision criteria are expected to become very important for possible future deep sea TLP designs, because they determine necessary spacings and pretensions in the risers.

The paper also describes a series of experiments to verify the collision criteria. The experiments consisted in towing tests with a model riser array, recording collisions by video observations.

1 INTRODUCTION

Practical evaluation of current forces on arrays of cylinders has until recently been based exclusively on experimental data, see for instance Refs.1 and 2. Considerable efforts have in recent years been spent on developing numerical procedures based on calculating the shedding and development of individual vortices in the wake flow and corresponding forces on cylinders. For survey of such methods see for instance Ref.3. In spite of recent progress these methods are still far from representing suitable tools in practical engineering work with TLP riser arrays, which may contain as many as 50 or more individual risers.

As a stage in between the purely empirical data and the vortex methods we have a group of methods based on considering the wake as a turbulent flow field, neglecting the effect of individual vortices. Such considerations have been applied in Refs.4 through 9 with considerable success, in some cases even for oscillatory flow.

The objective of the present publication is to present and verify a simple procedure for predicting the criteria for onset of collisions between individual risers in the riser system of deep sea tension leg platforms (TLPs). Such collisions may take place because the static current force on the

risers in downstream positions becomes smaller than the force on upstream ones due to wake interaction. The corresponding difference in their static deflections may lead to collisions between them. The numerical procedure for predicting such collisions is based on the theory of turbulent wakes together with momentum considerations. The basic principles of the procedure were first published in Ref.8, then limited to in-line force and rigidly supported cylinders in fixed positions. In Ref.9 the theory was extended to include calculation of the static deflection of each riser in the array. Typical applications are determination of necessary pretension and spacing to prevent collision between individual risers, optimization of the pretension of a drilling riser operating within an array of production risers, position stability of individual risers, etc. The present publication presents for the first time experimental data for verification of the collision prediction.

2 WAKE FIELD OF SINGLE CYLINDER

Schlichting (Ref.10) has solved the equations of motion in a wake by use of different mixing theories from L. Prandtl. By assuming two-dimensional motion, neglecting viscous stress and holding the pressure constant through the fluid, it is possible to set up expressions for the turbulent wake field as follows:

$$b = 0.25 \, (C_{d1} \, D_1 \, x_s)^{1/2} \qquad (1)$$

$$U_o = V_c \, (C_{d1} \, D_1 \, / \, x_s)^{1/2} \qquad (2)$$

$$u = U_o \, \exp(-0.693 \, (y/b)^2) \qquad (3)$$

Eqs. 1, 2, and 3 are expected to be valid only some distance downstream of the cylinder. Very close to the cylinder the above expressions will give a wake peak which is too high and narrow, which leads to erroneous results when trying to calculate the force on a second body placed in the wake. In order to correct for this the concept of "virtual source position" is introduced. This simply means that the distance x_s in the above equations is substituted by

$$x_s = x_v + x \qquad (4)$$

where x_v is the distance from cylinder to virtual source and x is the distance between the centres of the wake generating cylinder, C1, and the cylinder on which force is to be calculated, C2, see Fig.3.

Requiring that

$$b = D_1 \, / \, 2$$

at the position of cylinder C1 leads to

$$x_v = 4 \, D_1 \, / \, C_{d1} \qquad (5)$$

At large distances this correction does not make much difference when calculating the wake field. However, it makes a significant difference for the wake field close to cylinder C1.

3 FORCE ON SECOND CYLINDER

The in-line current force on the second cylinder C2 placed in the wake of C1 (Fig.3) can now be calculated as

$$F = 0.5 \, \rho \, D_2 \, C_{d2} \, (V_c - u)^2 \qquad (6)$$

A problem is now that u varies over the space occupied by the cylinder C2, such that the u value to be used in Eq.6 is not well defined. In the subsequent calculations this has been solved by using the u value from Eq.3 RMS averaged over the cylinder diameter.

4 FORCE ON ARRAY OF CYLINDERS

The force on each individual cylinder in an array of arbitrarily arranged cylinders at fixed positions can now be calculated by a small computer program as follows:

1) First the free stream current velocity V_c, the free stream direction α, and the coordinates, diameter and drag coefficient of each cylinder is read from an input file.

2) The cylinders are re-numbered in a sequence corresponding to their positions in the flow direction.

3) Starting with the cylinder far upstream, the total wake velocity u_n at the position of each cylinder in the array is calculated by RMS summation of the wake contributions generated by all upstream cylinders. The calculation of the wake contribution of each upstream cylinder is done by Eqs.1 through 4, substituting V_c in Eq.2 by V_c-u_n, the net inflow velocity to cylinder No.n, corrected for the wakes generated by all upstream cylinders.

4) Having now determined the total wake and thus the net inflow velocity of each cylinder in the array, the net drag force on each is calculated by

$$F = 0.5 \, \rho \, D \, C_d \, (V_c - u_n)^2 \qquad (7)$$

As a comment to item 3 above it should be emphasized that when summing up the wake contributions from all upstream cylinders one can not apply direct summation of wake velocities, but rather RMS summation. By that we mean summing the square of the wake velocities generated by the upstream cylinders, and then finally taking the square root of that sum. The proof of this principle can be easily obtained by applying wake and momentum considerations to a system of two or more cylinders as indicated in Fig.4. At position P1 close to the cylinders there is no overlapping of the two wakes and thus no summation to be done. At position P2 far downstream it is easily shown by the equations of Section 2 that in order to preserve a total momentum in the wake consistent with the sum of drag force on the two cylinders, RMS summation of the wakes must be applied.

The wake considerations described above represent hydrodynamic interaction between the cylinders in

56

the sense that the current force on a cylinder situated downstream of other cylinders is influenced by the wakes of the upstream ones, but not vice versa. However, even in a non-viscous fluid there is hydrodynamic interaction between the cylinders in an array. Each cylinder will surround itself with a flow field due to its displacement effect. As described in Ref.8 it is possible to correct for this effect by substituting each cylinder by a two-dimensional source/sink dipole.

In addition to the in-line force a riser situated in the wake of an upstream riser may also be subject to a static transverse force. Ref.9 gives some information also on this force component.

Results obtained with the above calculation procedure have been compared with experimental results. As described in Refs.8 and 9 the procedure predicts very accurately the static force on each cylinder in an array of rigid cylinders, as well as the force on the complete array.

5 METHOD FOR PREDICTING COLLISION BETWEEN RISERS

With the numerical tools described in the previous section we are now in a position to calculate the mean shape (current-induced static deflection) of each individual riser in the riser array of a TLP, and thus evaluate whether they will collide or not.

Input data to the calculation procedure are the horizontal coordinates of upper and lower end of each riser in the array, pretension at upper and lower end of each riser, diameter and drag coefficient of each riser, current velocity profile and direction.

The calculation procedure itself consists in doing the following calculations for each riser, starting with the far upstream riser, and continuing with each riser in the sequence of their position in downstream direction:

-   First the top end inclination angle of the riser is calculated under the assumption that the riser was sitting in undisturbed current.

-   Using the above inclination angle as a starting value the shape of the riser is calculated by Runge-Kutta integration of its curvature. The curvature at each level down the riser is calculated by the principles described in the above Sections 2 through 4, taking into account the wakes generated by all upstream risers.

-   The deviation of the resulting lower end point from the lower end point coordinate specified in the input is used to correct the inclination angle at the top end, and the integration along the riser is repeated. This iteration process is continued until one "hits the target", i.e. the specified lower end coordinates, or the riser collides with its upstream neighbour.

This calculation procedure can now be used to predict the performance of specific riser arrays, and also to obtain more general understanding of the behaviour of riser arrays.

6 STABILITY

An interesting result from the numerical calculations is that a riser with constant pretension, sitting in the wake of an upstream riser, can have more than one stable position. Fig.5 illustrates the physics of this type of interaction between two risers.

Let us consider what happens in the middle part of the risers. The wake field generated by the upstream riser is such that the drag on the downstream riser increases if it moves further downstream. This change in drag force with position can be considered as a negative restoring force. If the downstream riser is given a small displacement in or against the flow direction, the change in drag force will tend to increase this displacement, thus producing a de-stabilizing effect. The stabilizing effect, or positive restoring force, is produced by the combination of constant pretension and curvature of the riser. Depending on which of the two effects is the most pronounced, the downstream riser may represent a physically unstable system. In the numerical calculations it results in two or more different stable positions or camber values for the downstream riser. In Fig.5 this is indicated by the two dotted shapes.

Experience from the numerical calculations so far indicates that such instability can occur only when the camber of the upstream riser is considerably larger than the spacing at top and bottom. Furthermore, it occurs only for a limited range of pretension in the downstream riser.

If such instability should occur in real life it might well lead to galloping-like oscillations where the downstream riser oscillates between the two stable positions at a typical period corresponding to the first vibratory mode of the riser.

7 EXPERIMENTAL INVESTIGATION OF RISER COLLISIONS

7.1 Test set-up, model description

The tests were done in MARINTEK's Towing Tank No.III, which is 85 m long, 10.5 m wide, and 10 m deep. Fig.6 shows a principle sketch of the test set-up. An array of 5 riser models, A, B, C, D, and E, arranged in a single row, were tested. Each riser consisted of a stainless steel pipe. Their main (model scale) data were:

- length   8.25 m
- outer diameter   0.010 m
- inner diameter   0.008 m

The center-to-center spacing between neighbouring risers was 0.15 m (15 diameters) at top and bottom fixtures.

The top end of the risers were suspended from the towing carriage structure F by universal joints. At the lower end the risers were fixed to rods G, which were sliding freely up and down through teflon-lined guide holes in a heavy "bottom frame" H. This frame was suspended from the towing carriage by a wire I in each of its 4 corners. To the rods G were fitted lead weights J, serving the purpose of providing constant pretension in the risers. A tow-line K connected the bottom frame to a position far ahead on the towing carriage. A homogeneous inflow current of velocity V_c was produced by towing the whole test set-up along the tank.

7.2 Instrumentation

One of the main purposes of the tests was in fact to test various types of instrumentation for studying the behaviour of much longer model riser arrays in the open sea. The instrumentation used for producing the test results described in the present report consisted of:

- an underwater video camera looking at the risers in transverse direction at the level half-way down the risers

- accelerometers specially designed for fitting inside the riser models (inner diameter only 8 mm !), measuring in the two horizontal directions,

- an electronic system for measuring the distance between individual risers.

7.3 Test program

In this report is presented results from the following test program:

- the towing velocity was varied from 0.2 to 1.0 m/s in steps of 0.05 m/s,

- 3 different pretensions were tested, 80, 120, and 200 N. (referring to top end pretension, the lower end pretension being 17 N less.)

7.4 Test results

The accelerometer signals show the vibrations of the riser models due to the vortex shedding. By double integration of the signals the path travelled by the riser model is revealed. Fig.7 shows an example of such a path, recorded for riser A (the one far upstream).

Fig.8 shows an example of the distance measurement. Due to certain electronic interactions between the risers the calibration of the distance probes are not quite reliable. However, Fig.8 shows a very interesting phenomenon. During the first 8 seconds of the time trace the distance between the two upstream risers (A and B in Fig.6) is about 75 mm (5 diameters). Then suddenly it falls to about 30 mm and stays there. A possible explanation of this behaviour could be the instability mentioned in Section 6 above.

The most complete impression of the overall behaviour of the riser array, for instance the exact conditions for on-set of collision between the risers, was obtained by studying the video recordings. Interesting observations were:

- For those combinations of current and pretension where the risers were far from colliding the only mode of motion was the lock-in vibrations.

- At conditions close to or beyond the on-set of collision the motion pattern was dominated by apparently stochastic motions at large amplitudes and typical frequencies far below the vortex-shedding frequencies.

- Collisions first took place between the two most upstream risers (A and B in Fig.6).

The most important objective of the present report was to define the combinations of current an pretension that would cause risers to collide. The experimental data defining on-set of collision between risers A and B are given in Fig.9. Since the tests were done by varying the current by finite steps, and partly also due to a certain stochasticism in the on-set of collisions, the experimentally determined critical velocity is in Fig.9 indicated by a certain range of velocity. The experimental data in Fig.9 are based exclusively on the video recordings.

8 VERIFICATION OF NUMERICAL CALCULATIONS

In Fig.9 is also plotted the combinations of pretension and current velocity defining on-set of collision according to the numerical calculations described above. The calculations have been done with a drag coefficient $C_d = 1.3$ for all risers. Other input quantities are identical to the relevant test conditions, so that the experimental and numerical results in Fig.9 are directly comparable.

As can be seen the experiments verify the calculations very well at pretensions 80 and 120 N. At 200 N there is some deviation. The deviation can be explained as a consequence of uncertain input drag coefficient in the calculations. A sensitivity analysis shows that if a range of drag coefficients of 1.3 to 1.5 had been applied, then all 3 numerical results would have overlapped the experimental ones. Since the lock-in vibration amplitudes of the risers vary with varying pretension and current such variation of the drag coefficient is quite realistic.

It should be noted that when studying the video recordings one has defined it as a collision condition even if only one single collision occurred during the recording time of 30 seconds. Thus even for current and pretension combinations well above the curve in Fig.9 there could be long time intervals without collisions occurring. A hypothesis to explain this behaviour is as follows: The numerical calculations have been done for zero inflow angle, i.e. they are based on the assumption that the second riser is sitting exactly in the peak of the wake of the first (upstream) riser. In practice there are instabilities and oscillations in the system shifting the second riser laterally away from the wake peak, where it will be subjected to larger inflow velocity, and the clearance between the risers will increase.

In conclusion, the comparison with experimental results does not reveal any basic errors or short-

comings in the theory or computer program (RICOL) used in the calculations.

9 THE SKARNSUND EXPERIMENT

The riser array used in the present experiments is not a realistic modelling of a riser system of a TLP for very deep sea (e.g. 1000 to 2000 m or more). The main difference is that the modes of vibration excited by the current will be much higher in a realistic prototype system. And it is for such deep sea riser systems that the current-induced collisions become an important design consideration.

In order to have full confidence in the prediction procedure it should be verified by testing more realistic models. In fact such tests were done in May and June 1993 as a joint industry project at MARINTEK. The tests took place at Skarnsund in the fjord some 100 km north of Trondheim. This is a narrow sound with 200 m depth and a tidal current of up to 3 knots. The model set-up was 50 m tall. It was installed and operated from a bridge 50 m above sea level. The results are proprietary and will not be open for publication until 1996.

10 CONCLUSIONS

-   A calculation procedure, based on a combination of wake and momentum considerations together with potential flow calculations, has been developed for predicting the total current force on the array as well as the current force on each individual cylinder in an array of cylinders.

-   The procedure predicts forces which correlate very well with available experimental data on the mean in-line current force on rigid cylinders in fixed positions.

-   Using the above procedure to calculate the force, and thus the curvature at each level, numerical integration of the curvature along the risers gives the static shape and deflection of each riser in the array. Thus it also provides an approximate prediction of possible collision between individual risers.

-   A series of experiments in a model tank verifies the numerical procedure for collision prediction quite well.

ACKNOWLEDGEMENT

The author acknowledges the assistance of two students, Mr. Bart Eelen and Mr. Jan Bal of Katholieke Universiteit te Leuwen. They did their thesis at NTH/MARINTEK in 1992. Figs. 7 and 8 in the present paper has been taken from their report.

LIST OF SYMBOLS

b	half-width of wake, see Figs.1 and 2
C_d	drag coefficient
C_{d1}	drag coefficient of cylinder generating wake
C_{d2}	drag coefficient of second cylinder
D	diameter
D_1	diameter of cylinder generating wake
D_2	diameter of second cylinder, see Fig.3
F	force pr. unit length of cylinder
T	top end pretension
U_0	peak wake velocity
u	wake velocity profile
u_n	wake velocity RMS averaged over cylinder No.n
V_c	current velocity
V_{crit}	current velocity at on-set of collisions
x	coordinate in flow direction
x_s	distance from wake source
x_v	distance from cylinder center to virtual source
y	coordinate in transverse direction
α	inflow angle
ρ	density of water

REFERENCES

1 Bushnell, M. J. 1977. "Forces on Cylinder Arrays in Oscillating Flow", Paper No. 2903, Offshore Technology Conference, Houston.

2 Heideman, J. C. and Sarpkaya, T. 1985. "Hydrodynamic Forces on Dense Arrays of Cylinders", Paper No. 5008, Offshore Technology Conference, Houston.

3 Report of Ocean Engineering Commettee 1990. Proceedings International Towing Tank Conference, Madrid.

4 Huse, E. and Muren, P. 1987. "Drag in Oscillatory Flow Interpreted from Wake Considerations", Paper No.5370, Offshore Technology Conference, Houston.

5 Verley, R.L.P., Lambrakos, K.F., Reed, K. 1987. "Prediction of Hydrodynamic Forces on Seabed Pipelines", Paper No.5503, Offshore Technology Conference, Houston.

6 Huse, E. 1990. "Resonant Heave Damping of Tension Leg Platforms", Paper No.6317, Houston.

7 Cuffe, P.D., Finn, L.D., Lambrakos, K.F. 1990. "Compliant Tower Loading and Response Measurements", Paper No.6313, Offshore Technology Conference, Houston.

8 Huse, E. 1992. "Current Force on Individual Elements of Riser Arrays", Paper No.C5-83, Proceedings ISOPE, San Fransisco.

9 Huse, E. 1993. "Interaction in Deep-Sea Riser Arrays", Paper No.7237, Ofshore Technology Conference, Houston.

10 Schlichting, H. 1968. "Boundary Layer Theory", McGraw Hill Book Company Inc., New York.

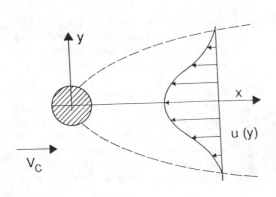

Fig.1. Wake profile behind a
------ cylinder in stationary flow.

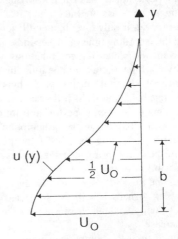

Fig.2. Definition of wake
------ half-width b.

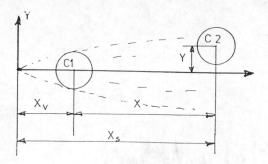

Fig.3. Definition of virtual
------ source position.

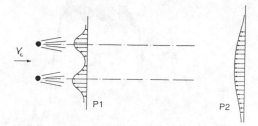

Fig.4. RMS summation of wakes.

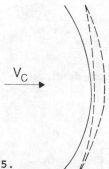

Fig.5.

Principle sketch showing shape
of stable upstream riser (solid
line) and two stable positions of
downstream riser ((broken lines).

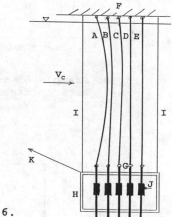

Fig.6.

Principle sketch of test set-up.

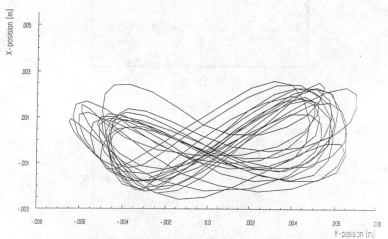

Fig.7. Example showing polar plot of vortex-
------ induced riser motion. Riser C, level
half-way down, V_c=0.5 m/s, T=120 N.

61

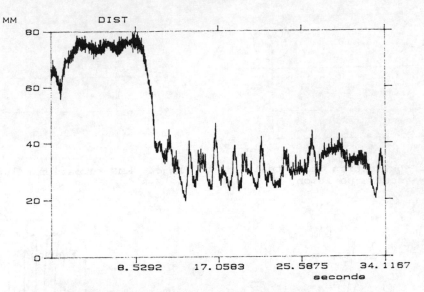

Fig.8. Example of recording showing distance
------ between risers A and B. Level half-way down,

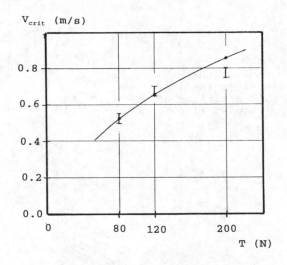

Fig.9. Combinations of pretension and
------ current defining on-set of collision
       between risers A and B.
       Solid line: numerical results
       Vertical bars: range of experimental
                        results.

Hydroelasticity in Marine Technology, Faltinsen et al. (eds) © 1994 Balkema, Rotterdam, ISBN 90 5410 387 6

Response calculation using an enhanced model for structural damping in flexible risers compared with full scale measurements

Tor D. Hanson & Audun Otteren
E&P Research Centre, Norsk Hydro a.s., Bergen, Norway

Nils Sødahl
MARINTEK, Trondheim, Norway

ABSTRACT: Flexible risers and hoses are some of the most important components in floating production units for oil and gas. Hence, comprehensive research efforts are going on in order to understand the nature of the dynamic response of such components.

Some important mechanisms in this context, are the cross-section formulations including bending stiffness and structural damping. A common analysis approach is the application of a stiffness proportional global damping matrix (Rayleigh damping). In the present paper also refined damping models are applied. The nonlinear bending stiffness of the riser and the internal friction between the different layers in the pipe structure are more realistically described in a hysteresis model. An enhanced Rayleigh damping model is also supplied where the contributions from axial- torsion- and bending-responses are separated. This shows a significant improvement of the simulation results which were compared to full scale measurements undertaken at the production and test ship "Petrojarl 1" while operating at the Oseberg and Troll Fields in the North Sea at 110 m and 330 m water depths respectively.

1 INTRODUCTION

Cross-section modelling of flexible pipes has been identified to be crucial to the computer prediction of dynamic curvature response. The scope of the present work is to perform comparison between simulations and full scale measurements, using data from the test and production ship "Petrojarl 1" at the Oseberg and Troll fields, with focus on an accurate modelling of stiffness and structural damping properties of flexible pipes.

The "Petrojarl 1" measurement concept is shown in Figure 1. Major system parameters both for the Oseberg and Troll fields are given in Table 1. Static and dynamic curvatures as well as governing environmental parameters were measured. The Oseberg measurements have been published by Otteren and Hanson (1990), and experiences from full scale measurements both at Oseberg and Troll are given by Hanson and Otteren (1992). The main

Table 1. Key field data

Production Test	Oseberg	Troll
Water Depth (m)	105	331
Flexible Riser	Coflexip	Coflexip
ID of riser (inch)	5	6
Length of riser (m)	165	480
Number of buoyancy elements	32	54
Static top tension of riser (kN)	60	250
Wellhead pressure max. (bar)	140	50
Wellhead temperature (°C)	85	60

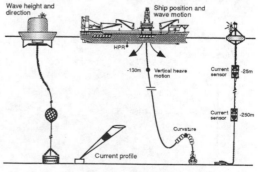

Figure 1. "Petrojarl 1" Riser configuration and instrumentation

conclusion from these studies was that simulations, using standard computational approaches and parameters, give an overprediction of the dynamic curvature when compared with the measurements.

2 CROSS-SECTIONAL PROPERTIES OF NONBONDED FLEXIBLE PIPES

A flexible pipe structure comprises of several different layers each having a very specific function. Due to the complex cross-section structure, it is convenient to describe the material stiffness properties of the pipe in terms of force-displacement relations such as axial force versus axial elongation and bending moment versus curvature.

The stiffness and structural damping properties of flexible pipes have been investigated by laboratory measurements on test specimen (Bech and Skallerud 1992, Bech et.al. 1992, Fang and Lyons 1992, Tan et.al. 1992). The main findings from these studies are summarized in the following .
- Different structural damping properties are related to axial, bending and torsional deformation.
- The axial and torsional stiffness can for all practical purposes be regarded as constant for a given internal pressure.
- Structural damping for axial and torsional deformations can be approximated by a viscous damping model. Equivalent viscous damping ratios of 1-3 % and 5% of critical are reported for axial and torsional deformations, respectively.
- A significant hysteretic behaviour is present in the bending-moment/ curvature relation for large amplitudes of bending deformation.
- An equivalent viscous damping ratio up to 25 - 30 % of critical damping can be established for large amplitudes of bending deformation with a fully developed bending moment - curvature hysteresis.
- The pipe structure behaves as a solid material for small amplitudes of bending deformation, i.e. no sliding, with a constant initial bending stiffness, EI_i , and a viscous damping ratio in the order of 5 % of critical.

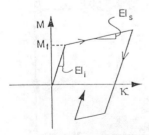

Figure 2. Idealized bending moment/curvature hysteresis from Sødahl et. al. (1992)

The characteristics of the bending hysteresis is governed by the material properties of each layer and the friction between the layers. The hysteresis can be idealized in terms of a friction moment M_f , an initial bending stiffness EI_i , and a sliding bending stiffness EI_s , as shown in Figure 2. The initial bending stiffness describes the stiffness before the static friction has been overcome. This stiffness is rather high because the flexible pipe structure behaves like a solid material as long as the friction moment is not exceeded. The sliding bending stiffness experienced when the friction moment has been exceeded, is much lower due to the relative movement between the layers in the cross-section. A stiffness ratio of $EI_i / EI_s \approx 10$ is estimated for the test specimen considered.

One should be aware that the available numerical data for stiffness and structural damping are based on a very limited number of test specimens. Stiffness and damping properties should therefore be provided by detailed numerical cross-section analysis or by laboratory measurements for each actual pipe considered. However, the reported physical cross-section behaviour is of general nature, and should therefore be adequately modelled in design analyses.

3 METHODS OF RISER ANALYSIS AND DAMPING FORMULATION

3.1 Equation of equilibrium

The most important nonlinear effects present in flexible riser systems are Morrison-type hydrodynamic loading, geometric stiffness, and material properties. All these effects can be properly described using a finite element method approach (FEM), and a nonlinear time domain step by step numerical integration of the incremental equilibrium equation.

The incremental dynamic equilibrium equation for a spatial discretized riser system can be expressed as:

$$M_T \Delta \ddot{r} + C_T \Delta \dot{r} + K_T \Delta r = \Delta Q \qquad (1)$$

where :

M_T , C_T , K_T - Tangential mass, damping and stiffness matrices.

$\Delta \ddot{r}$, $\Delta \dot{r}$, Δr - Incremental nodal acceleration, velocity and displacement vectors.

ΔQ - Incremental load vector

A complete nonlinear analysis is rather time consuming because an equilibrium iteration is performed at each time step. In many practical applications it is sufficient to use a more efficient linearized analysis without significant loss of accuracy. This approach is based on linearization of the inertia, damping, and stiffness forces at the static equilibrium position, which means that the mass, damping and stiffness matrices are kept constant throughout the analysis. However, nonlinear loading and relative motions between structure and water particles are included.

Alternative structural damping formulations which can be applied in FEM-analysis of flexible risers are discussed in the following sections.

3.2 The global Rayleigh damping model

The tangential damping matrix is in this case expressed as a linear combination of the tangential mass and stiffness matrices :

$$C_T = \alpha_1 M_T + \alpha_2 K_T \qquad (2)$$

The global Rayleigh damping model gives a modal damping ratio, λ_i , expressed as a function of the damping coefficients and the eigenfrequency, ω_i , :

$$\lambda_i = \frac{1}{2}\left(\frac{\alpha_1}{\omega_i} + \alpha_2 \omega_i\right) \qquad (3)$$

α_1 and α_2 are denoted the mass and stiffness proportional damping coefficients respectively. The mass proportional damping is in most situations neglected in order to avoid structural damping due to rigid body motions (i.e. $\alpha_1 = 0.$) (Bech and Skallerud 1992). The stiffness proportional damping will give a modal damping ratio which is proportional to the eigenfrequency. The reason for introducing the global Rayleigh formulation in analysis is mainly computational conveniences. The approach gives stable numerical computation both for nonlinear and linearized time domain analyses. In practical applications α_2 is selected to give realistic energy dissipation at the peak frequency of the loading spectrum. The major disadvantage using the global Rayleigh model is that the damping coefficients α_1 and α_2 apply to all global degrees of freedom. Hence it is impossible to specify different damping ratio to e.g. axial and bending deformations.

3.3 The separated Rayleigh damping model

A modified Rayleigh damping model which allows for separation of damping coefficients into specific coefficients in tension, torsion, and bending has been proposed by Bech et. al. (1992).

In this approach the separated coefficients are applied to the relevant degrees of freedom in the local mass and stiffness matrices before transformation and assembly into the global system matrices. The element damping matrix is expressed as :

$$c = \alpha_{1t} m_t + \alpha_{1tor} m_{tor} + \alpha_{1b} m_b$$
$$+ \alpha_{2t} k_t + \alpha_{2tor} k_{tor} + \alpha_{2b} k_b \qquad (4)$$

where subscript "t","tor", and "b" refer to tension, torsion, and bending contributions respectively. The matrices c, m and k are local element matrices, e.g. k_b includes all bending deformation terms in the local element stiffness matrix.

The equivalent viscous damping ratios measured for tension, torsion and bending can now be introduced directly in the numerical model by use of a proper selection of the respective damping coefficients. However, the separated Rayleigh model does not provide a global damping matrix with complete orthogonality characteristics. Hence, relating the damping ratio to eigenfrequencies according to eq. 3 will, strictly speaking, be an approximation. To obtain the accurate damping ratios, simulation of free vibration decays could be performed.

This model may be applied in combination with a representative average dynamic bending stiffness for large amplitude bending vibrations. The model is applicable both in nonlinear and linearized analyses. Use of mass proportional axial damping should still be avoided due to the influence from rigid body motion.

3.4 Hysteretic model

A more realistic physical cross-section model is obtained by introducing the measured hysteretic bending moment / curvature relations directly in the simulation model. (Fylling and Bech, 1991, Bech and Skallerud 1992).

This formulation will introduce an average dynamic bending stiffness which is larger than the sliding bending stiffness. The structural bending damping will be governed by the bending moment / curvature hysteresis.

Since the dynamic curvature normally shows a considerable variation along the riser, the hysteretic model will only give hysteretic damping in areas of the riser where the dynamic bending moment exceeds the friction moment. The hysteretic models should therefore always be used in combination with the separated Rayleigh model to account for damping related to axial, torsional and small amplitude bending deformations. The combination of these damping models will also provide a more stable numerical simulation. The disadvantage using

65

the hysteretic model is the rather large computing time due to the need of a fully nonlinear analysis and small time steps.

4 COMPUTER MODEL SIMULATION APPROACH

All numerical studies in the present work have been carried out using the RIFLEX computer program for static and dynamic finite element analysis of flexible riser systems (Sintef 1987). Nonlinear and linearized time domain analyses as well as all the damping formulations described in the former paragraph, are optionally available in this program system.

The riser models are build up using 3D beam elements. The equally spaced buoyancy elements were regarded as a continuously wrapped buoyancy. Hydrodynamic coefficients and cross-sectional data are mainly based on the pipe vendor's design specifications.

Measured vessel position and heading, wave spectra including wave direction, and current profiles were used as input in the numerical simulations. Vessel motion transfer functions have been derived using the panel analysis program WAMIT (1991).

The simulations are restricted to static (mean) riser profile (influence from current and vessel offset), and to dynamic riser response caused by direct wave excitation and by first order riser top-point (vessel) motions.

Timeseries for wave elevation and wave kinematics are obtained using an inverse fast fourier transform of the measured wave spectra. The simulated 1. order vessel response timeseries are based on measured wave spectra and computed motion transfer functions.

5 COMPARISON OF MEASUREMENTS AND COMPUTER ANALYSES AT OSEBERG

Measured time series with a duration of 10 minutes are available for the curvature response in the hog bend for the Oseberg system. The comparative studies are limited to mean values (i.e. static response) and standard deviations of the response.

5.1 Static curvature

Results from comparison of computed static curvature to mean values from full scale recordings are presented by Otteren and Hanson (1990). The curvature was close to 0.2 m^{-1} for all cases considered, and good agreements between measurements and computer analysis were obtained.

5.2 Dynamic curvature response using a global Rayleigh damping

Design analyses of flexible risers have traditionally been carried out using a global stiffness proportional Rayleigh damping model in combination with the sliding bending stiffness. This approach has been applied in comparative studies reported by Otteren and Hanson (1990) using a stiffness proportional damping of 20 % of critical at a eigenperiod of 10 seconds. It was found that simulations gave an overprediction of the dynamic curvature response with a factor of approximately 2 for the highest sea states in the range of 4 - 8 metres.

5.3 Dynamic curvature response using a hysteretic model

With background in the discrepancies reported by Otteren and Hanson (1990) additional case studies were carried out by Sødahl et al. (1992) using a hysteretic bending moment / curvature formulation. A stiffness ratio of $EI_i/EI_s = 10$, and a friction moment of $M_f = 0.5$ kNm were applied in all simulations. The hysteretic model was used in combination with a global stiffness proportional damping of 6 % of critical.

Analysis using the constant sliding bending stiffness and a global stiffness proportional damping of 6 % of critical, ($M_f = 0.0$ $\alpha_2 = 0.2$) , was included for comparison to the hysteretic approach . Results from comparative studies are presented in Figure 3. The combined effects of an increased average dynamic stiffness and hysteretic damping contribute to closer agreement with full scale measurements. A significant hysteretic damping was found for the highest sea state while the improvement observed

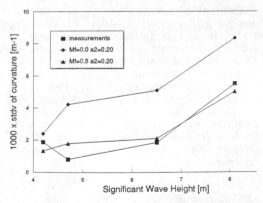

Figure 3. Dynamic curvature vs. significant wave height, Oseberg riser. Use of Hysteretic damping model

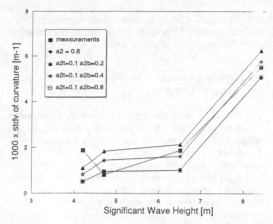

Figure 4. Dynamic curvature vs. significant wave height, Oseberg riser. Use of separated Rayleigh damping model

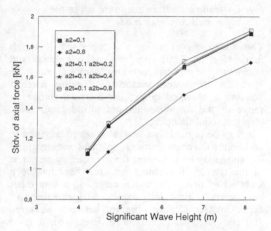

Figure 5. Dynamic top tension vs. significant wave height, Oseberg riser. Use of separated Rayleigh damping model

for the lower sea states is mainly due to an increased average dynamic bending stiffness (Sødahl et. al. 1992).

5.4 Dynamic curvature response using the separated Rayleigh model

Case studies have been carried out in order to investigate the benefit of using the separated Rayleigh damping model. The initial bending stiffness is used in all analysis due to the rather low overall dynamic response level observed. A stiffness proportional damping coefficient of $\alpha_{2t} = 0.1$ is used for tension in combination with a bending damping coefficient of $\alpha_{2b} = 0.2$, 0.4 and 0.8 respectively.

Analysis using a global stiffness proportional damping of 3 % and 25 % of critical (i.e. $\alpha_2 = 0.1$ and 0.8) have also been included for comparison to the separated damping model.

Standard deviation of curvature at the hog bend, and standard deviation of axial force at upper end of the riser are given as functions of significant wave height in Figures 4 and 5, respectively.

The tension response is found to be sensitive to the global stiffness proportional damping coefficient α_2, while it is seen that the tension response for all practical purposes is independent of the damping level used in bending in the separated damping model. Further, the tension response predicted using the separated damping model with $\alpha_{2t} = 0.1$ is in good agreement with results using a global Rayleigh model with $\alpha_2 = 0.1$ It is also seen that the curvature response found using the separated Rayleigh model with $\alpha_{2t} = 0.1$ and $\alpha_{2b} = 0.8$, agrees very well with results using the global Rayleigh model with $\alpha_2 = 0.8$ Hence, it has been demonstrated that the separated damping model is capable of describing different damping levels in bending and tension, and that the damping ratios for bending and tension can be estimated to close approximation by eq. 3 also while using a separated Rayleigh model.

Good agreement with full scale measurement is obtained using 12 % of critical damping in bending. It can therefore be concluded that the separated damping model represents a promising alternative for modelling of cross sectional properties of flexible pipes.

6 COMPARISON OF MEASUREMENTS AND COMPUTER ANALYSES AT TROLL

The static and dynamic curvatures in the hog bend were also addressed for the Troll riser. The length of the measured timeseries was the same as for the Oseberg case. Standard deviation and mean value of twenty-one measured timeseries curvatures are compared to the corresponding simulated realizations.

6.1 Static curvature

While the static bending radius at Oseberg was approximately 5 metres, the measured bending radius at Troll was close to 27 metres. The computed radius was equal to 20 metres. The measured static curvatures were quite stable. An example showing 7 different measurements is given in Figure 6.

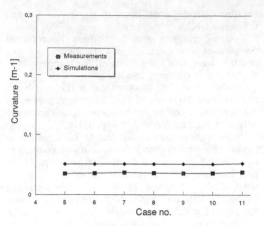

Figure 6. Static curvature of Troll riser

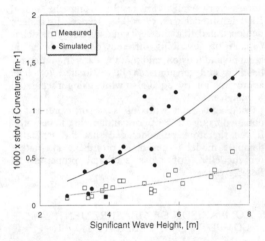

Figure 7. Dynamic curvature vs. significant wave height, Troll riser. Comparison of measured and simulated response

6.2 Dynamic curvature response using a global Rayleigh damping

A similar approach as the one described for Oseberg has been applied, i.e. the sliding bending stiffness was combined with a stiffness proportional damping of 20 % of critical at a period of 10 seconds. In Figure 7 the response is given as standard deviation of curvature versus significant wave height. The most severe sea states which were measured are close to 8 metres significant wave height. For these sea states the simulated responses are in average overpredicted by a factor of approximately 5 compared to the measurements. Using the global Rayleigh model and the sliding bending stiffness, it

will be impossible to obtain a close agreement with measurements. This without increasing the damping ratio far beyond a physical level.

6.3 Dynamic curvature response using a hysteretic model

For the Troll riser the dynamic curvature response is rather low. At the largest sea states (H_s in the range of 6 - 8 metres) the standard deviation of curvature is in the order of 10 to 30% of the response at Oseberg for similar wave heights. This response level is assumed to be below the threshold value for which sliding between layers develops. Hence, hysteretic damping will not be pronounced. Accordingly, the non-linear hysteretic damping model is not supposed to give any improvements in this case, and has not been applied for the Troll riser.

6.4 Dynamic curvature response using the separated Rayleigh model

The advantage of using the separated Rayleigh model for the Troll riser is that both mass- and stiffness proportional bending damping may be specified. Because a reasonable damping level in bending is obtained throughout the total frequency range of the response without introducing non-physical axial damping.

It may be argued that several other parameters in addition to the cross-sectional stiffness and structural damping may be important for the bending response of the risers. Therefore, the Troll riser has been selected to demonstrate the effect of a few other parameters. Especially, since a discrepancy of approximately 25 % is observed between measured and computed static curvature (Figure 6), it is of interest to investigate how parameters that govern the static profile, also may affect the dynamic response.

In the following a separated Rayleigh damping model is used to study the bending response as function of some input parameters which generally are of great importance while computing the response of flexible risers.

6.5 Drag Damping

The drag damping due to buoyancy devices could be significant if the riser motion is large. However, in this case the motion of the riser is very low both in longitudinal and transverse direction. Even though the quadratic damping coefficients especially in tangential direction may be taken extremely high due to very low KC-numbers, the total damping force is limited due to small motion amplitudes.

One case having a significant wave height of

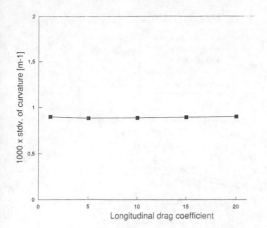

Figure 8. Dynamic curvature vs. longitudinal drag coefficient, Troll riser

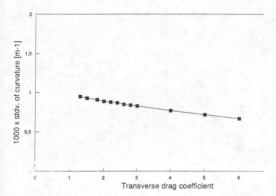

Figure 9. Dynamic curvature vs. transverse drag coefficient, Troll riser

approximately 7.5 metres, has been selected for sensitivity analyses. While the hydrodynamic parameters were varied, damping coefficients α_{b1} and α_{b2} were set to 0.1, and the bending stiffness was taken as 10 times the nominal value, which may be a reasonable choice of an initial bending stiffness.

The axial drag coefficient is taken in the range between 1.2 (nominal value) and 20. According to the low KC-numbers a value of 15 is not too high. Figure 8 shows that the response is unaffected by the axial drag damping.

The lateral drag coefficient is varied in the range between 1.3 and 6.0. as shown in Figure 9. No evidences have been identified which justify larger values than 2.0. Hence, the response is only slightly influenced by the lateral drag coefficient.

6.6 Bending stiffness

The given nominal bending stiffness which is provided by the vendor, is supposed to be the sliding stiffness which represents the stiffness of the pipe when the different pipe structure layers are allowed to slide against one another. However, theoretical computations using a computer program for capacity analysis of flexible pipes, indicate that the real sliding stiffness for the Troll riser may be twice as high as the nominal value. Hence, the real initial stiffness in this case may be 20 times the nominal value. In the buoyancy area of the riser the buoyancy clamping devices may cause the stiffness to be even higher. If each clamping device is assumed to be infinitely stiff compared to the nominal bending stiffness, the average bending stiffness in the buoyancy zone will be in the order of 30 times the nominal stiffness.

Figure 10 depicts the graph of static bending radius in the hog-bend as function of bending stiffness. The abscissa values are given as the ratio of nominal bending stiffness. If the bending stiffness increases above 10 times the nominal value, it is observed that the stiffness becomes important for the static riser configuration. The discrepancy between measured and computed static bending radius as given in Figure 6, could be explained through the uncertainty of the real bending stiffness of the pipe. The measured and the computed bending radii correspond if the bending stiffness is increased approximately 150 times the nominal value. However, this value is probably far above a reasonable bending stiffness even though contributions from buoyancy element clamping devices are included.

In order to study the importance of bending stiffness in combination with different damping levels, a case study having a significant wave height of 7.5 metres has been analyzed. The stiffness has been selected in the range from 0.1 to 150 times the nominal value. The damping is given as a mass-

Table 2. Damping coefficients, case studies

Case	α_{1b}	α_{1b}	Damping ratio % at 10 sec. eigen-period	Damping ratio % at 20 sec. eigen-period
A	0.	0.	0.	0.
B	0.05	0.05	5.6	8.7
C	0.1	0.05	9.5	16.7
D	0.2	0.05	17.5	32.6
E	0.4	0.05	33.4	64.5

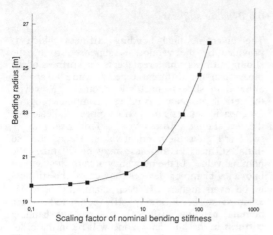

Figure 10. Static bending radius vs. scaling factor of nominal bending stiffness, Troll riser

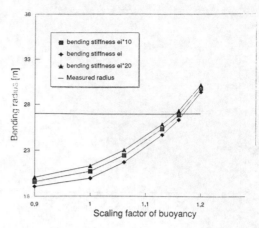

Figure 12. Static curvature vs. scaling factor of buoyancy weight, Troll riser

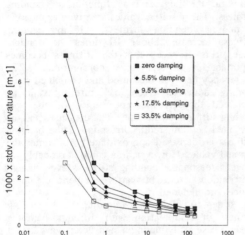

Figure 11. Dynamic curvature vs. scaling factor of nominal bending stiffness, Troll riser

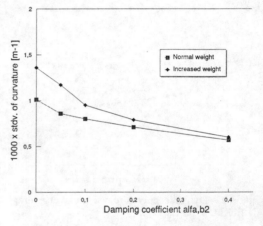

Figure 13. Dynamic curvature vs. damping coefficients, Troll riser

and stiffness- proportional bending damping according to the separated Rayleigh model. Hence, there is a total decoupling of bending, torsional and axial damping where in this case axial and torsional damping are neglected. The actual damping coefficients (α_{1b}, α_{2b}) are given in table 2.

Figure 11 gives the standard deviation of dynamic curvature versus bending stiffness for the different levels of damping ratio.

It is obvious that both bending stiffness and damping level influence the curvature response significantly. The smaller the bending stiffness, the larger is the bending response. Further, the smaller the bending stiffness, the more important becomes the damping ratio.

The measured dynamic response (standard

deviation of curvature) in this case is approximately $5.73 \cdot 10^{-4}$ m^{-1}. In order to obtain this response from the simulations, bending stiffness and damping ratios may be tuned. Figure 11 shows that a relatively moderate damped system (case B) in combination with a highly increased bending stiffness (150 times nominal value), or a heavily damped system (case E) combined with a realistic increase of bending stiffness (20 times nominal value), give both a response level close to the measured response. However, neither of these two sets of parameters are expected being within a reasonable physical range.

6.7 Weight of buoyancy elements

It is reasonable to assume that over some time the buoyancy elements could absorb water which cause the net buoyancy to decrease. In Figure 12 is depicted the static bending radius in the hog bend as function of buoyancy weight. Using a bending stiffness of 20 times the nominal value, a 15 % increase of buoyancy weight is sufficient to obtain correspondence between measured and calculated static bending radius.

The dynamic curvatures (standard deviations) are given as function of damping level in Figure 13. Both normal buoyancy weight and increased buoyancy weight are included. Unfortunately, the result shows that the weight increase also causes the response to increase. Hence, increase of buoyancy weight does not give an improved correspondence between measurements and simulations.

6.8 Top-point excitation and wave excitation

The influence from excitation sources is investigated for a case with bending stiffness 20 times the nominal value, adjusted buoyancy weight, and damping coefficients $\alpha_{1b} = 0.1$ and $\alpha_{2b} = 0.05$.

In order to study the main sources of riser excitation, a range of different excitation combinations were addressed. The results are given in table 3 :

Table 3. Response level of dynamic curvature given for alternative combination of excitation sources.

Case no	Description of case	Response $m^{-1} *10^{-3}$
1	Basic parameters	0.946
2	Direct wave excitation neglected	0.879
3	Horizontal top-point motion neglected	1.074
4	Vertical top-point motion neglected	0.244
5	Both direct wave excitation and horizontal top-point motion neglected	0.984
6	Current neglected	0.998
7	Modified wave spectrum shape (PM-spectrum)	1.089

It is obvious that none of these modifications, except ignoring vertical top-point motion, influence the response level dramatically. Hence, the main source of excitation seems to be the vertical top-point motion.

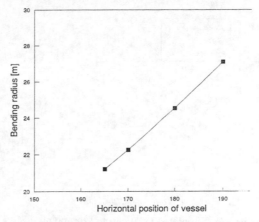

Figure 14. Static curvature vs. horizontal position of vessel, Troll riser

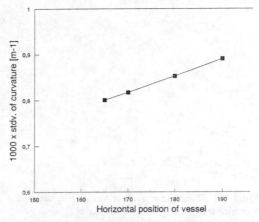

Figure 15. Dynamic curvature vs. horizontal position of vessel, Troll riser

6.9 Horizontal distance between riser end-terminations

The nominal average horizontal distance between well head and top of riser is 165 metres. Sensitivity analyses are carried out using different horizontal distances, and keeping the bending stiffness equal to 20 times the nominal value. As shown in Figure 14 it is necessary to increase the horizontal distance to 190 metres in order to obtain correspondence between measured and calculated static curvature, i.e change of bending radius from 20 to 27 metres.

In Figure 15 it is depicted how the variability of horizontal distance influences the dynamic curvature. Since the dynamic response is showing a moderate increase as the horizontal distance is enlarged, change of horizontal distance does not

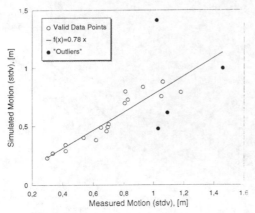

Figure 16. Simulated vs. measured top-point motion, Troll riser

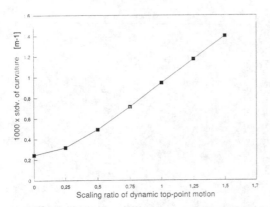

Figure 17. Dynamic curvature vs. scaling ratio of dynamic top-point motion, Troll riser

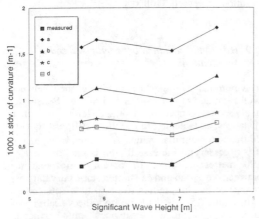

Figure 18. Dynamic curvature vs. significant wave height, Troll Riser. Final comparison to measurements

make any improvements to the correspondence between measurements and simulations.

6.10 Accuracy of Troll measurements

Curvature Sensors :
The theoretical accuracy of the measured static curvature as well as the accuracy of the standard deviation of dynamic curvature have been calculated, and found to be within an order of 3 - 4 %. However, if there is a systematical bias in the measurements, the error may increase correspondingly. Such a systematic bias may be that the curvature sensors are not in the anticipated position along the riser, or that the sensors are rotated relatively to the main axis of the riser. Based on the comparison of measured and calculated static curvature in the hog bend, a systematic bias of 25 % could be present.

Top-point motion of riser :
As shown in the preceding section, the vertical motion of the riser top gives the most important contribution to the bending response. Vertical motion of the upper part of the riser has been measured. This motion reflects the heave motion in the turret-area. Figure 16 depicts the correspondence between measured and simulated vertical motion. Except for a few "outliers", the trend is that the simulated standard deviations give approximately 80% of the measured response. The reason for this discrepancy is not fully identified, but an important factor may be sensitivity to wave heading, and that any difference between wind sea and swell wave direction is not included in the simulation model.

In Figure 17 it is shown how the curvature response is varied when the vertical top-point motion has been scaled. The correspondence between variation of vertical top-point motion and curvature is quite linear. Hence, if it is assumed that measured vertical top-point motion is correct, and this excitation is used as input to the simulations, a 25% increase of the simulated curvature response is anticipated.

The total effect of these "corrections" of measurements and simulations will almost cancel one another. Hence, these are not encountered in the comparison of measurements and simulations.

6.11 Final comparison between measurements and simulations at the Troll riser

Based on the previous sensitivity analyses all the various improvements of parameters have be included in the simulation of four different measur realizations. The simulations are carried out for several combinations of the stiffness and damping parameters in order to show the improvements of results :

a)   -Nominal bending stiffness, 3% of critical global Rayleigh damping, linear dynamic analysis.
b)   -Nominal bending stiffness, 20% of critical global Rayleigh damping, linear dynamic analysis.
c)   -Initial bending stiffness (20 times nominal), 6% of critical separated rayleigh damping, linear analysis
d)   -Initial bending stiffness adjusted for clamping devices (30 times nominal), 6% of critical separated Rayleigh damping, non-linear analysis.

The results are shown in Figure 18. Compared to the original simulations using the sliding bending stiffness and a global Rayleigh formulation for damping, case a and b, the improved simulations, case c and d, are approaching the level of the measured response.   For the case with highest response the correspondence between measurements and simulations has become quite close.   The ratio between simulated and measured response has been decreased from 5 to a ratio in the order of 2 for the three other realizations.   Taking into account the general low level of the response, the use of a separated Rayleigh model in combination with an adjusted bending stiffness shows a significant improvement of simulated response.

7 CONCLUSIONS

For the Oseberg riser a separated Rayleigh damping model has shown to be a relevant alternative for representing of cross-sectional and damping properties of the riser.   The advantage of the separated damping formulation is that different damping levels may be applied for axial, torsional and bending damping.   Further, this formulation can be used for a linearized time domain analysis which is much faster than a hysteretic damping formulation since the latter always require a full non-linear analysis.   Good agreement with measurements is obtained while using the initial bending stiffness in combination with an equivalent bending damping equal to 12 % of critical.

Parameter studies carried out for the Troll riser have revealed that bending stiffness, structural damping and vertical top-point motion is the dominant parameters governing the curvature response in the hog-bend. The separated Rayleigh damping formulation has shown to improve the results of simulations significantly also for the Troll riser. However, due to the rather low response level some discrepancy still remains.

The curvature response simulation has shown to be very sensitive to a correct cross-sectional modelling. This is believed to be most important for durability assessment of the riser including wear and fretting.

Taking into account all the uncertainties related to full scale measurements at an operating vessel, the overall conclusion is that computer simulations of curvature response of flexible risers prove acceptable agreement with measurements if a detailed formulation of the cross-sectional and damping properties is provided.

8 REFERENCES

Bech A. , et al , "Structural Damping in Design Analysis of Flexible Risers" Proc. of the First European Conference on Flexible Pipes, Umbilicals and Marine Cables, London 1992

Bech A. , Skallerud B. , "Structural Damping in Flexible Pipes: Comparisons Between Dynamic Tests and Numerical Simulations" Proc. of the Second International Offshore and Polar Engineering Conference, San Francisco, USA, 14-19 June 1992

Fang J. , Lyons G.J. , "Structural Damping Behaviour of Unbonded Flexible Risers", Marine Structures Vol. 5 No 2&3 1992

Fylling I. , Bech A. , "Effects of Internal Friction and Torque Stiffness on the Global Behavior of Flexible Risers and Umbilicals" Proc. of the 10. International Conference on Offshore Mechanics and Arctic Engineering (OMAE), Stavanger Norway, 23-28 June 1991

Hanson T.D. , Otteren A. , "Experiences Using Full-Scale Measurements for validation of Theoretical Models", Proc. of the International Seminar on Recent Research and Development within Flexible Pipe Technology, The Marine Technology Centre, Trondheim Norway, 18-20 February 1992

Otteren A. , Hanson T.D. , "Full Scale Measurements of Curvature and Motions on a Flexible Riser and Comparison with Computer Simulations" Proc. of the 9. International Conference on Offshore Mechanics and Arctic Engineering, (OMAE) Houston, Texas February 18-23, 1990

Sintef (1987) : RIFLEX - Flexible Riser System Analysis Program, User Manual, MARINTEK and Sintef, Div. of Struct. Engineering, Trondheim Norway

Tan Z. , et. al. , "On the Influence of Internal Slip Between Component Layers on The Dynamic Response of Unbonded Flexible Pipe" Proc. of the 10. International Conference on Offshore Mechanics and Arctic Engineering (OMAE), Stavanger Norway, 23-28 June 1991

WAMIT, Version 4.0. (1991) A Radiation-Diffraction Panel Program for Wave - Body Interactions. Dept. of Ocean Engineering, Massachusetts Inst. of Technology.

Hydroelasticity in Marine Technology, Faltinsen et al. (eds) © 1994 Balkema, Rotterdam, ISBN 90 5410 387 6

A study of the dynamics of hanging marine risers

K. Matsunaga
Ishikawajima-Harima Heavy Industries Co. Ltd, Tokyo, Japan

M. Ohkusu
Research Institute for Applied Mechanics, Kyushu University, Japan

ABSTRACT: The dynamics of a relatively short marine riser in hang-off mode were studied experimentally and numerically. Here we assumed the predominance of the vending deflection, demonstrating good agreement with each other. For a long hanging riser, numerical studies were performed to clarify the effect of the elongation. Incorporating the axial rigidity, our numerical method was developed to succeed in simulating the complex behavior of the elongation vibration.

1 INTRODUCTION

As offshore drilling operations develop into deeper waters, the requirements placed on the drilling riser become increasingly severe. In addition, a deep-sea drilling vessel system is now under the early stage of a design study in Japan. Retrieval of the drilling riser in the face of developing severe weather becomes critical for these risers. Long retrieval time will lead the riser be disconnected from the wellhead and suspended from the ship under storm conditions. When this long suspended riser is subjected to severe sea conditions, large axial and lateral forces may be induced in the riser and may cause riser damage (Long 1983, Wang 1983, Denison 1984).

The precise prediction of the dynamic behavior of a marine riser is of primary importance, for assuring the good performance of a marine riser itself and that of riser systems totally including a connecting body in waves. The characteristics of the dynamic behavior of a marine riser can be considered to depend on the length stronger than the diameter and the thickness. In this report, the dynamics of a rigid riser in hang-off mode, which would be the most severe design condition, are studied numerically and experimentally focusing the effect of the length.

The dynamics of a suspended short riser may resemble those of beams than cables. As the length of the riser becomes longer, the effect of the elongation should be incorporated together with the bending moment. In this case, it is afraid that the natural frequency of the elongation vibration may close to the range of the ocean waves (Sparks 1982). When the length of the riser becomes more longer, the dynamics may resemble that of cables than beams except at the low-tensional points (Howell 1992).

The mechanisms for a relatively short hanging riser's response to excitation, planar harmonic excitation at the top, are explored experimentally and numerically. An elastic model, which corresponds to an actual riser pipe with 400 m in length, is adopted for tank test, measuring the distributions of deflection and bending moment along the length with various exciting periods. Numerical simulations are also performed using discretized beam elements with bending stiffness, admitting finite amplitude effects of motion, and demonstrate good agreement with the measured data.

Secondary, the mechanisms for longer hanging risers are explored numerically. In these cases, the characteristics of the dynamic behavior resulting from the elongation may appear to be significant. As such, we seek to incorporate the effect of axial rigidity together with the bending stiffness. Two numerical approaches are developed to apply this particular complex problem, to clarify the characteristics of the dynamic behavior of a long marine riser.

2 DYNAMIC RESPONSE OF RELATIVELY SHORT HANGING RISER

In the beginnings, we treat a basic and rather simple problem in this section. Experimental and numerical studies are conducted to confirm the dynamic behaviors of a suspended rigid riser with relatively short length.

2.1 Numerical Simulation

There have been many researches conducted to predict

the dynamic behaviors of flexible risers (Paulling 1979, Dareing 1979). Two analytical ways have been applied to the problem. If one assume that a riser behaves like cables, the tensional force would govern the problem. On the other hand, if one assume that a riser behaves like beams, the vending deflection would be predominant. From physical intuition, the later may be fitted for the dynamics of a relatively short hanging riser.

Sol.0

The numerical method adopted here has already reported one of the current authors (Ohkusu 1992). Assuming that the effect of elongation on the bending and torsional deflections must be of higher order, the equations of the bending and torsional deflections can be formulated independently of the elongation.

The riser is divided into a number of segments that we assume do not deflect. This means that we discretize the riser as assembles of beam elements. The coordinate of the mass center (x_i, y_i, z_i) of each segment can be denoted using the motion of the upper end and the Eulerian angles $(\theta_i, \varphi_i, \psi_i)$. Here we define the Lagrangean of the total system and obtain the equations of the segments, as follows

$$
L = \sum_{i=1}^{N}\left[\frac{m}{2}\left(\dot{x}_i^2 + \dot{y}_i^2 + \dot{z}_i^2\right) + \frac{I_x}{2}\left(\dot{\theta}_i^2 + \dot{\varphi}_i^2 \sin^2\theta_i\right) \right.
$$
$$
+ \frac{I_z}{2}\left(\dot{\psi}_i^2 + \dot{\varphi}_i^2 \cos^2\theta_i + 2\dot{\psi}_i\dot{\varphi}_i\cos\theta_i\right)
$$
$$
+ \frac{\Delta m}{2}\left\{\dot{x}_i^2\left(\sin^2\varphi_i + \cos^2\varphi_i\cos^2\theta_i\right)\right.
$$
$$
+ \dot{y}_i^2\left(\cos^2\varphi_i + \sin^2\varphi_i\cos^2\theta_i\right) + \dot{z}_i^2\sin^2\theta_i
$$
$$
- \dot{x}_i\dot{y}_i\sin^2\theta_i\sin 2\varphi_i - \dot{y}_i\dot{z}_i\sin 2\theta_i\sin\varphi_i
$$
$$
\left. - \dot{z}_i\dot{x}_i\sin 2\theta_i\cos\varphi_i\right\} - (m - \Delta m)gz_i\Big]
$$
$$
- \sum_{i=1}^{N-1}\left[\frac{EI}{2l_i}\left\{\cos^{-1}(\sin\theta_i\cos\varphi_i\sin\theta_{i+1}\cos\varphi_{i+1}\right.\right.
$$
$$
\left.+ \sin\theta_i\sin\varphi_i\sin\theta_{i+1}\sin\varphi_{i+1} + \cos\theta_i\cos\theta_{i+1})\right\}^2
$$
$$
+ \frac{C}{2}\left\{\varphi_i + \psi_i - \varphi_{i+1} - \psi_{i+1}\right\}^2 \right] \tag{1}
$$

Where N denotes the total number of the segments, m and Δm represent the mass and the added mass of the segment respectively, I_x and I_z are the moment of inertia. In addition, EI denotes the bending stiffness and C is the torsinal rigidity. From this definition, differential equations of motion can be derived and are integrated step by step at the time step Δt with Newmark-β method. Hereafter we call this numerical method as Sol.0.

For example a trace of the deflection at every 1/16 period of the elastic riser model, which will be explained in the following chapter, is presented in

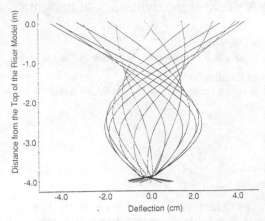

Fig.1 Simulated Deflection of the Elastic model (Pin-jointed, T=1.2 seconds)

Fig.1. In this numerical simulation, we used N=20 and Δt=0.025 seconds. The numerical computation was started from the initial rest condition, after 10 to 20 cycles the transient died out and steady oscillation would develop. The simulated profile of the deflection is fairly smooth as shown here.

2.2 *Experimental study*

Model test experiments using an elastic riser model were conducted to examine the dynamic behavior of a suspended rigid riser with relatively short length. We intended here to study the dynamic characteristics of the suspended rigid riser governed by bending experimentally, and intended to make code verification (Kitakouji 1993).

Elastic model

A 4.0m elastic model made by aluminum was selected here. Particulars of the model, accompanying that of the corresponding actual riser, are shown in Table 1. The model is constructed from 20 segments

Table 1 Paticulars of the Elastic Model

Principal Dimensions		Experimental Model	Actual Riser Pipes
	Scale	1/100	1
L	Length	4,000 mm	400m
l	Length of 1String	200mm	20m
do	Outer Diameter	4 mm	0.400 m
A	Sectional Area	12.57 mm²	0.0261 m²
I	Moment of Inertia	12.566368 mm⁴	4.68E-04m⁴
W	Mass per Unit Length*1	3.89E-09 kgsec²/mm²	2.09E-05 kgsec²/mm²
	Material Riser Pipes	Aliminum	Steel
	Bouy	Polyuretan Form	Syntactic Form
E	Modulus of Elasticity	7000.00 kg/mm²	21000.00 kg/mm²
EI		10.04E+04 kgmm²	9.8321E+12 kgmm²
EA		8.7965E+4 kg	5.4864E+8 kg

*1 : with bouy,mud water and lines

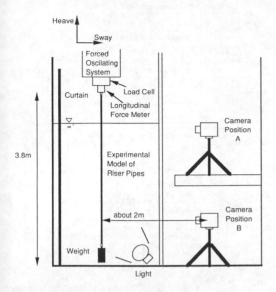

Fig. 2 Arrangement of Tank Test

connecting each other by a nut, and satisfies the similarity not only in outer geometry but also in bending stiffness. The data of the bending stiffness of the model shown here is experimentally measured.

Unfortunately we could not satisfy the similarity of the elastic model in axial rigidity and in weight . Each segment is rolled up by polyurethane form to make almost neutral buoyancy, which covers 95% of the total weight in the air as the actual riser. At the lower end of the riser model, a sinker (d=56mm, l=106mm, t=3mm, w=160g) is connected using a pin joint, modeling the BOP (Blowout Preventer).

Experimental Setup
An overview of the experimental setup is shown in Fig.2. The elastic riser model, hanging down into the water tank is forced to oscillate at its top end. The forced oscillation is intended to simulate the motion of a ship suspending the riser at sea. The conditions of the experiments are presented in Table 2. The amplitudes of the oscillation adopted here correspond to the maximum values expected in storm conditions.

We measured the distribution of the lateral displacements and the bending moments along the length as shown in Fig.3. Several nuts connecting

Table 2 Conditions of Experiments

Conditions	Pin、Fix
Forced Osclation Period	0.58~3.0sec.
Forced Osclation Mode	Heave & Surge Coupling
Amplitude of Oscilation	Heave : 46mm、Surge : 44mm
Phase	Heave=90deg. as Surge=0deg.

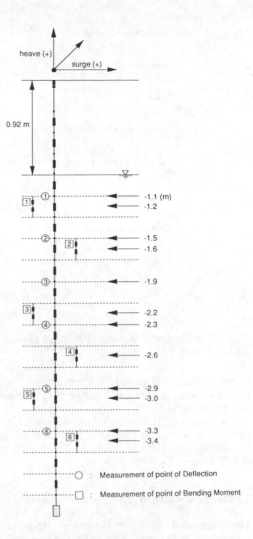

Fig. 3 Positions of Measurement

adjacent segments were illuminated in water, and their displacements were measured using X-Y tracking system. Strain measuring gauges were equipped on some segments to measure the bending moment.

The model was attached rigidly to the oscillator or attached with a pin joint, measuring the vertical and horizontal reacting forces. The axial force induced at the pin joint was also measured using a longitudinal force meter. Time series records for the oscillated period (fundamental period) T=1.2 and 0.6 seconds are presented in Figs.4 and 5, showing the experimentally obtained heave and surge oscillations, deflections and bending moments respectively.

When we oscillated the model with T=0.6 seconds, a strange motion with large amplitude was observed at the sinker. The period of this motion was about twice

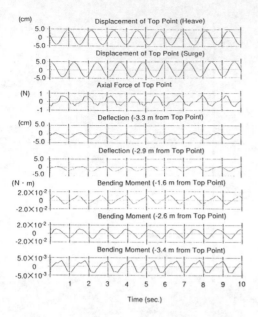

Fig. 4 Time Series of Measured Data (Pin-jointed, T=1.2 seconds)

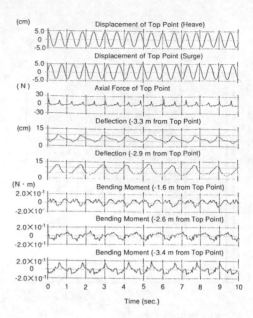

Fig. 5 Time Series of Measured Data (Pin-jointed, T=0.6 seconds)

as the fundamental period. The influence of this phenomenon on the deflection and bending moment appears clearly in Fig.5.

The drag coefficients used in the numerical simulation were also measured from pure surge oscillation tests. They were 1.42 for T=0.6 seconds and 0.92 for T=1.2 seconds.

2.3 *Computational and experimental results*

The measured data involve some frequency components as shown in Figs.4 and 5. The amplitudes belonging to the oscillated period were extracted from the time series data using a FFT analyzing system. Same processes are applied to the numerically simulated time series as well. Experiments were conducted with various frequencies, and here we give the results of typical three cases, T=0.6, 1.2 and 3.0 seconds respectively.

Deflection

In Fig.6, experimental and computational distributions of the deflection along the length are presented. The marks correspond to the measured horizontal displacements at each point. The numerical results denoted by lines have smooth and continuous curvature except at the lower end part of the riser. These kinks are caused by the motion of the sinker. There appear distinct nodes in the numerical simulation, but in the experimental results they are

not so clear. The numerical results demonstrate fairly well agreement with the measured data.

Bending moment

The distributions of the bending moment along the length are presented in Fig.7. The numerical simulation gives good prediction of the measured data

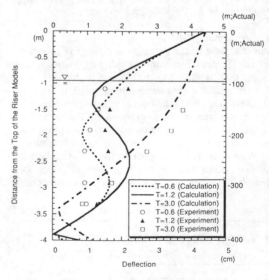

Fig. 6 Deflection of the model (Pin-jointed)

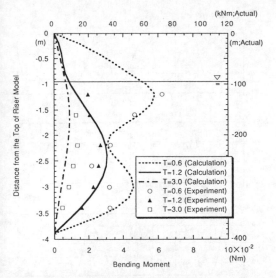

Fig.7 Bending Moment of the Model (Pin-jointed)

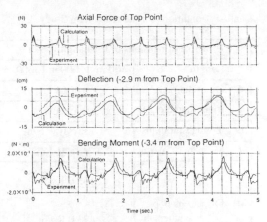

Fig. 8 Time Series of Measured and Simulated Result
(Pin-jointed, T=0.6 seconds)

in quantity. From these results, the maximum bending moment that would be induced in the actual rigid riser with 400 m in length can be estimated as 70 kN-m, and the maximum stress will be 23 kN/mm2.

Sinker motion
When the forced oscillation period was 0.6 seconds, a strange sinker motion with twice period was observed as mentioned before. In Fig.8, we present the time series of measured and simulated results. At the top end, a snap (impulse-like) load with the interval same as the oscillated period can be observed. On the contrary, the deflection and the bending moment vary with the period 1.2 seconds. The simulated results demonstrate good agreement with the measured time series, including the strange oscillation.

The cause of this oscillation can be considered as the dynamic coupling of the double pendulum. If we assume the sinker as a pendulum, the natural period is calculated as 1.2 seconds. The motion of the riser might happen to excite the oscillation of this component. When we changed the sinker to a steel rod that has different natural period, we could not observe the strange oscillation. This experimental result supports our conjecture.

From the time series of the axial force, we can consider that the inertia force is predominant in it. As such, the effect of the elongation on the bending moment and the bending deflection would be negligibly small in this case. We can conclude that our numerical simulation is applicable to the estimation of the dynamic behavior of a relatively short rigid riser.

3 THE EFFECT OF THE ELONGATION

Numerical methods to simulate the behaviors of a hanging riser are developed to include the effect of the elongation. Some numerical studies are performed to explore the dynamic behavior of long risers.

3.1 *Formulation*

The elongation of a riser can be divided into two components, one comes from the bending deflection and the other comes from the elongation vibration.

Sol.1
When we neglect the effect of the elongation on the bending and torsional deflections, the elongation can be determined on the quasi-steady deflected beam, in the form

$$dl_i = \frac{l_i}{2AE}(T_{i1} + T_{i2}) \qquad (2)$$

where A denotes the sectional area, T_{i1} and T_{i2} are tensional forces acting on the upper and lower ends of a segment respectively. If we have already known the tensional forces, we can compute the elongation.

The effect of the elongation on the potential energy of bending can be expressed as follows

$$U_1 = \frac{EI}{2l_i}\Theta^2\left(1 - \frac{dl_i}{l_i}\right) \qquad (3)$$

where Θ denotes the angle of adjacent segments. Using this expression, we can deform the equations of the riser motion including the elongated length.

In this numerical procedure, the elongation is calculated at each time step and the effect of the elongation is taken into account at next time step.

Table 3 Conditions for the Numerical Study

Principal dimensions of the hanging riser			
L Length	400 m	1000 m	3000 m
d_O Outer diameter	0.4064 m		
d_I Inner diameter	0.3744 m		
E Young's modulus	1.132×10^{11} N/m²		
I Moment of inertia	3.745×10^{-4} m⁴		
M_A Mass per unit length	545.0 Kg/m		
M_W Displacement per unit length	473.0 Kg/m		
End conditions			
Upper end	Ball joint		
Forced osallation	Circulr Motion		
amplitude	5 m		
period	10 sec		
Lower end	Free		
Current condition (linear distribution)			
Water depth 0 m	2.5 m/sec		
500 m	1.5 m/sec		
1000 m	0.75 m/sec		
2000 m	0.5 m/sec		
6000 m	0.5 m/sec		

Sol.2
In Sol.1, the simulated effect of the elongation comes from the last time step. More precise simulation can be accomplished if we deal the elongation, the bending and torsional deflections at same time.

Here we introduce the potential energy of the elongation, that is

$$U_2 = \frac{1}{2} AE \frac{\left(dl_i\right)^2}{l_i} \tag{4}$$

Adding the elongation terms to the equation (1) , we can define new Lagrangean including the effect of the elongation vibration.

3.2 *Numerical simulation*

Numerical simulations using Sol.0, Sol.1 and Sol.2

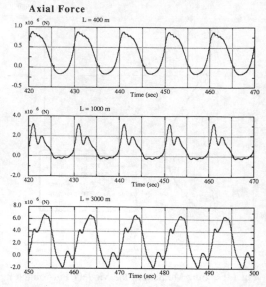

Axial Force

Fig. 9 Simulated Axial Force at Top of The Riser (Sol.2)

are performed and compared with one another. Here the dynamic behaviors of three hanging risers with different length are simulated. Table 3 gives the conditions for the calculations. The total numbers of the segments used in the simulations for 400 m, 1000 m and 3000 m risers are N=20, 40, and 60 respectively. The time steps are 0.001 seconds for Sol.0 and Sol.1, and we need smaller time step 0.0003 for Sol.2.

In Fig.9, we illustrate computed time series of the axial forces acting on the top end including the effects of the elongation vibrations (Sol.2). When the length of the riser becomes longer, we can clearly find the component coming from the elongation vibration. At that case, the natural frequency of the elongation vibration becomes low, and it effects significantly on the axial force.

The simulated results of the total elongation of the 3000 m riser using Sol.1 and Sol.2 are compared in Fig.10. We can find that the elongation vibration whose period is about 3 seconds is excited in the simulation. For a long rigid riser, the effect of the elongation vibration is troublesome, because it can enlarge the amplitude of the total elongation and may cause compressive stress in the riser.

In Figs.11 and 12, the time series of the bending moments of the 1000 m and 3000 m risers are presented. The effects of the elongation on the bending moment, coming from the bending deflection and the elongation vibration, are small for the 1000 m riser. The quasi-steady method (Sol.1) is applicable in this case.

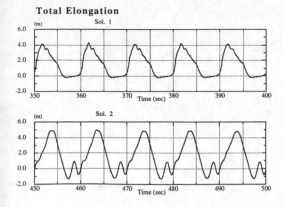

Fig. 10 Simulated Total Elongation of a 3000m Riser

Quite different time series is simulated for the 3000 m riser. Many modes of the bending and the elongation vibrations are excited in this case. The component of the harmonic forced oscillation with period 10 seconds is hidden by the excited bending vibrations of both lower and higher modes (Sol.0 and Sol.1). When we take into account the effect of the elongation vibration, the simulated time series turns to be more complicated (Sol.2).

Although the condition of this calculation is too hard, we can understand that the effect of the elongation vibration plays an important in long riser

dynamics. As such, we must seek to incorporate the effect of the elongation vibration on the design and operation stages of a long riser.

4 CONCLUDING REMARKS

The dynamics of a marine riser in hang-off mode are studied experimentally and numerically focusing the effect of the length in this report.

When the length of a riser is relatively short, the dynamic behavior is governed by the bending stiffness and the effect of the elongation vibration is negligibly small. The dynamics of a relatively short elastic riser hanging freely in water, driven harmonically at the top are studied experimentally. The deflection and the bending moment distributions along the length are measured to examine. The time series data and the amplitude of the oscillated component are compared with the results of the numerical simulation, demonstrating good agreement.

On the contrary, the elongation vibration would play a role for a long rigid riser. The code is developed to incorporate the effect of the elongation. One is a quasi-steady method and the other is a general unsteady method. A series of numerical simulations of three risers with different length is performed to evaluate the effect of the elongation, demonstrating the importance of the effect of the elongation vibration for a long hanging riser.

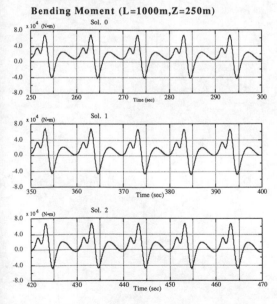

Fig. 11 Simulated Bending Moment of a 1,000m Riser

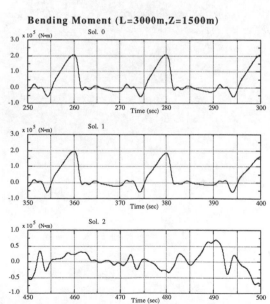

Fig. 12 Simulated Bending Moment of a 3,000m Riser

REFERENCES

Dareing, D.W., and Hwang, T. 1979. Marine Riser Vibration Response Determined by Modal Analysis. *J. Energy Resources Technology,* ASME, Vol. 101

Denison , E. B., Kolpak, M. M., and Garett, D. L. 1984. Comprehensive Approach to Deepwater Marine Riser Management. *Proc. 16th OTC*

Howell, C. T. 1992. *Investigation of the Dynamics of Low-Tension Cables*: Ph. D. Thesis, MIT

Kitakouji, Y. , et al. 1993. Dynamic Behavior of Riser System in Hang- off Mode. *Proc. Japan Society of Naval Architects*, Vol. 174

Long, J. R., Steddum, R., and Young, R. D. 1983. Analysis of a 6,000-ft Riser During Installation and Storm Hangoff. *Proc. 15th OTC*

Ohkusu, M. 1992. A Study of Behaviors of a Long and Flexible Pipe in the Water. *Proc. BOSS'92*

Paulling, J. R. 1979. Frequency Domain Analysis of OTEC CW Pipe and Platform Dynamics. *Proc. 11th OTC*

Sparks, C. P. , Cabillic, J. P. and Schawann J-C. 1982. Longitudinal Resonant Behavior of Very Deep Risers. *Proc. 14th OTC*

Wang, E. 1983. Analysis of Two 13,200-ft Riser Systems Using a Three -Dimensional Riser Program. *Proc. 15th OTC*

Risers cables and pipelines

Hydroelasticity in Marine Technology, Faltinsen et al. (eds) © 1994 Balkema, Rotterdam, ISBN 90 5410 387 6

Dynamic analysis of mooring underwater cable lines

J.Wauer
Institut für Technische Mechanik, Universität Karlsruhe, Germany

ABSTRACT: The modelling and the formulation of the the dynamic interaction of a two–point–fixed, sagging cable surrounded by an incompressible, viscous fluid are dealt with. Starting point is the governing boundary value problem for the longitudinal–transversal oscillations of an inextensible onedimensional continuum coupled to the Navier–Stokes equations of the fluid by corresponding transition conditions. Applying Galerkin's method, a boundary value problem averaged with respect to the axial direction is generated. Within a finite–term truncation, the general procedure to find the governing eigenvalue equation is described. To get quantitative results, the calculation is performed based on a one–term approximation. For a realistic, very small viscosity, the complex–valued eigenvalues determining the added mass effect and the hydrodynamic damping properties are computed by a (singular) perturbation analysis.

1 INTRODUCTION

In ocean engineering, underwater cable systems in a great variety are of importance. For many concrete applications, the interaction of the structural member and the surrounding fluid is a fundamental problem. An exact modelling of the dynamic coupling seems to be hopeless but a more accurate modelling than describing the added mass effect by addition of a certain mass quantified experimentally or the hydrodynamic damping by formulating it in the sense of the Morrison equation, for instance, is surely desired.

In the past, exactly such non–satisfactory approaches were usual. A short note by Kleczka and Kreuzer (1992) on remotely operated underwater vehicles points it out with reference to some other papers by Choo and Casarella (1973) and Hwang (1986), for instance, but for the modelling of the interaction, there is also not specified an innovation to overcome the difficulties.

The objective of the present contribution is to suggest now an accurate description of the dynamics in the sense of continuum mechanics for a two–point–fixed, sagging cable with a circular cross section surrounded by an incompressible, viscous fluid. Preliminary work which is helpful has

been published by Saxon/Cahn (1953), Ahmadi–Kashani (1989), Tadjbakhsh/Wang (1994) and Seemann/Wauer (1993, 1994). The first three papers concern a category of problems considering small vibrations of sagging chains or cables while the other two deal with the transverse vibrations of an elastically supported, rigid cylinder in a surrounding fluid. In contrast to the latter contributions by Seemann/Wauer (1993,1994) in which a purely planar problem is considered, here a spatial problem appears because the cable configuration is not straight neither in the equilibrium position nor in motion.

2 PHYSICAL MODEL

Fig. 1 shows the system under consideration. The structural member is a bending–flappy, inextensible cable (mass per unit length m) of circular cross section (radius R) and length ℓ. It is suspended from two points $L < \ell$ apart and having a difference in elevation of h. It is surrounded by an unbounded incompressible viscous fluid of viscosity μ and density ρ_F at rest. An appropriate global Cartesian X, Y, Z–coordinate system and two local reference frames placed in the plane of a general non–vibrating cable cross section are

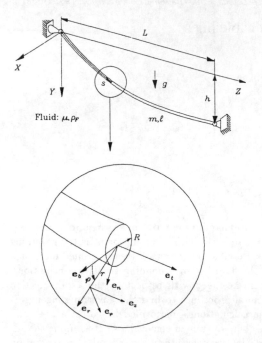

Fig. 1: Geometry of the mooring cable

introduced. The first triad of unit basis vectors $(\mathbf{e}_b, \mathbf{e}_n, \mathbf{e}_t)$ is attached at the centroid, the second one, $(\mathbf{e}_r, \mathbf{e}_\varphi, \mathbf{e}_z = \mathbf{e}_t)$, at an associated general point (distance r, angle φ). The direction of $\mathbf{e}_t$ is associated with the outside normal of the cross section. Small spatial oscillations

$$\Delta \mathbf{r}(s,t) = \Delta x(s,t)\mathbf{e}_b + \Delta y(s,t)\mathbf{e}_n + \Delta z(s,t)\mathbf{e}_t \quad (1)$$

about the equilibrium configuration

$$\mathbf{r}_0(s) = Y_0 \mathbf{e}_Y + Z_0 \mathbf{e}_Z \quad (2)$$

are studied where s is the arclength denoting the general cross section of the cable. They interact with the fluid motion which is characterized by the velocity components $u(r,\varphi,s,t)$, $v(r,\varphi,s,t)$, $w(r,\varphi,s,t)$ ($\mathbf{u} = u\mathbf{e}_r + v\mathbf{e}_\varphi + w\mathbf{e}_z$) and the pressure fluctuation $\Delta p(r,\varphi,s,t)$ of the equilibrium pressure $p_0(s)$. Boundary effects of the cable line supports on the fluid motion are neglected; the length of the cable and also the radius of curvature of the equilibrium catenary are sufficiently large with respect to the cross–sectional dimensions.

3 MATHEMATICAL FORMULATION

For the mathematical description, the continu-

ity equation and the the linearized Navier–Stokes equations for the fluid

$$\frac{\partial u}{\partial r} + \frac{u}{r} + \frac{1}{r}\frac{\partial v}{\partial \varphi} + \frac{\partial w}{\partial s} = 0, \quad (3)$$

$$\rho_F \frac{\partial u}{\partial t} + \frac{\partial \Delta p}{\partial r} = \mu \left(\Delta[u] - \frac{u}{r^2} - \frac{2}{r^2}\frac{\partial v}{\partial \varphi} \right),$$

$$\rho_F \frac{\partial v}{\partial t} + \frac{1}{r}\frac{\partial \Delta p}{\partial \varphi} = \mu \left(\Delta[v] - \frac{v}{r^2} + \frac{2}{r^2}\frac{\partial u}{\partial \varphi} \right),$$

$$\rho_F \frac{\partial w}{\partial t} + \frac{\partial \Delta p}{\partial s} = \mu \Delta[w] \quad (4)$$

together with the dynamic boundary conditions (force balance per unit length of the cable)

$$m\frac{\partial^2 \Delta x}{\partial t^2} - \frac{\partial}{\partial s}\left(T_0 \frac{\partial \Delta x}{\partial s} \right)$$

$$+ \int_0^{2\pi} (-\sigma_{rr}\cos\varphi + \sigma_{r\varphi}\sin\varphi)_{r=R} R d\varphi = 0,$$

$$m\frac{\partial^2 \Delta y}{\partial t^2} - \frac{\partial}{\partial s}\left[T_0 \left(\chi_0 \Delta z + \frac{\partial \Delta y}{\partial s} \right) \right] - \chi_0 \Delta T$$

$$+ \int_0^{2\pi} (\sigma_{rr}\sin\varphi + \sigma_{r\varphi}\cos\varphi)_{r=R} R d\varphi = 0,$$

$$m\frac{\partial^2 \Delta z}{\partial t^2} + T_0 \chi_0 \left(\chi_0 \Delta z + \frac{\partial \Delta y}{\partial s} \right) - \frac{\partial \Delta T}{\partial s}$$

$$+ \int_0^{2\pi} \sigma_{rs}|_{r=R} R d\varphi = 0, \quad (5)$$

and kinematical no–slip conditions

$$u(R,\varphi,s,t) = \frac{\partial \Delta x}{\partial t}\cos\varphi + \frac{\partial \Delta y}{\partial t}\sin\varphi,$$

$$v(R,\varphi,s,t) = -\frac{\partial \Delta x}{\partial t}\sin\varphi + \frac{\partial \Delta y}{\partial t}\cos\varphi,$$

$$w(R,\varphi,s,t) = \frac{\partial \Delta z}{\partial t} \quad (6)$$

at the contact surface between fluid and structure are the starting point where

$$\Delta[.] = \frac{\partial^2[.]}{\partial r^2} + \frac{1}{r}\frac{\partial[.]}{\partial r} + \frac{1}{r^2}\frac{\partial^2[.]}{\partial \varphi^2} + \frac{\partial^2[.]}{\partial s^2} \quad (7)$$

is the Laplacian operator and

$$\sigma_{rr} = -\Delta p + 2\mu \frac{\partial u}{\partial r},$$

$$\sigma_{r\varphi} = \mu \left(\frac{\partial v}{\partial r} - \frac{v}{r} + \frac{1}{r}\frac{\partial u}{\partial \varphi} \right),$$

$$\sigma_{rs} = \mu \left(\frac{\partial w}{\partial r} + \frac{\partial u}{\partial s} \right) \quad (8)$$

are the fluid stresses. T indicates the tension in the cable assumed as the sum of a static part T_0

and its change ΔT if the cable is in motion. Far away from the cable, e.g., at $r \to \infty$, the fluid velocities die out:

$$u(\infty, \varphi, s, t) = v(\infty, \varphi, s, t) = w(\infty, \varphi, s, t) = 0. \tag{9}$$

Additionally, a constraint equation

$$-\chi_0 \Delta y + \frac{\partial \Delta z}{\partial s} = 0 \tag{10}$$

denoting the inextensibility of the cable and boundary conditions at the suspension points of the cable

$$\Delta x(0, t) = \Delta y(0, t) = \Delta z(0, t) = 0,$$
$$\Delta x(\ell, , t) = \Delta y(\ell, t) = \Delta z(\ell, t) = 0 \tag{11}$$

have to be formulated. The relation (10) follows from the general constraint relation

$$\left(\frac{\partial \mathbf{r}}{\partial s}\right)^2 = 1 \tag{12}$$

for small vibrations $\Delta\mathbf{r}(s, t)$; χ_0 is the curvature of the equilibrium configuration of the cable.

Introducing a stream vector function $\boldsymbol{\psi} = \psi_r \mathbf{e}_r + \psi_\varphi \mathbf{e}_\varphi + \psi_z \mathbf{e}_z$ defined by

$$\mathbf{u}(r, \varphi, s, t) = \text{rot } \boldsymbol{\psi}(r, \varphi, s, t) \tag{13}$$

substitutes three decoupled equations

$$\Delta \left[\Delta[\psi_j] - \frac{1}{\mu} \frac{\partial \psi_j}{\partial t} \right] = 0, \quad j = r, \varphi, z \tag{14}$$

for the linear Navier–Stokes equations (3), (4) as an alternative.

4 ANALYSIS

To obtain the properties T_0, $\frac{dY_0}{ds}$, $\frac{dZ_0}{ds}$ and χ_0, the static shape of the cable subjected to weight and lift (surrounded by the fluid at rest completely) has to be determined first. Subsequently, small deviations from the static shape in form of coupled fluid–structural vibrations can be considered.

4.1 Static Problem

In equilibrium, the governing boundary value problem reduces to a purely structural one:

$$-\frac{d}{ds}\left(T_0 \frac{dY_0}{ds}\right) = mg,$$

$$-\frac{d}{ds}\left(T_0 \frac{dZ_0}{ds}\right) = 0, \tag{15}$$

$$Y_0(0) = Z_0(0) = 0,$$

$$Y_0(\ell) = L, \quad Z_0(\ell) = h, \tag{16}$$

$$\left(\frac{dY_0}{ds}\right)^2 + \left(\frac{dZ_0}{ds}\right)^2 = 1. \tag{17}$$

mg indicates the resulting external load. The complete solution satisfying (15)–(17) is given in the earlier mentioned paper by Tadjbakhsh and Wang (1994), for instance. One obtains

$$T_0 = H_0 \cosh \beta(Z_0 - c_1),$$
$$\frac{dY_0}{ds} = \frac{1}{\cosh \beta(Z_0 - c_1)},$$
$$\frac{dZ_0}{ds} = -\tanh \beta(Z_0 - c_1),$$
$$\chi_0 = \frac{-\beta}{\cosh^2 \beta(Z_0 - c_1)} \tag{18}$$

where c_1 and H_0 are constants. H_0 is the horizontal component of the tension T_0 in the cable, $\beta = \frac{mg}{H_0}$ and c_1 are determined from the boundary conditions (16) and the inextensibility condition (17) which yield

$$c_1 = \frac{1}{\beta} \cosh^{-1} \left\{ \left[\cosh^2 \frac{\beta L}{2} + \left(\frac{\beta h}{2}\right)^2 \right. \right.$$
$$\left. \left. \times \left(\frac{1}{\sinh^2 \frac{\beta L}{2}}\right) \right]^{\frac{1}{2}} + \frac{\beta h}{2} \right\} \tag{19}$$

explicitly and

$$s(Z_0) = \int_0^L \left[1 + \left(\frac{dY_0}{dZ_0}\right)^2 \right]^{\frac{1}{2}} dZ_0$$
$$= \frac{1}{\beta}[\sinh \beta(Z_0 - c_1) + \sinh \beta c_1]$$
$$\Rightarrow \ell = \frac{1}{\beta}[\sinh \beta(L - c_1) + \sinh \beta c_1] \tag{20}$$

as an implicit relationship.

4.2 Dynamic Analysis

Compared with the boundary value problem of the transverse vibrations of an elastically supported, rigid cylinder in a surrounding fluid, the variational equations (3)–(11) are more complicated because an additional space coordinate (the arclength s) is involved. To end in the same approach used by Seemann and Wauer (1993,1994), first of all, the s–dependence is averaged.

A preceding calculation step is to eliminate the change of Tension ΔT. For that purpose, differentiate $\frac{1}{\chi_0} \times (5)_2$ with respect to s and subtract $(5)_3$. In this way, one obtains

$$m\frac{\partial}{\partial s}\left(\frac{1}{\chi_0}\frac{\partial^2 \Delta y}{\partial t^2}\right) - \frac{\partial}{\partial s}\left\{\frac{1}{\chi_0}\frac{\partial}{\partial s}\left[T_0\left(\chi_0\Delta z\right.\right.\right.$$
$$\left.\left.\left. + \frac{\partial \Delta y}{\partial s}\right)\right]\right\} + \frac{\partial}{\partial s}\left[\frac{1}{\chi_0}\int_0^{2\pi}(\sigma_{rr}\sin\varphi\right.$$
$$\left. + \sigma_{r\varphi}\cos\varphi)_{r=R}R\mathrm{d}\varphi\right] - \int_0^{2\pi}\sigma_{rs}\big|_{r=R}R\mathrm{d}\varphi$$
$$- m\frac{\partial^2 \Delta z}{\partial t^2} - T_0\chi_0\left(\chi_0\Delta z + \frac{\partial \Delta y}{\partial s}\right) = 0. \quad (21)$$

The real averaging process is begun by representing Δq $(q \in x, y, z)$ as

$$\Delta x(s,t) = \sum_{k=1}^{\infty} x_k(t)\sin k\pi\frac{s}{\ell},$$
$$\Delta y(s,t) = \sum_{k=1}^{\infty} y_k(t)\sin k\pi\frac{s}{\ell},$$
$$\Delta z(s,t) = \sum_{k=1}^{\infty} y_k(t)R_k(s) \quad (22)$$

where

$$R_k(s) = \int_0^s \chi_0 \sin k\pi\frac{\hat{s}}{\ell}\mathrm{d}\hat{s}. \quad (23)$$

Thus the boundary conditions (11) and the constraint relation (10) are satisfied. For the fluid velocities and the pressure, a corresponding s–dependence is assumed:

$$u(r,\varphi,s,t) = \sum_{k=1}^{\infty} U_k(r,\varphi,t)\sin k\pi\frac{s}{\ell},$$
$$v(r,\varphi,s,t) = \sum_{k=1}^{\infty} V_k(r,\varphi,t)\sin k\pi\frac{s}{\ell},$$
$$w(r,\varphi,s,t) = \sum_{k=1}^{\infty} W_k(r,\varphi,t)\cos k\pi\frac{s}{\ell},$$
$$\Delta p(r,\varphi,s,t) = \sum_{k=1}^{\infty} P_k(r,\varphi,t)\sin k\pi\frac{s}{\ell} \quad (24)$$

or

$$\psi_j(r,\varphi,s,t) = \sum_{k=1}^{\infty} \Psi_{jk}(r,\varphi,t)\sin k\pi\frac{s}{\ell},$$
$$j = r, \varphi, z \quad (25)$$

Now, (22) and (23) and also (24) or (25) are substituted into the boundary value problem $(3),(4)$ or $(14),(5)_1,(21)$ and $(6),(9)$. Obviously, the s–dependence of the field equations $(3),(4)$ or (14) and

the boundary conditions $(6)_{1,2}$, (9) takes out without additional operations:

$$\frac{\partial U_l}{\partial r} + \frac{U_l}{r} + \frac{1}{r}\frac{\partial V_l}{\partial \varphi} - \frac{l\pi}{\ell}W_l = 0, \quad (26)$$

$$\rho_F\frac{\partial U_l}{\partial t} + \frac{\partial P_l}{\partial r} = \mu\left(\Delta_l[U_l] - \frac{U_l}{r^2} - \frac{2}{r^2}\frac{\partial V_l}{\partial \varphi}\right),$$

$$\rho_F\frac{\partial V_l}{\partial t} + \frac{1}{r}\frac{\partial P_l}{\partial \varphi} = \mu\left(\Delta_l[V_l] - \frac{V_l}{r^2} + \frac{2}{r^2}\frac{\partial U_l}{\partial \varphi}\right),$$

$$\rho_F\frac{\partial W_l}{\partial t} + \frac{l\pi}{\ell}P_l = \mu\Delta_l[W_l] \quad (27)$$

or

$$\Delta_l\left[\Delta_l[\Psi_{jl}] - \frac{1}{\mu}\frac{\partial \Psi_{jl}}{\partial t}\right] = 0$$
$$j = r, \varphi, z, (28)$$

$$U_l(R,\varphi,t) = \frac{\mathrm{d}x_l}{\mathrm{d}t}\cos\varphi + \frac{\mathrm{d}y_l}{\mathrm{d}t}\sin\varphi,$$

$$V_l(R,\varphi,t) = -\frac{\mathrm{d}x_l}{\mathrm{d}t}\sin\varphi + \frac{\mathrm{d}y_l}{\mathrm{d}t}\cos\varphi, \quad (29)$$

$$U_l(\infty,\varphi,t) = V_l(\infty,\varphi,t) = W_l(\infty,\varphi,t) = 0, \quad (30)$$

$$\Delta_l[.] = \frac{\partial^2[.]}{\partial r^2} + \frac{1}{r}\frac{\partial[.]}{\partial r} + \frac{1}{r^2}\frac{\partial^2[.]}{\partial \varphi^2} - \left(\frac{l\pi}{\ell}\right)^2[.], \quad (31)$$

$$l = 1, 2, \ldots, \infty.$$

Multiplying the differential equations $(5)_1,(21)$ by $\sin l\pi\frac{s}{\ell}$ and integrating with respect to s from zero to ℓ (Galerkin procedure for minimizing error) and doing the same for the remaining boundary condition $(6)_3$ where $\cos l\pi\frac{s}{\ell}$ is used leads to

$$m\frac{\mathrm{d}^2 x_l}{\mathrm{d}t^2} + \sum_{k=0}^{\infty} c_{kl}x_k + \frac{1}{\ell}\int_0^{2\pi}\big[-\Sigma_{rr\,l}\cos\varphi$$
$$+ \Sigma_{r\varphi\,l}\sin\varphi\big]_{r=R}\mathrm{d}R\varphi = 0,$$
$$\sum_{k=0}^{\infty}\left\{a_{kl}\frac{\mathrm{d}^2 y_k}{\mathrm{d}t^2} + b_{kl}y_k\right.$$
$$+ d_{kl}\int_0^{2\pi}\big[\Sigma_{rr\,k}\sin\varphi + \Sigma_{r\varphi\,k}\cos\varphi\big]_{r=R}\mathrm{d}R\varphi$$
$$\left. + e_{kl}\int_0^{2\pi}\Sigma_{rs\,k}\big|_{r=R}R\mathrm{d}\varphi\right\} = 0, \quad (32)$$

$$W_l(R,\varphi,t) = \sum_{k=0}^{\infty}\frac{\mathrm{d}y_k}{\mathrm{d}t}\int_0^{\ell}R_k\cos l\pi\frac{s}{\ell}\mathrm{d}s, \quad (33)$$

$$l = 1, 2, \ldots, \infty$$

where

$$\Sigma_{rr\,k} = -P_k + 2\mu\frac{\partial U_k}{\partial r},$$

88

$$\Sigma_{r\varphi k} = \mu\left(\frac{\partial V_k}{\partial r} - \frac{V_k}{r} + \frac{1}{r}\frac{\partial U_k}{\partial \varphi}\right),$$

$$\Sigma_{rsk} = \mu\left(\frac{\partial W_k}{\partial r} + \frac{l\pi}{\ell}U_k\right), \qquad (34)$$

and

$$c_{kl} = -\frac{2}{\ell}\int_0^\ell \frac{\partial}{\partial s}\left(T_0 \frac{k\pi}{\ell}\cos k\pi\frac{s}{\ell}\right)\sin l\pi\frac{s}{\ell}ds,$$

$$a_{kl} = m\int_0^\ell \left[\frac{\partial}{\partial s}\left(\frac{1}{\chi_0}\sin k\pi\frac{s}{\ell}\right)\right.$$
$$\left. - R_k\right]\sin l\pi\frac{s}{\ell}ds$$

$$b_{kl} = -\int_0^\ell \left\{\frac{\partial}{\partial s}\left[\frac{1}{\chi_0}\frac{\partial}{\partial s}(T_0 B_k)\right.\right.$$
$$\left.\left. + T_0\chi_0 B_k\right]\right\}\sin l\pi\frac{s}{\ell}ds,$$

$$d_{kl} = \int_0^\ell \frac{\partial}{\partial s}\left(\frac{1}{\chi_0}\sin k\pi\frac{s}{\ell}\right)\sin l\pi\frac{s}{\ell}ds,$$

$$e_{kl} = -\int_0^\ell \cos k\pi\frac{s}{\ell}\sin l\pi\frac{s}{\ell}ds, \qquad (35)$$

$$B_k(s) = \chi_0 R_k + \frac{k\pi}{\ell}\cos k\pi\frac{s}{\ell}. \qquad (36)$$

In order to express all governing equations of motion in dimensionless form, nondimensional variables

$$\bar{q} = \frac{q}{\ell}\ (q \in X, Y, Z, s), \quad \bar{r} = \frac{r}{R},$$

$$\bar{t} = \sqrt{\frac{H_0}{m\ell^2}}t,$$

$$\bar{x}_k = \frac{x_k}{R}, \ \bar{y}_k = \frac{y_k}{R},$$

$$\bar{U}_k = \frac{U_k}{R\sqrt{\frac{H_0}{m\ell^2}}}, \ \bar{V}_k = \frac{V_k}{R\sqrt{\frac{H_0}{m\ell^2}}}, \ \bar{W}_k = \frac{W_k}{R\sqrt{\frac{H_0}{m\ell^2}}},$$

$$\bar{P}_k = \frac{P_k m\ell^2}{\rho_F R^2 H_0}, \ \bar{T}_0 = \frac{T_0}{H_0}, \ \bar{\chi}_0 = \chi_0\ell \qquad (37)$$

or

$$\bar{\Psi}_k = \frac{\Psi_k}{\sqrt{\frac{H_0}{m\ell^2}}} \qquad (38)$$

and parameters

$$\mathrm{Re} = \frac{\rho_F\sqrt{\frac{H_0}{m\ell^2}}R^2}{\mu}, \quad \alpha = \frac{\rho_F \pi R^2}{m}, \quad \gamma = \frac{R}{\ell} \qquad (39)$$

are introduced. Under this rescaling, the boundary value problem governing the coupled cable–fluid

oscillations becomes after dropping all overbars for convenience

$$\frac{\partial U_l}{\partial r} + \frac{U_l}{r} + \frac{1}{r}\frac{\partial V_l}{\partial \varphi} - l\pi\gamma W_l = 0, \quad (40)$$

$$\frac{\partial U_l}{\partial t} + \frac{\partial P_l}{\partial r} = \frac{1}{\mathrm{Re}}\left(\Delta_l[U_l] - \frac{U_l}{r^2} - \frac{2}{r^2}\frac{\partial V_l}{\partial \varphi}\right),$$

$$\frac{\partial V_l}{\partial t} + \frac{1}{r}\frac{\partial P_l}{\partial \varphi} = \frac{1}{\mathrm{Re}}\left(\Delta_l[V_l] - \frac{V_l}{r^2} + \frac{2}{r^2}\frac{\partial U_l}{\partial \varphi}\right),$$

$$\frac{\partial W_l}{\partial t} + l\pi\gamma P_l = \frac{1}{\mathrm{Re}}\Delta_l[W_l] \quad (41)$$

or

$$\Delta_l\left[\Delta_l[\Psi_{jl}] - \mathrm{Re}\frac{\partial \Psi_{jl}}{\partial t}\right] = 0$$
$$j = r, \varphi, z, (42)$$

$$U_l(1,\varphi,t) = \frac{dx_l}{dt}\cos\varphi + \frac{dy_l}{dt}\sin\varphi,$$

$$V_l(1,\varphi,t) = -\frac{dx_l}{dt}\sin\varphi + \frac{dy_l}{dt}\cos\varphi,$$

$$W_l(1,\varphi,t) = \sum_{k=0}^\infty \frac{dy_k}{dt}\int_0^1 R_k\cos l\pi s\,ds, \quad (43)$$

$$U_l(\infty,\varphi,t) = V_l(\infty,\varphi,t) = W_l(\infty,\varphi,t) = 0, \quad (44)$$

$$l = 1, 2, \ldots, \infty,$$

$$\frac{d^2 x_l}{dt^2} + \sum_{k=0}^\infty c_{kl}x_k + \frac{\alpha}{\pi}\int_0^{2\pi}\left[-\Sigma_{rr\,l}\cos\varphi\right.$$
$$\left. + \Sigma_{r\varphi l}\sin\varphi\right]_{r=1}d\varphi = 0,$$

$$\sum_{k=0}^\infty\left\{a_{kl}\frac{d^2 y_k}{dt^2} + b_{kl}y_k\right.$$

$$+ d_{kl}\int_0^{2\pi}\left[\Sigma_{rr\,k}\sin\varphi + \Sigma_{r\varphi k}\cos\varphi\right]_{r=1}d\varphi$$

$$\left. + e_{kl}\int_0^{2\pi}\Sigma_{rs\,k}\big|_{r=1}d\varphi\right\} = 0 \quad (45)$$

where

$$\Delta_l[.] = \frac{\partial^2[.]}{\partial r^2} + \frac{1}{r}\frac{\partial[.]}{\partial r} + \frac{1}{r^2}\frac{\partial^2[.]}{\partial \varphi^2} - (l\pi\gamma)^2[.], \quad (46)$$

$$\Sigma_{rr\,k} = -P_k + \frac{2}{\mathrm{Re}}\frac{\partial U_k}{\partial r},$$

$$\Sigma_{r\varphi k} = \frac{1}{\mathrm{Re}}\left(\frac{\partial V_k}{\partial r} - \frac{V_k}{r} + \frac{1}{r}\frac{\partial U_k}{\partial \varphi}\right),$$

$$\Sigma_{rs\,k} = \frac{1}{\mathrm{Re}}\left(\frac{\partial W_k}{\partial r} + k\pi\gamma U_k\right) \quad (47)$$

and

$$c_{kl} = -2\int_0^1 \frac{\partial}{\partial s}\left(T_0 k\pi\cos k\pi s\right)\sin l\pi s\,ds,$$

$$a_{kl} = \int_0^1 \left[\frac{\partial}{\partial s} \left(\frac{1}{\chi_0} \sin k\pi s \right) - R_k \right] \sin l\pi s \, ds$$

$$b_{kl} = -\int_0^1 \left\{ \frac{\partial}{\partial s} \left[\frac{1}{\chi_0} \frac{\partial}{\partial s} (T_0 B_k) \right. \right.$$
$$\left. \left. + T_0 \chi_0 B_k \right] \right\} \sin l\pi s \, ds,$$

$$d_{kl} = \frac{\alpha}{\pi} \int_0^1 \frac{\partial}{\partial s} \left(\frac{1}{\chi_0} \sin k\pi s \right) \sin l\pi s \, ds,$$

$$e_{kl} = -\int_0^1 \cos k\pi s \sin l\pi s \, ds, \quad (48)$$

$$R_k(s) = \int_0^s \chi_0 \sin k\pi \hat{s} \, d\hat{s},$$
$$B_k(s) = \chi_0 R_k + k\pi \cos k\pi s. \quad (49)$$

Allowing k and l have the range

$$k = 1, 2, \ldots, N, \quad l = 1, 2, \ldots, N \quad (50)$$

for a fixed N, (40), (41) or (42) and (43)–(45) compose a truncated subset of equations as an approximation.

Considering the three decoupled field equations (42) using the stream functions Ψ_{jk}, their solutions can be given as

$$\Psi_{jk}(r, \varphi, t) = \left[a_{jk} J_1(\kappa r) + b_{jk}(\kappa r) + c_{jk} r \right.$$
$$\left. + \frac{d_{jk}}{r} \right] \left(1 + e^{i\varphi} \right) e^{i\lambda t} \quad (51)$$

where

$$\kappa = \sqrt{i\lambda \mathrm{Re}}. \quad (52)$$

Expressing velocities and strains by Ψ_{jk} and assuming corresponding solutions

$$x_k(t) = C_k e^{i\lambda t}, \quad y_k(t) = \bar{C}_k e^{i\lambda t} \quad (53)$$

for the vibrational coordinates of the cable, fitting the general solutions (51) to the boundary conditions (43)–(45) is possible. This procedure yields for the φ–dependent parts of the solution the governing characteristic equation in form of a vanishing determinant to determine the complex–valued eigenvalues λ. But due to the implicit form, numerical results can only be evaluated for moderate viscosities. For the practically realistic case of very low viscosities numerical difficulties arise because the coefficient of the highest derivative of the governing differential equation (28) or (42) is proportional to μ or $\frac{1}{\mathrm{Re}}$, respectively. Also the influence of the parameters on the eigenvalues can not be seen directly.

As an alternative, useful also for very small viscosities, a singular perturbation analysis can be applied. Then, it is of advantage to remain in the equations of motion (40), (41) for the velocity components and the pressure together with the boundary conditions (43)–(45) and to introduce a small parameter

$$\varepsilon^2 = \frac{1}{\mathrm{Re}}. \quad (54)$$

Similar to (51), now corresponding solutions

$$U_k(r, \varphi, t) = \left[\bar{X}_k(r) \sin \varphi + X_k(r) \cos \varphi \right.$$
$$\left. + X_{k0} \right] e^{i\lambda t},$$
$$V_k(r, \varphi, t) = \left[Y_k(r) \sin \varphi + \bar{Y}_k(r) \cos \varphi \right.$$
$$\left. + Y_{k0} \right] e^{i\lambda t},$$
$$W_k(r, \varphi, t) = Z_{k0}(r) e^{i\lambda t},$$
$$P_k(r, \varphi, t) = \left[\bar{Q}_k(r) \sin \varphi + Q_k(r) \cos \varphi \right.$$
$$\left. + Q_{k0}(r) \right] e^{i\lambda t} \quad (55)$$

for velocities and pressure are assumed to separate the φ– and t–dependence. Introducing them into the boundary value problem (40), (41) and (43)–(45) yields two decoupled eigenvalue problems for $\mathbf{X}_k = (X_k, Y_k, Q_k)^T$ and C_k or $\bar{\mathbf{X}}_k = (\bar{X}_k, \bar{Y}_k, \bar{Q}_k)^T$ and $\bar{C}_k$ not written down explicitly.

In addition, there appears a boundary problem for $\mathbf{X}_{k0} = (X_{k0}, Y_{k0}, Z_{k0}, Q_{k0})^T$ which is coupled to the eigenvalue problems by the magnitude $\bar{C}_k$ of the vertical cable vibrations $y_k(t)$. If the eigenvalue problems are solved, the solution of the latter boundary value problem describing a secondary, φ–independent fluid oscillation is straightforward. Subsequently, it is clear now that for a complete solution $\Psi_{jk}(r, \varphi, t)$ a φ–independent term is necessary as shown in (51).

An important detail is that disregarding the differing coefficients c_{kl} and a_{kl}–e_{kl} the second eigenvalue problem originates from the first one if $\mathbf{X}_k$ and C_k are replaced by $\bar{\mathbf{X}}_k$ and $\bar{C}_k$, respectively, and several signs are changed. The procedure to calculate the eigenvalues is identically the same.

According to singular perturbation theory, the solution $\mathbf{X}_k$ (for instance) is computed in a region outside the boundary, called outer solution, and in a small range close to the boundary, called the boundary layer solution. For both the outer and the inner solution, the eigenvalue λ is assumed to be an asymptotic expansion

$$\lambda = \lambda_0 + \varepsilon \lambda_1 + \varepsilon^2 \lambda_2 + \ldots \quad (56)$$

of the small parameter ε. The outer solution $\hat{\mathbf{X}}_k(r)$

is expressed in the form

$$\hat{X}_k = \hat{X}_{k0} + \varepsilon \hat{X}_{k1} + \varepsilon^2 \hat{X}_{k2} + \dots \quad (57)$$

while for the inner solution near $r = 1$ — in a pre–step the radial coordinate r is transformed to

$$\tilde{r} = \frac{1}{\varepsilon}(r - 1) \quad (58)$$

— the solution $\tilde{X}_k$ in the boundary layer is expanded as

$$\tilde{X}_k = \tilde{X}_{k0} + \varepsilon \tilde{X}_{k1} + \varepsilon^2 \tilde{X}_{k2} + \dots \quad (59)$$

Introducing all expressions into the governing eigenvalue problems and collecting equal powers of ε leads to a sequence of eigenvalue problems which can be solved recursively. For larger upper limits N, the calculation can only be performed by computer–aided formel manipulation systems.

5 EVALUATION

In order to be able to follow the computation in all details, a one–term approximation is presented. Only the approach applying singular perturbation theory is followed up. Starting point is the boundary value problem $(40), (41)$ with (43)–(45) for $N = 1$. Introducing (55) and dropping the index k for convenience yields the eigenvalue problems

$$X' + \frac{1}{r}X + \frac{1}{r}Y = 0,$$

$$i\lambda X + Q'$$
$$= \frac{1}{\text{Re}}\left[X'' + \frac{1}{r}X' - \frac{2}{r^2}X - (\pi\gamma)^2 X - \frac{2}{r^2}Y\right],$$

$$i\lambda Y - \frac{1}{r}Q$$
$$= \frac{1}{\text{Re}}\left[Y'' + \frac{1}{r}Y' - \frac{2}{r^2}Y - (\pi\gamma)^2 Y - \frac{2}{r^2}X\right],$$

$$X(1) = i\lambda C, \quad Y(1) = -i\lambda C,$$
$$X(\infty) = 0, \quad Y(\infty) = 0,$$

$$(c - \lambda^2)C + \alpha\left[Q(1) - \frac{2}{\text{Re}}X'(1)\right.$$
$$\left. + \frac{1}{\text{Re}}\left(Y'(1) - \frac{1}{r}Y(1) - \frac{1}{r}X(1)\right)\right] = 0 \quad (60)$$

and

$$\bar{X}' + \frac{1}{r}\bar{X} - \frac{1}{r}\bar{Y} = 0,$$

$$i\lambda\bar{X} + \bar{Q}'$$
$$= \frac{1}{\text{Re}}\left[\bar{X}'' + \frac{1}{r}\bar{X}' - \frac{2}{r^2}\bar{X} - (\pi\gamma)^2\bar{X} + \frac{2}{r^2}\bar{Y}\right],$$

$$i\lambda\bar{Y} - \frac{1}{r}\bar{Q}$$
$$= \frac{1}{\text{Re}}\left[\bar{Y}'' + \frac{1}{r}\bar{Y}' - \frac{2}{r^2}\bar{Y} - (\pi\gamma)^2\bar{Y} + \frac{2}{r^2}\bar{X}\right],$$

$$\bar{X}(1) = i\lambda\bar{C}, \quad \bar{Y}(1) = -i\lambda\bar{C},$$
$$\bar{X}(\infty) = 0, \quad \bar{Y}(\infty) = 0,$$

$$(b - \lambda^2 a)\bar{C} + d\pi\left[\bar{Q}(1) - \frac{2}{\text{Re}}\bar{X}'(1)\right.$$
$$\left. + \frac{1}{\text{Re}}\left(\bar{Y}'(1) - \frac{1}{r}\bar{Y}(1) - \frac{1}{r}\bar{X}(1)\right)\right] = 0 \quad (61)$$

in λ for $\mathbf{X} = (X, Y, Q)^T$, C and $\bar{\mathbf{X}} = (\bar{X}, \bar{Y}, \bar{Q})^T$, $\bar{C}$, respectively, and the boundary value problem

$$X_0' + \frac{1}{r}X_0 - \frac{1}{r}Y_0 - \pi\gamma Z_0 = 0,$$

$$i\lambda X_0 + Q_0' = \frac{1}{\text{Re}}\left[X_0'' + \frac{1}{r}X_0' - (\pi\gamma)^2\right],$$

$$i\lambda Y_0 = \frac{1}{\text{Re}}\left[Y_0'' + \frac{1}{r}Y_0' - (\pi\gamma)^2 Y_0\right],$$

$$i\lambda Z_0 + \pi\gamma Q_0 = \frac{1}{\text{Re}}\left[Z_0'' + \frac{1}{r}Z_0' - (\pi\gamma)^2\right],$$

$$X_0(1) = 0, \quad Y_0(0) = 0,$$

$$Z_0(1) = i\lambda\bar{C}\int_0^1\left(\int_0^s \chi_0\sin\pi\hat{s}\,d\hat{s}\right)\cos\pi s\,ds,$$

$$X_0(\infty) = Y_0(\infty) = Z_0(\infty) = 0 \quad (62)$$

for $\mathbf{X}_0 = (X_0, Y_0, Z_0, Q_0)^T$ where $(.)'$ indicates derivatives with respect to r.

As predicted, there are two decoupled eigenvalue problems for $(X, Y, Q)^T$, C and $(\bar{X}, \bar{Y}, \bar{Q})^T$, $\bar{C}$ and an additional boundary value problem for $(X_0, Y_0, Z_0, Q_0)^T$ coupled to the eigenvalue problems by the source term $\bar{C}$. Of particular interest is only the discussion of the eigenvalue problems; they characterize the fluid–structure interaction above all.

Since the eigenvalue problems (60) and (61) are decoupled, it is sufficient to deal with the first one, for example. Eliminating Y and Q in the form

$$Y(r) = -X(r) - rX'(r),$$
$$Q(r) = -i\lambda r[X(r) + rX'(r)]$$
$$+ \frac{1}{\text{Re}}(r^2 X''' + 4rX'') \quad (63)$$

where here and in the following the summand $(\pi\gamma)^2(rX' + X)$ is neglected because $\gamma^2 \ll 1$, the remaining shortened eigenvalue problem reads

$$i\lambda(r^3 X'' + 3r^2 X')$$
$$- \frac{1}{\text{Re}}(r^3 X'''' + 6r^2 X''' + 3rX'' - 3X') = 0,$$

91

$$X(1) = X(1) + X'(1) = i\lambda C,$$
$$X(\infty) = X'(\infty) = 0,$$
$$(c - \lambda^2)C - \alpha\left\{i\lambda[X'(1) + X(1)] + \right.$$
$$\left. + \frac{1}{\text{Re}}[3X'(1) - 3X''(1) - X'''(1)]\right\} = 0. \quad (64)$$

5.1 Perturbation Analysis

Following (57), the outer solution $\hat{X}$ is represented as
$$\hat{X}_0 + \varepsilon\hat{X}_1 + \varepsilon^2\hat{X}_2 + \dots \quad (65)$$
The differential equation for the outer solution is exactly the original one, $(64)_1$. After introducing λ and $\hat{U}$ according to (56) and (65), respectively into this differential equation and ordering equal powers of ε, a system of differential equations
$$3\hat{X}_0' + r\hat{X}_0'' = 0,$$
$$3\hat{X}_1' + r\hat{X}_1'' = -\frac{\lambda_1}{\lambda_0}(3\hat{X}_0' + r\hat{X}_0''),$$
$$3\hat{X}_2' + r\hat{X}_2'' = -\frac{\lambda_2}{\lambda_0}(3\hat{X}_0' + \hat{X}_0'')$$
$$- \frac{\lambda_1}{\lambda_0}(3\hat{X}_1' + r\hat{X}_1'')$$
$$+ \frac{i}{\lambda_0 r^2}(3\hat{X}_0' - 3r\hat{X}_0''$$
$$- 6r^2\hat{X}_0''' - r^3\hat{X}_0''''), \quad (66)$$
$$\vdots$$
results. The solution of $(66)_1$ is given by
$$\hat{X}_0 = C_{10} + \frac{C_{20}}{r^2}. \quad (67)$$
Introducing the solution $\hat{X}_0$ into equation $(66)_2$ yields
$$3r\hat{X}_1' + r^2\hat{X}_1'' = 0 \quad (68)$$
with the solution
$$\hat{X}_1 = C_{11} + \frac{C_{21}}{r^2}. \quad (69)$$
A recursive investigation shows that all right–hand sides of $(66)_j$ $(j > 1)$ vanish due to solutions $\hat{X}_{j-1}$ calculated before. Therefore, all solutions $\hat{X}_j$ $(j > 1)$ are also given by
$$\hat{X}_j = C_{1j} + \frac{C_{2j}}{r^2}, \quad j = 2, 3, \dots \quad (70)$$
Applying the boundary condition $(64)_3$ forces that every C_{1j} $(j = 0, 1, \dots)$ vanishes and the relationship
$$\left.\frac{C_{2j}}{r^2}\right|_{r\to\infty} = 0, \quad j = 0, 1, \dots \quad (71)$$

has to be fulfilled.

Using the transformation (58) and based on (59), the solution $\tilde{X}$ in the boundary layer is expanded as
$$\tilde{X} = \tilde{X}_0 + \varepsilon\tilde{X}_1 + \varepsilon^2\tilde{X}_2 + \dots \quad (72)$$
The same argumentation as before leads to a system of differential equations
$$-\tilde{X}_0'''' + i\lambda_0\tilde{X}_0'' = 0,$$
$$-\tilde{X}_1'''' + i\lambda_0\tilde{X}_1'' = 3\tilde{r}\tilde{X}_0'''' + 6\tilde{X}_0''' - 3\tilde{r}i\lambda_0\tilde{X}_0''$$
$$- i\lambda_1\tilde{X}_0'' - 3i\lambda_0\tilde{X}_0',$$
$$-\tilde{X}_2'''' + i\lambda_0\tilde{X}_2'' = 3\tilde{r}\tilde{X}_1'''' + 6\tilde{X}_1''' + 12\tilde{X}_0'''$$
$$+ 3\tilde{X}_0'' - i\lambda_2\tilde{X}_0'' - i\lambda_1\tilde{X}_1''$$
$$- 3i\lambda_1\tilde{r}\tilde{X}_0'' - 3i\lambda_0\tilde{r}\tilde{X}_1''$$
$$- 3i\lambda_1\tilde{X}_0' - 3i\lambda_0\tilde{X}_1'$$
$$- 6i\lambda_0\tilde{r}\tilde{X}_0' + 3\tilde{r}^2\tilde{X}_0''''$$
$$- 3i\lambda_0\tilde{r}^2\tilde{X}_0'', \quad (73)$$
$$\vdots$$
where now $(.)' = \frac{d(.)}{d\tilde{r}}$. In addition, the boundary conditions at $\tilde{r} = 0$ — ordered by collecting equal powers of ε — are
$$\tilde{X}_0(0) = Ci\lambda_0,$$
$$\tilde{X}_1(0) = Ci\lambda_1,$$
$$\tilde{X}_2(0) = Ci\lambda_2, \quad (74)$$
$$\vdots$$
$$\tilde{X}_0'(0) = 0,$$
$$\tilde{X}_0(0) + \tilde{X}_1'(0) = Ci\lambda_0,$$
$$\tilde{X}_1(0) + \tilde{X}_2'(0) = Ci\lambda_1, \quad (75)$$
$$\vdots$$
$$\alpha[\tilde{X}_0'''(0) - i\lambda_0\tilde{X}_0'(0)] = 0,$$
$$\alpha[\tilde{X}_1'''(0) - i\lambda_0\tilde{X}_1'(0)] = (\lambda_0^2 - c)C$$
$$+ \alpha[i\lambda_0\tilde{X}_0(0)$$
$$+ i\lambda_1\tilde{X}_0'(0)$$
$$- 3\tilde{X}_0''(0)],$$
$$\alpha[\tilde{X}_2'''(0) - i\lambda_0\tilde{X}_2'(0)] = 2\lambda_0\lambda_1 C + \alpha[i\lambda_0\tilde{X}_1(0)$$
$$+ i\lambda_1\tilde{X}_0(0)$$
$$+ i\lambda_1\tilde{X}_1'(0)$$
$$+ i\lambda_2\tilde{X}_0'(0) + 3\tilde{X}_0'(0)$$
$$- \tilde{X}_1''(0)], \quad (76)$$
$$\vdots$$
The general solution of $(72)_1$ is given by
$$\tilde{X}_0(\tilde{r}) = K_{10} + K_{20}\tilde{r} + K_{30}e^{-\kappa\tilde{r}} + K_{40}e^{\kappa\tilde{r}} \quad (77)$$

with the parameter

$$\kappa = \sqrt{i\lambda}. \tag{78}$$

Applying the boundary conditions $(74)_1,(75)_1$ and $(76)_1$ leads to

$$\tilde{X}_0(\tilde{r}) = C\kappa^2 - 2K_{40} + K_{40}e^{-\kappa\tilde{r}} + K_{40}e^{\kappa\tilde{r}}. \tag{79}$$

The differential equation for $\tilde{X}_1$ then is

$$\tilde{X}_1'''' - \kappa^2\tilde{X}_1'' = \left(\frac{\lambda_1}{\lambda_0}\kappa^4 - 3\kappa^2\right)K_{40}e^{-\kappa\tilde{r}}$$
$$+ \left(\frac{\lambda_1}{\lambda_0} - 3\kappa^2\right)K_{40}e^{\kappa\tilde{r}} \tag{80}$$

with the solution satisfying the boundary conditions $(74)_2,(75)_2$ and $(76)_2$

$$\tilde{X}_1(\tilde{r}) = \left(\kappa^2\frac{\lambda_1}{\lambda_0} - \frac{c-\lambda_0^2}{\alpha\kappa^2} + \kappa\right)C$$
$$+ \left(2\frac{\lambda_1}{\lambda_0} - \frac{3}{\kappa}\right)K_{40} - 2K_{41}$$
$$+ \left(\frac{c-\lambda_0^2}{\alpha\kappa^2} - \kappa^2\right)C\tilde{r}$$
$$+ \left[\left(\frac{c-\lambda_0^2}{\alpha\kappa^2} - \kappa\right)C\right.$$
$$\left. + \frac{3}{\kappa}K_{40} + K_{41}\right]e^{-\kappa\tilde{r}}$$
$$+ K_{41}e^{\kappa\tilde{r}} - \frac{1}{2}\left(\frac{\lambda_1}{\lambda_0} - 3\right)$$
$$\times \left(\frac{2}{\kappa} + \tilde{r}\right)K_{40}e^{-\kappa\tilde{r}} + \left(\frac{\lambda_1}{\lambda_0}\kappa - 3\right)$$
$$\times \left(\tilde{r} - \frac{2}{\kappa}\right)K_{40}e^{\kappa\tilde{r}}. \tag{81}$$

The solution for $\tilde{X}_2$ can be found in an analogeous manner; the final result depending also from C, K_{40}, K_{41} and K_{40} is lengthy and is not written down explicitly. These integration constants have to be determined during the matching process which is performed subsequently.

Between the boundary layer close to $r = 1$ and the outer region, the solutions are assumed to be equal at a point

$$r = 1 + \eta r_\eta \tag{82}$$

or expressed in $\tilde{r}$ at

$$\tilde{r}_\eta = \frac{\eta}{\varepsilon}r_\eta \tag{83}$$

where

$$\varepsilon \ll \eta \ll 1. \tag{84}$$

Because of (83), the quotient $\frac{\eta}{\varepsilon}$ tends to infinity. Therefore, $e^{\kappa\tilde{r}_\eta}$ tends to infinity, too, and $e^{-\kappa\tilde{r}_\eta}$ is exponentially small if the real part of κ is greater than zero. Thus a bounded solution solution is possible only if all integration constants K_{4j} $(j = 0, 1, \ldots)$ vanish. The remaining solution in the boundary layer at r_η is

$$\tilde{X}(r_\eta) = K_{10} + K_{20}\frac{\eta}{\varepsilon}r_\eta + \varepsilon\left[\left(\kappa^2\frac{\lambda_1}{\lambda_0}\right.\right.$$
$$\left. - \frac{c-\lambda_0^2}{\alpha\kappa^3} + \kappa\right)C + \left(\frac{c-\lambda_0^2}{\alpha\kappa^2}\right)C\frac{\eta}{\varepsilon}r_\eta\right]$$
$$+ \varepsilon^2\left[-\frac{3}{2} + \frac{3}{2}\frac{C}{\alpha\kappa^4}(c-\lambda_0^2) + \frac{\lambda_1}{\lambda_0}\frac{C}{\alpha\kappa^3}\right.$$
$$\times \left(\frac{3}{2} + \frac{1}{2}\lambda_0^2\right) + \frac{1}{2}\frac{\lambda_1}{\lambda_0}\kappa C + \frac{\lambda_2}{\lambda_0}\kappa^2 C$$
$$- \frac{\lambda_1}{\lambda_0}\left(\frac{c+\lambda_0^2}{\alpha\kappa^2} + \kappa^2\right)C\frac{\eta}{\varepsilon}r_\eta + \left(3\kappa^4 C\right.$$
$$\left.\left. - \frac{3C}{\alpha}(c-\lambda_0^2)\right)\frac{1}{2\kappa^2}\left(\frac{\eta}{\varepsilon}\right)^2 x_\eta^2\right]$$
$$+ O(\eta^3, \eta^2\varepsilon, \eta\varepsilon^2, \varepsilon^3). \tag{85}$$

The outer solution at r_η is

$$\hat{X}(r\eta) = \frac{C_{20}}{(1+\eta r_\eta)^2} + \varepsilon\frac{C_{21}}{(1+\eta r_\eta)^2}$$
$$+ \varepsilon^2\frac{C_{22}}{(1+\eta r_\eta)^2} + \ldots \tag{86}$$

Sorting both solutions at r_η for equal powers of ε, η and products of both gives

	outer solution	inner solution
$\varepsilon^0:$	C_{20}	$\kappa^2 C,$
$\varepsilon^1:$	C_{21}	$\left(\frac{\lambda_1}{\lambda_0}\kappa^2 - \frac{c-\lambda_0^2}{\alpha\kappa^2} + \kappa\right)C,$
$\eta^1:$	$-2C_{20}r_\eta$	$\left(\frac{c-\lambda_0^2}{\alpha\kappa^2}\right)Cr_\eta,$
$\varepsilon^2:$	C_{22}	$-\frac{3}{2}C + \frac{3}{2}\frac{C}{\alpha\kappa^4}(c-\lambda_0^2)$
		$+ \frac{1}{2}\frac{\lambda_1}{\lambda_0}\kappa C + \frac{\lambda_1}{\lambda_0}\frac{C}{\alpha\kappa^3}$
		$\times\left(\frac{3}{2} + \frac{1}{2}\lambda_0^2\right) + \frac{\lambda_2}{\lambda_0}\kappa^2 C,$
$\varepsilon\eta:$	$-2C_{21}r_\eta$	$-\frac{\lambda_1}{\lambda_0}\left(\frac{c+\lambda_0^2}{\alpha\kappa^2} + \kappa^2\right)Cr_\eta,$
$\eta^2:$	$3C_{20}r_\eta^2$	$\left(3\kappa^4 - \frac{3}{\alpha}(c-\lambda_0^2)\right)\frac{C}{2\kappa^2}r_\eta^2$
$\vdots$		

$$\tag{87}$$

For a correct matching, both columns have to be equal for every power of ε and η.

5.2 Eigenvalues

Based on the calculations performed till to this

93

point, both the eigenvalues and the eigenfunctions can be determined. Here, attention is focussed to the eigenvalues only; for practical applications, they are much more interesting than the modes.

Equations $(87)_1$ and $(87)_3$ form a set of two homogeneous equations for C_{20} and C with nontrivial solutions if the corresponding determinant vanishes:

$$\begin{vmatrix} 1 & -\kappa^2 \\ -2r_\eta & -\left(\frac{c-\lambda_0^2}{\alpha\kappa^2} - \kappa^2\right) r_\eta \end{vmatrix} = 0. \qquad (88)$$

Taking into consideration that $\kappa^2 = i\lambda_0$ leads to

$$\lambda_0^2 = \frac{c}{1+\alpha} \qquad (89)$$

which is the lowest normalized frequency of the cable vibrating in an incompressible, inviscid fluid. Obviously, there is an added mass effect, e.g., if there is a surrounding fluid of finite density ($\alpha > 0$), this frequency is smaller than the corresponding value, $\lambda_0 = c$, if the cable vibrates in vacuum. Eqs $(87)_3$ and $(87)_6$ are linearly dependent so that eqs $(87)_2$ and $(87)_5$ constitute a set of two homogeneous equations for C_{21} and C which has nontrivial solutions for special values of λ_1. The characteristic equation is

$$\begin{vmatrix} 1 & \frac{\lambda_1}{\lambda_0}\kappa^2 - \frac{c-\lambda_0^2}{\alpha\kappa^2} + \kappa \\ -2r_\eta & -\frac{\lambda_1}{\lambda_0}\left(\frac{c+\lambda_0^2}{\alpha\kappa^2} + \kappa^2\right) r_\eta \end{vmatrix} = 0 \qquad (90)$$

with the root

$$\frac{\lambda_1}{\lambda_0} = -\frac{2}{\kappa}\frac{c - \lambda_0^2 + \alpha\lambda_0^2}{c + \lambda_0^2 + \alpha\lambda_0^2}. \qquad (91)$$

This formula can be simplified introducing λ_0^2 (89),

$$\frac{\lambda_1}{\lambda_0} = -\frac{2}{\kappa}\frac{\alpha}{1+\alpha}, \qquad (92)$$

so that finally, the approximate solution λ is given as

$$\frac{\lambda}{\lambda_0} = c\left[1 - \frac{1}{\sqrt{\mathrm{Re}}}\frac{2}{\sqrt{i\lambda_0}}\frac{\alpha}{1+\alpha} + O\left(\frac{1}{\mathrm{Re}}\right)\right]. \qquad (93)$$

The square root of $i\lambda_0$ in the denominator can be expressed in the form

$$\frac{1}{\sqrt{i\lambda_0}} = \frac{1}{\sqrt{2\lambda_0}}(1-i) \qquad (94)$$

which means that the absolute value of the real und imaginary part are equal. Obviously, the added mass effect (slightly) intensifies if the fluid is

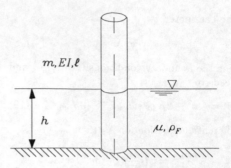

Fig. 2: Pile partially submerged into the surrounding fluid

viscous and the vibrations are really damped because the imaginary part of λ is positive.

6 MODIFICATIONS AND GENERALIZATIONS

The presented approach considering the coupled fluid–structural vibrations of mooring underwater cable lines can be applied to similar problems, e.g., beam–shaped piles (partially) submerged into the surrounding water (see Fig. 2) or interaction problems where a cross–flow (velocity u_∞) of the fluid appears (see Fig. 3).

In the first case, there is also a spatial problem but in a sufficient approximation, the axial fluid velocity component vanishes. Therefore, the Navier–Stokes equations fundamentally unchanged simplify and they are valid only in the interval $0 < s < h$ where h is the height of the fluid. In this interval, also kinematical no–slip conditions at the surface between beam and fluid are satisfied. The equations of motion for the vibrating structure are the well–known differential equations of a Bernoulli–Euler (bending stiffness EI, mass per unit length m), for instance, coupled in the interval $0 < s < h$ to the Navier–Stokes equations by loading due to the fluid stresses on the beam surface. Outside this interval, the beam–shaped pile performs free oscillations if an additional exitation by wind etc. is neglected. To complete the mathematical formulation, boundary conditions for the beam at its suspension points and for the fluid at its lower and upper surface have to be selected. While for the beam a fixed and a free end at the bottom $s = 0$ and the top $s = \ell$ (ℓ length of the pile), respectively, are an appropriate choice, a no–slip condition at the lower level and a free upper surface for the liquid are mostly found in prac-

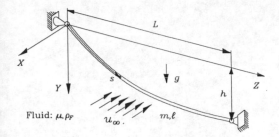

Fig. 3: Cable line in a steady cross–flow

tice. The calculation procedure experiences only one slight modification: The averaging process has to be performed over two intervals, from $s = 0$ to $s = h$ and from $s = h$ to $s = \ell$.

For the second example, the only but complicating change of the governing boundary value problem is that now the fully nonlinear Navier–Stokes equations have to be used, all other relations remain identically the same. The consequences become evident during the real calculation: It must be performed step–wise. Following an earlier paper by Wauer (1990), for instance, the steady flow past the flexible structure statically deformed in the direction of the flow has to be analyzed first. Subsequently, small coupled fluid–structural oscillations about this stationary basic state can be considered. Assuming a solution $\mathbf{q} = \mathbf{q}_0 + \Delta \mathbf{q}$, substituting into the governing nonlinear boundary value problem and linearizing in the $\Delta \mathbf{q}$–quantities leads to a linear boundary value problem again which can be dealt with according to the same approach presented here. The complications are that the calculation of the basic state is more expensive and there are additional space–dependent coefficients (as a result of the pre–computed stationary state).

CONCLUSIONS

The fluid–structural interaction of mooring cable lines surrounded by a viscous fluid have been examined. Both the structural member and the fluid have been modelled without significant restrictions in the sense of continuum mechanics. While the first one has been assumed as an inextensible, bending–flappy, onedimensional cable with circular cross section, the fluid has been described as an incompressible Newtonian medium.

It has been shown that for an unbounded fluid region as assumed, the change of the natural frequencies of the cable due to the surrounding fluid

can be calculated analytically. Both the damping factor due to the viscosity of the fluid and the added mass effect as a result of the vibrating fluid can be determined. As expected, the natural frequncies of the cable vibrating in vacuum decrease (dependent in a characteristic manner from the density ratio α) and the oscillations are damped now where the magnitude of the damping measure is also influenced by α. To limit the calculation expense, here a one–term truncation applying a singular perturbation analysis has been presented. It clearly demonstrates that even this rough computation exhibits all essential influence parameters on the vibrational characteristics. Remarkable is that within the presented approximation the magnitude of the damping measure is the same as the frequency change.

Finally, two other examples have been touched for which the same methods can be used with success. They emphasize that the proposed approaches can be applied to a broad class of dynamic fluid–structure interaction problems.

REFERENCES

Ahmadi–Kashani, K. 1989. Vibration of Hanging Cables. *Computers & Structures* 31: 699–715.

Kleczka, W. & Kreuzer, E. 1992. Zur Modellierung von Unterwasserrobotern. *Z. Angew. Math. Mech.* 72: T53–55.

Choo, Y. & Casarella, M.–J. 1973. A Survey of Analytical Methods for Dynamic Simulation of Cable–Body Systems. *J. Hydronautics* 7: 137–144.

Hwang, Y.–L. 1986. Nonlinear Dynamic Analysis of Mooring Lines. In *Proc. of OMAE Symp., Vol. 3, No. 5:* 499–506.

Seemann W. & Wauer, J. 1993. Fluid–Structural Coupling of Vibrating Bodies in Contact with a Fluid. In *Proc. of German–Polish Workshop on Nonlinear Dynamics.* To be published.

Seeman, W. & Wauer, J. 1994. Fluid–Structural Coupling of Vibrating Bodies in a Surrounding Confined Liquid. *J. Fluids Structures.* To be published.

Tadjbakhsh, I.G. & Wang, Y.-M. 1994. Transient Vibrations of a Taut Inclined Cable with a Riding Accelerating Mass, *Nonlinear Dynamics* 4. To be published.

Wauer J. 1991. On the dynamics of a viscous, incompressible flow past an elastically supported, circular cylinder at small Reynolds numbers. In *Structural Dynamics (W.B. Krätzig et al., eds.), Vol 2*. Balkema, Rotterdam: 827–836.

Hydroelasticity in Marine Technology, Faltinsen et al. (eds) © 1994 Balkema, Rotterdam, ISBN 90 5410 387 6

Vortex shedding induced oscillations during pipe laying

Roberto Bruschi, Luca Ercoli Malacari, Enrico Torseletti & Luigino Vitali
Snamprogetti S.p.A., Offshore Division, Fano, Italy

ABSTRACT: In the framework of submarine pipeline technology, laying is one of the most delicate phases, especially when operating in deep waters. Indeed, during laying a considerable pipe length is free spanning from the lay barge to the sea bottom. Then pipe submerged weight yields pipe bending which is controlled pulling the pipeline on the laying ramp by means of a tensioning machine. When operating in unfavourable environments, due to waves and currents, the additional dynamic excitation gives rise to over-imposed bending cycles, potentially triggering unacceptable permanent strains or even buckling. In geographical areas where new pipes are envisaged to be laid in the near future, severe current profiles impacting on the pipelay free span are expected to be encountered. These may cause resonant oscillations due to vortex shedding involving one or more natural frequencies of the spanning length. In this paper, the physics of the vortex shedding-induced oscillation of pipelay free spans is introduced. The structural model used to determine the response to permanent and external loads and therefore to assess the pipe integrity, is presented. Then some typical applications are discussed with the aim to point out areas of concern. Finally, considerations are made with regard to further improvements of the response model and associated experimental surveys to calibrate relevant parameters and to improve the understanding of the concerned phenomena.

1 INTRODUCTION

In the 70's, the offshore pipeline industry gave an impressive impulse to tackle pipe laying in deep waters, notably with a view to the at-that-time challenging projects in the Gulf of Mexico and in the Mediterranean Sea, Shell (1974-1977). The R&D efforts made during those years contributed to the successful implementation in the early 80's of the most advanced pipeline systems ever made, Albano et al. (1992) and Burattini et al. (1993). The satisfactory performance of these systems stands in witness of that, Iovenitti et al. (1994). The most advanced lay-barges for S-lay were constructed in the same period and they are still the most important heritage which actual plans for future pipelines are relying on, Anselmi and Bruschi (1993).

The structural modelling of the pipe laying operation is one of the first subjects developed within the offshore pipeline engineering, Anselmi et al. (1987) and Clauss and Weede (1989). Nevertheless, the aspects concerning the lay criteria, mainly related to the acceptability of the evaluated static and dynamic response with respect to an allowable state of stress and strain, have not been fully clarified, Sriskandarajah and Mahendran (1992). The minimum requirements to evaluate the response and the approximation of the obtained results in a given specific scenario are not yet univocally defined. Indeed, in many circumstances the acceptance of a proposed pipelay strictly depends on the laying scenario (shallow or deep waters, small or large diameter pipe, light or heavy lines), on the strength criteria assumed and on the structural integrity assessment procedure, Bruschi et al. (1994). At present, several studies are being carried out to determine, through computer-aided laying simulations, which parameters, e.g. length and geometrical configuration of the stinger, pulling force at the tensioner etc., are suitable to undergo upgrading changes that would improve the lay-barge capabilities, Langner and Ayers (1985).

The most critical lay simulations are the ones regarding the dynamic response of the pipeline to the encountered seastates. The loading process on the pipe is both direct (inertial and drag forces) and induced by the lay-barge response to wave action. As to the latter, oscillations in pitch (and heave) and surge (shallow waters) are the most critical ones. The final scope of a dynamic pipelay analysis is usually to predict a limit sea state for which the limit pipe strength capacity, as defined by the accepted criteria for a given pipelay, is reached. From these analyses the worthiness of current laying equipment can be evaluated. These topics are relevant in case of large diameter pipelines to be laid in deep waters, especially if the lay-barge has no supplementary pulling reserves to help the pipeline to provide a better dynamic response. Therefore possible upgrading works to extend the lay barge capacities can be addressed, Bruschi et al. (1994). The limit sea state should indicate the circumstances when the line is to be smoothly abandoned on the seabed to be recovered after the storm. Actually, the decision to release laying because of severe environmental conditions is assisted by monitoring the pipe at the stinger exit. Other parameters may also affect the decision, such as operational limits of the tugs managing anchor movements or waves affecting directly the working area on the firing ramp.

Besides, dynamic effects may also arise due to vortex shedding phenomena caused by marine currents which can have a significant effect on the pipeline in some situations. In 1979, when laying the second line crossing the Strait of Messina, in absolute calm days with a weak evidence of surface current, the pipeline experienced considerable oscillations and this was evidenced by the camera monitoring the pipe over the last roller. In some cases this oscillation caused the pipe to impact against the roller or to lift from it giving the impression of leaving the ramp. At that time this was diagnosed as due to vortex shedding and the pipe laying was stopped. In a later stage slow oscillations of the current profile due to large eddies were measured at that specific site. This allowed to conclude that the concurrence of slow oscillation of the lay span perpendicular to the lay vertical plane and of VIV in the vertical plane was responsible of such large oscillations.

The offshore industry has often tackled the problem of static and dynamic response of slender tubular members of considerable length exposed to vortex shedding induced vibrations (VIV). This is the case of drilling and production risers, tethers of tension leg platforms, collecting pipe for deep ocean mining, slender members of jackets, free spans of submarine pipelines, etc., as documented in the subject literature and recently reviewed by Pantazopoulos (1994). As for the structural response, different approaches have been developed, modelling the member deformation either in the two-dimensional or in the three-dimensional space, considering the actual material behaviour and the large displacement-rotation theory in different ways, taking into account different force models, etc. Bernitas (1982). Sometimes the static and dynamic analyses are combined but in many

circumstances the most common and used approach (slow versus high frequency oscillations) consists in the assessment of the quasi-static equilibrium configuration and in the evaluation of the dynamic response to the cyclic loads acting on the structure in the proximity of the static equilibrium configuration. Kim and Tryantafyllou (1984). This is the case of pipe laying, Bernitas and Vlahopoulos (1990).

As for the vortex shedding-induced oscillations, major research efforts have focused on slender members subjected to uniform flow along their length, Griffin (1982), often disregarding that in these special applications the incident flow is variable along the component axis. This is the case of tubular members going from the sea surface to the sea bottom, for instance subjected to stationary marine current due to wind, from a maximum in the proximity of the sea surface to a minimum at the sea bottom, the so-called shear flows. In this case to describe the interaction between the fluid and the tubular component is extremely hard, as vortex shedding occurs in separate cells and there are cells involving only a part of the span where the wake separation is regular. The subsequent cyclic force on the cylindrical section can excite the natural modes whose natural frequencies are close to the vortex shedding frequency, Griffin (1985). In the last two decades the research efforts, focused on the investigation of the mechanisms which control the separation of the cells and the interaction between adjacent cells, Rooney and Peltzer (1982), Stansby (1976) and Maull and Young (1973), and on the development of analysis procedures to determine the member response and the subsequent fatigue damage, Whitney and Nikkel (1983), Vandiver (1985), Brooks (1987) and Lyons and Patel (1989). These studies pointed out the extremely complex nature of the vortex separation in case of shear flows and indicated that the velocity gradient is the main VIV controlling parameter. High values of the velocity gradient generate a multi-modal excitation where the tubular member shows both excited and damped zones. For the assessment of the fluidodynamic damping in the non-synchronised areas which affects the amplitude of oscillation, data is few and the topic is still being studied, Vandiver and Chung (1987) and Humphries (1988).

The scope of this paper is to deal with vortex shedding induced oscillations on to pipelay free spans. This aspect is scarcely documented in the subject literature and nevertheless may result topical in certain projects next to implementation. The nature and heading of the predominant hydrodynamic field with respect to the pipelay bending plane in calm waters, may deeply influence lay criteria and lay ability from most lay barges currently operating without considering upgrading works. This paper presents the structural model describing the lay span in calm water, together with the approach to calculate the response to vortex shedding in the presence of currents varying across the depth. The loading model is based on the state of-the-art methods proposed by other authors with minor modification based on the authors' experience on pipeline free spans. The proposed model is two-dimensional as it is meant to calculate dynamic stresses to be superimposed to the state of bending envisaged in the S or J lay configurations.

2. VIV IN SHEAR FLOWS: BACKGROUND.

The main parameter characterising the vortex separation in linear shear flows is the velocity gradient, expressed in non dimensional form as:

$$\beta = (D/U_m) \, dU/dx; \qquad (2.1)$$

where: D is the outer diameter of the tubular, U_m is the average current velocity along the axis and dU/dx the velocity gradient.

Experiments show that the separation occurs in separate cells for velocity gradients, β, larger than 10^{-2}. It has been verified that the cross-flow lock-in regime occurs within a reduced velocity range and with an amplitude comparable with the ones occurring in uniform flows for velocity gradients, β, less than $2 \cdot 10^{-4}$. However, most experimental evidences are limited to test set ups characterised by a slenderness (length over diameter ratio) lower than 50. For very slender tubulars and high β it is expected that a multi-modal excitation occurs, involving both excited and damped zones.

Predictive models proposed in literature are generally concentrated on VIV induced by shear currents with uniform velocity gradients. The cyclic loading from vortex shedding is assumed as organised in separated cells characterised by a Strouhal frequency possibly captured by the natural frequency of the tubular member closest to this. The boundaries of each cell are in someway linked to the velocity gradient and to an assumed lock-in range around the Strouhal frequency ($\pm 40\%$ according to Walker and King, 1988). For long members and steep velocity gradients, a multi-modal response may occur: for each excited mode, it results an active length where cyclic pressures due to regular vortex shedding work on the structure increasing the oscillation amplitude and passive length where separation is not organised and the oscillation induced by active length is counteracted by drag forces. The amplitude of the modal oscillation associated to the excitation of a single cell can be assessed through the typical relations used for uniform flows, that is a function of the modal shape of the oscillation and of the stability parameter. Then, the overall multi-modal response can be obtained by weighting each different modal response in some way. Some researchers tend to previlege the hydrodynamic aspects, that is consider that mainly one mode is excited by vortex shedding, while others, mainly interested in associated cyclic stresses and fatigue, consider as topical a multi-modal response . It should be noted that the dynamic behaviour of the tubular member is assumed as fully linear in this discipline: the envisaged amplitude of VIV is such as not to trigger any dynamic stiffening or softening requiring a non linear analysis.

In a critical review of the predictive models e.g. proposed by Whitney (1983) and Vandiver (1985), main steps may be more or less considered as the ones aforementioned. In a companion paper, Bruschi et al. (1989) present a part of an extensive study activity carried out on VIV, Vitali (1986), dealing with a comparison of the predictive models of Whitney and Vandiver with a few experimental results obtained at BHRA, Humphries (1988) and Humphries and Walker (1988). The following conclusions were reached:

- The model of Whitney is quite approximate as the elastic response of the component, obtained as RMS of each single modal amplitude in the presence of simultaneous excitation of modes affected by the various synchronisation fields along the axis, is not able to provide evidence either of some prevailing modes rather than others or the localisation of the response and consequent mechanical stresses. In general the response is over-.estimated

- The model of Vandiver, as it was proposed in its first formulation, could be considered rather simplified as it produces evidence of the response of only one mode, thus neglecting the response due to the possible synchronisation of other modes associated to a higher stability parameter. From the hydrodynamic point of view, this assumption seems reasonable. However, this is in contrast with some experimental results which clearly show the multi-modal nature of the response.

It is to be pointed out that the Vandiver's model seems to be aimed at defining a stability parameter to be associated to the possible response modes rather than formulating a response model.

The model proposed in this paper is an integration of the previous ones: the dynamic response is calculated as the sum of transversal deflections associated to the separate synchronisation of the various cells on the natural frequencies considered, weighted through an equivalent stability parameter. This model can be called multi-modal weighted model. However, the comparison with some experimental result shows a general over-prediction of the response for all these models, which may be due to a weak calibration of the dissipation mechanism or to an over-estimate of the capacity from each active cells to work on the tubular member.

It should be mentioned that Lyons and Patel (1989) proposed a new model, based on a semi-empirical formulation to be used within a time domain calculation, which for a certain extent adopted the concept of multi-modal response or multi-cell excitation. It was based on an early model proposed by Iwan (1981) comprising the effects both of limited spatial extent of lock-in and fluid damping of inactive elements.

Irrespective of the research efforts during these last years, current perspectives in deep water are calling for new study activities. Interpretative models as regards the extent and the force of active cells and the role of the passive lengths are still questionable. As an example, fluid dynamic drag for such small oscillations in presence of a wake more or less perpendicular to the direction of free oscillation should be re-addressed. Moreover, in predictive models currently in force there is no mention of the equivalent mass expressing the ratio between the kinetic energy of the oscillating tubular, active plus passive lengths, and the kinetic energy of the passive length.

The VIV amplitudes at lock-in of a given mode and the relevant weight in a multi-model excitation to give a cumulative response envelope is also topical.

From the structural modelling point of view, a more thorough analysis of some aspects related to some antagonistic mechanisms influencing the dominant response taking origin from the nature of the structural response, Triantafyllou (1991) and Patrikalakis and Chryssostomidis (1987), is also recommended.

3. PIPE LAY FREE SPAN: STATICS

With the conventional laying techniques (mainly S and J lay) entailing the assembly of the pipe string on board a lay barge, the pipeline can undergo severe bending loads before reaching the seabed, fig. 1. With the S-lay method the pipeline is supported by the barge ramp/stinger for the first length aft of the tensioners (overbend) then it lifts off and spans up to the touch down point (sagbend). Consequently two different equilibrium conditions take place, namely a displacement controlled condition in the overbend and a load controlled condition from the lift-off point to the touch down point. The geometrical configuration and levels of stress are primarily governed by the ramp length and curvature, the pipeline self-weight and stiffness, the hydrostatic pressure, the water depth and the pull force applied by the tensioners. From another point a view, the pull force results in a fictitious reduction of the pipeline weight. In addition the level of stress can be affected to a large extent by environmental loads due to waves and currents acting directly on the suspended portion and environment induced barge motions resulting in displacements imposed to the pipeline string. Finally accidental loads can occur due to the erroneous ballasting of the stinger, loss of tension at the tensioners, barge movement due to failure of anchor lines, etc. When laying in deep water with the available third generation barges the overbend is the most critical location. In fact due to the long suspended span and associated weight, (the high ratio between the suspended span and the ramp/stinger length) and the available tension capacities, high load reactions and bending can take place at the lift-off point. Apart from the overbend region, a major role is played by the pull force which, coupled with the pipeline curvature, has the effect to support the pipeline span thus reducing the pipeline bending. When laying in shallow waters, the sagbend is the most critical location due to the large pipeline weight required to achieve on-bottom stability under severe environmental loads. Finally as far as the cyclic stresses ranges induced by the laybarge motions are concerned, it can be observed that they are usually negligible in the area of the overbend closer to the tensioners and have peaks both at the lift-off area and in the sagbend.

The structural behaviour of a pipelay free span is analysed in accordance with the large displacement and rotation theory of deflected beams. Special attention is given to the acting loads, such as the hydrostatic pressure. The effects of external pressure on a small pipe length is evidenced by formulating the equilibrium of a volume of fluid of equal size and in the same position as the small pipe section shall be considered, fig. 1. The forces acting on the volume of fluid are its weight and the hydrostatic pressure on the lateral and on the end surfaces. The contribution of the hydrostatic pressure on the lateral surface is defined through a surface integral. The equilibrium in the normal direction leads to:

$$ds \int p \, dc \, \underline{n} \cdot \underline{n} - ((p + dp) \, A \, d\vartheta/2 +$$
$$pA \, d\vartheta/2) \, \underline{n} \cdot \underline{n} - \gamma_w \, A \, \underline{j} \cdot \underline{n} = 0 \qquad (3.1)$$

where: p, external hydrostatic pressure; g_w, specific gravity of fluid; A, cross section of pipe element; dc, arc element. It is assumed that $d\vartheta/2 \sim \sin d\vartheta/2$. From the line geometry the following is recalled:

$$\underline{t} = (dx/ds) \, \underline{i} + (dy/ds) \, \underline{j};$$
$$\underline{n} = (-dy/ds) \, \underline{i} + (dx/ds) \, \underline{j}; \qquad (3.2)$$
$$d\vartheta/ds = 1/R$$

Neglecting the 2nd order terms and introducing the equation (3.2) in the equation (3.1), it results:

$$\int p \, dc = (pA/R) + \gamma_w \, A \, (dx/ds) \qquad (3.3)$$

This gives the resultant of hydrostatic pressure on the pipe element. The force is normal to the pipe axis and is relevant to the pipe unit length.

Fig. 1 shows the behaviour of a pipe subject to its dead weight, pressure loads and tension in S-lay. The equilibrium in the vertical direction leads to the following:

$$dV/dx = -wds/dx + pA/R + \gamma_w \, A \, dx/ds \qquad (3.4)$$

where w is the pipe submerged weight.
In the horizontal direction:

$$dH/dx = (pA/R + \gamma_w \, A \, dx/ds) \, dy/dx \qquad (3.5)$$

The rotation equilibrium requires that:

$$dM/dx = V + Hdy/dx \qquad (3.6)$$

Deriving equation (3.6) and introducing equations (3.4) and (3.5), it is obtained:

$$M'' = -wds/dx + pA/R + \gamma_w \, Adx/ds +$$
$$Hy'' + y'^2 \, (pA/R + \gamma_w \, Adx/ds) \qquad (3.7)$$

where the apex stands for the derivative of the variable in question with respect to x.

Again the following is recalled from geometry considerations:

$$ds/dx = (1 + y'^2)^{1/2};$$
$$dy/ds = y' \, (1 + y'^2)^{1/2}; \qquad (3.8)$$
$$1/R = y''/(1 + y'^2)^{3/2} .$$

Introducing the equation (3.8) in (3.7), it results:

$$M'' = (H+pA/(1+y'^2)^{1/2})y'' =$$
$$(-w+\gamma_w A) \, (1+y'^2)^{1/2} \qquad (3.9)$$

The term M could be written as a function of curvature through:

$$M = EJ/R = EJ \, y''/(1 + y'^2)^{3/2} \qquad (3.10)$$

which relates the curvature to the bending moment and is valid if the pipe stress does not exceed the elastic limit.
Equation (3.6) becomes:

$$(EJ \, y''/(1 + y'^2)^{3/2})'' - (H + pA/(1 + y'^2)^{1/2})y'' =$$
$$(- w + \gamma_w A) \, (1 + y'^2)^{1/2} \qquad (3.11)$$

which describes the flexural behaviour of the pipe in the vertical plane.
The horizontal force is related to the configuration and the pressure loads through the equilibrium condition:

$$dH/dx = (pAy''/(1 + y'^2)^{3/2} + \gamma_w A/(1 + y'^2)^{1/2})y'$$
$$(3.12)$$

The two equations, plus boundary conditions allow to solve the problem.
The term in brackets multiplying y'' in equation (3.9) seems a very complex function of pressure and configuration but it can be easily computed. Differentiating this term it is found:

$$dH_E/dx = dH/dx + (Adp/dx/(1+y'^2)^{1/2} +$$
$$pA \, (-(1+y'^2)^{-3/2})y'y'') \qquad (3.13)$$

in which:

$$dp/dx = -\gamma_w \, dy/dx = - \gamma_w \, y' \qquad (3.14)$$

It gives:

$dH_E/dx = dH/dx - (\gamma_w A/(1 + y'^2)^{1/2} +$
$pAy''/(1 + y'^2)^{3/2})y'$ (3.15)

which combined with equation (3.12) gives:

$$dH_E/dx = 0 \qquad (3.16)$$

The effective horizontal force H_E which governs the flexural behaviour of the pipe is a constant except in the event of concentrated actions while the horizontal pulling force H (on steel) changes with the law given by equation (3.12).
Therefore, equation (3.9) becomes:

$(EJ\ y''/(1 + y'^2)^{3/2})'' - H_E y'' =$
$(-w+\gamma_w A)\ (1+y'^2)^{1/2}$ (3.17)

in which H_E is a constant for a long section of the lay pipe span beyond the stinger.

Equations (3.12), (3.16) and (3.17), plus typical boundary conditions, will fully describe the mathematical model of the pipelay free span. The configuration $y(x)$ and the horizontal force on the steel section $H(x)$ are the unknowns of the problem. The equations which govern the problem are non-linear due to the first term and to the second member of equation (3.17).

The proposed method uses a finite element description of the lay free span and solves the non-linear problem by successive calculations of conveniently linearized systems.
Equation (3.17) can be considered linear in a short pipe element:

$$\overline{EJ}\ y^{IV} - H_E y'' = -(\overline{w - \gamma_w A}) \qquad (3.18)$$

where the marked terms represent the medium value considered as constant in the element.
Moreover:

$$\overline{EJ} = EJ\ /\ (1+\overline{y}'^2)^{3/2} \qquad (3.19)$$

$$\overline{w - \gamma_w A} = (w - \gamma_w A)(1+\overline{y}'^2)^{1/2} \qquad (3.20)$$

$\overline{y}'$ is the medium value of the slope taken from the previously performed calculation.
These approximations will be corrected on the basis of the new configurations until the convergence.

The horizontal force is known as it is the horizontal component of lay tension and the solution does not depend on the horizontal displacements. The only unknowns are the vertical displacements and rotation of each node. The term H_E in equation (3.18)

couples the vertical configuration to the horizontal equilibrium: H_E does not change for pressure loads but only due to the horizontal components of roller reactions, at rollers, see fig. 3.

The monolateral contact between rollers and pipe is simulated in the vertical plane. A frictionless behaviour is assumed and a tentative constraint is imposed at rollers and corrected where the condition is not verified. The correction of the constraint causes a variation in the horizontal forces due to the connection of roller reaction component and an updating of the effective horizontal force.

The internal actions written in the tangent reference system are linked by the following:

$dN/ds = -T/R - q_t\ ;$
$dT/ds = -N/R - q_n\ ;$ (3.21)
$dM/ds = T$

in which N, T, M represent the internal actions, q_t is the tangent component of loads, q_n the shearing component and R the bending radius. Considering that the axial component of load is $q_t = w\ dy/ds$, equation (3.21) can be integrated and gives:

$$N_2 = N_1 + w(y_2 - y_1) - (M_2^2 - M_1^2)/2J \qquad (3.22)$$

where $y_1, N_1, M_1, y_2, N_2, M_2$ are the internal action and the coordinate at two freely chosen pipeline points.
The equation (3.22) can be used to check the pipelay configuration.

4. PIPE LAY FREE SPANS: DYNAMICS

The dynamic configuration of the pipeline as a Bernoulli beam is defined by the radius vector $\mathbf{r}(s,t)$ of the pipe axis and the torsion angle $\chi(s,t)$ as functions of cord lengths s and time t, Love (1927). The initial equations are derived from the dynamic force and moment equilibrium of the pipe and from the constitutive equations of bending and torsion.
If it is assumed that:

- the pipe static configuration due to the pipe weight in water and to the pulling force applied at the tensioner is on a vertical plane;
- the curvature radius of the pipeline in the static configuration is high;
- the dynamic loads act on the vertical plane and orthogonally to the pipe axis;
- the vertical vibration takes place around the pipeline static configuration;

102

- the vibration amplitudes are small;
- the dynamic variation of the axial force is neglected;
- the material has a linear behaviour.

the vertical dynamic equilibrium equation becomes:

$$\frac{\partial^2}{\partial s^2}\left[EJ\frac{\partial^2 v}{\partial s^2}\right] - \frac{\partial}{\partial s}\left[N_e(s)\frac{\partial v}{\partial s}\right] + c\frac{\partial v}{\partial t} + m\frac{\partial^2 v}{\partial t^2} = f(s,t)$$

(4.1)

where:

s   = curvilinear abscissa;
v   = transversal displacement in every point of abscissa s orthogonal to vector **t** tangent to the static elastic line;
EJ   = flexural stiffness;
$N_e(s)$ = equivalent axial force, including the effect of the external pressure;
m   = total mass per unit length including the added mass;
c   = structural damping per unit length;
f(s,t) = hydrodynamic forces acting on the structure due to vortex shedding.

It is worthwhile to remember that when the displacements are such as to cause sensible variations of the dynamic pull with respect to the static pull, a term shall be added to equation (4.1), accounting for the increase of pull due to dynamic deflection. This term makes the problem non-linear and is not taken into account in this analysis. In order to verify the validity of the linear solution the maximum value of this increase is evaluated once the dynamic linear response of the structure has been calculated.

Eq. (4.1) with the associated boundary conditions is integrated using the finite element technique in the spatial domain and the modal superposition technique for the time integration. In particular it is assumed that the dynamic response of the pipe component is given by the superimposition of the modal shapes $\Phi_i(s)$ of the undamped system, multiplied by suitable weight functions, $q_i(t)$:

$$v(s,t) = \sum_{i=1}^{\infty}\phi_i(s)q_i(t)$$

(4.2)

The modal shapes $\Phi_i(s)$ and the correspondent natural frequencies f_i are determined by the eigenvalue analysis of the laying static configuration. The eigenvalue analysis was perfomed considering both the axial and flexural inertia and stiffness

characteristics of the pipe, while the proposed formulation does not take into account the effect of the axial inertia in the evaluation of the maximum amplitude of oscillation.

Replacing the equation (4.2) in the (4.1) and considering the orthogonality property of the modal shapes, equation (4.1) becomes as follows:

$$M_i\frac{d^2q_i(t)}{dt^2} + C_i\frac{dq_i(t)}{dt} + K_iq_i(t) = f_i(t), \text{ for } i = 1,2,3...\infty$$

(4.3)

In the differential equation system (4.3), M_i, C_i, K_i and f_i are respectively the modal mass, the modal damping, the modal stiffness and the modal force and are expressed by the following terms:

$$M_i = \int_L m\phi_i^2(s)ds$$

(4.4)

$$C_i = \int_L c\phi_i^2(s)ds$$

(4.5)

$$K_i = M_i(2\pi f_i)^2$$

(4.6)

$$f_i(t) = \int_L f(s,t)\phi_i(s)ds$$

(4.7)

where L is the pipeline length and f_i is the i-th natural frequency of the tubular.

In the hypothesis that the pipe is subjected to the action of vortex shedding loads caused by a sheared flow, a cyclic action, having a frequency varying between a cell and the other one, will occur along the structure. If the vortex shedding frequencies are close to the natural frequencies of the slender member, they will synchronise on them. The number of pipe stretches where the fluid is able to transfer energy to the structure depends on the slenderness of the component and the velocity gradient; increasing the pipe length and the velocity gradient, the number of the pipe stretches, where synchronisation occurs, increases. The following assumptions are made:

- the structure can be divided in cells where the vortex shedding frequency is constant and is synchronised with the natural frequency of the structure, fig. 4;
- the limits of the cells are defined considering that the vortex shedding frequency, for a uniform flow, locks on the natural frequency of the structure in the field of the reduced velocities

$$4 \leq V_m = U(s)/f_n D \leq 10 \qquad (4.8)$$

where

$U(s)$ = current velocity at the curvilinear abscissa s perpendicular to the vertical plane containing the pipeline static configuration;

f_n = n-th vertical natural frequency of the pipe in the static equilibrium configuration;

D = pipe outer diameter;

V_{rn} = reduced velocity corresponding to the n-th natural frequency, f_n.

For each natural frequency it is possible to calculate by equation (4.8) a peak velocity and a minimum velocity, as follows: $U_p = 6.0\, f_n D$; $U_{min} = 4.0\, f_n D$. Peak and minimum velocities define the velocity range of synchronisation of the n-th frequency and the cell length.

The synchronisation zones are defined starting from the top of the pipeline, where the velocity is higher. The natural mode with U_p closer to the current velocity at the sea surface is considered synchronised in the first cell, where its upper edge is the sea surface and the lower one corresponds to the water depth where $U_{min} = U(s)$. The natural mode excited in the successive cell is the higher mode having a U_p value lower than the U_{min} value corresponding to the lower limit of the upper cell; the lower limit for this cell is still defined equating the U_{min} value, relevant to this mode, to the current velocity, relevant to different water depths.

This process continues until the whole length of the pipeline is analysed. Afterwards the number N of the locked-in cells is determined.

In the following the forcing term, $f(s,t)$ is explained.

For each natural mode synchronisation, there is an active cell, denoted by R_n, where the energy is transferred to the structure and a passive cell, denoted by $(L-R_n)$, where the energy is dissipated by the hydrodynamic damping.

The active force is modelled as follows:

$$f_{CLn}(s,t) = 0.5\rho D V^2(s) C_{Ln}\phi_n(s)\sin(2\pi f_n t) \qquad (4.9)$$

Instead the passive force is modelled considering that the drag force per unit length is given by:

$$f_{Dn}(s,t) = 0.5\rho \cdot C_D \cdot D |\overline{U}_R| \overline{U}_R \qquad (4.10)$$

where $U_r = (U(s), \partial v/\partial t)$ is the relative velocity between the flow and the structure.
For $\partial v/\partial t << U(s)$, the component of f_{Dn} in the lateral direction can be written as:

$$f_{Dn}(s,t) = -0.5\rho \cdot C_D \cdot DU(s)\frac{\partial v}{\partial t} \qquad (4.11)$$

The total force $f_n(s,t)$ acting along the pipe, due to the synchronisation in the n-th cell, is:

$$f_n(s,t) = \begin{cases} 0.5\rho DU^2(s)C_{Ln}\phi_n(s)\sin(2\pi f_n t), \; for\; s \in R_n \\ \qquad\qquad\qquad\qquad\qquad\qquad\qquad (4.12) \\ -0.5\rho C_D DU(s)\dfrac{\partial v}{\partial t}, \qquad for\; s \in (L-R_n) \end{cases}$$
$$(4.13)$$

where C_D is the drag coefficient and C_{Ln} is the lift coefficient.

The vibration amplitudes, due to the synchronisation in a defined cell, are calculated by substituting equations (4.12-4.13) into equations (4.3). Since the differential equation system is linear, the response of the structure is proportional to the lift coefficient C_{Ln} and, in particular the system solution is:

$$q_{ni}(t) = q_{ni0}(t) + q_{niG}(t), \; for\; i=1, 2, 3,.\infty \qquad (4.14)$$

where:

$$q_{no}(t) = e^{-2\pi f_i t}\left(A_{ni}\sin\left(\left(2\pi f_i\sqrt{1-\xi_i^2}\right)t\right) + B_{ni}\cos\left(\left(2\pi f_i\sqrt{1-\xi_i^2}\right)t\right)\right)$$
$$(4.15)$$

$$q_{niG}(t) = C_{Ln}F_{ni}\sin(2\pi f_i t - \theta_{ni})/(2\pi f_i)^2((1-(f_n/f_i)^2)^2 + (2\xi_i f_n/f_i)^2)^{\frac{1}{2}}$$
$$(4.16)$$

$$\vartheta_{ni} = \tan^{-1}(2\xi_i f_n/(f_i\cdot(1-(f_n/f_i)^2))) \qquad (4.17)$$

$$\xi_i = \left(C_i + 0.5\rho DC_D \int\limits_{(L-Rn)} U(s)\phi_i^2(s)ds\right)/(4\pi f_i M_i) \qquad (4.18)$$

$$F_{ni} = (0.5\rho D \int\limits_{Rn} U^2(s)\phi_n(s)\phi_i(s)ds)/M_i \qquad (4.19)$$

The terms A_{ni} and B_{ni} are integration constants obtained by imposing the initial conditions that, since the structure is assumed as still, are $q_{ni}(0)=dq_{ni}(0)/dt=0$.

On the contrary, if the regime response of the system is considered, $q_{nio}(t)$ tends to zero as the time t increases, and the total displacement due to the synchronisation in the cell R_n is then given by:

$$v_n(s,t) = C_{Ln} \sum_{i=1}^{\infty} q_{ni}(t)\phi_i(s) \tag{4.20}$$

where

$$q_{ni}(t) = \overline{q}_{ni} \cdot \sin(2\pi f_i t - \vartheta_{ni}) \tag{4.21}$$

$$\overline{q}_{ni} = F_{ni} / ((2\pi f_i)^2 ((1-(f_n/f_i)^2)^2 + (2\xi_i f_n/f_i)^2)^{\frac{1}{2}}) \tag{4.22}$$

As a general rule, the response models developed for the fluidoelastic oscillation induced by the vortex shedding are imposed displacement models; as a consequence the value of C_{Ln} is obtained imposing a maximum vibration amplitude equal to the amplitude obtained through the following relationship, valid for uniform flow and suggested by Blevins (1990):

$$(v_n/D)_{max} = 0.32\gamma_n / (0.06 + S_{Gen}^2)^{\frac{1}{2}} \tag{4.23}$$

where

$$\gamma_n = |\phi_n|_{max} \left(\int_L \phi_n^2(s)ds / \int_L \phi_n^4(s)ds \right)^{\frac{1}{2}} \tag{4.24}$$

and S_{Gen} is the equivalent response parameter, as defined by Vandiver (1985). S_{Gen} is calculated as follows:

$$S_{Gen} = S_{Gu}(P_u/P_s)(R_s/R_u) \tag{4.25}$$

where

$$P_u = \int_L \phi_n^2 ds \tag{4.26}$$

$$P_s = \int_{Rn} \phi_n^2(s)ds \tag{4.27}$$

$$R_u = \int_L c\phi_n^2(s)ds \tag{4.28}$$

$$R_s = R_u + 0.5\rho DC_D \int_{(L-Rn)} U(s)\phi_n^2(s)ds \tag{4.29}$$

$$S_{GU} = \frac{2\pi S_{tn}^2 R_u}{\rho D^2 f_n \int_L \phi_n^2(s)ds} \tag{4.30}$$

The term S_{tn}^2, appearing in equation (4.30) is the Strouhal number and is assumed equal to 0.2.

S_{Gu} and S_{Gen} are the response parameters for the vortex shedding synchronisation respectively along the whole structure and on a limited portion of it (R_n).

S_{Gen} is always higher than S_{Gu} as when the flow is not uniform there is an area where the fluid acts as damper causing lower oscillation amplitude than in the corresponding case with uniform flow (see equation 4.23).

The member oscillation, due to the vortex shedding synchronisation in the different cells, is obtained by the above described procedure. Afterwards the total pipe response is evaluated through a "weighted" sum of the response due to the synchronisation in the various cells, where the "weight" is given by the inverse of Vandiver equivalent response parameter S_{Gen}, as the vibration amplitudes are inversely proportional to this parameter:

$$v(s,t) = \left(\sum_{n=1}^{N} v_n(s,t)/S_{Gen} \right) / \sum_{n=1}^{N} (1/S_{Gen}) \tag{4.31}$$

5. STRENGTH EVALUATION

On the basis of the response model, previously described, the total stress distribution along the laying span and the pipe cross section is known. Therefore the evaluation of the circumferential crack growth can be performed on the basis of a fracture mechanics approach, once the time interval necessary to a cross section to pass from the barge to the sea bottom, the initial dimensions of the maximum circumferential defect (length and thickness) and its starting position both along the laying span and the cross section are known.

The defect length is compared with the critical defect length in order to determine the cumulative fatigue damage of the pipe during laying operation. Hereinafter, a simplified methodology was adopted, based on the following assumptions:

-   The defect is located in the most stressed point of the pipe cross section.
-   The total time, T, during which the circumferential crack is exposed to dynamic stress is equal to the time necessary for the involved pipe section to

pass from the laying ramp to the touch-down point on the sea bed.

The fundamental steps are:

- Calculation of the time-average stress acting on particular points along the pipeline. These points usually coincide with the nodes of the finite element mesh.

- Calculation of the dynamic stress in the various selected points through the equation (4.31), which becomes:

$$\sigma(s_i, t) = E \frac{\partial^2 v(s_i, t)}{\partial s^2} \frac{D}{2} \qquad (5.1)$$

- Considering that the lock in frequencies are N, with N greater than 1, the stress state of the node is neither a harmonic nor a periodical function. Through the relation (4.20) the stress status due to the lock in of n-th frequency is calculated in every node.
In particular the difference between the maximum and the minimum values, indicated by $\Delta\sigma_{in}$, are determined. The index "i" refers to the node in question and the index "n" to the considered locked in frequency.
- Calculation of the time elapsing for the passage from node "i" to node "i+1" through the relation $T_i = L_i/V$, where L_i is the distance between the two nodes "i" and "i+1" and V is the laying velocity.
- Assuming that the state of stress keeps constant during the time interval T_i and equal to that occurring on node "i", the number of cycles corresponding to each $\Delta\sigma_{in}$ which the defects is exposed to, is calculate by $n_{in} = T_i f_n$.
- Assessment of the characteristic parameters of the S-N curve, C and m ($\Delta\sigma^m N = C$). In particular, three distinct S-N curves are determined in each node: the first two are evaluated by a simplified procedure based on the fracture mechanics according to the provisions of BS-PD6493-1991, one corresponding to the propagation of a superficial defect, the second relevant to an internal defect; the third curve is equal to the quality curve X (API-RP2A), as the weld is assumed of a very high quality being obtained by the butt-welding method.
- Calculation of the number of limit cycles corresponding to each $\Delta\sigma_{ij}$, by using the above calculated S-N curves: $N_{in} = C_i/\Delta\sigma_{in}{}^{m_i}$.

- Calculation of the partial cumulative damage when passing from node "i" to node "i+1", through the Palmgren-Miner law:

$$d_i = \sum_{n=1}^{N} n_{in} / N_{in} = \sum_{h=1}^{N} T_i f_n \Delta\sigma_{in}{}^{m_i} / C_i \qquad (5.2)$$

- Calculation of the total cumulative damage, during the laying operations, through the sum of the partial damages obtained by the equation (4.34) utilizing the three S-N curves:

$$d_T = \sum_{i=1}^{Nodi} d_i = \sum_{n=1}^{N} T_i / C_i \sum_{j=1}^{N} f_n \Delta\sigma_{in}{}^{m_i} \qquad (5.3)$$

The total cumulative damage, calculated by the equation (5.3), is to be considered an upper limit value as it is estimated on the basis of the envelope of the maximum dynamic stress range acting along the pipe axis. A less conservative estimation can be performed on the basis of the root mean square (RMS) of the dynamic stress variation envelope: where the RMS is evaluated considering the laying time interval necessary to a joint to pass from the stinger to the sea bottom.

6. APPLICATION

Different laying scenarios were been considered covering both medium and large diameter pipelines and both deep and very deep waters. The final scope was to evidentiate the criticality of vortex-induced oscillations during the laying operation if lock-in occurs at the higher natural frequencies of the pipelay free span.
The analysis has been carried out in the following steps:

- static analysis in order to determine the pipe static configuration;
- eigenvalue/eigenvector analysis in order to determine the pipe dynamic response characteristics;
- evaluation of the dynamic stress range associated to the dynamic stress range;
- pipe fatigue analysis in order to determine the damage accumulated during the laying operation.

The laybarge considered within the present application is a third generation laybarge, CASTORO SEI, with the following main characteristics (from in house data): maximum pulling capacity of 360 t

106

Tab. 1 - Laying scenarios.

Case	Outside steel diameter (m)	Specific gravity	Mass (kg/m)	Water depth (m)	Laying method
L1	0.508	1.2	317	350	S
L2	0.508	1.4	401	350	S
L3	1.016	1.2	1227	350	S
L4	0.508	1.2	317	700	S
L5	0.508	1.2	317	700	J

Tab. 2 - Laying static analysis results.

Case	Stinger radius (m)	Tensioner pulling force (kN)	Free span length (m)	Max. strain over-bend (%)	Max. stress sag-bend (MPa)
L1	137	414.6	695	0.207	156.5
L2	160	1153.7	737	0.207	124.3
L3	240	3467.8	1010	0.237	113.1
L4	150	1048.0	1422	0.213	132.1
L5	-	469.9	870	-	323.0

(assuming that the tensioner is upgraded); a welding ramp slope of 9.2 deg; a stinger with the overall length of 91m and 10 rollers; a trim of 1.5 deg; an assumed laying rate of about 170 joints per day.

Within the laying scenarios considered, a 20" diameter pipeline was selected as representative for the medium diameter category and a 40" diameter pipeline was selected for the large diameter category. The following water depths were considered: 350m and 700m for the 20" pipe and 350m for the 40" pipe. A diameter to thickness ratio (D/t) of 35.5 was chosen for all cases. This D/t ratio was selected by consideration concerning both local buckling and maximum allowable stress, DNV (1981), thus resulting in the following pipe dimensions:

	Outside diameter (m)	Wall thickness (m)
20"	0.508	0.0143
40"	1.016	0.0286

The concrete coating thickness was selected in order to have a pipeline specific gravity of 1.2 for the 20"

Tab. 3 - Response amplitudes. The value in brackets is the maximum non dimensional amplitude of oscillation.

Case	Current C01			Current C02			Current 03		
	Excited mode	Freq. (Hz)	v_{max}/D	Excited mode	Freq. (Hz)	v_{max}/D	Excited mode	Freq. (Hz)	v_{max}/D
L1	9 4 2	0.263 0.098 0.052	1.310 0.182 0.178 (1.140)	12 9	0.406 0.263	0.994 0.417 (0.789)	8 5 3	0.225 0.125 0.074	0.666 0.569 0.130 (0.473)
L2	9	0.260	0.951 (0.951)	13 8	0.365 0.230	0.907 0.419 (0.75)	8 5 3	0.230 0.144 0.093	0.602 0.498 0.182 (0.450)
L3	3 1	0.123 0.058	1.160 0.222 (1.110)	5 3	0.197 0.123	0.949 0.370 (0.730)	3 1	0.123 0.058	0.623 0.476 (0.506)
L4	20 13	0.272 0.175	1.210 0.121 (1.096)	20	0.292	0.723 (0.723)	19 12 8	0.254 0.162 0.108	0.681 0.538 0.321 (0.474)
L5	5 3 1	0.264 0.156 0.071	1.500 0.142 0.160 (1.439)	7	0.391	1.290 (1.290)	4 2 1	0.203 0.116 0.071	0.833 0.595 0.214 (0.238)

Tab. 4 - Cumulative fatigue damage.

Case	Cumulative	fatigue	damage			
	Current C01		Current C02		Current C03	
	Fracture mechanics approach	S-N curve	Fracture mechanics approach	S-N curve	Fracture mechanics approach	S-N curve
L1	0.340	0.022	0.556	0.037	0.004	0
L2	0.060	0.002	0.187	0.008	0.011	0
L3	0.008	0	0.018	0	0	0
L4	0.220	0.008	0.058	0.001	0.003	0
L5	0.006	0	0.019	0	0	0

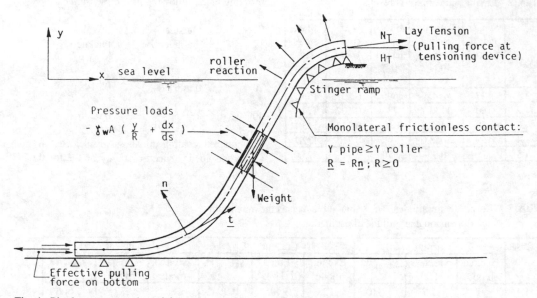

Fig. 1 - Pipelay - structural model.

and 40" pipe and 1.4 for the 20" pipe. The specific gravity is expressed as the ratio between the pipeline weight (in air) and the weight of the displaced water.

Afterwards both the S-lay and the J-lay methodologies were analysed. Tab. 1 shows the considered laying scenarios.

Static Analysis

The pipelaying static analysis was carried out in order to define the stinger radius and the lay pull at the tensioner. These were calculated in accordance to a maximum allowable longitudinal strain in the overbend for the S-lay and to a maximum equivalent stress in the sagbend for the J-lay, Bruschi and al.

(1994). The results of the static analysis are summarized in tab. 2.

Eigenvalues Analysis

The eigenvalues analysis was carried out assuming that the pipe ends are clamped. It is obvious that the boundary conditions do not significantly affect the dynamic behaviour of such a long free span, but they strongly affect the dynamic stress at the ends, causing very high stress if compared with the ones relevant to the free span section far from the two ends. Therefore their effect is important as regards the cumulative fatigue damage. In this present work no sensitivity analysis relevant to their effect is described.

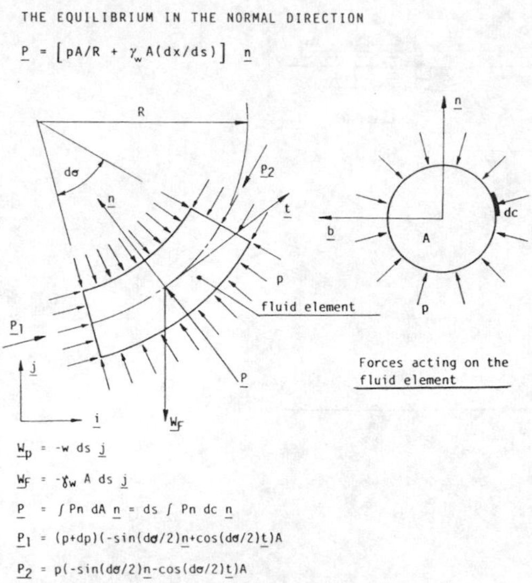

THE EQUILIBRIUM IN THE NORMAL DIRECTION

$$\underline{P} = \left[pA/R + \gamma_w A(dx/ds) \right] \underline{n}$$

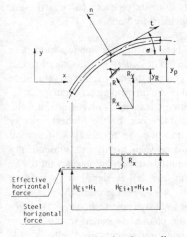

$$\underline{W}_p = -w \ ds \ \underline{j}$$

$$\underline{W}_F = -\gamma_w A \ ds \ \underline{j}$$

$$\underline{P} = \int Pn \ dA \ \underline{n} = ds \int Pn \ dc \ \underline{n}$$

$$\underline{P}_1 = (p+dp)(-\sin(d\sigma/2)\underline{n}+\cos(d\sigma/2)\underline{t})A$$

$$\underline{P}_2 = p(-\sin(d\sigma/2)\underline{n}-\cos(d\sigma/2)\underline{t})A$$

Fig. 2 - Effect of external pressure.

Fig. 3 - Effect of friction less roller on the horizontal force.

For the S-laying at the water depth of 350m, the pipe natural frequencies increase for increasing specific gravity. This effect is mainly due to the fact that the pipe behaves as a cable and the stiffening due to the pulling force prevails on the softening due to the increasing span length and increasing mass per unit length. Furthermore higher natural frequencies correspond to the larger diameter pipeline. Increasing the water depth, the natural frequencies decrease as the effect of the increasing span length prevails. Finally the analysis shows that the laying

configuration strongly affects the dynamic characteristics of the pipelay free span: in the J-lay case, the lowest natural frequencies are approximately three times higher than the ones relevant to the correspondent S-lay case.

Dynamic Response and Fatigue Analysis

Three different current profiles were considered acting on the pipeline. The velocity distribution of the profile, indicated as C01, is typical of wind generated current, characterised by the following expression:

$$U(z)=U_0((h-z)/h)^{1/7}$$

where: U_0 is the current velocity, at still water level, assumed equal to 1 m/s; h is the water depth of the site and z is the distance from the surface.
The current profiles C02 and C03 are shown in fig. 5; these profiles could be generated by the effect of the tidal currents combined with currents due to water density gradients. The above mentioned current profiles were combined with the laying scenarios described in tab.1.
Tab.3 is a summary of the pipe dynamic response analysis results in terms of:

- lock-on natural frequencies and associated natural modes;
- maximum amplitudes of oscillation relevant to each natural frequencies, calculated according to the equation (4.23);
- maximum amplitude of oscillation relevant to the total response of the pipe, according to the equation (4.31).

If the current profile C01 is associated to the different laying scenarios, multiple modal synchronisation occurs in all the cases, except the L2 scenario. The total pipe response is governed by the excitation from the upper cell, while the other modes scarcely contribute to the overall pipe response. As far as the laying scenarios at 350 m water depth is concerned, it can be pointed out that the 40" pipeline (case L4) synchronises at lower natural modes. Moreover, for the laying scenarios in 700 m water depth, the synchronised natural modes relevant to J-lay are lower than the ones relevant to the S-lay. For both cases, this significantly reduces the cumulative fatigue damage during laying, see tab. 4.

The comparison of the corresponding pipe responses due to the action of the current profile C02 with the one relevant to the current profile C01,

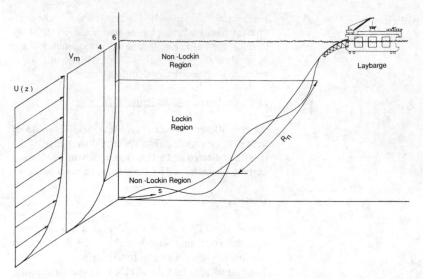

Fig. 4 - Third mode partial lockin in a sheared flow.

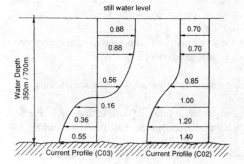

Fig. 5 - Current profiles (m/s).

shows that the maximum amplitude of oscillation decreases and the synchronisation of higher modes occur. The former is due the higher value of the velocity gradient along the current profile, while the latter is caused by the increase of the maximum current velocity. Both of them have an opposite effect on the cumulative fatigue damage: the fatigue damage showed in tab.4 evidences that the latter is predominant.

Figs. 6a, b and c show respectively:

- the pipe static equilibrium configuration and the modal shapes correspondent to the excited vortex shedding frequency;
- the time envelope of the transversal displacements along the pipe axis calculated using the proposed model

- the time envelope of the transversal displacements along the pipe axis calculated using the model suggested by Vandiver (1985).

The comparison between the two models evidences that, due to the contribution of the multi-modal excitation, light changes of the maximum amplitude of oscillation and of the time envelope occur.

Between the three considered current profiles, the profile C03 is characterised by the highest velocity gradient: this causes the synchronisation of several modes along the free span length and subsequently the maximum amplitude of oscillation relevant to the synchronisation in one cell is significantly reduced with respect to the others. Furthermore the maximum value of the pipe total dynamic response is furtherly reduced due to the phases associated to the forcing mechanism in the different synchronised cells.

These effects justify the low value of the cumulative fatigue damage calculated using the current C03 with respect to the others, see tab.4. As for the previous case fig.7a, b and c show the modal shapes correspondent to the excited frequencies and the time envelope of the transversal displacements along the pipe axis according to both the proposed and the Vandiver's models.

7. CONCLUSIONS

New pipelines in increasingly difficult environments are currently being discussed or planned for the

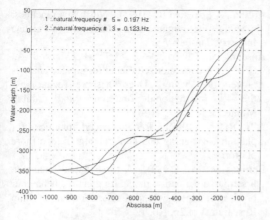

a) lock-in natural modes

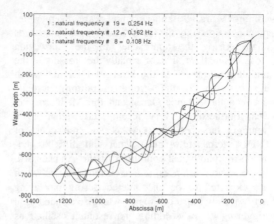

a) lock-in natural modes

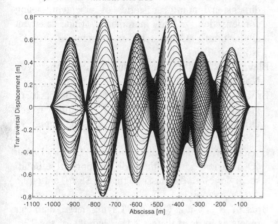

b) transversal displacement by the proposed model

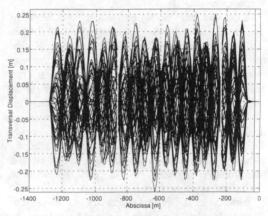

b) transversal displacement by the proposed model

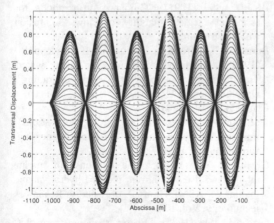

c) transversal displacement by Vandiver's model

Fig. 6 - 40" diameter pipe; specific gravity = 1.2; water depth= 350 m; S-lay (L3 case). Current profile C02.

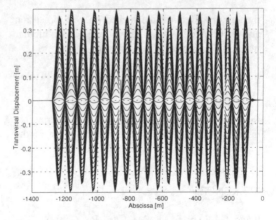

c) transversal displacement by Vandiver's model

Fig. 7 - 20" diameter pipe; specific gravity = 1.2; water depth= 700 m; S-lay (L4 case). Current profile C03.

coming years. Striking examples are deep water pipelines crossing the rocky-sharp peaked seabeds characterising the approaches to the Norwegian Coast inside the fjords, or actually the crossing of the undulating seabeds of the Strait of Gibraltar which are exposed to the strong on-bottom currents between the Atlantic Ocean and the Mediterranean Sea. Furthermore, very recent study relate to a strategic link between OMAN and the coast of INDIA across ultra deep water and difficult geo-sismo-morphological conditions of the Arabian Sea. In these circumstances, severe current profiles which are expected to be encountered during pipelaying may severely impact on the very long suspended span triggering vortex shedding and resonant oscillations involving one or more natural frequency of the spanning length.

The excitation based on a kind of response amplitude mechanisms, the involvement of higher modes and (consequently to the response amplitude mechanism) the resulting high stress ranges, the time interval for a pipe joint leaving the ramp to reach the seabottom, all together may contribute to cause an unacceptable cumulative damage on pipe joint and on girth welds. This is evidenced in a series of applications presented and discussed here above: Moreover the contribution of different factors influencing the cumulative damage, including the interpretative models of VIV in shear flow conditions, is pointed out.

From these it appears necessary to investigate thoroughly the excitation mechanism related to vortex shedding in current profile, notably:

- the factors affecting the locked on cell dimension and distribution along the axis of slender tubular members subjected to shear flows;

- the fluid dynamic damping on the pipe section where synchronisation does not occur;

- the effect of the synchronisation occurring in different cells on the multi-modal response of the pipe.

REFERENCES

Albano G., Lehrizi M., De Caro R., (1992): "The TRANSMED: a technological milestone now under further expansion", Proceedings of the 11th Int. Conf. on OMAE, Vol.1 part B, Calgary.

Anselmi A., Bruschi R, (1993): "North Sea pipelines: meeting the engineering challenges", Pipeline Industry, Jan-Feb-Apr.

Anselmi A., Cimbali W., Orselli B., Mazzoli A., (1987): "A computerized model for pipelay analysis in very deep water", Deep Offshore Technology, the 4th International Conference and Exhibit, Monte Carlo.

API RP2A (1989): Recommended practice for planning, designing and constructing fixed offhore platforms.

Bernitas M. M., (1982): "A three dimensional non linear large-deflection model for dynamic behaviour of risers", Pipelines and Cables. Journal of Ship Research. Vol.26, No. 1.

Bernitas M. M., Vlahopoulos N., (1990): "Three dimensional nonlinear statics of pipelaying using condensation in an incremental finite element algorithmm", Computers & structures. Vol.35, No.3, pp. 195-214.

Blevins R., (1990) : "Flow induced oscillations"; Van Nostrand Reinhold, 2nd Ed.

Brooks I.H. (1987): "A pragmatic approach to vortex - induced vibrations of a drilling riser", Offshore Technology Conference, OTC Paper No 5522, Houston - TX, USA.

Bruschi R., Vitali L., Tomassini E.P., (1989): "Oscillazioni idroelastiche di componenti tubolari snelli investiti da flussi tagliati", Sesto colloquio AIOM, Napoli.

Bruschi R. et al., (1994): "Laying large diameter pipelines in deep waters", Proceedings of the 13th Int. Conf. on OMAE, Houston.

BS-PD 6493 (1991): "Guidance on methods for assessing the acceptability of flows in fusion welded structures".

Burattini E., Lorenzetti L., Raffaeli E., Radenti R., (1993): "New achievements in deep water: Transmediterranean pipeline system upgrading", Deep Offshore Technology Conference, Monaco.

Clauss G.F., Weede H. E. W., (1989): "Dynamics of offshore pipelines during laying operations", International Symposium on Offshore Engineering. Rio de Janiero, Brazil.

Griffin O. M., (1982): "The Response of marine tubular and risers to current-induced hydrodynamic loadings", ASME Journal of Energy Resources technology. Vol. 104.

Griffin O. M., (1985): "Vortex shedding from bluff bodies in shear flow: A review", Transactions of ASME, Journal of Fluid Engineering, Vol. 107, September.

Humphries J. A. (1988): "Comparison between theoretical predictions for vortex shedding in shear flow and experiments", the 7th Int. Conf. on OMAE, Houston-TX, USA.

Humphries J. A., Walker D. H. (1989): "Vortex excited response of large scale cylinders in sheared flows", the 9th Int. Conf. on OMAE, Houston-TX, USA.

Kim Y. C., Triantafyllou M. S. (1984): "The nonlinear dynamics of long slender cylinders", Offshore Mechanics and Arctic Engineering Symposium", New Orleans, La.

Iovenitti L., Venturi M., Albano G., Tonisi E. H. (1994): "Submarine pipeline inspection: the 12 years experience of TRANSMED and future developments", the 13th Int. Conf. on OMAE; Houston.

Iwan W.D., (1981): "The vortex-induced oscillation of non-uniform structural systems", J. Sound and Vibration, Vol. 79 Part 2.

Langer C.G., Ayers R.R, (1985): "The feasibility of laying pipelines in deep waters", Proc. of the 4th Int. Conf. of Offshore Mechanics and Artic Engineering, ASME, New Orleans.

Love A.E.H. (1927): "A treatise on the mathematical theory of elasticity"; Dover Publications.

Lyons G.J., Patel M. H., (1989): "Application of a general technique for the prediction of riser vortex-induced vibration in waves and current", Journal of Offshore Mechanics and Artic Engineering. Vol. 111, pp. 82-91.

Maull D. J., Young R.A. (1973): "Vortex shedding from bluff bodies in a shear flow", J. Fluid Mech., Vol. 60 Part 2.

Pantazopoulos M. (1994): "Vortex-induced vibration parameters: critical review"; the 13th Int. Conf. on OMAE, Houston - TX, USA.

Patrikalakis N. M., Chryssostomidis C., (1987): "Vortex induced response of a flexible cylinder in a sheared current", the 7th Int. Conf. on OMAE, Dallas -Texas, USA.

Rooney D. M., Peltzer R. D., (1982): "The effects of roughness and shear on vortex shedding cell length behind a circular cylinder", Transactions of the ASME, Vol. 104, March.

Shell Development Company: "Deep water pipeline feasibility study", Thirthy Nine World Wide

Sriskandarajah T., Mahendran I. K. (1992): "Critique of pipelay criteria and effects on pipeline design", Offshore Technology Conference, OTC Paper No. 6847, Houston-TX. USA.

Stansby P. K. (1976): "The locking-on of vortex shedding due to the cross-stream vibration of circular cylinders in uniform and shear flows", Journal of Fluid Mechanics.

Triantafyllou M. S. (1991): " Influence of amplitude modulation of the fluid forces acting on a vibrating cylinder in cross-flow", 1st International and Polar Engineering Conference. Edinburgh, United Kingdom.

Vandiver J. K., (1985): "The prediction of lock in vibration on flexible cylinders in a sheared flow", OTC 5006. Houston, Texas.

Vandiver J. K. and Chung T.Y. (1987): "Hydrodynamic damping on flexible cylinders in sheared flows", Offshore Technology Conference, OTC Paper No 5524, Houston - TX, USA.

Vitali L., (1986): "Hydroelastic problems in the offshore engineering: the vortex shedding induced oscillations of risers and submarine pipelines", tesi di Laurea, Università di Ancona - Facoltà di Ingegneria - Dipartimento di Meccanica.

Whitney A. K., Nikkel K.G. (1983): "Effects of shear flow on vortex shedding-induced vibration of marine risers", OTC 4595. Houston, Texas.

Walker D. King R. (1988): "Vortex excited vibrations of tapered and stepped cylinders", the 7th Int. Conf. on OMAE, Houston-TX, USA.

Hydroelasticity in Marine Technology, Faltinsen et al. (eds) © 1994 Balkema, Rotterdam, ISBN 90 5410 387 6

Vortex shedding induced oscillations on pipelines resting on very uneven seabeds: Predictions and countermeasures

Roberto Bruschi, Paolo Simantiras & Luigino Vitali
Snamprogetti S.p.A., Offshore Division, Fano, Italy

Vagner Jacobsen
Danish Hydraulic Institute, Ports & Offshore Division, Hørsholm, Denmark

ABSTRACT: The offshore pipeline industry is currently embarking upon new challenging deep water crossings characterised by undulating rocky seabeds. It is therefore of increasing interest to reckon upon an unambiguous and recognised approach for the acceptance of pipe lengths in suspension under vortex shedding induced vibrations (VIV). In this discipline, noteworthy advances have been attained these last years. Specifically, a number of Joint Industry Research (JIR) projects have been carried out to improve the knowledge on peculiarities of VIV on submarine pipelines in the near-seabed scenario. The free span distribution and the structural-installation-functional conditions for the achieved state of elastic equilibrium, the boundary conditions of each free span, the initial as-laid deflection and the response to functional loads, the coupling of adjacent spans, the occurrence of VIV and the excitation regimes, etc., have been closely investigated theoretically and, in many circumstances, supported by extensive experimental campaigns both in laboratory and in-field. In this paper some aspects of the hydroelastic excitation and response are examined, focusing upon prediction tools and the implications from these through some specific evidences. In addition, traditional and advanced countermeasures are discussed, particularly looking into the applicability of hydrodynamic devices as temporary or permanent measures to tackle VIV in a more cost-effective way than in the recent past. Recent experimental data obtained in laboratory on slender tubulars, both bare and with VIV suppression devices, are briefly discussed.

1 INTRODUCTION

These last years the attention of the offshore pipeline industry strongly focused on submarine pipelines crossing very uneven seabeds. In these circumstances the as-laid pipeline is expected to be in contact with the seabed for short lengths and essentially free spanning for the remaining stretches, Bruschi et al. (1991-a). The assessment of the structural integrity of this configuration in the short and in the long term is based on the assessment of the maximum achievable stresses at the free span shoulders and, in the presence of persistent and considerable on-bottom currents, on the evaluation of the cumulative damage induced by VIV. In certain conditions, the compliance with strength and fatigue criteria may lead to extensive seabed preparation and free span correction works, deeply influencing the technical-economical effectiveness of a project. This aspect is particularly outstanding in deep waters as for the new challenging projects planned in the 90's: shore approaches crossing the rocky seabeds at the entrance of Norwegian fjords, deep water crossings

of the continental slope of the Gulf of Mexico, the Strait of Gibraltar and new Mediterranean Sea crossings, the Oman-India and South East Asia links, etc. The developments in the coming years are expected to encourage the efforts from Oil Companies and R&D bodies to move towards advanced concepts for such pipeline systems in order to properly ensure the transportation availability in the long term, and, at the same time, to improve the cost effectiveness of such projects, Bruschi and Vitali (1994).

Since the early days of the offshore pipeline technology, it was recognised that the structural integrity of an offshore pipeline might be jeopardised when pipeline lengths were suspended for a considerable extent and exposed to significant near-bottom cross currents, Mes (1976). However, the complexity of the problem was beyond the realm of the offshore R&D of those days. A tentative design guideline in the field of VIV of free standing piles exposed to river currents, was developed during the construction of the Immingham Oil Terminal in the UK, Wootton et al. (1974). The subsequent in-situ investigation provided a considerable amount of data for direct application to the proposed jetty design. This well-organised set of data represented the base for the development of the CIRIA design guidelines, Hallam et al. (1978), which was later incorporated by Det Norske Veritas in the "Rules for Submarine Pipeline Systems", in 1981. In this first formulation of a design guideline, a salient role was also played by the results of the contemporary research efforts at BHRA-UK, King (1977), and in USA, Sarpkaya (1979).

The problem of VIV in general and suspended spans, became important during the late 70's and early 80's for certain government bodies, Griffin et al. (1975) and Jacobsen et al. (1984), and major oil companies because of their specialised and future needs of exploration and production not only in harsher environments but also "deeper" waters, Tsahalis (1984). The full scale tests performed in the Lagoon of Venice by Snam before embarking on the Transmed Project, Bruschi et al. (1982), represent the very first sign of strong interest of the offshore pipeline industry to the problem. From this, a number of JIR projects followed during the 80's and notable examples are:
- Development of Guidelines for the Assessment of Submarine Pipeline Spans, Raven (1986), carried out by J.P. Kenney on behalf of the Department of Energy, UK. It included both laboratory and full scale field tests. The field tests were conducted with a free span exposed to tidal current in order to study dynamic effects. The laboratory tests were aimed at determining the effect of wall-seabed proximity on VIV.
- Pipeline Span Evaluation Manual, Bryndum et al. (1989-a), performed by the Danish Hydraulic Institute for a group of six oil companies and the Department of Energy, UK. The study was aimed at assessing the fatigue life of scour induced free spans. The assessment was based on laboratory test data base, covering the hydrodynamic aspects and on theoretically evaluated soil-pipe interaction.
- Submarine Pipeline Vortex Shedding (SVS) Project, Tassini et al. (1989), carried out by Snamprogetti, Italy. The project was partly funded by the EEC (European Economic Community) and ENI (Italy's National Hydrocarbon Corporation) and partly by four Italian Companies. It was basically an experimental project and a very large number of laboratory and field tests was carried out. The effects of irregular waves and/or currents, damping, pipe roughness and seabed proximity were studied among other issues.

Despite these comprehensive efforts, in the late 80's the state of affairs of treating the VIV of suspended spans was not satisfactory because sound and generally accepted guidelines for the design of suspended spans against VIV, i.e. determining the VIV and associated fatigue life, did not yet exist. This lack was remedied by the Gudesp Project, the basic objective of which was the drafting of a guideline to be issued jointly with Det Norske Veritas, as anticipated by Nielsen (1993). This project dealt in a very comprehensive way with the outstanding aspects involving the assessment of the structural integrity of pipeline free spans, Tura et al. (1994). Further, the need to tackle increasingly uneven seabeds and deep waters recently prompted the industry to look more specifically into seabed macro-roughness and the near-bedline hydrodynamic field, the VIV in such an environment and alternative approaches to handle it. In the Maspus Project, the main findings of which can be found in Bruschi et al. (1993-a, b), new tests were carried out to study the pipeline response in the presence of considerable roughness, and new calculation tools were implemented to tackle such scenarios in a safe and cost effective way. As an alternative approach to traditional seabed preparation and free span correction works, new concepts, based on the control of response amplitude of the exposed free spans, came to the forefront. This control can be achieved through the installation of proper "suppression" devices on the pipe. Actually, the tests of such devices on submarine pipelines were first conceived during the full scale test performed in the Lagoon of Venice both in 1978 (umbrella shaped dampers and three-start helically wrapped rope), Snamprogetti (1988), and repeated during the field tests of SVS Project (fringe type), Snamprogetti (1989). At the moment, the use of hydrodynamic devices in new challenging projects is becoming very attractive due to the improved project flexibility and potential cost benefits that such measures seem to warrant. Because of this, new investigations are currently being carried out in Norway in the framework of the Pinties Project, as recently discussed by Bruschi and Vitali (1994).

In the above only the key achievements relevant to VIV in the offshore pipeline technology have been mentioned. It should be noted that in the field of VIV an extensive basic research activity has been carried out in the last thirty years. This work has been recently reviewed by Pantazopoulos (1994) as regards offshore tubulars in general, and by Sumer and Fredsøe (1994) specifically for submarine pipelines.

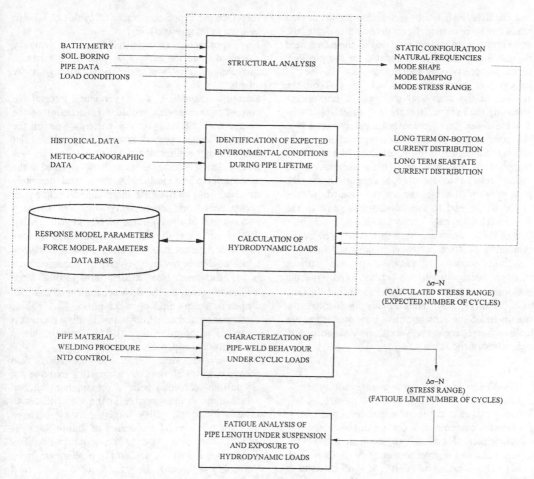

Fig.1 - Disciplines and relevant links in the free span assessment.

The scope of this paper is to present the crucial aspects affecting the assessment of the structural integrity of pipelines resting on uneven seabeds. First, predictions and contradictions in the current practice are discussed. They mainly concern the identification of the loading processes, the modelling of load effects and response of the free spans, basic principles for the formulation of design criteria for structural integrity and safety factors to be applied in different loading conditions. It is presented how the free span correction requirements and related uncertainties are usually tackled by the state-of-practice technology, i.e. additional supporting of the pipeline either with gravel sleepers or with mechanical trestles. The possibility for the introduction of countermeasures of hydrodynamic type, as extensively adopted for risers, is then stressed. In the authors' opinion, the considerable

economic implications of the novel application of an "old" concept will inspire the offshore pipeline industry to pursue such solutions in coming years.

2 PREDICTIONS AND CONTRADICTIONS

The prediction of the occurrence and of the amplitude of VIV on a sequence of free span lengths is mainly based on the following:
- The prediction of the near-seabed hydrodynamic field acting on the pipeline along the route in the short and in the long term. In deep waters it is mainly due to steady currents (varying with long periods), e.g. tidal or due to barotrophic effects sometimes induced by density stratification.
- The definition of the most probable state of equilibrium achieved by the pipeline on the uneven

seabed after installation and under the additional loads due to operation. For a given water depth, the equilibrium state is strictly related to the submerged weight and stiffness characteristics of the pipeline, to the adopted laying technology (e.g. "S" or "J" laying) and consequent residual lay pull on the seabed, to the scale and the type of unevenness involving the bedline, Bruschi et al. (1991-b).

- The identification of hydroelastic parameters which characterise the synchronisation regime between the regular shedding of vortices and the dynamic response of the pipeline. These parameters are mainly related to the free span pattern, to the proximity of the pipeline to the seabed, to the modal shape and to the interaction regime in the lock-in velocity range.

Most of these topics are more or less covered by the engineering practice, which is summarised in Fig.1. Therefore, some contradictions of the current practice which need a further and specific development are discussed in the following. In fact, these factors may considerably influence the assessment of the structural integrity of deep water pipelines free spanning on very uneven seabeds for a large extent of the route.

2.1 *On-bottom current long term distribution*

The design basis is in general a long term distribution of a bottom currents, at a certain distance from the seabed, either directional or projected over the perpendicular to the pipeline alignment, Bruschi et al. (1984). The long term distribution of currents is assessed either from measurements, numerical modelling (hindcasts) or a combination. In either case mean currents within certain time intervals - from 10 min to 1 hour - are used, neglecting any time variance with shorter period. The data resulting from measurements and/or hindcasts are evaluated and analysed to provide the long term distribution based on extreme value statistics. The long term distribution is then applied:

- To define a characteristic near seabed current corresponding to a specified annual probability of exceedence. This current in turn is used to calculate a cut-off frequency of free oscillations for the anticipated free span pattern, Vitali et al. (1993). The natural frequencies of the pipeline free span shall be greater than the cut-off frequency for a no-VIV criterion (under the envisaged load conditions, e.g. empty and operating).
- To estimate cumulative time interval of different lock-in regimes for the natural frequencies characterising the free span pattern, which can be expected during the operating lifetime (and therefore the expected number of cycles of VIV to be used in fatigue analysis).

This approach, recommended and generally recognised since the early 80's, presents a series of contradictions notably in applications to rough rocky seabeds:

- Location, sampling and averaging interval of current measurements should take account of the nature both of near seabed currents and of the seabed. The near seabed currents for an almost flat bedline may be different from the ones flowing over an undulating bedline (as the one encountered in the areas affected by iceberg scours), and definitely different from an irregular sharp-edged bedline with rocky peaks and deep depressions (as the one encountered at the entrance of Norwegian fjords). The pipeline may be fully immersed in a viscous sub-layer characterised by mild flow fluctuations and a negligible average velocity or impacted by a highly turbulent free stream transporting large eddies triggered by the flow separation at downhill slopes or sharp edges of rocky peaks. Under these circumstances, the hydrodynamic loading process is expected to be other than the VIV predictable in the traditional approach.

- The limited time span for typical field measurements allows to anticipate extrema of on-bottom currents with a confidence interval which may be questionable in the view of design criteria based on safety targets ranging between 10^{-2} and 10^{-4} annual probability of failure, Lawless (1978). This could lead to the adoption of safety factors apparently illogical to the pipeline engineer. A solution could be to couple the field measurement with the study of the flow in a hydraulic basin or in a 3-D numerical model to estimate some sort of physical limits for the processes which control the near seabed currents.

- Sometimes the most critical pipe lengths in suspension are caused by local features which also affect the local hydrodynamic field. Accelerated flows over convex morphologies or flow tunnelling within the walls of submarine canyons on one hand and stagnating flows inside deep depressions on the other may totally mislead any prediction based on generic location of current meters. In many circumstances, the limited number of current meter stations and unfortunate locations of these can lead to dramatically overestimate or underestimate the process affecting a specific area.

In general, the engineering practice overlooks this kind of topics and perhaps also the possibility of excessive conservatism of certain long term predictions from oceanographers. The considerable implications on project costs and pipeline reliability in the long term are obvious.

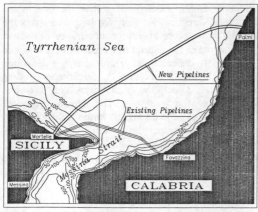

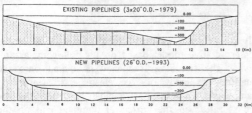

steel type	API 5LX-X65	
outer steel pipe diameter	0.508	m
wall thickness	17.48	mm
polyethylene thickness	4.0	mm
concrete thickness	50	mm
pipe hydraulic diameter	0.615	m
weight	420	kg/m
submerged weight	126	kg/m

Fig.2 - Strait of Messina crossing routings and pipe data.

2.2 *Natural frequencies and modal shapes*

This topic received great impetus after the introduction of the fatigue analysis for the in-line lock-in in the design practice. It occurred in the early 80's after the issue of Det Norske Veritas '81 rules. In those years the structural characterisation of free spanning pipelines was based on the assumption of isolated free spans behaving like a rectilinear beam pinned at its ends, sometimes including the effect of the equivalent axial force, Bruschi et al. (1982). The recognised practice recommended that the maximum free span length should be the one characterised by a natural frequency larger than that for the onset of cross-flow oscillations according to 100 years return period near seabed current. Fatigue assessment for in-line oscillation or, at worst, for no in-line oscillation criteria lead to definitely restrictive

requirements for free span corrections. More realistic boundary conditions for free span assessment of scour induced free spans were being incorporated in the Statpipe System in 1984, Bruschi et al. (1986) and Eide et al. (1988). Then, as a further refinement, the deflected shape of the as laid pipeline and the load history (permanent and functional loads) were accounted for, Bruschi and Vitali (1991). Finally, for free spanning configurations characterised by a random sequence of free span lengths interrupted by short stretches in contact with the seabed, the eigenvalue analysis of the entire sequence of free span lengths has been introduced in the late 80's, Blaker and Bruschi (1990). Advanced topics, as near buckling configuration and oscillation amplitude dependent frequency, have also been introduced, Bruschi and Vitali (1991).

The concept of maximum allowable free span length for a random sequence of free span lengths was substituted by a new format which compares the lowest natural frequency of the structural configuration with the cut-off frequency of cross-flow oscillations, referring to a characteristic value of the near seabed current, Bruschi et al. (1993-a) and Vitali et al. (1993). Obviously, all these efforts were aimed at reducing the free span correction works through a more accurate calculation of the dynamic properties of the configuration.

For the in-line oscillations, the work carried out to model the pipe-soil interaction at the free span shoulders and thereby to quantify the amount of energy dissipated by the soil during a cyclic excitation for the pipe was very valuable, Bernetti et al. (1991). However, this appears to be an excessive sophistication for an approach that has inherent uncertainties as regards the prediction of the actual equilibrium state on the discontinuous supports points and within the deepest and longest depressions. Indeed, the non-linearity of the reaction for the soil and the effect of initial deflection which triggers it, are examples of the complexity and the uncertainty to be carefully handled at the design stage, Vitali et al. (1993). Recalling an episode (unpublished) that occurred during the survey of the pipelines in the Strait of Messina in the early 80's, might be interesting, Benedini (1984). During the assistance works to the laying of the 2nd line of the crossing of the Strait of Messina, the oscillation amplitude in the vertical direction of a 51 m long free span (aspect ratio of about 100) between the pipe joints 497 and 504 was measured. This measurement was carried out using an echo sounder mounted on a saddle placed on the pipe at midspan. The signals were transmitted via wire to the Intersub's PC1602 submarine, where it was recorded on magnetic tape.

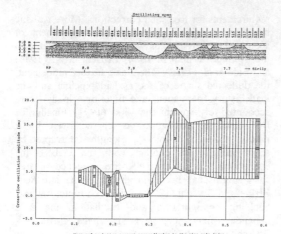

Fig.3 - Span location and recorded cross-flow oscillation amplitude versus the measured current normal to the pipe axis.

The principal pipe characteristics in the concerned zone are summarised in Fig.2. The pipe was resting on a sandy bank in 225 m water depth, with a 2.5% mean slope and with an azimuth of 105°. The estimated residual lay pull was 300 kN. During the survey, the flow field ranged between 0.1 to 0.6 m/s in amplitude and between -65° to 40° directions with respect to the pipe axis. The maximum measured amplitude of oscillation in the vertical plane was 6 cm (~0.1 times the hydraulic diameter). It is important to note that this oscillation occurred with a "zero position" 10-12 cm higher than the static equilibrium position of the pipeline. Indeed, all the oscillations occurred with a movement above the static equilibrium position, as can be seen from Fig.3. This example comprises and demonstrates the complexity of predicting VIV on pipeline free spans, in particular for the slenderest and the most compliant ones to hydrodynamic loads. Besides the importance of having recorded field VIV on a real pipeline, this experience was valuable from the viewpoint of the calibration of predictive models. It was, however, a bit depressing to realize the discordance between desk prediction and the real behaviour. What was responsible for pipe lifting, an hydrodynamic force or a dynamic stiffening in the subject free span or from adjacent ones? Obviously, this evidence triggered new studies.

A rectilinear pipe is characterised by the same natural frequencies in the plane parallel to the flow direction and in the transverse one. When synchronisation occurs, the pipe axis describes a figure "eight" pattern with an amplitude in the in-line

direction usually much smaller than the one in the transverse direction. Actually for long free spans the initially deflected equilibrium configuration of the laid pipeline makes the natural frequencies in the two said planes more different the greater the pipe deflection is, Bruschi and Vitali (1991). Superimposed drag forces may affect the equilibrium and the free oscillations of a free span. The dynamic response under lock-in conditions is completely different, indeed. Furthermore, the presence of the internal pressure may generate in a long span a near buckling condition, and, as a consequence, the response could be even chaotic. In such circumstances any prediction model becomes questionable.

2.3 Hydroelastic regimes

It is current practice to look into VIV using an interpretative model based on the following axiom: when the shedding frequency is locked onto the natural frequency of free elastic oscillations, the amplitude of the response exclusively depends on the mass ratio, on the structural damping and on the modal shape, for a given interaction regime. Many experimental evidences support this interpretative model, which can be classified as a Response Amplitude Model. In the authors' opinion force models, like the ones based on the correlation length, may still be valuable to explain the effect of the pipe oscillation for increasing amplitudes (~0.3 diameter), but will not be attractive to interpret the forcing process at large amplitudes of oscillation (~1.0 diameter) within the cross-flow lock-in range. Indeed, a negligible spanwise correlation in the lock-in range, was measured in many circumstances, see Fig.4 and Bryndum et al. (1989-b). On the other hand, the wide lock-in range of cross-flow VIV in water (V_r, reduced velocity from 4.5 to 10) and the contrasting narrower one for VIV in air, is a key evidence to explain that even a short correlated shedding properly tuned to the oscillation is capable to sustain a resonant oscillation in such lightly damped systems. The different stability parameter in water and in air is obviously the cause of such a difference. This aspect is also strikingly important to interpret experiments on the effectiveness of hydrodynamic devices applied to suppress vortex induced oscillations, as discussed below.

In general, the occurrence of three lock-in regimes is detected, notably the in-line lock-in regimes with symmetric and with alternating shedding of vortices from top and bottom sectors, and the cross-flow lock-in with alternating shedding of vortices. The symmetric shedding of vortices is a kind of instability regime as it may be triggered by an external action

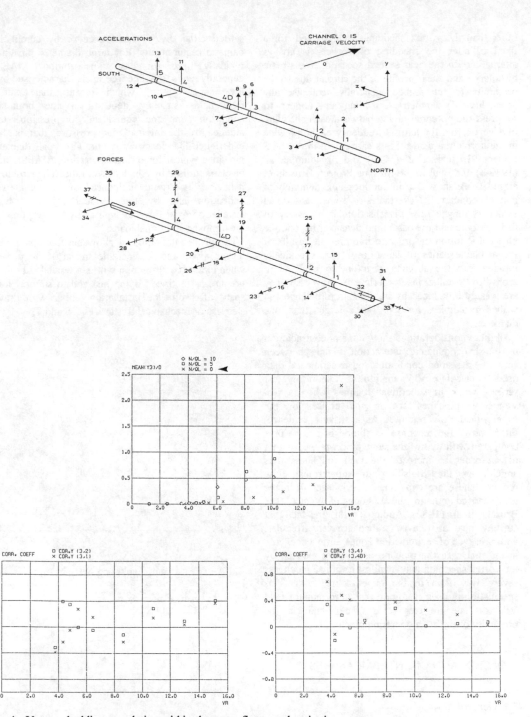

Fig.4 - Vortex shedding correlation within the cross-flow synchronisation range.

other than the vortex shedding itself. Indeed, for a fixed cylinder the shedding of vortices is always alternating! In the near seabed scenario, free stream turbulence and shear profile of the current due to the proximity to the seabed, generally neutralise any possibility of symmetric shedding and appear to suppress the potential tuning between each shedding of a vortex and the natural frequency of the pipeline in the horizontal plane. Tests carried out in the SVS Project, Bruschi et al. (1989) and Bryndum et al. (1989-b) and, and in the Maspus Project, Bruschi et al. (1993-a), show that in-line lock-in is definitely of minor concern. Nevertheless, some additional evidence, maybe flow visualisation, is necessary to further document the basic fluid dynamics beyond the physical intuition, in order to interpret the results of the available series of laboratory tests, specifically looking into the absence of in-line lock-in. As to cross-flow oscillations, experiments show that they are weakly influenced by free stream turbulence and seabed proximity, for gap ratios larger than one diameter.

All these considerations show a major contradiction in the hydrodynamic interaction regime between actual near seabed conditions and experimental data used to calculate VIV amplitudes. Laboratory test set-ups work in subcritical regimes whereas real submarine pipelines are in the critical or the supercritical flow regimes! As a striking evidence, Fig.5 shows the response amplitude of a bare pipe compared with that of the same pipe wrapped with a net in order to increase the wall roughness thus anticipating the transition to supercritical flow regime. These are experimental evidences from the tests carried out in the Lagoon of Venice, see Bruschi et al. (1982). Adding to this the fact that currents near an uneven seabed are very irregular, with presence of recirculation zones, turbulence, etc., in the real case the pipeline oscillations induced by the vortex shedding may manifest itself in a much less severe way than in the experimental tests. This speculation in itself is a good reason to monitor field VIV on specific test set-ups, e.g. built in the proximity of or on an operating pipeline.

3 DESIGN AND COUNTERMEASURES

3.1 General

The conventional approach adopted by the offshore pipeline industry is based on the assessment of the maximum allowable free span length (MAFSL), for both static and durability criteria. As regards VIV, the maximum allowable free span length can be settled either by no occurrence or by cumulative damage requirements. For large diameter pipelines, typically used for gas transportation, MAFSL generally tends to define free spans characterised by a low aspect ratio, which means that natural frequencies strongly depend on the boundary conditions and the equivalent compression state induced by the internal pipe pressure, that is they experience like-beam behaviour. For small diameter pipelines typical for oil transportation, MAFSL may be characterised by a high aspect ratio. The structural behaviour is strongly influenced by the initial pipe deflection and by the residual lay pull from installation axial force as well as by the response to the imposed additional loads.

The proper definition of allowable free spans may have a direct and considerable impact on the project when extensive correction works are needed both to prepare a "rectified" laying bed and to support long spans after pipeline installation, either with gravel sleepers or mechanical trestles (Figs.6 and 7).

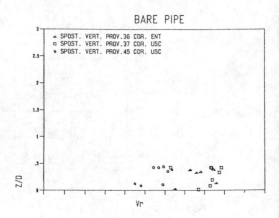

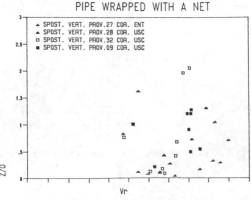

Fig.5 - Comparison of cross-flow response amplitudes of a free span: bare and net wrapped pipe.

Fig.6 - Gravel sleeper supporting the Transmed pipeline in the Sicilian Channel.

Fig.7 - Mechanical trestle supporting the Transmed pipeline in the Sicilian Channel.

The following sections will deal with new pipeline design criteria, as well as possible countermeasures to be adopted to satisfy such criteria, particularly focusing on one category of these countermeasures, that is the hydrodynamic suppression devices.

3.2 *New design criteria*

The free spanning configuration achieved by the empty pipe significantly changes depending on the pipeline diameter and submerged weight, the tension used during the installation and the degree of irregularity of the seabed. This has a great impact on the required intervention works, Bruschi et al. (1993-b). If the resulting configuration is free spanning, the evaluation of extreme cross-flow responses or cumulative fatigue damage possibly induced by hydrodynamic loads is to be carried out in order to identify proper correction works both to prepare a "rectified" laying bed and to support long free spans after pipeline installation. Extreme stresses due to the largest lateral and vertical oscillations and

fatigue damage predicted on the basis of a hypothetical stress history are then compared with allowance criteria. When the equilibrium configuration achieved by a pipeline on an uneven seabed shows a sequence of stretches in suspension, separated by short lengths, i.e. a few pipe diameters where the pipe touches the seabed, the dynamic behaviour of each free span is strictly linked to the adjacent span lengths. This means that the lowest natural frequencies of free spans in different locations but otherwise characterised by approximately equal lengths could span a considerable range, Vitali et al. (1993). In other words, the concept of maximum allowable free span length for the overall stretch conventionally applied in the offshore pipeline technology may be misleading in case of very uneven seabeds. Therefore, on the basis of the state of the equilibrium achieved by the pipeline just after installation, a full eigenvalue analysis should be carried out in order to correctly determine natural frequencies and modal shapes of the line extended to the entire stretch. On this basis:
- In compliance with a no-oscillation criterion, for the natural frequencies lower than the cut-off frequency previously defined, the corresponding free spans showing the maximum amplitude of the modal responses will be identified. In this way the location of required corrective measures will also be identified. Further eigenvalue analysis is then required to confirm that the resulting lowest natural frequency of the corrected configuration will be higher than the cut-off frequency.
- In compliance with a maximum acceptable damage, the fatigue cumulated at each joint from VIV will be calculated. One or more modes can affect each joint more or less seriously in accordance with the resulting stress range and the number of cycles estimated in the lock-in range. The location of the corrective measures will be identified. Further analysis is then required to document that the corrected configuration will comply with allowable cumulative damage criteria.
It is obvious that this is a new design format different from the traditional one and therefore necessitating sometimes to be accepted and then enter the engineering practice.

3.3 *Countermeasures*

The approach to satisfy the design criteria is based on the control of the response amplitude of the exposed free spans to the vortex shedding induced excitation. To this purpose, three types of action may be considered:

- changing the ratio of vortex shedding frequency to natural frequency so as to avoid resonance,
- increasing the stability parameter, that is increasing the system damping, so that large amplitude response is avoided,
- avoiding the formation of vortices or reducing their impact by disorganising the shedding.

In the current practice for offshore pipelines, an action of the first type is normally taken, namely controlling the free span length by both preparing the seabed to reduce free spanning and providing additional supporting after pipeline installation. This is accomplished using gravel sleepers dumped before or after laying, or mechanical trestles locked on the pipe, possibly in an active way (forcing on it by hydraulic jacks). These reduce the free span length and therefore increase the natural frequency of free oscillations.

Another way to achieve an action of this type could be increasing the pipe outer diameter with a material possibly neutral-buoyant and lasting, in order to decrease the shedding frequency and consequently the cut-off frequency, therefore allowing the acceptance of lower natural frequencies in the calculated configuration.

The second type of action, consisting in the increase of damping, seems less attractive, since damping is only effective when the structure moves, so oscillations can be reduced but cannot be fully avoided. This type of approach was proved successfully in the field of wind engineering, Sause et al. (1992) and Hart et al. (1992).

It is to be noted that the two types of action involve structural changes to the system. As the structural characteristics of a system depend on other specific design parameters they are not always likely to be modified and not to a level capable of reducing vibration amplitudes to the desired values.

In areas affected by rocky peaks and deep depressions, technical feasibility of rectification works may be very critical, and the last way seems to be the most attractive, namely to minimise the vortex shedding or disorganise the creation and shedding mechanism to a degree where the associated loads are diminished and uncorrelated. Indeed, in the current practice, in all the VIV suffering fields except offshore pipelines, e.g. risers and wind engineering technologies, this is the most widely adopted approach and the one supported by most experimental results. The next section provides a review of the most significant devices for VIV reduction and new original data from a recent experimental program.

3.4 *Hydrodynamic devices*

As already mentioned in the previous section, the way mostly adopted to control vortex shedding vibrations is to mitigate the shedding oscillating forces, both by preventing vortex formation and by reducing their correlation through the modification of the shedding mechanism. As a general rule, the different approaches can be grouped as follows:
- use of surface protrusions such as strakes, wires, fins, studs, spheres, etc., which affect the separation of the boundary layer;
- application of shrouds, usually perforated but also gauze, axial rods and axial slats which limit the growth of vortices by affecting the entrainment layer, i.e. the layer which supplies irrotational fluid necessary for the growth of vortices, in addition to the rotational fluid, in separated shear layers;
- use of near-wake stabilisers as saw-tooth plates, splitter plates, guide plates, guide vanes, base bleed and various compliant near-wake stabilisers, which affect the position of the confluence point, i.e. the point where the two entrainment layers, coming from the opposite sides of the cylinder, meet and interact;
- application of streamlined fairings which minimise shedding by providing the cylinder with an hydrodynamic shape.

Some researchers group "near wake stabilisers" and "streamlined fairings" devices in a general category called "fairings".

The large variety of helical strakes, which probably represent the most widely used vortex shedding reduction device, are mainly characterised by:
- number and pitch of helixes,
- height of strakes,
- coverage of cylinder.

While there is a general consensus on the optimal configuration which should have three helical sharpened strakes with height of 0.1 of the diameter of the cylinder (to be increased to 0.12 of a diameter for low damped structures), the optimal pitch of the helixes is a very discussed matter. Most investigations recommend a pitch of 5 diameters while recent studies show that a pitch of 15 diameters is more effective, Jones and Lamb (1992), though these tests shall be evaluated with some reservations, as they are conducted in critical flow conditions. For the cylinder with strakes configuration, the drag coefficient is not dependent on the Reynolds number. It keeps unchanged when going from sub critical flow conditions to critical and super critical conditions, and is equal to 1.35 for a strake height of 0.06 of a diameter and to 1.45 for a height of 0.12 of a diameter, Cowdrey and Lawes (1959). A 25% coverage of the cylinder is found to be inadequate whereas both a 50% and a 80% coverage gave large reductions of response amplitude, Vickery and Watkins (1962). Experimental tests have showed that the effectiveness of helical strakes for large values of the reduced velocity decreases, Every and King (1979). In addition, turbulence in the incoming flow was found to reduce the effectiveness of helical strakes : from a reduction of over 80% in amplitude for low turbulence levels only 50% reduction in amplitude was found for a turbulence level of 14%, Gartshore et al. (1978). Furthermore the test results show that these devices can reduce vibrations both underwater and in air, Vickery and Watkins (1962) and Every et al. (1982).

A device similar to the helical strake consists of circular wires helically wrapped around the cylinder, that is "helical wires". The diameter of these wires may range from 0.004, Nakagawa et al. (1963), to 0.10, Halkyard and Grote (1987), of the cylinder diameter although the value of 0.05 diameter is the most typical one. Tests have revealed that this device is extremely sensitive to the configuration as some configurations were found to be very efficient while others even increased the response amplitude, Nakagawa et al. (1963).

Shrouds have been tested with both circular and square holes. The latter appears to provide the best solution as regards reducing vortex induced oscillations. Typical shrouds diameters are 1.25 times the cylinder diameter whereas the porosity may vary from 20% to 40%. Circular holes have diameters from 0.05 to 0.1 of the cylinder diameter whereas square holes have side lengths from 0.05 to 0.07 of the cylinder diameter. Both circular and square hole shrouds proved efficient in reducing vortex induced vibrations not only at low values of reduced velocities, that is between 5 and 10, but also for reduced velocities up to 30, Zdravkovich and Volk (1972). It should be noted that these results were obtained at low (subcritical) Reynolds numbers and for very high values of the stability parameter, which is typical of aerodynamic structures. It is not fully clear whether the shrouds could produce similar effects at high Reynolds numbers and for low values of the stability parameter, which is typical of structures exposed to water flow, such as pipelines.

A relatively common method of preventing motion of drilling risers, mooring lines and similar systems consists in fitting streamlined fairings. These devices are generally the best method to prevent vortex shedding oscillations, although their installation is a nuisance during deployment (and recovery, if any). Indeed, it is essential that these fairings rotate freely to match changes in the approach flow direction and do so without inducing torque on the main structure.

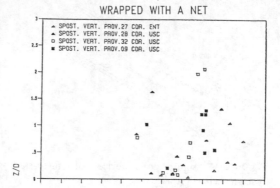

WRAPPED WITH A NET

PLUS THREE-START HELICALLY WRAPPED ROPE

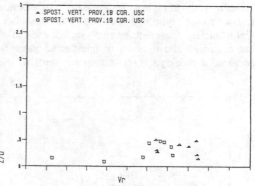

Fig.8 - Comparison of cross-flow response amplitudes of a free span (wrapped with a net): with and without reduction device of three-start helically wrapped rope type.

Furthermore, as they are the only devices capable of reducing the drag force, they can be applied to this purpose as well. Indeed the drag coefficient of a practical application of this device, assumed to be uninterrupted along the whole cylinder axis, is about 0.3, Gardner and Cole (1982). In the typical application for vortex shedding reduction, consisting of short sections spaced by uncovered sections of the cylinder, the drag coefficient can reach a value of about 1.0.

During the in-field experimental survey carried out in 1979 in the Lagoon of Venice by Snam, in collaboration with Ismes, Snamprogetti and the University of Pisa, the application of a three-start helically wrapped rope device to the pipe experienced a marked reduction of the response amplitude, as it can be seen for the cross-flow oscillations in Fig.8.

During the SVS Project, three types of devices

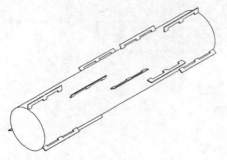

Fig.9 - VIV suppression device of three-start "pseudo-helical" strakes type.

were also in-field tested, namely a three-start helically wrapped rope, a "Japanese net" made device and a "fringe" type compliant nearwake stabiliser. Unfortunately, these tests were planned at the end of the campaign and a storm of unexpected intensity occurred damaging the test set up and most of the

stored information, whereby it was impossible for the few data available to completely assess the effectiveness of any of the three devices, Snamprogetti (1989).

As mentioned, the helical strakes result from bibliography among the most effective devices. Therefore, three devices approximating them, but not having the same construction and installation disadvantages, were designed in collaboration with the Danish Hydraulic Institute, Bruschi et al. (1994). A laboratory test programme was undertaken to determine the performance of the devices, both in terms of capacity to reduce the level of vibration, and hydrodynamic drag. During the test execution, the device efficacy was verified for different incident angles between the flow direction and the model axis and for high reduced velocities, where the second and the third modes lock-in, for which no data are available. The experimental results showed that:

- the effectiveness of the tested devices is comparable to the helical strakes, as far as helical

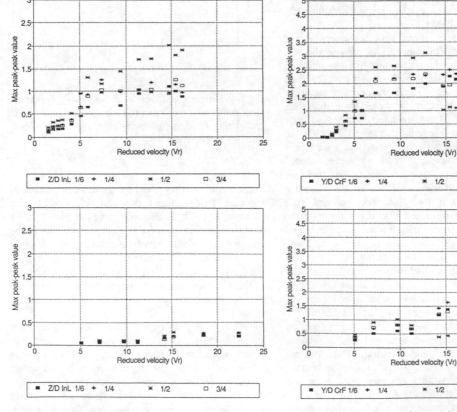

Fig.10 - Comparison of maximum in-line response amplitudes: bare pipe and pipe with the reduction device shown in Fig.9.

Fig.11 - Comparison of maximum cross-flow response amplitudes: bare pipe and pipe with the reduction device shown in Fig.9.

strakes data are available (low reduced velocities),
- all the devices are less effective at higher reduced velocities, where the second and third modes lock-in,
- the performance of the devices to reduce the level of vibration decreases as the incident angle decreases.

For example, for an incident current normal to the pipe axis, the achieved reductions for the device type of Fig.9 are shown in Figs.10 and 11, respectively for the in-line and the cross-flow motions.

The presence of the helical strakes on the outer surface of the pipe, in connection with the laying technology and consequential procedures, gives rise to complications that shall be overcome, either by installing the devices after the transit of the pipe through the tensioner or by bounding the device application on just the upper side of the pipe, to avoid any interference with the tensioner. As the latter way appears to be much simpler, an intense investigation program is necessary to value and optimise the residual mitigation efficacy of such "reduced" device.

4 CONCLUSIONS

The sophisticated approaches mentioned above, as far as VIV and their implications on the design are concerned, would lead to think that in many circumstances the submarine pipelines were damaged by this phenomenon up to failure. However, failure statistics, even the most up-dated ones, do not show any trace of this, or better, if some indications of concerns are found, it is clear that there was always time to take action and solve the critical situation. This aspect, that the presence of VIV does not cause failure, would indicate that the strength capacity of the steel pipes is far better than foreseen by the present technology. Former projects have adopted extensive and sometimes excessive preventive measures which were extremely expensive.

Should these criteria be applied to the new challenges in deep water and in uneven seabeds to be tackled in the second half of this decade, the economical feasibility on one hand and, once construction has started, the cost of contingencies associated both with installation in these seabeds and with restrictive free span correction criteria, would become quite crucial and in the worst case lead to prohibitively expensive measures. This is the reason why these topics are now being developed by the current R&D with a special care for the measures of the near seabed hydrodynamic field and interaction regime discussed in this paper. But the most important innovative aspect in connection with free

span correction technology, and which comes from the positive experiences acquired in other fields, is the application of the VIV suppression devices. Replacing the seabed preparation works with hydrodynamic devices, at least for temporary conditions, might involve considerable economical advantages in many circumstances

To continue aggressively the development and refinement of this old concept for application in the novel contexts of deepwater pipelines is therefore a mandatory step in proving the technical and economical feasibility of future projects that otherwise would be rejected by operators and certifying agencies based upon today's state of practice.

REFERENCES

Anon.: "Rules for Submarine Pipeline Systems", Det Norske Veritas, 1981.

Benedini A.: "Misure effettuate su una campata libera oscillante presente sulla linea n° 2 dell'attraversamento dello stretto di Messina il 1.3.1979", private communication.

Bernetti R., Bruschi R., Curti G., Simantiras P.: "Theoretical and Experimental Analysis of Soil-Pipe Interaction at Free-Span Shoulders for Oscillating Pipelines", European Offshore Mechanics Seminar, Trondheim, Norway, 1991.

Blaker F. and Bruschi R.: "Statoil Study Confirms Advanced Design for Condensate Pipeline", *Oil & Gas Journal*, Jan. 1990.

Bruschi R., Buresti G., Castoldi A., Migliavacca E: "Vortex Shedding Oscillations for Submarine Pipeline: Comparison between Full Scale Experiments and Analytical Models", Offshore Technology Conference, OTC paper No. 4232, Houston, Texas, 1982.

Bruschi R., Celant M., Matteeelli R., Mazzoli A.: "An Application of Structural-Hydroelastic Fracture Models to the Safe Life Design of a Submarine Pipeline", Proc. Int. Conf. on OMAE, New Orleans, 1984.

Bruschi R., Cimbali W., Ragaglia R., Vincenzi M.: "Scour Induced Free Span Analysis", the 6th Int. Conf. on OMAE, Tokio, Japan, 1986.

Bruschi R., Montesi M., Tura F., Vitali L., Accerboni E.: "Field Tests With Pipeline Free Spans Exposed to Wave Flow and Steady Current", Proc. 21st Offshore Technology Conference, Paper OTC 6152, pp. 301-316, Houston, Texas, May 1989.

Bruschi R., Curti G., Tura F.: "Free Spanning Pipelines - A Review", the 1st Int. ISOPE Conf., Edinburgh, 1991 (-a).

Bruschi R., Curti G., Marchesani F., Torselletti E.: "Limit States Approach to Crossing of Uneven Seabed Areas", the 10th Int. Conf. on OMAE, Stavanger, Norway, 1991 (-b).

Bruschi R. and Vitali L.: "Large-Amplitude Oscillations of Geometrically Nonlinear Elastic Beams Subjected to Hydrodynamic Excitation", *Journal of Offshore Mechanics and Arctic Engineering*, Vol. 113, pp. 92-104, May 1991.

Bruschi R., Curti G., Marchesani F., Torselletti E.: "The Maspus Project - Methodologies for Adapting Submarine Pipelines to Very Uneven Seabeds", the 12th Int. Conf. on OMAE, Glasgow, 1993 (-a).

Bruschi R., Curti G., Marchesani F., Torselletti E.: "The Maspus Project: Upgrade for Deep Water Pipelines on Very Uneven Seabeds", DOT Conference, Montecarlo, 1993 (-b).

Bruschi R. and Vitali L.: "Recent Advances in the Offshore Pipeline Technology", the 4th Int. Offshore and Polar Engineering Conf., Osaka, Japan, 1994.

Bruschi R., Simantiras P., Vitali L., Jacobsen V.: "Suppression devices of vortex-induced oscillations for very long free span", 10th Offshore South East Asia Conference, Singapore, December 1994.

Bryndum M. B. et al.: "Improved Design Methods for Spanning of Pipelines", Offshore Pipeline Technology, European Seminar, Amsterdam, 1989 (-a).

Bryndum M. B., Bonde C., Smitt L. W., Tura F., Montesi M.: "Long Free Spans Exposed to Current and Waves: Model Tests", Proc. 21st Offshore Technology Conference, Paper OTC 6153, pp. 317-336, Houston, Texas, May 1989 (-b).

Cowdrey C. F. and Lawes J. A.: "Drag measurements at high Reynolds numbers of a circular cylinder fitted with three helical strakes", Natl. Phys. Lab. (UK), Aero Rep. 384, 1959.

Eide L. O., Bruschi R., Leopardi G.: "The Experience from the Statpipe System on Free Span Development and Analysis", OPT Seminar, Stavanger, Norway, 1988.

Every M. J. and King R.: "Suppressing flow induced vibrations - An experimental comparison of clamp-on devices", BHRA report RR 1576, November 1979.

Every M. J., King R., Weaver D. S.: "Vortex-excited vibrations of cylinders and cables and their suppression", *Ocean Engineering*, 9, pp. 135-157, 1982.

Gardner T. N. and Cole N. W. J.: "Deepwater drilling in high current environment", Proc. 14th Offshore Technology Conference, Paper OTC 4316, pp. 177-201, Houston, Texas, May 1982.

Gartshore I. S., Khanna J., Laccinole S.: "The effectiveness of vortex spoilers on a circular cylinder in smooth and turbulent flow", Proc. 5th Int. Conf. on Wind Engineering, Fort Collins, CO, 1978.

Griffin O. M., Skop R. A. and Ramberg S. E.: "The Vortex-Excited Resonant Vibrations of Structures and Cable Systems", Offshore Technology Conference, OTC Paper NO. 2319, Houston, Texas, 1975.

Halkyard J. E. and Grote P. B.: "Vortex-induced response of a pipe at supercritical Reynolds numbers", 19th Offshore Technology Conference, Houston, Texas, April 1987.

Hallam M. G., Heaf N. J., Wootton L. R.: "Dynamics of Marine Structures - Methods of calculating the dynamic response of fixed structures subjected to wave and current action", CIRIA Report UR 8, 1978.

Hart J. D., Sause R., Wyche Ford G., Row D. G.: "Mitigation of wind-induced vibration of Arctic pipelines Systems", the 11th Int. Conf. on OMAE, 1992.

Jacobsen V., Bryndum M. B., Nielsen R. N., Fines S.: "Vibrations of Offshore Pipelines Exposed to Wave and Current Motions", Proc. of the 3rd Int. Symp. on OMAE, Vol. 1, pp. 291-299, New Orleans,

Jones G. S. and Lamb W. S.: "The use of helical strakes to suppress vortex induced vibration", Proc. 6th Int. Conf. on Behaviour of Offshore Structures, London, 1992.

King R.: "A Review of Vortex Shedding Research and its Applications", *Ocean Engineering*, Vol. 4, pp. 141-172, 1977.

Lawless J. F.: "Confidence Intervals Estimation for the Weibull and Extreme Value Distributions", *Technometrics*, Vol. 20 No. 4, 1978.

Mes M. J.: "Vortex Shedding Can Cause Pipelines to Break", *Pipeline and Gas Journal*, August 1976.

Nakagawa K., Fujino T., Arita Y., Shima T.: "An experimental study of aerodynamic devices for reducing wind-induced oscillatory tendencies of stacks", Symp. Natl. Phys. Lab. (UK), Teddington, 1963, H.M.S.O., pp. 774-795, 1965.

Nielsen N. J. R.: "Update of DnV 1981 Submarine Pipeline Rules", oral presentation at the 13th Int. Conf. on OMAE, ASME, Glasgow, 1993.

Pantazopoulos S.: "Vortex-Induced Vibration Parameters: Critical Review", Offshore Technology Conference, Vol. 1, pp. 199-255, Houston, Texas, March 1994.

Raven P. W. J.: "The Development of Guidelines for the Assessment of Submarine Pipeline Spans - Overall Summary Report", Dept. of Energy, OTH 86 231, HMSO, 1986.

Sarpkaya T.: "Vortex-Induced Oscillations A Selective Review", *Journal of Applied Mechanics*, Vol. 46, June 1979.

Sause R., Hart J. D., Wyche Ford G.: "Evaluation of wind-induced vibration of Arctic pipelines", Proc. 2nd Int. Offshore and Polar Engineering Conf., Paper No. ISPE-92-B2-4, San Francisco, June 1992.

Snamprogetti: "Critical Review of the Results of the Experimental Campaign in the Lagoon of Venice - October 1978, April 1979", SVS Project, Internal Report, June 1988.

Snamprogetti: "Risultati della campagna acquisizione dati in sito Laguna di Venezia", Internal Report, Rev. 0, March 1989.

Sumer B. M. and Fredsøe J.: "A Review on Vibrations of Marine Pipelines", the 4th Int. Offshore and Polar Engineering Conf., Osaka, Japan, April 1994.

Tassini P. et al.: "The Submarine Vortex Shedding Project: Background, Overview and Future Fall-out on Pipeline Design, Offshore Technology Conference, OTC Paper NO. 6157, Houston, Texas, 1989.

Tsahalis D. T.: "Vortex-Induced Vibrations of a Flexible Cylinder near a Plane Boundary Exposed to Steady and Wave-Induced Currents", *ASME Journal of Energy Resources Technology*, Vol. 106, June 1984.

Tura F., Dumitrescu A., Bryndum M. B., Smeed P. F.: "Guidelines for Free Spanning Pipelines: the Gudesp Project", the 13th Int. Conf. on OMAE, Houston, Texas, 1994.

Vitali L., Marchesani F., Curti G., Bruschi R.; "Dynamic excitation of offshore pipelines resting on very uneven seabeds", Proc. of the 2nd European Conf. on Structural Dynamics, Vol. 2, pp. 1181-1190, Trondheim, Norway, June 1993.

Vickery B. J. and Watkins R. D.: "Flow-induced vibration of cylindrical structures", Proc. 1st Australian Conf. on Hydraulics and Fluid Mechanics, pp. 213-241, Pergamon Press, Oxford, 1962.

Wootton L. R., Warner M. H., Sainsbury R. N., Cooper D. H.: "Oscillation of Piles in Marine Structures - A description of the full-scale experiments at Immingham", CIRIA Technical Note 40, 1974.

Zdravkovich M. M. and Volk J. R.: "Effect of shroud geometry on the pressure distribution around a circular cylinder", *J. Sound Vib.*, 20, pp. 451-455, 1972.

Hydroelasticity in Marine Technology, Faltinsen et al. (eds) © 1994 Balkema, Rotterdam, ISBN 90 5410 387 6

Vortex-induced vibrations of structural members in unsteady winds

Chen-Yang Fei & J. Kim Vandiver
Department of Ocean Engineering, Massachusetts Institute of Technology, Cambridge, Mass., USA

ABSTRACT: Natural wind-induced vibrations of structural members have been the source of fatigue damage to offshore platforms during fabrication and transportation and to flarebooms during in-service conditions. It is shown that natural unsteady fluctuations in the mean wind speed typically prevent vortex-excited vibrations from reaching steady state amplitudes. A probabilistic model is proposed in which the duration of visit by the mean wind speed is expressed in terms of the probability density function and spectrum of the wind speed. The vibration response amplitude and rate of fatigue damage are predicted in terms of the ratio of the duration of visit of the mean wind speed in the critical velocity interval to the rise time of the vibration response. A worked example is presented to illustrate the proposed probabilistic model.

1 Introduction

Natural wind-induced vibrations of structural members have been the source of fatigue damage to offshore platforms during fabrication and transportation and to flarebooms during in-service conditions. To avoid failures, it is important for designers to be able to predict such vibrations as well as the resulting fatigue damage.

Current response prediction methods generally assume that when the mean wind speed is within the critical wind speed range for a given structural member, then it is adequate to compute the steady state response of the member. However, practical experience has revealed [1] that these methods *over-predict* the response, and, predict structural failures *too* frequently.

One explanation of the over-conservatism was suggested by Rudge et. al [2]. They found that the frequency of occurrence of the critical wind velocity is greatly *over-estimated*, since it is assumed that ideal conditions, which require low turbulence and increasing wind speeds, allow lock-in to develop to its fullest extent at which peak amplitudes are seen. This explanation was later supported by the results of wind tunnel experiments.

In the wind tunnel experiments described here, the wind speed was varied continuously with time, as shown in Figure 1. In this figure, the wind speed is expressed in terms of reduced velocity; $f_n = 32.375$ Hz, $D = 0.0483$ m and $V_r = \frac{V}{f_n D}$. Cross-flow oscillations of a carbon-fiber tube were measured. The tube, which spanned the test section of the tunnel, was pinned at both ends and allowed to vibrate freely in response to vortex shedding. Throughout the testing at various wind speeds, the tube responded primarily in its first mode. The envelope of the measured transient motion of the tube at its mid-span is shown in Figure 2. This response time history is simultaneous with the wind speed shown in Figure 1. The main parameters of the tube were documented in Rudge et. al [3] and are summarized briefly in Table 1.

It is clear from Figure 2 that, although the mean wind speed was ideal for lock-in conditions, unsteady fluctuations in the wind speed typically prevented vortex-excited vibrations from reaching steady state amplitudes. The Brown & Root formula [4] predicted that the maximum steady state vibration amplitude would be 0.193 diameters.

0.191 diameters was the measured steady state vibration amplitude. In other words, the duration of time that the wind speed stayed within the critical velocity range for the member was less than the transient buildup time for this lightly damped vibration mode. The damping was 0.35% for this member and the reduced damping, K_s, was 8.54. Under ideal steady state lock-in conditions, one would expect mid-span vibration amplitudes of 0.19 diameters.

In this paper, a probabilistic model is proposed in which the duration of a visit by the mean wind speed is expressed in terms of the probability density function and spectrum of the wind speed. The vibration response amplitude is predicted in terms of the ratio of the duration of a visit of the mean wind speed in the critical velocity interval to the rise time of the vibration response. The proposed model is illustrated by a worked example.

2 Time Scales

There are two time scales that determine the vibration amplitude and the fatigue damage rate. One is the duration of visit by the wind speed to the critical velocity interval. The other is the rise time of the structural vibration response.

2.1 Duration of a Visit by the Wind Speed to an Interval

The duration of a visit by the wind speed to an interval $[a, b]$ is defined as the undisrupted length of time that the wind speed spends between levels a and b. The definition of the duration of a visit by the wind speed to an interval can be illustrated in Figure 3. The duration of a visit to an interval starts with either an upcrossing of the windspeed at level a or a downcrossing of windspeed at level b, and ends with either an upcrossing at b or a downcrossing at a.

In the case of random wind, the duration of a visit is a random variable that depends on the mean rates of crossings by the wind speed at levels a and b. The exact distribution of the duration of a visit by the wind speed to an interval is not known except for very few random processes [5]. However, it can be shown that in general the exact

mean can be calculated as follows, provided that the wind speed is a stationary random process:

$$E[T_{[a,b]}] = \frac{F_V(b) - F_V(a)}{\nu_a^+ + \nu_b^+} \qquad (1)$$

Where $F_V(c)$ is the cumulative probability distribution function, which specifies the probability that the wind speed is less than or equal to c; ν_c^+ is the mean rate of crossing the level $V(t) = c$ with positive slopes, and:

$$F_V(c) = \int_0^c p_V(v)dv \qquad (2)$$

$$\nu_c^+ = \frac{1}{2}\int_{-\infty}^{\infty} |\dot{v}| \, p_{V\dot{V}}(c, \dot{v})d\dot{v} \qquad (3)$$

Where $p_V(v)$ is the PDF of the wind speed and $p_{V\dot{V}}(v, \dot{v})$ is the joint PDF of the wind speed and its time derivative. Usually lower case symbols are used as arguments of PDF's. Upper case symbols are used as real time dependent variables.

It is clear from equation 1 that the mean duration of an undisrupted visit to a critical velocity interval depends not only on the probability distribution of the wind speed $V(t)$, but also on the properties of its time derivative process, $\dot{V}(t)$, due to the dependence of equation 1 on the mean rates of crossings. Mean upcrossing rates of non-Gaussian random processes can be determined from related Gaussian processes, through a univariate, nonlinear transformation [6] as follows.

Suppose $F_V(v)$ and $p_V(v)$ are the CDF and the PDF of the wind speed respectively, and σ_V and $\sigma_{\dot{V}}$ are the standard deviations of the wind speed and the time derivative of the wind speed respectively. Our objective is to derive the mean upcrossing rate of the process $V(t)$ in terms of the above quantities.

First, a Gaussian random process $\tilde{Y}(t)$ is derived from the wind speed process $V(t)$ through a non-linear transformation. The mean upcrossing rate of the process $V(t)$ at level $V(t) = c$ can be obtained from that of the derived Gaussian process $\tilde{Y}(t)$ at a corresponding level. Let $\tilde{Y}$ be a Gaussian random process of the same sampling rate and total length as $V(t)$, which consists of random variables $\tilde{Y}_i$ of zero mean and unit variance. Then there exists a real function $h(\tilde{Y}_i)$ such that

$$V_i = h(\tilde{Y}_i) = F_V^{-1}(\Phi(\tilde{Y}_i)) \qquad (4)$$

where Φ is the CDF of $\tilde{Y}_i$ and $\Phi(\tilde{Y}_i) = \frac{1}{\sqrt{2\pi}} \int_{-\infty}^{\tilde{Y}_i} \exp(-0.5\tilde{y}^2) d\tilde{y}$. Since both F_V and Φ are monotones, h is guaranteed to possess one to one mapping. The mean rate of crossing the level c by the process $V(t)$ can be obtained from the mean rate of crossing the level $\tilde{y} = h^{-1}(c)$ by the Gaussian process $\tilde{Y}(t)$, since $V(t)$ and $\tilde{Y}(t)$ upcross the level c and $\tilde{y}$ respectively at the same instant. Thus the mean rate of crossing the level c by $V(t)$ can be expressed as the mean rate of crossing the level $\tilde{y} = h^{-1}(c)$ by the Gaussian process $\tilde{Y}(t)$

$$\nu_c^+ = \frac{\sigma_{\dot{\tilde{Y}}}}{\sqrt{2\pi}} \phi(h^{-1}(c)) \qquad (5)$$

where ϕ is the PDF of $\tilde{Y}_i$ and $\phi(\tilde{Y}_i) = \frac{1}{\sqrt{2\pi}} \exp(-0.5\tilde{Y}_i^2)$.

To calculate ν_c^+ requires finding the value of $\sigma_{\dot{\tilde{Y}}}$. Next $\sigma_{\dot{\tilde{Y}}}$ is expressed in terms of the statistics of the original processes $V(t)$ and $\dot{V}(t)$. First a new random process $\tilde{V}(t)$ is defined that can be derived from the process $V(t)$ as follows

$$\tilde{V}(t) = \frac{V(t) - E[V]}{\sigma_V} \qquad (6)$$

where $E[V]$ and σ_V are the mean and the standard deviation of the random process $V(t)$. The derived process $\tilde{V}(t)$ has a zero mean and a unit variance. Furthermore, the following equations hold for $\tilde{V}(t)$:

$$F_{\tilde{V}}(\tilde{v}) = F_V(v) \qquad (7)$$
$$p_{\tilde{V}}(\tilde{v}) = \sigma_V p_V(v) \qquad (8)$$
$$\sigma_{\dot{\tilde{V}}} = \frac{\sigma_{\dot{V}}}{\sigma_V} \qquad (9)$$

where $F_{\tilde{V}}(\tilde{v})$ and $p_{\tilde{V}}(\tilde{v})$ are the CDF and the PDF of the random process $\tilde{V}(t)$, and $\sigma_{\dot{\tilde{V}}}$ is the standard deviation of the process $\dot{\tilde{V}}(t)$.

Combining the equations 4 and 7, the process $\tilde{V}(t)$ can be related to the process $\tilde{Y}(t)$:

$$\tilde{V}(t) = h^*(\tilde{Y}(t)) = F_{\tilde{V}}^{-1}(\Phi(\tilde{Y})) \qquad (10)$$

Taking the time derivative of the above equation, the process $\dot{\tilde{V}}(t)$ can be related to the process $\dot{\tilde{Y}}(t)$:

$$\dot{\tilde{V}}(t) = \frac{dh^*(\tilde{Y})}{d\tilde{Y}} \dot{\tilde{Y}}(t) \qquad (11)$$

Since $\tilde{Y}(t)$ and $\dot{\tilde{Y}}(t)$ are independent, as stationary Gaussian processes, the process $\dot{\tilde{V}}(t)$ has zero mean and variance given by the equation

$$\sigma_{\dot{\tilde{V}}}^2 = E\left\{ \left[\frac{h^*(\tilde{Y})}{d\tilde{Y}} \right]^2 \right\} E\left\{ \left[\dot{\tilde{Y}}(t) \right]^2 \right\} \qquad (12)$$

That yields $\sigma_{\dot{\tilde{V}}}^2 = \sigma_{\dot{\tilde{Y}}}^2 \eta^2$, where η is given by:

$$\eta = \sqrt{\int_{-\infty}^{\infty} \frac{\phi^3(\xi) d\xi}{\left\{ p_{\tilde{V}}[F_{\tilde{V}}^{-1}(\Phi(\xi))] \right\}^2}} \qquad (13)$$

Finally, the mean upcrossing rate by the wind speed process $V(t)$ at level c can be expressed more explicitly as

$$\nu_c^+ = \frac{\eta}{2\pi} \frac{\sigma_{\dot{V}}}{\sigma_V} e^{-0.5\tilde{y}^2} \qquad (14)$$

where $\tilde{y}$ is given by:

$$\tilde{y} = \Phi^{-1}\left(F_{\tilde{V}}\left(\frac{c - E[V]}{\sigma_V} \right) \right) \qquad (15)$$

σ_V and $\sigma_{\dot{V}}$ can be obtained directly from the power spectral density function of the wind speed $S_V(\omega)$:

$$\sigma_V = \sqrt{\int_0^{\infty} S_V(\omega) d\omega} \qquad (16)$$

$$\sigma_{\dot{V}} = \sqrt{\int_0^{\infty} \omega^2 S_V(\omega) d\omega} \qquad (17)$$

In reality, the upper limit of the integral should be replaced by a constant beyond which the wind speed spectrum can be regarded as zero.

$$\sigma_V = \sqrt{\int_0^{f_\infty} S_V(\omega)d\omega} \qquad (18)$$

$$\sigma_{\dot{V}} = \sqrt{\int_0^{f_\infty} \omega^2 S_V(\omega)d\omega} \qquad (19)$$

where f_∞ is the cut-off frequency beyond which the wind speed spectrum can be regarded as zero.

In summary, for a given PDF and power spectral density function of the wind speed, the calculation of the mean upcrossing rate by the non-Gaussian process $V(t)$ involves the following steps:

- The first step is to derive wind statistics σ_V and $\sigma_{\dot{V}}$ from the power spectral density function of the wind speed, using equations 16 and 17.

- The second step is to define a Gaussian process $\tilde{Y}(t)$ from the wind speed process $V(t)$, through a non-linear transformation. Since the CDFs of both processes are monotone, the transformation function h^* is guaranteed to possess one to one mapping. In general, h^* can only be derived numerically.

- The third step is to calculate η and $\tilde{y}$ numerically by equations 13 and 15. Then the mean upcrossing rate can be calculated by equation 14.

The calculated mean upcrossing rates by the wind speed can be used directly to calculate the mean duration of a visit by the wind speed to the critical velocity interval, by equation 1. The mean duration of a visit by the wind speed to the critical velocity interval will be shown later to determine the fatigue damage of a structure in random winds.

2.2 Rise Time of Structural Response

As the excitation force is switched on, the structure needs a finite amount of time to build up its vibration amplitude towards a steady state value. The transient response envelope of a single degree of freedom oscillator with a constant sinusoidal excitation at the natural frequency f_n is approximated by the following equation

$$A(t) = A_\infty(1 - e^{-\zeta\omega_n t}) \qquad (20)$$

where $A(t)$ is the instantaneous vibration amplitude envelope at time t after the excitation has begun and A_∞ is the steady state vibration amplitude. ζ is the structural damping ratio measured in still air. ω_n is the natural frequency of the nth mode in radians per second.

The rise time of structural response is defined as the duration required to build up to 63.2% of the steady state value, from an initial stationary state. From this definition:

$$t_r = -\frac{1}{\zeta\omega_n}\ln(1 - 0.632) \qquad (21)$$

$$= \frac{1}{\zeta\omega_n} = \frac{T}{2\pi\zeta} \qquad (22)$$

where T is the undamped vibration period. Clearly, the rise time of structural response depends on the structural damping ratio and the natural frequency of the structure. For lightly damped structures, such as members on oil production platforms, the rise time can exceed 100 or more periods of vibration.

3 Predictions of Response and Fatigue Damage Rate

3.1 Instant Rise Time Model

The instant rise time model is the model for predicting the response and fatigue damage rate assuming the rise time of structural response is instantaneous, i.e., that the structure instantly reaches steady state response to excitation. Such a model provides a simple but overly conservative closed-form solution for predicting the response and fatigue damage for a given probabilistic description of wind speed. However, it provides a useful starting point for a more realistic prediction model.

3.1.1 Prediction of Response

Assuming an instant rise time model, it is possible to establish mappings between the instantaneous wind speed and the corresponding steady state VIV response as expressed below, where we define A^* as the ratio of A, the steady state response amplitude at a particular reduced velocity, to A_{max}, the maximum response amplitude at the critical wind speed, V_{crit}.

$$A^* = \frac{A}{A_{max}} = f(V_r(t_i)) \qquad (23)$$

$$= f(\frac{V(t_i)}{f_n D}) \qquad (24)$$

A_{max} is the maximum vibration amplitude, occurring at the critical wind speed V_{crit}. A_{max} can be predicted using various VIV design methodologies, such as **DnV**, **BS 8100**, **ESDU 85038** and **Brown & Root**, etc. A complete review of these methodologies was documentated in [3]. $V(t_i)$ and $V_r(t_i)$ are the instantaneous values of wind speed and reduced velocity respectively. D is the diameter of the structural member and f_n is the natural frequency of the cylinder in hertz. $f(V_r)$ is a real function that relates the reduced velocity to the steady state vibration amplitude of the structure that is excited by the flow at that speed. It can be determined based on experimental evidence [2] [7]. Different expressions of this function resulting from the models proposed by various authors are given below.

• **DnV** [8] [9]

$$f(V_r) = \begin{cases} 1 & 4.7 \leq V_r \leq 8.0 \\ 0 & \text{otherwise} \end{cases} \qquad (25)$$

• **BS8100** [10]

$$f(V_r) = \begin{cases} \frac{(3.6 - 0.52V_r)V_r^2}{25} & 3.85 \leq V_r \leq 6.90 \\ 0 & \text{otherwise} \end{cases} \qquad (26)$$

• **ESDU 85038** [7] **and Brown & Root** [4]

$$f(V_r) = \begin{cases} e^{-a(1 - \frac{V_r}{(V_r)_{crit}})^2} & 4.25 \leq V_r \leq 5.25 \\ 0 & \text{otherwise} \end{cases} \qquad (27)$$

where $a = 104.5(\frac{m\zeta}{\rho D^2})^{1.8}$.

• **Proposed Model**

$$f(V_r) = \begin{cases} V_r - 5.0 & 5.0 \leq V_r \leq 6.0 \\ 2(6.5 - V_r) & 6.0 \leq V_r \leq 6.5 \\ 0 & \text{otherwise} \end{cases} \qquad (28)$$

If f possesses one-to-one mapping, then,

$$V_r = f^{-1}(A^*) = g(A^*) \qquad (29)$$

where g is the inverse function of f and $A^* = \frac{A}{A_{max}}$. Based on the theory of random variables, the probabilistic structure of the response can be derived from that of the wind speed as follows.

$$p_A(a) = p_{V_r}(g(A^*)) \mid \frac{dg(A^*)}{dA^*} \mid \frac{1}{A_{max}} \qquad (30)$$

where $p_A(a)$ is the PDF of vibration amplitudes and $p_{V_r}(v_r)$ is the PDF of the wind speed expressed in terms of reduced velocity.

If $f(V_r)$ is not a monotone, the range of variations of V_r can be partitioned into segments within each of which the function is a monotone and

$$p_A(a) = \sum_l p_{V_r}(g_l(A^*)) \mid \frac{dg_l(A^*)}{dA^*} \mid \frac{1}{A_{max}} \qquad (31)$$

where $g_l(A^*)$ is the inverse mapping in the lth segment. For the proposed model of $f(V_r)$ there are two such non-zero monotonic segments.

135

3.1.2 Prediction of fatigue damage rate

If the response is deterministic and cyclic, the response and the number of cycles to fatigue failure can be defined by an S-N curve

$$NS^m = c \qquad (32)$$

where S is the cyclic stress amplitude. N is the number of cycles to fatigue failure at the stress amplitude S. m and c are positive constants that relate to material properties.

Let $\mathcal{D}(t)$ denote the fraction of damage accumulated per unit time due to a random stress $S(t)$. According to Lin [11], the expectation of $\mathcal{D}(t)$ can be expressed as

$$E[\mathcal{D}(t)] = c^{-1} E[M_T(t)] \int_0^\infty \sigma^m p_\Sigma(\sigma,t) d\sigma \qquad (33)$$

where $\sigma(t)$ is the peak of the random stress $S(t)$, $p_\Sigma(\sigma,t)$ is the probability density function of stress peaks at time t. $E[M_T(t)]$ is the expected number of peaks of the stress per unit time.

If the random stress $S(t)$ is a stationary, narrow-band process, then the number of positive stress peaks per unit time is the center frequency of the stress spectrum, f_c.

$$E[M_T(t)] = f_c \qquad (34)$$

And the expected fatigue damage rate is

$$E[\mathcal{D}] = f_c c^{-1} \int_0^\infty \sigma^m p_\Sigma(\sigma) d\sigma \qquad (35)$$

For a flexible cylinder with pinned ends vibrating at its first mode, the response amplitude spectrum will be narrow band with a peak at the natural frequency. Therefore, $f_c = f_n$. Furthermore, the maximum bending stress can be related to the maximum displacement response A as

$$\Sigma = Ek^2 \frac{D}{2} A \qquad (36)$$

where k is the wave number and $k = \frac{\pi}{L}$.

Because they are linearly related, the PDF of the peaks of random stresses, $p_\Sigma(\sigma)$, can be calculated from the PDF of the peaks of the displacement response, $p_A(a)$, through equation 36

$$p_\Sigma(\sigma) = \frac{1}{Ek^2 \frac{D}{2}} p_A\left(\frac{\sigma}{Ek^2 \frac{D}{2}}\right) \qquad (37)$$

Combining equations 35 and 37, the fatigue damage rate can be expressed in terms of the distribution of the peaks of the displacement response:

$$
\begin{aligned}
E[\mathcal{D}] &= f_n c^{-1} \int_0^\infty \sigma^m p_\Sigma(\sigma) d\sigma \\
&= f_n c^{-1} \left(E \frac{\pi^2}{L^2} \frac{D}{2}\right)^m \int_0^\infty a^m p_A(a) da \quad (38)
\end{aligned}
$$

The expected fatigue damage rate assuming instant rise time, $E[\mathcal{D}_0]$, can be further expressed in terms of the PDF of the wind speed expressed in terms of reduced velocity, after using equation 31 in evaluating equation 38:

$$
\begin{aligned}
E[\mathcal{D}_0] &= \int_0^\infty A^m \sum_l p_{V_r}(g_l(A^*)) \left| \frac{dg_l(A^*)}{dA^*} \right| dA \\
&\times \frac{1}{A_{max}} f_n c^{-1} \left(E \frac{\pi^2}{L^2} \frac{D}{2}\right)^m \quad (39)
\end{aligned}
$$

Equation 39 is used to calculate the expected fatigue damage rate of a structural member assuming instant rise to steady state amplitudes, for a given PDF of the instantaneous wind speed. The calculation of the expected fatigue damage rate of a structural member assuming instant rise is a required step in the estimation of the actual fatigue damage rate in random winds.

3.2 Finite Rise Time Model

The natural fluctuation of the wind speed often does not allow fully developed vibration. The actual response amplitude and fatigue damage rate in natural winds, therefore, depend on the relative length of the duration of visit to the rise time of a structural response mode. It is postulated that the fatigue damage rate due to VIV in random winds is a function of the ratio of the duration of visit to the rise time

$$\frac{E[\mathcal{D}]}{E[\mathcal{D}_0]} = q(m, \zeta\omega_n E[\mathcal{T}_{[a,b]}]) \qquad (40)$$

where $\zeta\omega_n E[\mathcal{T}_{[a,b]}]$ is the ratio of $E[\mathcal{T}_{[a,b]}]$, the mean duration of a visit by the wind speed in the interval $[a,b]$, to $\frac{1}{\zeta\omega_n}$, the rise time. This ratio is defined as r. $\mathcal{D}$ is the resulting fatigue damage rate of the structural member excited by random wind after the effect of finite rise time has been included and $\mathcal{D}_0$ is the fatigue damage rate assuming instant rise time. $q(m,r)$ is a real function that relates the ratio of fatigue damage rates to r, the ratio of the duration of a visit by the wind speed to rise time. m is the exponent from the S-N curve expression in equation 32.

The function q can be estimated by Monte-Carlo simulations using the following steps:

- **Generating Wind Speed.** A time history of a Gaussianly distributed wind speed with a specified spectrum [1] can be generated by providing Gaussianly distributed white noise, as input to an optimum AutoRegressive MovingAverage filter, which best simulates the target spectrum [13]. The generated wind speed is denoted by $V(t)$. This simulation may of course be replaced by using real wind data.

- **Calculating the Excitation Force and Response.** Due to single mode dominance of VIV response, the vibration of the cylinder is regarded as the same as that of an equivalent single degree of freedom oscillator which has been obtained using the techniques of modal analysis.

$$\ddot{x} + 2\zeta\omega_n\dot{x} + \omega_n^2 x = f_a \cos\omega_n t \qquad (42)$$

where $x(t)$ is the mid-span vibration response amplitude of the cylinder to the

given excitation. ζ is the measured structural damping ratio of the cylinder. ω_n is the natural frequency of the nth mode in radians per second and f_a is the amplitude of the modal excitation force per unit modal mass of the oscillator and is to be determined from the generated wind speed. Assuming that the wind speed is held at a constant value of $V(t_i)$, the vibration of the cylinder would eventually reach a steady state given by

$$x(t_i) = A(t_i)\sin\omega_n t = \frac{1}{2\zeta\omega_n^2}f_a(t_i)\sin\omega_n t \qquad (43)$$

where $A(t_i)$ is the steady state vibration amplitude, and is given by

$$A(t_i) = \frac{f_a(t_i)}{2\zeta\omega_n^2} \qquad (44)$$

The $\sin\omega_n t$ term accounts for the periodicity of the lift force and is assumed to be independent and uncoupled from the amplitude modulation caused by variations in $V(t_i)$. At any given time t_i, the amplitude of the excitation force, $f_a(t_i)$, can be derived from the value of the instantaneous wind speed $V(t_i)$ in a way such that the steady state vibration amplitude resulting from the derived excitation could be predicted by equation 28.

$$A(t_i) = \frac{f_a(t_i)}{2\zeta\omega_n^2} = A_{max}f(V_r(t_i)) \qquad (45)$$

or, equivalently

$$f_a(t_i) = 2\zeta\omega_n^2 A_{max}f(V_r(t_i)) \qquad (46)$$

---

[1]The windspeed spectrum was defined as [12]:

$$S_V(f) = \frac{320 \times (0.1\overline{V}_{10})^2 \times (0.1Z)^{0.45}}{(1+\tilde{f}^n)^{\frac{5}{3n}}} \qquad (41)$$

where $\tilde{f} = 172 \times f \times (0.1Z)^{\frac{2}{3}} \times (0.1\overline{V}_{10})^{-0.75}$; $n = 0.468$; $S_V(f)$ is the power spectral density at frequency f in Hertz; Z is the height above sea level in meters; $\overline{V}_{10}$ is the 1 hour mean wind speed at 10 meters above sea level. This formula results from extensive windspeed measurements at Sletringen, Norway, and has been proposed as a model spectrum for design of North Sea structures.

At this point an important approximation is necessary which has proven to be quite accurate. The excitation $f_a(t_i)$ evaluated above is derived from the the steady state magnitude of the periodic lift force which would be required to drive the cylinder to the steady state response amplitude corresponding to the reduced velocity. Since the lift coefficient changes with response amplitudes, the approximation is made here that as the cylinder vibration rises toward the steady state value, the periodic excitation force magnitude stays constant. In other words, during finite rise time, the lift coefficient is assumed to be constant at the value which would correspond to the final steady state response amplitude. With this approximation we may estimate the excitation which corresponds to any wind speed $V(t)$.

Due to changes in wind speed $V(t_i)$ and therefore changes in reduced velocity $V_r(t_i)$, the vibration response will be modulated in amplitude. These modulations in response may be estimated by a standard convolution integral of the time varying excitation force, $f_a(t_i)$, and the impulse response function for the oscillator, as follows.

$$x(t) = \frac{1}{\omega_d} \int_0^t f_a(\tau) \cos \omega_n \tau e^{-\zeta \omega_n(t-\tau)} \times \sin \omega_d(t-\tau) d\tau \qquad (47)$$

where ω_d is the damped natural frequency and $\omega_d = \omega_n \sqrt{1-\zeta^2}$.

When this was done to estimate the response at mid-span of the carbon fiber tube in the wind tunnel, excited by the wind shown in Figure 1, the resulting response amplitude envelope is as shown in Figure 2. It is a very close approximation. The convolution accounts for the finite rise time.

- **Calculating the fatigue damage rates**. Vibrations of an elastic cylinder cause cyclic bending stresses, which result in fatigue damage. The fatigue damage, Δ_i, resulting from the ith cycle of stresses, can be

expressed as below.

$$\Delta_i = \frac{1}{N_i} \qquad (48)$$

where N_i is the number of cycles to fatigue failure at the stress peak σ_i. The total fatigue damage accumulated over the duration of stresses, Δ, could be expressed as the sum of Δ_i over the total number of the applied cyclic stresses, n, as follows.

$$\Delta = \sum_{i=1}^{n} \Delta_i \qquad (49)$$

$$= \sum_{i=1}^{n} \frac{1}{N_i} \qquad (50)$$

$$= c^{-1} \sum_{i=1}^{n} \sigma_i{}^m \qquad (51)$$

where c and m are the constants of the S-N curve that were shown in equation 32. By virtue of the relationship between the peaks of vibrations and the peaks of stresses shown in equation 36, Δ could be further expressed in terms of the peaks of vibrations as follows

$$\Delta = c^{-1} (Ek^2 \frac{D}{2})^m \sum_{i=1}^{n} X_i^m \qquad (52)$$

where $X(t)$ is the envelope of the transient vibration $x(t)$ that was derived from equation 47. X_i is the discrete sequence of $X(t)$ sampled at f_n, the natural frequency of the member.

The fatigue damage rate, $\mathcal{D}$, can be calculated as below.

$$\mathcal{D} = \frac{\Delta}{n} f_n = \frac{f_n}{nc} (Ek^2 \frac{D}{2})^m \sum_{i=1}^{n} X_i^m \qquad (53)$$

The fatigue damage rate assuming instant rise time, $\mathcal{D}_0$, can also be calculated as below.

$$\mathcal{D}_0 = \frac{f_n}{nc}(Ek^2\frac{D}{2})^m \sum_{i=1}^{n} A_i^m \qquad (54)$$

where A_i is the steady state vibration amplitude envelope at the wind speed V_i.

The ratio of the fatigue damage rates between finite rise time and instant rise time can be expressed as follows.

$$\frac{\mathcal{D}}{\mathcal{D}_0} = \frac{\sum_{i=1}^{n} X_i^m}{\sum_{i=1}^{n} A_i^m} \qquad (55)$$

- **Calculating the Time Scales.**

A time history of a Gaussianly distributed wind speed with a target spectrum was created as the output of an ARMA filter, as previously described. The mean duration of a visit by the wind speed to the critical velocity interval and the rise time of the first mode were also calculated from equations 1 and 22 respectively, based on the wind statistics derived from the simulated wind speed record. The ratio between the two time scales as defined previously as $r = \zeta\omega_n E[\mathcal{T}_{[a,b]}]$ was computed.

For each simulated time history of Gaussianly distributed wind speed, a single pair of values $(\frac{\mathcal{D}}{\mathcal{D}_0}, r)$ was formed. This pair relates the ratio of actual fatigue damage rate to that assuming instant rise and the ratio of the mean duration of a visit by the wind speed to the critical velocity interval to rise time.

The above steps were repeated to generate wind speed records with different hourly means, and to calculate the fatigue damage rates of a single structural member and the durations of a visit by the wind speed to the critical velocity interval, until a considerable number of such pairs $(\frac{\mathcal{D}}{\mathcal{D}_0}, r)$ were formed. An empirical relationship between $\frac{\mathcal{D}}{\mathcal{D}_0}$ and r was identified as the curve which minimizes the least square error to the results of the Monte-Carlo simulations.

The following empirical expressions were identified as the best fit to the results of the Monte-Carlo simulations based on the least square error technique.

$$\frac{E[\mathcal{D}]}{E[\mathcal{D}_0]} = \begin{cases} 1 - \exp\left(-0.7093r^{0.2859}\right) & m = 3.74 \\ 1 - \exp\left(-0.5718r^{0.3085}\right) & m = 4.38 \end{cases}$$
$$(56)$$

where $r = \zeta\omega_n E[\mathcal{T}_{[a,b]}]$.

Equation 56 was derived for a single structural member. Strictly, it is only valid for that member at a single critical velocity. Research is going on to investigate if such formulae can be generalized to different members with different critical velocities. Equation 56 is plotted in Figure 4 with the results of the numerical simulations.

4   A Worked Example

An example is given to demonstrate the implementation of the proposed finite rise time model. The fatigue damage rate of a flexible cylinder is predicted based on the finite rise time model. The cylinder, a tube made of carbon-fiber, is assumed to have pinned-pinned ends. Structural parameters of the cylinder, such as the total length, diameter and the natural frequency, are described in Table 1.

For the purpose of illustrating the use of the proposed finite rise time model, we assume that both the PDF and the power spectral density function of the wind speed are given. The PDF of the wind speed is defined to be Gaussian with a PDF $p_V(v)$, a mean $\overline{V}$ and a variance σ_V^2:

$$p_V(v) = \frac{1}{\sqrt{2\pi}\sigma_V} \exp\left(-\frac{(v-\overline{V})^2}{2\sigma_V^2}\right) \qquad (57)$$

where $\overline{V}$ is the mean wind speed, and σ_V is the standard deviation of the wind speed. Values of both quantities will be assigned below in **Step 1**.

The wind speed spectrum is defined by the equation 41. Reiterating

$$S_V(f) = \frac{320 \times (0.1\overline{V}_{10})^2 \times (0.1Z)^{0.45}}{(1+\tilde{f}^n)^{\frac{5}{3n}}}$$

where $\tilde{f} = 172 \times f \times (0.1Z)^{\frac{2}{3}} \times (0.1\overline{V}_{10})^{-0.75}$; $n = 0.468$; $S_V(f)$ is the power spectral density

Table 1: Summary of a worked example of the proposed finite rise time model

section	variable name	variable symbol	variable value	SI units
input	L	total length	2.0955	meter
	D	outside diameter	0.0483	meter
	t	wall thickness	0.0023	meter
	f_1	natural frequency	32.375	second^{-1}
	ρ_m	mass density	1597.674	kg/meter^3
	ζ	structural damping ratio	0.35%	
	E	Young's modulus	5.3×10^{10}	Pascal
	K_s	reduced damping	8.54	
	A_{max}/D	Brown & Root prediction	0.1928	
	Z	height above sea level	10	meter
step 1	$\overline{V}$	mean windspeed	9.38	meter/sec.
	σ_V	standard deviation of windspeed	0.9232	meter/sec.
	$\sigma_{\dot{V}}$	standard deviation of acceleration	0.9727	meter/sec.^{2}
	a	lower bound of the critical reduced velocity interval	5.0	
	b	upper bound of the critical reduced velocity interval	6.5	
	$F_V(a)$	CDF of windspeed at a	0.0554	
	$F_V(b)$	CDF of windspeed at b	0.7874	
step 2	η	non-normality factor	1.00	
	ν_a^+	mean upcrossing rate at a	0.0467	second^{-1}
	ν_b^+	mean upcrossing rate at b	0.1219	second^{-1}
	$E[\mathcal{T}_{[a,b]}]$	mean duration of visit to interval $[a, b]$	4.3427	second
	$\zeta \omega_1 E[\mathcal{T}_{[a,b]}]$	ratio of mean duration of visit to rise time	3.0918	
step 3	m	exponent of S-N curve	3.74	
	$\frac{E[\mathcal{D}]}{E[\mathcal{D}_0]}$	ratio of fatigue damage rate	0.6245	
	$E[\mathcal{D}_0]$	fatigue damage rate (instant rise)	0.0725	day^{-1}
	$E[\mathcal{D}]$	fatigue damage rate (finite rise)	0.0453	day^{-1}
	$E[\mathcal{D}_{s.s.}]$	fatigue damage rate with constant mean windspeed	0.3295	day^{-1}

at frequency f in Hertz; Z is the height above sea level in meters; $\overline{V}_{10}$ is the 1 hour mean wind speed at 10 meters above sea level. The member is assumed to be at 10 meters above sea level, thus $Z=10$ meters and $\overline{V}_{10} = \overline{V}$.

It is important to recognize that wind speeds may behave quite differently in different geographic areas. The use of Gaussian PDF's and the wind speed spectrum defined by equation 41 may not generalize. To develop an appropriate PDF of the wind speed which adequately characterize the effect of the local wind environment on transient VIV is a current research topic at MIT. Comments on this subject will be made briefly in the next section.

Step 1: To Derive σ_V and $\sigma_{\dot{V}}$

To derive the wind statistics requires calculating the mean duration of a visit by the wind speed to the critical velocity interval. The required wind statistics are the standard deviation of the wind speed and the standard deviation of the acceleration. Both quantities are calculated from the power spectral density function given in equation 41, after a value of $\overline{V}_{10}$ is assigned.

The mean wind speed is assumed to be the value which corresponds to a reduced velocity value of 6, at which the maximum steady state vibration of the tube would be achieved based on the observation of the results from the wind tunnel experiments.

$$\overline{V} = \overline{V}_{10} = 6.0 f_n D = 9.38 \text{ meters/second} \quad (58)$$

The standard deviation of the wind speed, σ_V, and the standard deviation of the acceleration, $\sigma_{\dot{V}}$, are calculated using equations 18 and 19. The cut-off frequency is chosen to be 1 Hertz. The power spectral density function at 1 Hertz is roughly 0.013% of the value at 0 Hertz. (where the peak of the spectrum is located). *i.e.*,

$$\frac{S_V(f=1)}{S_V(f=0)} = 1.3 \times 10^{-4} \quad (59)$$

Such a choice of the cut-off frequency is arbitrary in the example, only for the purpose of illustration. A more rational choice should depend on the frequency resolution of the wind speed records on which the wind speed spectrum is based, the rise time of the structural response, and the natural frequency of the structure.

After the cut-off frequency is defined, both σ_V and $\sigma_{\dot{V}}$ can be calculated as follows.

$$\sigma_V = \sqrt{\int_0^1 S_V(f)df} = 0.9232 \text{ meters/sec.}$$

$$\sigma_{\dot{V}} = 2\pi\sqrt{\int_0^1 f^2 S_V(f)df} = 0.9727 \text{ meters/sec.}^2$$

The above integrations were implemented numerically using the *trapezoidal* rule. The frequency resolution was chosen to be 1/2000 Hertz.

The critical reduced velocity interval was defined as [5, 6.5]. The corresponding lower and

upper bounds of the critical velocity interval were 7.82 and 10.16 meters per second respectively. The cumulative density function at the lower and upper bounds were calculated to be 0.0554 and 0.7874 respectively.

Step 2: To Calculate $E[\mathcal{T}_{[a,b]}]$ and $\frac{1}{\zeta \omega_n}$

To calculate $E[\mathcal{T}_{[a,b]}]$, the mean duration of a visit by the wind speed to the critical velocity interval, requires calculation of the mean upcrossing rates by the wind speed at levels a and b respectively. Since the wind speed is assumed to be Gaussian, the mean upcrossing rate at a level c is known as follows:

$$\nu_c^+ = \frac{1}{2\pi}\frac{\sigma_{\dot{V}}}{\sigma_V}\exp\left(-0.5c^2\right) \quad (60)$$

Using the above equation, the mean upcrossing rates at levels a and b are 0.0467 and 0.1219 per second respectively.

In general, the PDF of the wind speed may not be Gaussian. The calculation of the mean upcrossing rates by a non-Gaussian process is more complicated, and can only be done numerically, as indicated earlier by equations 13, 14 and 15.

Having obtained all of its determinants, the mean duration of a visit by the wind speed to the critical velocity interval can be calculated as follows.

$$
\begin{aligned}
E[\mathcal{T}_{[a,b]}] &= \frac{F_V(b) - F_V(a)}{\nu_a^+ + \nu_b^+} \\
&= \frac{0.7874 - 0.0554}{0.0467 + 0.1219} \\
&= 4.3427 \text{ seconds}
\end{aligned}
$$

Another key time scale is the rise time of the structural response, and it can be calculated directly from the data as follows.

$$
\begin{aligned}
\frac{1}{\zeta \omega_n} &= \frac{1}{0.35\% \times 2 \times 3.1416 \times 32.375} \\
&= 1.4045 \text{ seconds}
\end{aligned}
$$

141

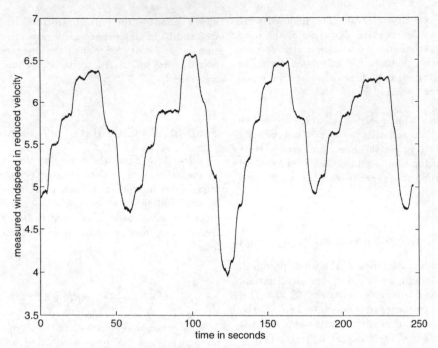

Figure 1: Time history of the measured windspeed expressed in terms of reduced velocity (first mode)

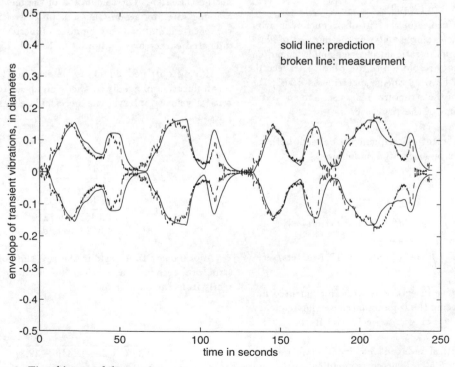

Figure 2: Time history of the envelope of transient vibrations at the mid-span of the cylinder, in diameters

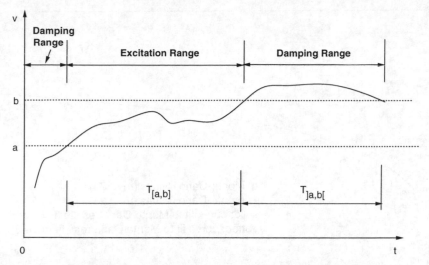

Figure 3: Duration of visit by wind speed to an interval $[a, b]$

The ratio of the two time scales is $r = \zeta \omega_n E[\mathcal{T}_{[a,b]}] = 4.3427/1.4045 = 3.0918$.

Step 3: To Predict $E[\mathcal{D}]$

To predict $E[\mathcal{D}]$, the actual fatigue damage rate in random winds, we need to calculate the following two quantities. The first quantity is the ratio of the fatigue damage rates between finite rise time model and instant rise time model, which takes into account of the finite rise time. This ratio was assumed to depend only on r, the ratio of the mean duration to rise time, and can be calculated by equation 56. Assuming the exponent of the $S - N$ curve to be 3.74 (This is a typical value for steel, not carbon fiber, but is used here for example purposes only), the ratio of fatigue damage rates may be calculated as follows.

$$
\begin{aligned}
\frac{E[\mathcal{D}]}{E[\mathcal{D}_0]} &= 1 - \exp\left(-0.7093 r^{0.2859}\right) \\
&= 0.6245
\end{aligned}
$$

Readers should be aware that the above formula was derived for a different critical velocity. Research is going on to verify if such a formula is also valid for any member.

The second quantity that needs to be calculated is $E[\mathcal{D}_0]$, the expected fatigue damage rate assuming instant rise time. This quantity accounts for the unsteadiness of the instantaneous winds. It can be calculated from the PDF of the wind speed by equation 39. Based on the assumed PDF of the wind speed and $m = 3.74$, $E[\mathcal{D}_0]$ is calculated as follows.

$$
\begin{aligned}
E[\mathcal{D}_0] &= \int_0^\infty A^m \sum_l p_{V_r}(g_l(A^*)) \mid \frac{dg_l(A^*)}{dA^*} \mid dA \\
&\quad \times \frac{1}{A_{max}} f_n c^{-1} (E \frac{\pi^2}{L^2} \frac{D}{2})^m \\
&= \int_0^{A_{max}} A^m (p_{V_r}(A^* + 5.0) + \\
&\quad 0.5 p_{V_r}(6.5 - 0.5A^*)) dA \times \\
&\quad \frac{1}{A_{max}} f_n c^{-1} (E \frac{\pi^2}{L^2} \frac{D}{2})^m \\
&= 0.0725 \text{ per day}
\end{aligned}
$$

where $A^* = \frac{A}{A_{max}}$ is the dimensionless vibration amplitude.

The above integration is implemented numerically using the *rectangular* rule over 30 stations evenly dividing response amplitude from 0 to A_{max}.

143

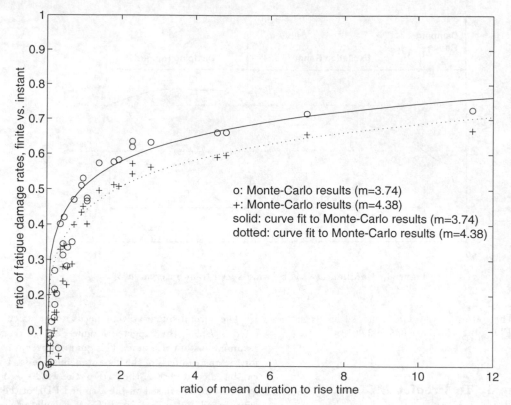

Figure 4: Variation of the ratio of the fatigue damage rates with the ratio of the mean duration of visits to structural response rise time

The actual fatigue damage rate in random winds is the product of the above two quantities and can be predicted as follows.

$$
\begin{aligned}
E[\mathcal{D}] &= \frac{E[\mathcal{D}]}{E[\mathcal{D}_0]} \times E[\mathcal{D}_0] \\
&= 0.6245 \times 0.0725 \\
&= 0.0453 \text{ per day}
\end{aligned}
$$

The result shows that the predicted fatigue damage rate based on the finite rise time model is 0.0453 per day, *i.e.*, the model would fail after a 22 day exposure to such a wind environment. It is worth mentioning that the conventional fatigue damage prediction based on mean wind speed and steady state response yields a fatigue damage rate of 0.3295 per day, 7 times the predicted value.

Table 1 summaries the use of the three step implementation of the finite rise time model.

5 Commentary on the PDF of the Wind Speed

To date, most available PDFs of the wind speeds are from hourly averaged wind speed measurements. PDFs based on 1 hour mean wind speed characterize the long-term behavior of mean wind speeds, but exclude the short term transient winds which prevent vortex-excited vibrations from reaching steady state amplitudes. It is also inappropriate to use the PDFs based on the 1 hour mean wind speed to calculate the mean upcrossing rates and the duration of visits, since the short term transient wind speed may behave quite differently than the mean wind speed.

The desirable PDF of the wind speed that characterizes transient vibrations must be derived from the instantaneous wind speed records with high sampling rates. Such fine quality wind speed records rarely exist, and when available, may often have too few samples to derive statistically reliable PDFs. Appropriate averaging techniques need to be employed to process the measured wind speed data. The objective of an on-going research project on wind-induced VIV at MIT is to study short-term high resolution wind speed records and to develop PDFs of the wind speed which can be used by design engineers as an input to the proposed probabilistic model.

6 Acknowledgements

This work was sponsored by the American Petroleum Institute and by an industry consortium research project . Sponsoring companies were: Amoco, British Petroleum, Chevron, Conoco, Exxon Production Research, Mobil, Petrobras, and Shell Development Company.

References

[1] B. L. Grundmeier, R. B. Campbell, and B. D. Wesselink. OTC 6174: A Solution for Wind-Induced Vortex-Shedding Vibration of the Heritage and Harmony Platforms During Transpacific Tow. In *Offshore Technology Conference*, 1989.

[2] D. Rudge, C. Fei, S. Nicholls, and J. K. Vandiver. OTC 6902: The Design of Fatigue-Resistant Structural Members. In *Offshore Technology Conference*, May 1992.

[3] D. Rudge and C. Fei. Response of structural members to wind-induced vortex shedding. Master's thesis, Massachusetts Institute of Technology, September 1991.

[4] R. W. Robinson and J. Hamilton. *A Revised Criterion for Assessing Wind Induced Vortex Vibrations in Wind Sensitive Structures*. Brown and Root Limited, August 1991. Revision of Report Number: 0T0 88021.

[5] Ove Ditlevsen. Duration of visit to critical set by gaussian process. *Probabilistic Engineering Mechanics*, 1(2), 1986.

[6] M. Grigoriu. Crossings of non-gaussian translation processes. *Journal of Engineering Mechanics*, 110(4), April 1982.

[7] ESDU International Plc. *ESDU: Item numbers 85038/39: Circular-cylindrical Structures: Dynamic Response to Vortex Shedding*, December 1985. Amendments A to C, March 1990.

[8] Det Norske Veritas. *Rules for Submarine Pipeline Systems (1981)*.

[9] Det Norske Veritas. *Classification Note No. 30.5, Environmental Conditions and Environmental Loads (March,1991)*.

[10] British Standards Institution. *BS8100:Parts 1 and 2:1986*.

[11] Y. K. Lin. *Probabilistic Theory of Structural Dynamics*. McGraw-Hill, 1967.

[12] Norwegian Potroleum Directorate. *Acts, Regulations and Provisions for the Petroleum Activity. Volumn 2. Guidelines concerning loads and load effects*, 1994.

[13] M. G. Rivero. Random wave simulation using a.r.m.a. models. Master's thesis, Massachusetts Institute of Technology, February 1987.

Ships

Hydroelasticity in Marine Technology, Faltinsen et al. (eds) © 1994 Balkema, Rotterdam, ISBN 90 5410 387 6

Bending moment of ship hull girder caused by pulsating bubble of underwater explosion

B. Smiljanić & N. Bobanac
Brodarski Institute, Zagreb, Croatia

I. Senjanović
University of Zagreb, Croatia

ABSTRACT: The theory of interaction of an underwater explosion gas bubble with a nearby ship hull structure is outlined. The excitation due to bubble pulsation is determined for two types of explosive on the basis of the bubble pulsation theory, using the pulsation period measurement results. The numerically obtained dependence of the maximum bending moment on the mass and depth of explosive for two different ship hulls is discussed.

1. INTRODUCTION

This paper considers a method for the estimation of hull girder bending moment in the case of hull girder whipping due to the pulsation of the bubble caused by a nearby underwater explosion.

Recently the ship hull girder response calculation has been widely used in stress analysis during the ship hull design stage. This calculation is based on the theory of bubble pulsation described by Cole (1965) and the theory of transient response of ship hull girder caused by nearby underwater explosion (Chertock 1953, 1970). A.N. Hicks (1986) adapted these theories for computer implementation.

The first effect of an underwater explosion is the shock wave which transfers about 50% the explosive energy to the surrounding water. The gaseous products of the explosion form a bubble which induces an unsteady field of pressure in the surrounding water. A structure immersed in the vicinity of the underwater explosion is exposed thus to the effects of two types of time-dependent load:

a) The shock wave effect, which lasts for a short time, and can be in the beginning approximated by the interaction between the elastic structure and the acoustic wave in water. This effect later on gets continuously transformed into the reaction due to the acceleration of the added mass of water according to E.H.Kennard (Cole 1965), J.H.Haywood (1953) and T.L.Geers (1971).

b) The pulsating effect of the gaseous bubble dominated by the reaction of the added mass acceleration, which is manifested as a periodic change of the buoyancy of the structure caused by the surrounding water acceleration (G.Chertock 1953, 1970), (Hicks 1986).

The shock wave progression can be described as an outward radial propagation of a relatively thin sphere of compression in the water. The particles involved, close to the immersed structure, are exposed to a short-time acceleration in different moments, which causes local pressure impulses on the structure.

Unlike the shock wave, the gas bubble pulsation instantaneously accelerates the whole water volume surrounding the structure, and this effect can be described as the global oscillation of buoyancy.

The pressure-time dependence at a distance of 30m from the centre of explosion is shown in Fig.1.1. The load spectrum acting on the immersed structure is dominated by low-frequency components caused by the pulsating bubble and high-frequency components due to the shock wave effect. If a ship hull of a usual construction exposed to the effects of an underwater explosion is described as a linear elastic system, then in its low-frequency response, the first natural form of the ship hull girder vibration is dominant. Due to differences in impedance, the low-frequency response of the hull to the shock wave is negligible, and therefore the influence of the shock wave on the bending moment of the ship hull girder is small. Thus, the effects of shock wave and pulsating bubble on the ship hull can be observed separately.

This paper considers the main features of global loading of the ship hull structure exposed to a nearby explosion. The basic assumptions and conclusions important for the description of pulsation and interaction are presented.

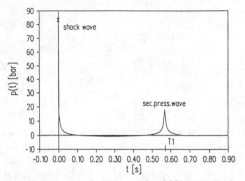

Fig 1.1 Pressure-time curve for 200kg TNT charge fired 30m below water surface

The calculated natural frequencies and modes are compared with the experimentally obtained results for two ship hulls. The hull responses due to pulsation are calculated for different TNT charges and depths. The calculation procedure for hull response of other different explosives is shown. The differences in the effect of pulsating gas bubble and shock wave on the ship hull due to underwater explosion are discussed.

2. GAS BUBBLE PULSATION

After the shock wave emission, approximately 40% of explosion energy remains in the gas bubble in the form of internal gas energy. The bubble expands and the energy transforms into kinetic and potential energy of the surrounding water. In the moment when the bubble reaches the maximum radius a_m, the gas pressure falls below the surrounding hydrostatic pressure p_o, causing thus the bubble contraction. During the contraction, a large amount of kinetic and potential energy is again transformed into the bubble internal gas energy. In the moment when the minimum gas volume is reached, the surrounding water deceleration generates the secondary pressure wave, and the next pulsation is started.

The gas bubble pulsation theory is based on the following assumptions:

a) The explosion occurs far enough from the water surface and bottom, thus the sphere bubble geometry of radius $a=a(t)$ can be assumed.

b) The pressures induced by bubble pulsation are much lower than the shock wave pressures, therefore the water compression is assumed to be negligible.

c) The gravitation effect is neglected and the absolute pressure in the surrounding water p_o is assumed to be constant.

d) Pulsating frequency is high enough so that the heat exchange between the gas in the bubble and the surrounding water can be neglected.

f) The influence of viscous forces on the bubble pulsation is negligible too.

Under these assumptions, the gas bubble pulsation in water can be modelled as a non-stationary potential flow around source having a strength $4\pi a^2 \dot{a}$, described by the energy conservation equation:

$$E_k(a) + E_p(a) + E_i(a) = E_b \qquad (2.1)$$

where E_b is explosion energy which remains after the shock wave emission, $E_k(a)$ is kinetic energy of the surrounding water, $E_p(a)$ is potential energy of the surrounding water, and $E_i(a)$ is gas bubble internal energy.

The kinetic energy is described as:

$$E_k(a) = -\frac{1}{2} \rho_o \oint_s \varphi_i \frac{\partial \varphi_i}{\partial n} \, dS \qquad (2.2)$$

and the velocity-potential flow around the pulsating bubble φ_i being:

$$\varphi_i = \frac{a^2}{r} \frac{da}{dt} \qquad (2.3)$$

where ρ_o is water density, S is bubble surface area, n is bubble surface normal, t is time, and r is point radius.

The flow potential and gas bubble internal energy are:

$$E_p(a) = \int_0^{V(a)} p_o dV \qquad (2.4)$$

$$E_i(a) = \int_{V(a)}^{\infty} p dV = \frac{k_e m_e}{(\kappa-1)} \left(\frac{m_e}{V(a)} \right)^{\kappa-1} \qquad (2.5)$$

where $V(a)$ is bubble volume, m_e is explosive mass, κ is adiabatic gas constant, and k_e is explosive dependent constant.

The first approximation of the pulsation period (Willis formula) and the maximum radius of the gas bubble are obtained by neglecting the internal energy $E_i(a)$ because it is very small during the predominant pulsation period. Neglecting the internal energy, the first pulsation period T_o and the bubble gas maximum radius a_{mo} are obtained as:

$$T_o = K \frac{E_b^{1/3}}{p_o^{5/6}} \qquad (2.6)$$

$$a_{mo} = \left(\frac{3 E_b}{4 \pi p_o} \right)^{1/3} \qquad (2.7)$$

where K is fluid density dependent constant, $p_o = \rho_o g Z_o$ is surrounding fluid pressure at the explosion depth.

Inserting (2.2),(2.3),(2.4) and (2.5) into (2.1), introducing non-dimensional variables $\alpha = a/a_{mo}$ and $\tau = t/[a_{mo}(3\rho_o/2p_o)^{1/2}]$ according to Friedman and Shiffman (Cole 1965), and normalizing the equation by E_b from (2.1), the following differential equation, dependent on $\alpha(\tau)$, is obtained:

$$\alpha^3 \dot{\alpha}^2 + \alpha^3 = 1 - k_{be}\alpha^{-3(\kappa-1)} \qquad (2.8)$$

In the developed equation, the internal energy member is dependent on the type of explosive through constants k_{be} and κ, and does not satisfy similarity laws for variable explosion depths Z_o.

The initial condition, $\alpha(0)$, necessary for solving the equation is obtained from the condition of minimum $\alpha(\tau)$, i.e. $d\alpha/d\tau=0$, under the assumption of $\alpha(\tau)$ curve symmetry, i.e. $\alpha(0)=\alpha(T)$, where T is pulsation period.

The numerical solution of the equation (2.8) is obtained by routines from the IMSL Differential Equation Solver for CDC computer. The gas bubble radius a(t) and the bubble pulsation pressure p(t), for the explosion of 10kg of TNT at 10m depth and 10m distance from the explosion centre are presented in Fig 2.1.

In reality the gas bubble in the gravitation field is affected by buoyancy that accelerates the bubble towards the free surface, by surface and bottom vicinity and by viscous forces of the surrounding water. These effects become important at the end of the first pulsation period. The vertical migration velocity influences the pulsation pressure at the minimum bubble radius. The field measurement have shown that the secondary pressure wave impulse remains approximately the same (Cole 1965). According to the numerical calculation results, the vicinity of boundary surfaces tends to reduce the migration velocity. The field measure-

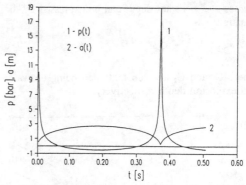

Fig.2.1 Time dependent gas bubble radius a(t) and bubble pulsation pressure p(t)

ments have shown that the numerical results are somewhat overestimated.

The viscous forces effect is due to irregular and unstable bubble geometry during the contraction, manifested as a reduction of bubble migration velocity and as an increase of gas heat dissipation. In the case of an explosion below the ship hull, the bubble buoyancy at the end of the first pulsation period significantly moves the bubble towards the hull and at the same time the bubble energy dissipation gets strongly increased. None of these effects is taken into consideration in equation (2.8), but fortunately these effects are mostly compensated among themselves.

The numerical constants used in (2.8), given by Hicks (1986), are valid for TNT explosive only. The Willis formula, however, shows that the pulsation period is mainly dependent on explosive energy E_b. Under the assumption that the thermodynamic properties of explosion products of different explosives are similar to those of TNT, the equation (2.8) can be also applied for other explosives if the corresponding E_b is used. The explosive gas bubble energy can be obtained by the measurement of bubble pulsation period (Bjarnholt 1980). The pulsating period for an unknown explosive X is obtained from the time dependent pressure record. The results of the measurements performed for a number of X and TNT explosions in the mass span from 1 to 20kg at a sea depth of 20m are presented in Fig.2.2 (Smiljanić 1990).

The obtained numerical results, according to Friedman, Shiffman and Hering (Cole 1965), show that the influence of the boundaries on the pulsation period is small for the distances between the bubble centre and the boundary greater than $3a_m$. In this case the first pulsation period for different explosive masses at a constant depth can be expressed by the equation:

$$T_1 = C_1 E_b^{1/3} + C_2 E_b^{2/3} \qquad (2.9)$$

where C_1 and C_2 are constants depending on water and explosion depth respectively.

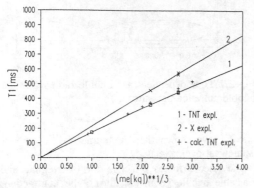

Fig. 2.2 Calculated and measured first pulsation periods for TNT and unknown explosive X

Since the explosive energy is mass proportional, the equation (2.9) could be expressed as:

$$T_1 = a m_e^{1/3} + b m_e^{2/3} \qquad (2.10)$$

The constant C_1 from the Willis formula (2.6) is $1.135 \rho_o^{1/2}/p_o^{5/6}$. By fitting the equation (2.10) to the measurement results, the coefficients a and b are obtained. Eliminating C_2, the equation for the gas bubble energy E_b of an unknown explosive reads:

$$E_b = \frac{a^2}{8\,C_1^3 b}\left[\sqrt{1+\frac{4b}{a^2}T_1} - 1 \right]^3 \qquad (2.11)$$

3. BUBBLE INDUCED EFFECTS

For the estimation of the effects induced by the gas bubble on the ship hull, the following assumptions are taken:

a) The ship hull is considered as an elastic girder of variable cross-section floating on the calm free surface.

b) The explosion takes place in the vertical plane of symmetry at a depth $h > 2.5 B_m$, where B_m is maximum waterline breadth.

c) The ship hull is represented by a number of strips with constant geometrical properties.

d) hull does not affect velocity-potential around the gas bubble.

The surface boundary conditions, ($\varphi = 0$ and p = const.), are approximately satisfied by introducing sink O', which is with respect to free surface symmetrical with the source representing the gas bubble O, Fig.3.1. In this way the ship hull floating on the free surface is represented by a fully immersed body symmetrical to the waterline. The velocity-potentials at each particular point of the surrounding water is equal to the sum of the potentials of the source O and sink O'.

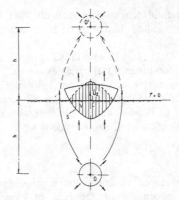

Fig.3.1 Ship hull and gas bubble pulsation model in the vicinity of free surface

The bubble pulsation induced force on the hull strip is:

$$\oint_S p\,dS = \oint_V \nabla p\,dV \qquad (3.1)$$

where S is strip wetted surface and V is immersed strip volume.

The bubble pulsation induced pressure on the strip is:

$$p = \rho_0 \frac{\partial \varphi}{\partial t} - \frac{1}{2}\rho_0 (\nabla\varphi)^2 \qquad (3.2)$$

where φ is velocity-potential in the vicinity of the ship hull.

The second right-hand member of (3.2) is steady stagnation pressure which is proportional to $(a/r)^4$. Thus, by increasing the distance r, its value is rapidly diminished, and therefore it can be neglected in the vicinity of the ship hull. The (∇p) components are cancelled in the horizontal plane and are summed up in the vertical direction. Considering thus that the mean value of the vertical component

152

(∇p) is approximately equal to the vertical component in point C, Fig.3.1, the equation (3.1) can be written:

$$\oint_s p \, dS \approx \rho_o V \left[\nabla \left(\frac{\partial \varphi}{\partial t} \right) \right]_C \cdot \vec{k} \qquad (3.3)$$

where ∇ is Hamilton operator and $\vec{k}$ is vertical unit vector.

Inserting (2.3) into (3.3), the segment force due to bubble pulsation is obtained:

$$f_a = \rho_o V \frac{2h}{r^3} (2a\dot{a}^2 + a^2 \ddot{a}) = m_d \dot{u}_z \qquad (3.4)$$

where $a = a(t)$ is gas bubble radius obtained by numerical integration of (2.8), m_d is displacement mass, $\dot{u}_z$ is water acceleration vertical component due to gas bubble pulsation, r is strip point C distance from explosion centre, and h is explosion depth.

Each strip is also affected by the reaction inertial force of the strip and added mass of the surrounding water. Since $\ddot{y}$ and $\dot{u}_z$ are the vertical acceleration of the strip and the surrounding water respectively, the inertial reaction strip force can be written as:

$$f_r = m_b \ddot{y} + m_a (\ddot{y} - \dot{u}_z) \qquad (3.5)$$

here m_b is hull strip mass, $\ddot{y}$ is strip vertical acceleration, and m_a is added mass.

The hull strip is simultaneously affected by the restoring buoyancy force due to the strip draught changes:

$$f_u = \rho_o g B y \, l \qquad (3.6)$$

where B is segment waterline breadth, y is strip vertical translation and l strip length.

Finally, the strip ends are also affected by internal elastic forces f_{el}, and therefore the strip equilibrium equation is:

$$f_{el} = f_a - f_r - f_u \qquad (3.7)$$

Considering each hull strip as a beam finite element, the hull equilibrium equation in the vertical plane can be expressed as:

$$([K] + \lceil D \rfloor) \{\delta\} + ([M_b] + [M_a]) \{\ddot{\delta}\} = \\ = ([M_d] + [M_a]) \{\dot{u}_z\} \qquad (3.8)$$

where [K] is hull stiffness matrix, $\lceil D \rfloor$ is matrix of restoring buoyancy forces, $[M_b]$ hull mass matrix, $[M_a]$ added mass matrix, $[M_d]$ is buoyancy mass matrix and $\{\delta\}$ is node displacement vector.

The node displacement vector has two degrees of freedom per node - vertical translation y (including bending and shear deflection) and rotation ϑ in vertical plane.

The equation (3.8) represents the equation of hull motion due to gas bubble pulsation. The both sides of the equation depend on displacement changes, however predominant is short duration excitation (during the secondary pressure wave), and thus, the displacement mass can be considered as constant during that time. The diagonal matrix of restoring buoyancy forces is hull form dependent and can be considered as constant (vertical hull side approximation) for small displacement changes.

In spite of the involved assumptions, the presented approach offers a useful engineering tool for estimation of global bending moments caused by a nearby underwater explosion (Chertock 1970, Hicks 1986).

For solution of the equation of motion (3.8), the adapted DIANA 88 computer program (Senjanović et al. 1989) was employed. This computer program is intended for solving the equation of ship hull forced vibration by the mode superposition method in space domain, and harmonic acceleration method (Senjanović 1984) in time domain.

In this paper the response is obtained by the summation of two forms of rigid body displacement in vertical plane and four lowest vibration modes.

4. ILLUSTRATIVE EXAMPLE

On the basis of the analyses of the shock wave and gas bubble pulsation influence on the global bending moment of the hull girder, the excitation diagram presented in Fig.4.1 is obtained. The critical bending moment usually occurs shortly after the secondary shock wave (Hicks 1986). Due to the fact that the gas bubble energy content after the first pulsation is less than 15% of the total explosive energy, the excitation after the second shock wave could be neglected.

Some response results for two steel ship hulls of the following features are presented in the paper:

Hull A: $L_{OA} = 43$m, displacement $D = 215$t, transversely framed,

Hull B: $L_{OA} = 45$m, displacement $D = 195$t, longitudinally framed.

In all calculations of response the ship hull at rest was considered in the moment $t = 0$. The explosive

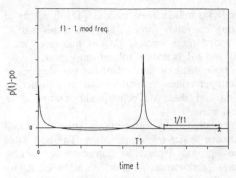

Fig.4.1 Time dependent pressure excitation

mass and explosion depths were chosen to satisfy the assumptions of the theory of pulsation and the theory of interaction.

Both hulls were subjected to deep water vibration excitation tests by means of electromechanical vibration exciter to obtain natural frequencies and forms of vibration. The measurement results are presented in Table 4.1.

Table 4.1 Natural frequencies of vibration, Hz

mode	A hull		B hull	
	calc.	meas.	calc.	meas.
1.	4.48	4.38	5.26	5.2
2.	9.61	8.72	10.8	9.0
3.	15.23	13.0	16.9	14.0
4.	20.74	-:-	22.8	20.5

The position of explosion with respect to the ship hull is defined by the depth h and the distance of the centre of explosion from the aft perpendicular x_e, Fig.4.2. The obtained time dependence of response is presented for the hull cross-section at a distance x from the aft perpendicular, Fig.4.2.

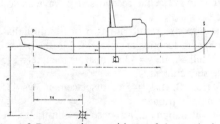

Fig. 4.2 Presentation positions of the explosion and response

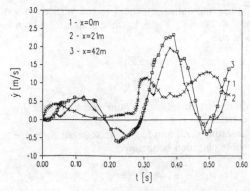

Fig 4.3 Velocity at stern, midship and bow hull cross-sections during the explosion of 12 kg of TNT at 30m depth and at distance $x_e=21$m

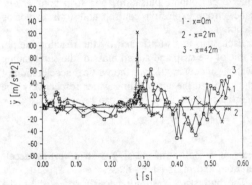

Fig 4.4 Acceleration at stern, midship and bow hull cross-sections during the explosion of 12 kg of TNT at 30m depth and at distance $x_e=21$m

The integral presentation of the calculated hull response includes displacements, velocities, accelerations, bending moments and shear forces. The time dependent velocities and accelerations of the hull A are given for illustration in Figs. 4.3 and 4.4.

The maximum bending moment is obtained during the midship explosion. The results of a number of bending moment calculations for midship explosions are presented in Fig 4.5. For each particular explosion the maximum sagging and hogging moments were obtained. The sagging moments divided by the Rule bending moment, for each particular hull are given in Fig.4.5. The Rule bending moment is the sum of still water moment and wave bending moment, for each hull, obtained from Bureau Veritas (1990).

The "humps" on the maximum moment curves are the result of the coincidence of predominant

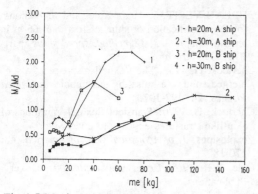

Fig 4.5 Maximum bending moment dependence on explosive mass and explosion depth for hulls A and B

excitation harmonics with the frequency of the first natural mode of hull vibration. The low frequency content of the pulsating bubble induced pressure spectrum for two TNT charges is shown in Fig. 4.6. The before mentioned coincidences of frequencies is clearly visible in the same figure.

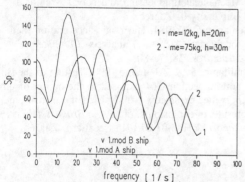

Fig 4.6 Pulsating bubble low frequencies for 12kg of TNT at 20m and for 75kg of TNT at 30m depth

A higher pulsation frequency is obtained by smaller amounts of explosive. Since the hull B has higher natural frequencies, its resonance response occurs in the case of smaller amounts of explosive. Besides, lower buoyancy means lower excitation force, and therefore the bending moments due to these both facts are lower. The final result is that small amounts of explosive can produce the bending moments equal to the Rule bending moment.

The relationship of the maximum bending

moments divided by the Rule bending moment and the keel shock factor Q_k for the hulls A and B is shown in Fig.4.7. The keel shock factor is defined by the following expression:

$$Q_k = \frac{\sqrt{m_e}}{R}\left(\frac{1+\cos\theta}{2}\right) \qquad (4.1)$$

where θ and R are shown in Fig.4.8.

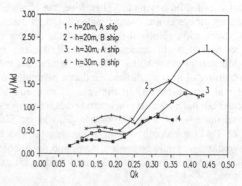

Fig.4.7 Maximum bending moment dependence on keel Q factor

The local elastic response of the hull structure is approximately linearly proportional to the shock factor. A general opinion is that serious local structure damage due to shock wave occurs at shock factor 0.45 or greater.

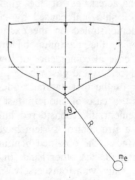

Fig.4.8 Keel shock factor variables

The global ship hull response, excited by the bubble pulsation of smaller amounts of explosive at lower depths, can produce higher damage in the case of resonance at a lower shock factor, see Fig.4.7. It is also visible from Fig.4.7 that the hull response to bubble pulsation is rapidly reduced with the increase of explosion depth.

The bending moments, Figs. 4.5 and 4.7, are in favour of the hull B. The advantages of the hull B over the hull A are specially pronounced if comparing the critical deck and bottom buckling stresses, which is caused by differences in the construction of the hulls. This topic, however, is beyond the scope of this paper.

With the increase of the explosive mass, the error of the presented model of interaction gets increased too, because the assumption of constant displacement is not valid any more. Therefore, the obtained bending moments are overestimated for such cases. The calculation results also show that the explosive mass necessary to cause serious hull damage is within narrow boundaries. This fact diminishes the probability of heavy hull damage due to whipping in the real circumstances. In the case of exposure of such small hulls like the considered ones, to conventional underwater mine explosions (300 kg or 500 kg of TNT) at a depth of 35m or greater, the shock wave damage would be predominant, and the bubble pulsation would result in hull rigid body vertical motion.

The hull response due to different types of explosive types can be calculated by equivalent TNT mass from equation (2.11), under the assumption that the bubble pulsation excitation is equivalent to that of TNT.

5. CONCLUSION

The basic characteristics of the global bending moment of the ship hull caused by the pulsating bubble are demonstrated on the example of two hulls. Generally, the maximum bending moment occurs in the case of coincidence of low frequency excitation induced by the gas bubble pulsation and the frequency of the first natural mode of the hull vibration. In the case of modern fast combatants (200t displacement, like considered hull B) which because of the fire control system have extremely high hull stiffness, the obtained bending moments are not critical. On the other hand, in the case of older types of ships with transversely framed hulls, which have relatively low stiffness (considered hull A), the bending moments caused by the bubble pulsation can produce deck and bottom stresses which exceed the critical buckling stresses.

REFERENCES

Bjarnholt, G. 1980. Underwater explosion test, *Propellants and Explosives*.5: 67-74.

BUREAU VERITAS, 1990. Rules and regulations for the classification of ships of less than 65 m in length, *Chapter 13, Light high speed ships,* part II-B, Hull structure.

Chertock, G. 1953. The flexural response of a submerged solid to a pulsating gas bubble. *J.Applied Physics*. 24: 192-197.

Chertock, G. 1970. Transient flexural vibrations of ship-like structures exposed to underwater explosions. *J. of Acoustical Soc. of America*. 48: 170-180.

Cole, R.H. 1965. *Underwater explosions*. Dover Publications

Geers T.L. 1971. Residual potential and approximate methods for three-dimensional fluid-structure interaction problems. *J.Acoust.Soc. Amer*.49: 1505-1510.

Haywood J.H. 1958. Response of an elastic cylindrical shell to a pressure pulse. *Quart. J. Mech. App. Math*. 11: 129-141.

Hicks, A.N. 1986. Explosion induced hull whipping. *Proceedings of Int.Conf. Advances in Marine Structures*. Dunfermline.

Senjanovic,I., V.Ćorić, T.Agustinović, Y.Fan 1989. Manual for DIANA computer program for ships hull vibration analysis. (in Croatian) Faculty of Mechanical Engineering and Naval Architecture, University of Zagreb.

Senjanović, I. 1984. Harmonic acceleration method for dynamic structural analysis. *Computer & Structures*. 18: 71-80.

Smiljanic, B. & Ž.Đuračić 1990. Testing of the ship bottom section Č64 exposed to the action of underwater explosion - analysis of the measurement results of explosion parameters, *Brodarski Institute Report*. YH02-01-013 (in Croatian), Zagreb.

Hydroelasticity in Marine Technology, Faltinsen et al. (eds) © 1994 Balkema, Rotterdam, ISBN 90 5410 387 6

Speed dependence of the natural modes of an elastically scaled ship model

K. Riska
Helsinki University of Technology, Arctic Offshore Research Centre, Finland

T. Kukkanen
Helsinki University of Technology, Ship Laboratory, Finland

ABSTRACT: In order to investigate ships that ram massive floating ice features in model tests, a backbone ship model was constructed. The model consists of nine segments attached to a backbone giving the proper stiffness in bending and in shear. With the construction it was possible to scale the lowest hull bending natural modes in addition to the rigid body features. The main topic of the paper is an investigation of the speed dependency of natural frequencies and modes. This factor is important in theoretical simulation of the ramming process. The test was done by carrying out an exciter test on the model at different speeds up to 19 knots (full scale). The natural frequencies show a marked decrease with increasing forward speed. The values of modal amplitudes also decrease with increasing speeds. The results are discussed and commented in the paper.

1 INTRODUCTION

The elastic response of a ship hull is dynamic when impact excitation is acting on the hull. This external excitation can be, for example, slamming impact in waves, or ramming an ice floe. When modelling theoretically the dynamical behaviour of the hull girder, the mass distribution and the structural properties of the ship have to be determined. The surrounding water induces an additional inertial force that may be taken into account as a hydrodynamic added mass. The calculation models used in dynamic analysis can be divided to two or three dimensional methods. In the two dimensional method the ship is modelled as a beam and the hydrodynamic added mass is calculated by two dimensional sink-source or conformal mapping. For the lowest hull girder vibration modes the beam model is sufficient, but when higher modes or more accurate calculations are needed, three dimensional methods are necessary. The added mass may be determined by three dimensional sink-source method or by finite element method. The common methods for calculating added mass in ship vibration analysis assume zero speed, which means that the ship forward speed effects are ignored.

An experiment was carried out to study the relationship between the natural modes of vibration of the ship's hull and the forward speed at which she sails. Model tests were done on the Canadian bulk carrier M.V. Arctic during the summer of 1991. The model used in testing is an elastic model made up of nine separate segments that are connected to a backbone beam. The model was originally build for ship ramming tests that have been carried out earlier. Details of the construction and calibration of the model and the backbone beam can be found in Riska (1988). There also exist full scale data of natural frequencies and damping factors from an exciter tests which has been conducted for the ship M.V. Arctic. This makes it possible to verify the success of the model.

This report discusses the open water towing tests and the results obtained from them. Theoretical calculations were conducted by finite element method and are compared with the model test results. The main subject of the investigation was to study the forward speed effect on the hydrodynamic added mass and what are the relative differences in natural frequency and mode shapes compared to zero speed values. The model test was carried out in still water. Hence rigid body motions, heave and pitch, and wave exciting forces do not exist in the measurements and are not included in calculations.

1.1 Theoretical background

The total deflection $w(x,t)$ of a Timoshenko beam is composed of two parts; one caused by bending and the other by shear. So the slope of deflection curve at the point x can be written

$$\frac{dw(x,t)}{dx} = \theta(x,t) + \gamma(x,t), \tag{1}$$

where γ is angle due to shear and θ is angle due to bending (Bishop & Price 1979). The shear force and bending moment is, ignoring structural damping

$$Q(x,t) = kGA(x)\gamma(x,t) \tag{2}$$

$$M(x,t) = EI(x)\frac{\partial\theta(x,t)}{\partial x}, \tag{3}$$

where $I(x)$ the moment of inertia, $A(x)$ is cross-section area, k is shear area reduction factor, E is Young's modulus and G is shear modulus. The equations of equilibrium for shear and bending are, if F is the external force, the following

$$\frac{\partial Q}{\partial x} + F = m(x)\frac{\partial^2 w(x,t)}{\partial t^2} \tag{4}$$

$$\frac{\partial M}{\partial x} + Q = I_m\frac{\partial^2\theta(x,t)}{\partial t^2}, \tag{5}$$

where $m(x)$ is the mass of the beam per unit length and $I_m(x)$ is rotary inertia per unit length about a normal that is perpendicular to x-z plane.

When the ship is vibrating the surrounding water induces additional pressure forces that have to be taken into account. In still water the cross sectional hydrodynamic force is commonly divided into three parts; one proportional to acceleration, one proportional to velocity, and one proportional to displacement. The proportionality factors are the cross sectional added mass a_{33}, the cross sectional hydrodynamic damping b_{33}, and cross sectional hydrodynamic restoring force $\rho gB(x)$. The hydrodynamic damping is negligible in the frequency range of ship hull vibration. Then the hydrodynamic force related to vibration may be given as

$$f_h = -a_{33}\frac{\partial^2 w}{\partial t^2} - \rho gB(x)w. \tag{6}$$

External force may be divided into two parts; the hydrodynamic f_h part and the external force f as

$$F = f + f_h. \tag{7}$$

The first step in modal analysis is to solve the eigenvalue problem

$$\begin{cases} \dfrac{\partial}{\partial x}\left(kGA\left(\dfrac{\partial w_r}{\partial x} - \theta_r\right)\right) - \rho gB(x)w_r = -\omega_r^2 w_r\left(m + a_{33}\right) \\ \dfrac{\partial}{\partial x}\left(EI\dfrac{\partial\theta_r}{\partial x}\right) + kGA\left(\dfrac{\partial w_r}{\partial x} - \theta_r\right) = -\omega_r^2\theta_r I_m \end{cases}$$
$$\tag{8}$$

for natural frequencies ω_r and mode shapes w_r and θ_r.

It has been shown that the mode shapes fulfil two orthogonality conditions (Bishop & Price 1979). Taking into account these orthogonality conditions, the equations of motions are independent for different modes and the generalized coordinate q_r can be calculated from equation

$$\ddot{q}_r + 2\zeta_r\omega_r\dot{q}_r + \omega_r^2 q_r = \frac{1}{m_r}\int_0^L fw_r dx, \tag{9}$$

where the damping is included as modal damping $2\zeta_r\omega_r$, and ζ_r is modal damping factor for mode r. m_r is the generalized mass, which is equal to one $m_r = 1$ if the mode shapes are normalized with respect to mass.

When the generalized coordinates q_r are solved from equation (9), then forced response analysis can be done for the structure and all the values of interest can be calculated. The displacement w in location x is

$$w(x,t) = \sum_{r=1}^{N} w_r(x)q_r(t), \tag{10}$$

where w_r is the modal displacement. The shear force and bending moment can be calculated from equations

$$Q(x,t) = \sum_{r=1}^{\infty} Q_r(x)q_r(t) \tag{11}$$

$$M(x,t) = \sum_{r=1}^{\infty} M_r(x)q_r(t), \tag{12}$$

where Q_r is the modal shear force and M_r is the modal bending moment.

The modes $w_r(x)$, $r = 1, 2, \dots$ are sometimes called the 'wet' modes of the ship hull as the hydrostatic restoring force and added mass is included in equations (8). Then the mode number 1 corresponds to heave mode and mode number 2 to pitch mode. Starting from the third mode, elastic forces dominate over the restoring forces $\rho gB(x)$. This formulation of the 'wet' modes contains inherent difficulties, even if damping is ignored. Sometimes the 'dry' modes are solved first, and the hydrodynamic components are treated as external force (Bishop & Price 1979). The topic in this investigation is to investigate some of the problems in the basic formulation given above.

2 TEST ARRANGEMENT AND PROGRAM

2.1 *General*

The investigation of the influence of the forward speed on the natural modes of the hull beam was done by towing a ship model with a constant speed and applying an impact force on it. The test was done by hitting the model at bow with an instrumented hammer a number of times during each run in the towing tank. The response of the ship model was measured. The simultaneous measurement of the exciting force and response made it possible to calculate the transfer function and to determine the natural modes.

2.2 *Model*

The model used in the tests is a 1:40 scale wooden model of the Canadian arctic bulk carrier M.V. Arctic. The dimensions of the ship and model are presented in table 1.

The model consists of nine separate segments were connected by a backbone beam (see figure 1). Rubber straps were attached between each of the segments, which makes it water-tight while making a negligible change to the properties of the model.

The backbone beam was rectangular, hollow and made of fibreglass. There were two different cross-sectional areas of the beam, a smaller cross-section in the fore and aft and a larger one in the mid-body of the model. The thicknesses of the walls of the beam are uniform at 13 mm for the fore and aft, while the horizontal thickness of 13 mm is coupled with a vertical thickness of 8 mm in the mid-body (see figure 1). The characteristics of the beam are in table 2. The dimensions were selected in order to enhance the shear deformation.

2.3 *Measuring equipment*

The bending stress was evaluated by ten strain gauges placed at the top and at the bottom of the beam. The shear stress was evaluated by eight separate strain gauges placed on port and starboard sides of the beam. The strain gauges were attached to the beam at the joints between the segments. The bending gauges were connected so that normal stress was eliminated in all but two gauges that measured separately the stresses from top and bottom of the beam (gauges 2A, 2Y and 5A, 5Y). This makes it possible to check the magnitude of the normal stress.

When the measurements were analysed the normal stress was found to increase slightly only with higher modes of vibration.

The shear gauges were arranged so that the effect of horizontal bending stress was avoided. In addition, 5 displacement transducers were place on the top of the beam to measure the deflection of the beam during the tests.

For the testing in the towing tank, the model had two DC amplifiers attached to it in the middle segment number 5. These were used to take the voltage readings from the strain gauges. To record the responses from various devices during testing, two 14 channel tape recorders were used. The impact hammer was recorded by both tape recorders to synchronise the tapes during analysis. The set up of the recording equipment was that the recorders were placed on the carriage next to the model with the wires running from the amplifiers to them.

2.4 *Calibration*

The strain gauges were first calibrated in known loading situations, with the responses recorded as voltages from the strain gauges. The calibration was carried out by placing the ship model on simple supports and hanging weights from the model. The results of these loadings were then analysed and compared to a theoretical calculation of the load cases. This theoretical calculation gave the shear force and bending moment of the beam and corresponding bending and shear stress (Riska 1988). The calibration results were compared with the theoretical data to obtain calibration coefficients between measured voltages and shear forces and bending moments. These calibration coefficients were then used in the final analysis to determine the natural frequencies, mode shapes and damping factors from the strain gauge data.

The second step in the calibration was two swing tests, lateral and longitudinal. These calibrations were done to check the vertical centre of gravity and the longitudinal radius of gyration. From this, the radius of gyration of the model was found to be 1.18 m and the vertical centre of gravity was 0.216 m. These values were the same as the values obtained by Riska (1988) for the same model.

2.5 *Towing tests*

The tests were conducted in the 120 m long open water towing tank at Helsinki University of Technology. Tests simulated full-scale speeds between 0 to 19 knots (table 3). For the tests, the model was fitted with a towing bar across section 4. It extended out from both sides of the model and

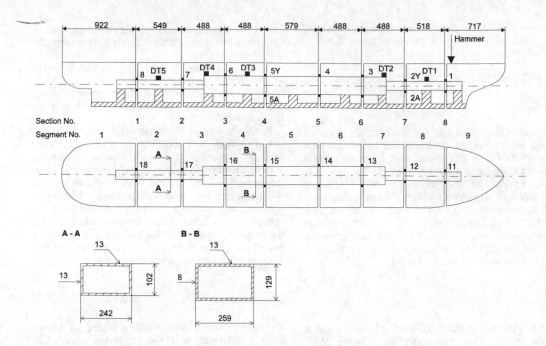

Figure 1. Schematic view of the model used in the towing test. The bending stress gauges are numbered from 1 to 8 and the shear stress gauges are from 11 to 18. DT means displacement transducer.

Table 1. The main dimensions of M.V. Arctic and the model.

		Ship	Model
Length, overall	L_{OA}	209.6 m	5.24 m
Length between perpendiculars	L_{PP}	196.6 m	4.91 m
Breadth	B	22.9 m	0.57 m
Draft	T	11.0 m	0.27 m
Displacement	Δ	38 030 tn	589 kg
Block coefficient	C_B	0.76	

Table 2. The characteristic of the backbone beam.

		Fore and aft	Mid-body
Section modulus	W_y	270 cm^3	380 cm^3
Shear area	$k_z A$	22.1 cm^2	16.3 cm^2
Young's Modulus	E	8.23 GPa	
Shear Modulus	G	3.10 GPa	

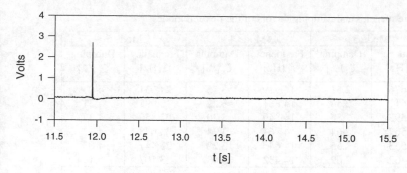

a) Impact hammer force at bow.

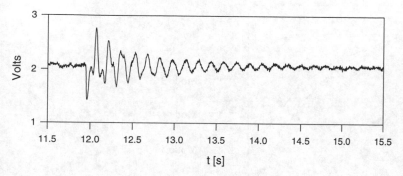

b) Bending moment at section 6 (strain gauge 3).

Figure 2. Detailed time histories from towing test. The model speed was 0.949 m/s corresponding Froude number Fn = 0.137.

ropes from it were attached to the front and back of the carriage to keep the model steady during testing. The set up allowed the heave and pitch motions. The model was towed at constant speed by the carriage and during each run an impact force was applied to bow of the model at intervals of approximately ten seconds. A typical time history from a towing test is shown in figure 2. The impacts were made after the model position had settled in each velocity.

Table 3. The speeds of the model in the towing test.

Ship [kn]	Froude number $Fn = \dfrac{U}{\sqrt{gL_{pp}}}$
0.0	0.0
3.9	0.046
7.8	0.091
11.7	0.137
15.6	0.182
19.4	0.228

3 ANALYSIS OF TEST DATA

The natural frequencies, damping factors and mode shapes can be obtained from frequency response function $H(\omega)$

$$H(\omega) = \frac{X(\omega)}{F(\omega)} \qquad (13)$$

where $X(\omega)$ is the measured response spectrum and $F(\omega)$ is excitation spectrum, obtained by Fourier transform of the time trace.

The test data were analysed at the Laboratory for Strength of Materials at Helsinki University of Technology (Kantola 1993, Porkka 1994). From the measured signals the frequency response functions were derived by spectrum analyser. The results are shown in table 4, where the scaled full scale values are also shown in the last row. Non-dimensionalised natural frequencies and damping factors as a function of speed are presented in figures 3 and 4 respectively.

Table 4. Measured natural frequencies and damping factors of the flexural modes.

Model	Mode 3		Mode 4		Mode 5	
Speed [Fn]	Frequency f_3 [Hz]	Damping ζ_3 [%]	Frequency f_4 [Hz]	Damping ζ_4 [%]	Frequency f_5 [Hz]	Damping ζ_5 [%]
0.0	6.626	2.51	14.929	3.15	24.031	4.03
0.046	6.611	2.56	14.881	3.22	23.997	4.31
0.091	6.593	2.58	14.849	3.20	23.850	4.37
0.137	6.567	2.64	14.762	3.34	23.678	4.32
0.182	6.533	2.56	14.640	3.22	23.441	4.05
0.228	6.462	2.69	14.432	3.43	23.107	5.13
Full scale $Fn = 0.0$	5.9	2.0	12.5	2.5	19.7	2.8

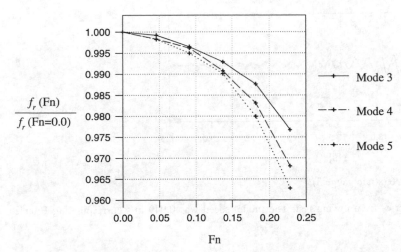

Figure 3. Non-dimensionalised natural frequencies as a function of forward speed.

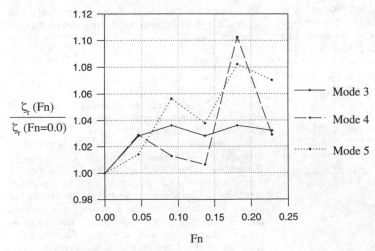

Figure 4. Non-dimensionalised damping factors as a function of forward speed.

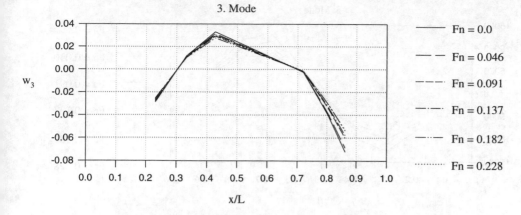

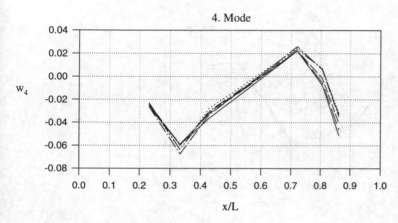

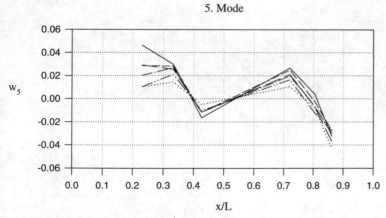

Figure 5. Modal displacements ($\frac{1}{\sqrt{kg}}$).

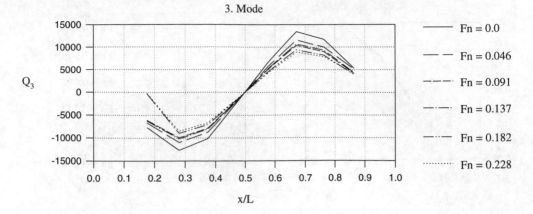

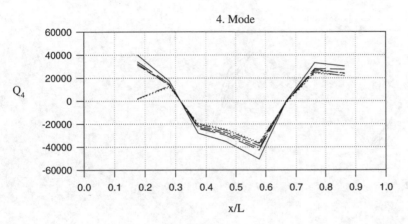

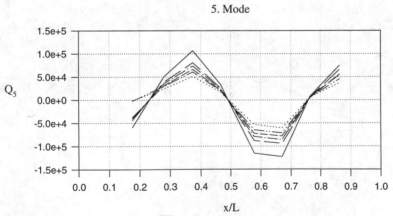

Figure 6. Modal shear forces ($\frac{\sqrt{kg}}{s^2}$).

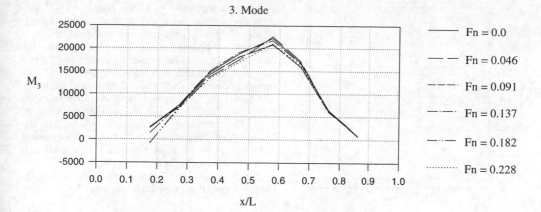

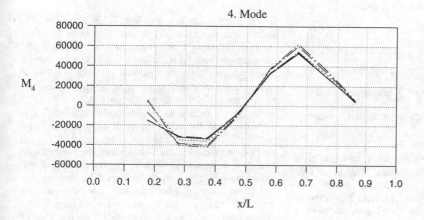

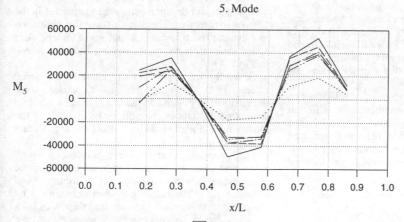

Figure 7. Modal bending moments ($\frac{\sqrt{kg}}{s^2}m$).

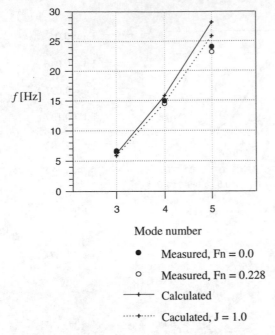

●   Measured, Fn = 0.0

○   Measured, Fn = 0.228

─────── Calculated

·····+····· Caculated, J = 1.0

Figure 8. Calculated and measured natural frequency.

From figure 3 it can be seen that when forward speed increases the natural frequencies decrease. This means that added mass increases, because the stiffness should be speed independent. The forward speed has more effect on the higher modes of vibration.

Forward speed influence on damping is not so evident as it was in the case of natural frequencies. For example, modal damping factor for mode 3 is almost constant at different forward speeds. In the analysing procedure the damping factors are the most inaccurate values and that partly explains the large deviations in them.

The measured mode shapes of displacements, shear forces, and bending moments for different forward speeds are shown in figures 5, 6, and 7. The modes have been normalized so that the generalized mass is one. For all modes the main tendency is that the mode shape values decrease when the forward speed increases.

4 THEORETICAL CALCULATIONS

The main parameters that influence the calculated natural frequencies and mode shapes are the stiffness of the beam and the mass, including hydrodynamic added mass. The stiffness includes bending rigidity and shear area and these are independent of the forward speed. The stiffness can be calculated quite

accurately for the backbone beam. The only difficulty is the beam foundations in the model, which might reduce the accuracy of the calculation.

More complicated is to take into account the added mass, that is the induced pressure force when the ship is vibrating in water. Integration of this pressure force yields the added mass and damping. The flow may be described by the velocity potential ϕ_k if it is assumed that linear potential theory can be used. The governing equations for calculating the heave added mass with two and three dimensional methods can be found in Newman (1978), Salvesen et. al. (1970) and Inglis & Price (1982).

The boundary condition at the free surface is usually approximated by high frequency limit

$$\phi_k = 0 \quad \text{on } z = 0 \text{ when } \omega \to \infty, \tag{14}$$

in the ship vibration problem. This approach has been used, for example, in the 3D Green function method (Hylarides & Vorus 1982) and 3D finite element method (Hakala 1986). It has been shown that by using three dimensional methods the modal values are given with quite good accuracy even for higher modes of vibration (Hakala 1986).

In two dimensional methods the hull is approximated by series of two dimensional cross sections. The flow field is determined for each section separately and therefore the interaction between different sections is neglected. The basic assumption is that ship is slender. This means that the three dimensional normal vectors can be approximated by two dimensional ones. The consequence of this assumption is that the rate of change of hydrodynamic pressure in longitudinal direction should be small compared to changes in the transverse direction. The two most common methods to calculate the two dimensional speed independent added mass are conformal mapping and sink-source method.

Generally in ship vibration analysis all speed effects have been ignored in two and three dimensional methods. In addition to this, in two dimensional methods the hull shape changes in the longitudinal direction do not affect to added mass computation.

The measured modes are compared with calculated ones in order to gain insight into the relationship between added mass distribution and natural modes. The calculations were carried out by finite element method, where the ship hull was modelled by 25 Timoshenko beam elements and the added mass was calculated by Frank close-fit method (Frank 1967).

In vibration analysis the two dimensional added mass is usually corrected by J-factor to take account of the three dimensional effects. In the present calculation the following empirical formula was used (Townsin 1967)

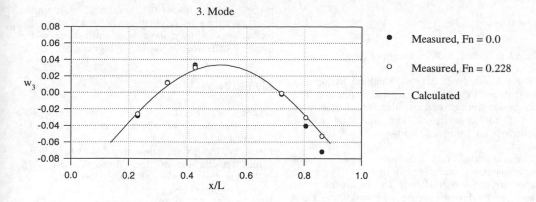

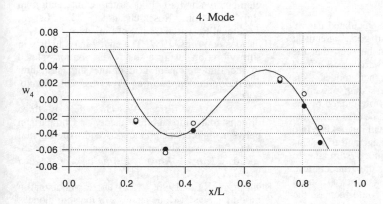

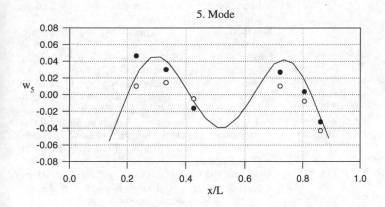

Figure 9. Mode shapes of displacements w_r.

$$J_r = 1.02 - 3\frac{B}{L}\left(1.2 - \frac{1}{r-1}\right), \tag{15}$$

where B is breadth and L is length of the ship and $r = 3,4,5$ is the mode number.

The calculated and measured natural frequencies for the three lowest modes are given in figure 8. The calculations were carried out also for J-factor value $J_r = 1$ for comparison. The measured natural frequencies are for Froude numbers $Fn = 0.0$ and $Fn = 0.228$.

The calculated and measured vertical modal displacements w_r are given in figure 9. The mode shapes are mass normalized. The measured mode shapes are for Froude numbers $Fn = 0.0$ and $Fn = 0.228$.

For the lowest bending mode, the calculated natural frequency and mode shape are in good agreement with the measured results. It can be seen that the three dimensional effects and forward speed effects become more important for higher modes. The influence of forward speed is less than the effect from three dimensionality of the flow judged by the variation of the J-factor. It should be also noted that the added mass was calculated without taking into account wave profile and ship sinkage and trim. This means that wave elevation around a ship due to forward speed is ignored. In the model test, especially the bow wave grew when forward speed of the model was increased. This could increase the added mass and therefore decrease the natural frequency.

5 CONCLUSIONS

In the vibration tests described in this paper the natural frequencies and modal values were derived for the elastic model of the bulk carrier M.V. Arctic. From the results of the measurements it can be seen that natural frequency decreases when the forward speed increases. The modal amplitudes of, displacement, shear force, and bending moment decrease with increasing forward speed.

The natural frequencies and mode shapes were calculated by two dimensional methods for zero forward speed. The calculated values corresponds satisfactorily to measured ones. The difference between calculated and measured values were greater than the forward speed effects in measurements. In added mass calculations, three dimensional effects are more important than forward speed, particularly for higher modes. The forward speed effects, including the wave elevation around a ship, are not taken into account in added mass calculations.

When the forward speed is one of the main parameters, such as in slamming or ramming situations then in calculations of added mass the forward speed may have some influence on the dynamic behaviour of the ship hull girder.

REFERENCES

Bishop, R. E. D. & Price, W. G. 1979. *Hydroelasticity of ships*. London: Cambridge University Press.

Frank, W. 1967. Oscillation of cylinders in or below the free surface of deep fluids, Report 2375. Washington DC: Naval Research and Development Center.

Hakala, M., K. 1986. Application of the finite element method to fluid structure interaction in ship vibration, Research notes 433. Espoo: Technical Research Centre of Finland.

Hylarides & Vorus. 1982. The added mass matrix in ship vibration, using a source distribution related to the finite element grid of the ship structure. *International Shipbuilding Progress*, Vol. 29, No. 330.

Inglis, R. B. & Price, W. G. 1982. A three dimensional ship motion theory - Comparison between theoretical predictions and experimental data of the hydrodynamic coefficient with forward speed. *Trans. Royal Institution of Naval Architects*, Vol. 124.

Kantola, K. 1993. Laivan pienoismallin moodianalyysi. (Mode analysis of the ship model). Espoo: Helsinki University of Technology, Laboratory for Strength of Materials, Unpublished, (in Finnish)

Newman, J. N. 1978. The theory of ship motion. *Advances in Applied Mechanics* 18.

Porkka, E. 1994. Laivan pienoismallin moodianalyysi (Siirtymät). (Mode analysis of the ship model (Displacements)). Espoo: Helsinki University of Technology, Laboratory for Strength of Materials, Unpublished, (in Finnish)

Riska, K. 1987. On the mechanics of the ramming interaction between ship and massive ice floe, Publications 43. Espoo: Technical Research Centre of Finland.

Riska, K. 1988. Ship ramming multi-year ice floes, Model test results, Research notes 818. Espoo: Technical Research Centre of Finland.

Salvesen, N., Tuck, E. O. & Faltinsen, O. M. 1970. Ship motions and sea loads. *Trans. Society of Naval Architects and Marine Engineers*, Vol. 78.

Townsin, R. L. 1969. Virtual mass reduction factors. 'J' values for ship vibration calculations derived from tests with beams including ellipsoids and ship models. *Trans. Royal Institution of Naval Architects*, Vol. 111.

Hydroelasticity in Marine Technology, Faltinsen et al. (eds) © 1994 Balkema, Rotterdam, ISBN 90 5410 387 6

Characteristics of hydrodynamic loads data for a naval combatant

Bill Hay
Carderock Division-Naval Surface Warfare Center, USA

Jim Bourne, Allen Engle & Rick Rubel
Naval Sea Systems Command, Washington, D.C., USA ·

ABSTRACT: This paper discusses recent efforts by the U.S. Navy in the area of hydrodynamic loads research. Specific areas covered include both model test and full scale trials data. Included within our growing data base, are ordinary wave induced loads and combined wave plus whipping response for both vertical and lateral loads. The methods that have been used to collect this data and its subsequent analysis will be discussed in detail. Recommendations, based upon our findings, for future work will also be presented.

1 INTRODUCTION

The current USN design criteria utilizes a "standard wave" for determining primary stresses. Developed over forty years ago, this approach was established at a time when high speed computers were not available nor was our understanding of physical oceanography or applied statistics as advanced as they are today. Similarly, the methods available for predicting structural response (e.g., fatigue strength and fracture performance) were not available.

This "standard wave" approach determines the design bending moment by statically balancing the ship on a trochoidal wave whose length is equal to the ships length and whose height is equal to $1.1\sqrt{LBP}$. The stresses derived from this bending moment are then compared with allowable values and adjusted on a "trial and error" basis, to reflect past experiences with ships already in operation. Although this approach has worked well, this "standard wave" approach does not specifically account for the effects of transient loads (e.g., whipping, green seas, wave slap), fatigue or their effects on longitudinal distribution of bending moments other than by empirical "rules of thumb". In addition, lateral loads, torsion and associated effects are not addressed.

As a result of these uncertainties, the designer has been forced to apply a generous safety margin, particularly at stations forward of midships, to account for effects of slamming. In addition, this design methodology applies only to ships that are within the historical database. We are now beginning to use new structural materials (e.g., high strength steels, composites), develop unconventional ship designs (e.g., SWATH, SES, Advanced Double Hull) and anticipate the need to improve our ships' capability to operate at higher speeds and severe environments for longer durations. Furthermore, there is an ever present demand for lighter, more efficient structures. Although extrapolations of current design methods are possible, there exists a level of uncertainty when one takes an empirically based design procedure and applies it to different ship types, displacements or operational requirements.

With the advent of finite element methods the naval architect has the capability to assess these variations in design and/or materials. However, while these analytic techniques can help one evaluate the ability of specific structural members to resist a given load (and hence the consequence of failure of that member) the designer can be lulled into a false sense of security as the probability of failure cannot be determined. As structural safety, and hence an acceptable level of risk, is defined as the probability of failure times the consequence of failure, it is clear that an alternate structural design criteria must be developed in order for the naval architect to have a quantitative basis from which appropriate safety levels can be determined.

In response to the need to design structurally efficient, and affordable yet innovative ships, the US Navy is pursuing revisions to their current structural design criteria and to incorporate reliability based design methodologies within these new procedures. This is a major undertaking, requiring research in the areas of hydrodynamic loads, structural response and reliability theory. To date, our major efforts have been in the collection of model test and full scale hydrodynamic loads data. Included within our database are ordinary wave induced loads and combined wave induced plus whipping loads for both vertical and lateral loads. What follows is a discussion of the methods that have been used to collect this data and its subsequent analysis.

2.0 FULL SCALE TRIALS

2.1 Trial objectives

In accordance with our program plan (NAVSEA 1992), a series of full scale trials, both of short and long term duration, were performed on a ship of the CG 47 Class. The hull lines for the CG 47 Class are shown in Figure 1; the main attributes being a large bow mounted sonar dome, pronounced bow flare and a transom stern.

The short term, heavy weather trial, was performed in the North Atlantic from 14 January to 19 January 1991. Weather forecasting, furnished by the Naval Eastern Ocean Center and Bendix Field Engineering Corporation, provided vectoring information for the desired sea conditions.

The objectives of the short term, heavy weather trial were to allow for the collection of primary ship structural hull girder response, ship motions and in-situ wave spectral data. The test plan (AEGIS 1991) required these measurements to be made through Sea State 6, so as to provide the opportunity to collect sufficient slam induced whipping data. Data were collected in 30 minute segments during which ship speed and heading remained constant. The resulting measurements were converted into vertical and lateral bending

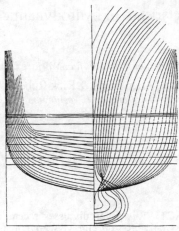

FIG. 1 CG 47 HULL FORM GEOMETRY

moments and used to develop Response Amplitude Operators (RAO's). These measurements also provide a database to validate existing scaling laws through correlation model tests (Note; it is expected that these data will also be of use in validating emergent non-linear time domain loads and motions computer programs). Upon completion of the short term trial, the data acquisition system was left onboard in order to collect long term bending moment magnitude distribution data.

2.2 Instrumentation

2.2.1 Strain gages

A total of 28 strain gage bridges were installed, at 5 locations longitudinally, for measurements of overall longitudinal vertical and lateral hull girder bending. Primary bending response gages were placed on the 01 level, outboard, and as near the keel as could practically be made accessible. Theses gages were located approximately at stations 3.5, 5, 6.5, 10, and 15 (figure 2). The majority of the gages were located in the forward portion of the ship to provide additional data on the response of the structure to slamming. The strain gages for these bridges were installed on the stiffener webs at the local neutral axis of the plate stiffener combination (figure 3). This is to ensure that the effects of the local secondary bending is negated and the strain measurements

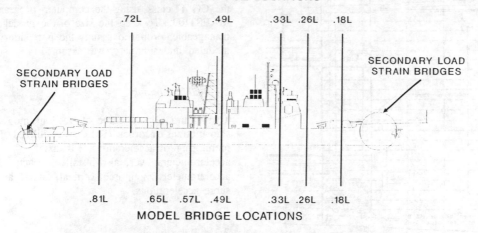

FULL SCALE BRIDGE LOCATIONS

.72L   .49L   .33L .26L .18L

SECONDARY LOAD
STRAIN BRIDGES

SECONDARY LOAD
STRAIN BRIDGES

.81L   .65L .57L .49L   .33L .26L   .18L

MODEL BRIDGE LOCATIONS

Figure 2 General Arrangement of USS MONTEREY (CG-61)
Strain Gage Instrumentation

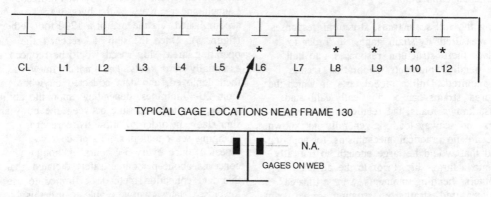

CL   L1   L2   L3   L4   L5   L6   L7   L8   L9   L10   L12

TYPICAL GAGE LOCATIONS NEAR FRAME 130

N.A.

GAGES ON WEB

FIGURE 3 CROSS SECTION OF 01 LEVEL SHOWING FULL SCALE TRIALS GAGE LOCATIONS

are from primary bending only. It should be noted that the hull girder bending channels were not wired as vertical or lateral bending bridges, they were wired and recorded as deck edge response channels, to provide data on how vertical and lateral bending combine to produce a deck edge response.

Strain gages were also installed to provide information on the local structural response of the foredeck region structure (figure 4). Ten strain gage bridges were mounted near the deckhouse front to provide data on uplifting forces from the deckhouse. The hull girder torsional responses were measured at frame 130.

The transverse distribution of the strains induced by the primary bending moment were measured by 7 strain gage bridges located on the 01 level stiffeners at frame 130. These bridges provided data on in plane stress magnitudes and the effects of shear lag.

To provide data on local secondary responses due to wave impact, a number of strain gage bridges were installed on the bow region stiffeners. The area between frames 32 and 40 was heavily instrumented to measure local response to bow flare slamming and to measure 01 level and bulwark response to green sea loads.

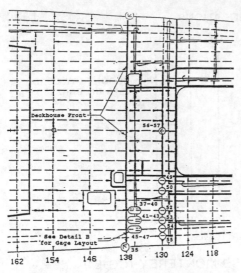

FIGURE 4 - Details of Gage Locations

2.2.2 Loads calibration

During the full scale trials, structural responses were measured by strain gages. In order to translate these structural responses into hull girder bending moments, calibration factors were required. Unlike model tests, in which the measured strains can be physically calibrated against known loads, the relationship between strains and seaway loads is generally not known. There are no practical mechanisms for applying a load that would be large enough to statically calibrate a large ship. Prior to the advent of computers, bending moments were estimated from measured strains by assuming simple beam theory and using the section modulus from the ships strength drawings. These strength drawings, however, neglected the contribution of the deckhouse and conservatively estimated the ineffective area which resulted in bending moments that could not be expected to be completely accurate.

However, by using a whole ship finite element model, a numerical calibration can be performed (Sikora et al 1991). A known load is applied to the numerical model and the resulting strains calculated. Calibrations factors are then obtained at selected locations corresponding to the location of the strain gages on the ship. The strain gage readings can then be translated into corresponding hull girder bending moments. In this case a coarse mesh,

full ship finite element model was developed for the CG 47 class, using the computer program MAESTRO. To keep the size of the model to manageable proportions only the port side was modeled and symmetry was assumed.

2.2.3 Ship motions

To measure all six degree of freedom motions a ship motions recorder (SMR) was utilized. For monitoring of accelerations 3 tri-axial accelerometers were also installed. Each accelerometer monitored vertical, lateral and surge accelerations.

2.2.4 Wave height

Two types of wave buoys were used for collection of in-situ sea state data. For measurement of directional wave spectra an Endeco directional wave buoy was utilized. Launch and recovery of the buoy was accomplished via the use of a 1200 foot tether (figure 5). Once the ship had reached the operating area, ship speed would be reduced to essentially zero knots. The Endeco buoy was then deployed, and data collected anywhere from 20-40 minutes, depending upon the ability to keep the tethered line between the buoy and ship slack. In order to allow us to record changing sea conditions, this process was repeated once every 3-4 hours. Although concerns about personnel safety dictated that this data collection method be limited to use in lower sea states, we were able to use this method during selected sea state 6 conditions.

When environmental conditions would not allow us to use the Endeco buoy, Mark II disposable wave buoys were utilized. During these situations a buoy would be launched prior to the start of a test. Wave data would then be continuously recorded throughout the prescribed run. As the battery life of these buoys are approximately 20 hours, it was often possible to use one buoy for multiple test conditions.

2.2.5 Data acquisition

All primary hull girder structural response and motion data were digitally recorded by means of a 128 channel data acquisition system (DAS).

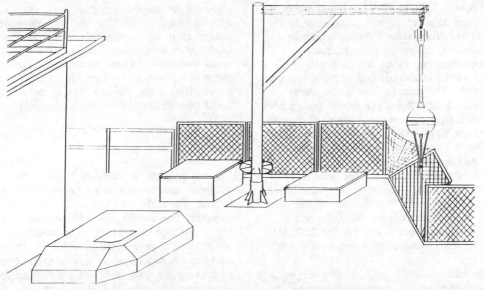

Figure 5 - Retrievable Buoy Station

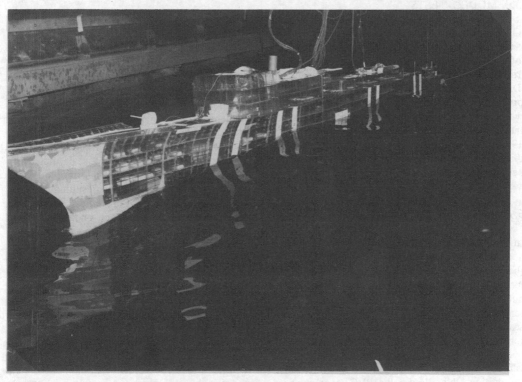

Figure 6 - PVC Model of CG 47

The system utilized 10 Hz low pass filters to effectively eliminate electrical noise at 60 Hz and higher frequencies. With a maximum unattenuated frequency of 5 Hz, data corresponding up to the second mode of longitudinal vertical bending of the ship were recorded. The data for this system were digitized at 20 samples per second per channel. At this sample rate, approximately 20 samples define the first mode vibratory response. The maximum through-put rate was approximately 2000 samples per second.

When the cruiser encountered seas in which slamming was expected, the above stated sample rate was insufficient to adequately define local panel stiffener responses. Accordingly, an analog system, capable of recording 2500 Hz, was utilized to record wave impact response data.

Information such as ships heading angle, ship speed, wind speed and direction were recorded directly from ship's gyro.

3.0 MODEL TESTS

Once the trial was completed, two sets of model tests were performed. The first phase consisted of a set of correlation model tests, where the results were used to validate scaling laws. During the second phase, a comprehensive test program was accomplished, where the model was subjected to conditions ranging from moderate to extreme sea states (Bishop et al 1992). The model was physically calibrated to directly relate measured responses and applied bending moment.

3.1 Model construction

In order to measure both local (shear lag and deck edge stresses) responses and global (bending moment) loads, a structurally scaled model was constructed (Rodd et al 1992). Built out of rigid poly-vinyl-chloride (PVC), the 1/25 scale model was continuous and structurally scaled so as to have the same load paths as the full scale ship (see figure 6). As a result, the overall stiffness and neutral axis for both vertical and lateral bending were scaled exactly. In addition, this allowed the PVC model to be instrumented to directly measure not only loads but structural response as well. Phasing of all loads could also be determined.

The model was self propelled with various systems (gages, propulsion and steering motors) connected to the carriage via cables. Installation of the propulsion system required the use of wood foundation blocks. Accordingly, section properties in the area of the blocks could not be scaled. Hence, the midship area of the model could provide bending moment data only.

3.2 Model test program

The test program is shown in Table 1. For all irregular wave tests, a minimum of 30 minutes of data (full scale) was recorded. Where severe slamming was present, one hour of data (full scale) was recorded. The regular wave tests were run at a wave steepness ratio of 1/50. Variation in hull response was determined by performing the 15 knot bow sea condition at a wave steepness ratio of 1/24.

3.2 Model instrumentation

3.2.1 Strain gages

Strain gages were installed on the model to provide structural responses due to hull girder loads and secondary wave impact loads. To obtain sufficient data to define the longitudinal distribution of vertical and lateral bending moments, seven frames were instrumented with strain gage bridges to monitor hull girder responses. Most of the frames chosen correspond to those of the full scale, however exceptions were made aft of midships due to the placement of the motors and resulting motor supports in the model. For vertical bending, the strain gages were installed on or near the CVK and on the 01 level. Lateral bending bridges were installed on the 01 level at the deck edges.

3.2.2 Loads calibration

The model was calibrated as a loads sensor by suspending it and applying static loads at known specific locations to induce a static bending moment. Loads calibration were performed for longitudinal vertical and lateral bending as well as torsion. For the vertical calibration the model was upright and suspended at frames 220 and 300, while static loads were applied at bulkheads 58, 138, and 506. The lateral calibration was similar to the vertical, except

Table 1 Experimental Test Conditions
Test Conditions In Regular (R) and Irregular Waves Representing Fully Developed
Bretschneider Energy Density Waves for Sea State 5 (5), Sea State 6 (6), Sea State 7 (7),
and Storm Waves (S)

SHIP SPEED	0 KTS	5 KTS	10 KTS	15 KTS	20 KTS	25 KTS
REL WAVE HDNG						
0 DEG	567SR	567S	567SR	567	56R	5
15 DEG	567S	567S	567S			
30 DEG	567SR	567S	567SR	6SR		
45 DEG	567S	567S	567S			
60 DEG	567S	567S	567S			
90 DEG	567	567S				

that the model was suspended with its starboard side up. Static loads were then hung at the same frame as reported above.

4.0 LOADS DETERMINATION

Structural responses from full scale at sea trials and irregular wave model tests were analyzed to provide the design community with a seaway induced loads data base for surface ship design. From time domain analyses both vertical and lateral bending moment distributions along the length of the ship were determined, as were the phase angle relationships between vertical and lateral ordinary wave induced loads and whipping. Weibull analyses of selected primary bending channels were also performed to provide probability distributions for slamming induced whipping moment magnitudes. Frequency domain analyses were performed to provide Response Amplitude Operators (RAO's) and phase angle relationships for vertical and lateral bending. Based on these RAO's a lifetime loads analysis was then performed to provide estimates of lifetime maximum seaway induced loads.

4.1 Bending moment distributions

Bending moment distributions were determined for longitudinal vertical and lateral bending due to: ordinary wave (i.e. weight minus buoyancy), whipping, and combined ordinary wave plus whipping. For the ordinary wave (i.e. weight minus buoyancy) moment distribution, the time histories of the load channels at each instrumented frame were examined from runs without whipping, as well as from low pass filtered time histories of runs where whipping was present. The results are plotted in Figures 7 and 8 for vertical and lateral bending, respectively. Included in each figure is a traditional 1-Cosine moment distribution curve currently assumed for vertical bending in U.S. Navy designs. Although current U.S. Navy design practice does not consider lateral bending, the 1-Cosine curve is shown for comparison purposes.

The longitudinal distributions of moments due to whipping were derived from high pass filtered time histories of the load channels for those runs that exhibited a significant number of whipping events. In this manner the quasi-static wave component was eliminated, leaving only the whipping responses. Plots of the whipping moment distribution are given in Figures 9 and 10 for vertical and lateral bending, respectively. The figures show that a trapezoidal distribution with a constant maximum value between 40 and 60 percent of ship length seems to provide the best fit to both vertical and lateral moment distributions.

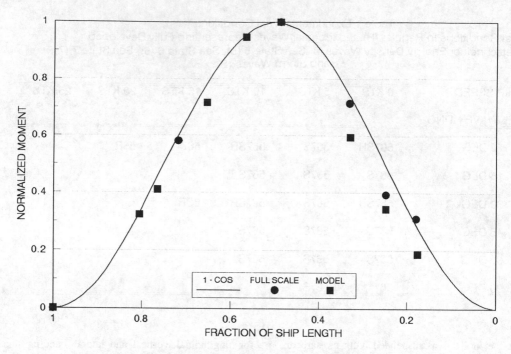

FIGURE 7 ORDINARY WAVE VERTICAL BENDING MOMENT DISTRIBUTION

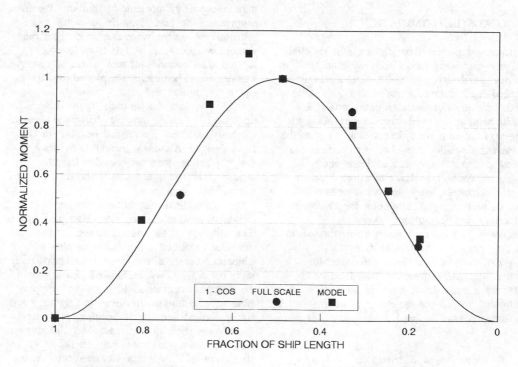

FIGURE 8 ORDINARY WAVE LATERAL BENDING MOMENT DISTRIBUTION

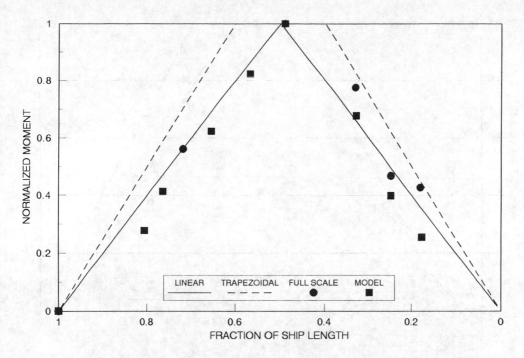

FIGURE 9 VERTICAL BENDING MOMENT DISTRIBUTION FOR WHIPPING ONLY

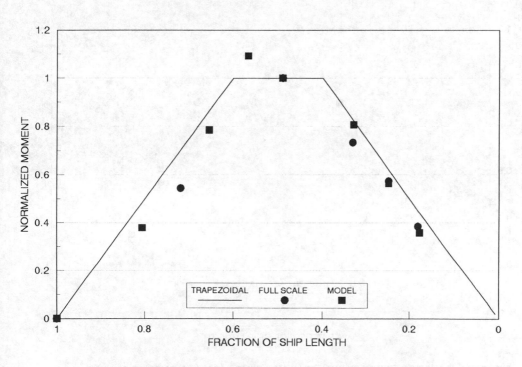

FIGURE 10 LATERAL BENDING MOMENT DISTRIBUTION FOR WHIPPING ONLY

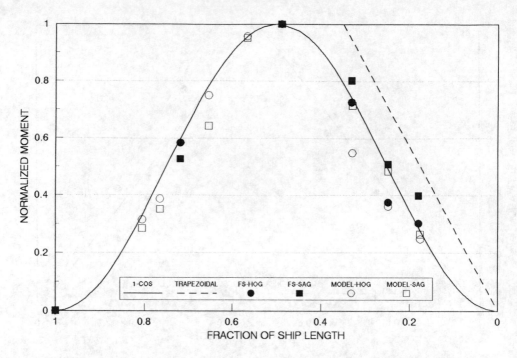

FIGURE 11 COMBINED ORDINARY WAVE PLUS WHIPPING VERTICAL MOMENT DISTRIBUTION

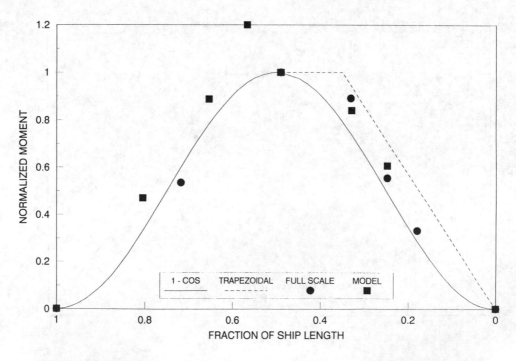

FIGURE 12 COMBINED ORDINARY WAVE PLUS WHIPPING LATERAL MOMENT DISTRIBUTION

From the unfiltered time histories of those runs with significant whipping events, the ratios of the maximum moments at each instrumented frame were derived with respect to midships so as to determine the envelope of maximum combined ordinary wave plus whip moments along the length of the ship. The resulting moment distribution for vertical bending is presented in Figure 11, which includes the 1-Cosine curve and a trapezoidal design moment curve. The corresponding longitudinal distribution of combined lateral bending moments is shown in Figure 12. For comparison, this figure includes both the 1-Cosine distribution curve as well as the trapezoidal distribution.

4.2 Phase angle relationship between ordinary wave and whipping

Whipping increases maximum and minimum strain values by superimposing a first and/or second mode vibratory response on top of the ordinary wave induced response cycles, see Figure 13. As was discussed in the previous section, ordinary wave and whipping responses were isolated by high and low pass filtering of the time histories. An analysis of the filtered time histories was performed for a number of sea and ship configurations from the full scale trials and model test data to provide the phase angle for the onset of whipping with respect to the ordinary wave cycle. This analysis showed that the wave impacts were occurring at 217 to 223 degrees into the ordinary wave cycle, see Figure 13. This phase angle is believed to be the result of bow flare slamming as opposed to hull bottom slamming. The major consequence of bow flare slamming is that it occurs further into the sag cycle, resulting in greater combined moments, as is shown in the figure.

4.3 Weibull analyses of hull girder whipping

In order to improve upon the Navy's current methodology for predicting lifetime maximum seaway loads, a statistical analysis of the variables associated with slam induced whipping of the hull girder was performed (Sikora 1992). To provide the means of extrapolating whipping loads it was first necessary to determine the probability distribution of the measured whipping events. For this analysis it was

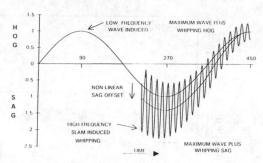

Figure 13 - Phase Angle between Hull Girder Bending Due to Vertical Ordinary Wave and Whipping

assumed that the distribution of the magnitudes of the initial whipping peaks could best be described by a three parameter Weibull distribution. The Weibull distribution may be expressed as:

$$P(M) = 1 - e^{-(\frac{x-x_0}{\theta-x_0})^\beta}$$

where, P(M) = of probability of exceedance; x = the whipping moment; x_0 = the threshold value, below which there were no measured data; β = the Weibull shape parameter or slope; and θ = the characteristic value, which corresponds to 63.2 percentile of the distribution.

The threshold, slope, and characteristic values (x_0, β, θ) are the three parameters that define a specific Weibull distribution. It should be noted that a Weibull shape parameter of 1.0 is an exponential distribution, and a shape parameter of 2.0 corresponds to a Rayleigh distribution. Details as to the determination of the Weibull parameters is given in reference (Richardson 1987).

Typical Weibull data plots are shown as Figure 14 for both vertical and lateral whipping for a number of trial conditions. Note that in these plots the data have been non-dimensionalized by dividing the magnitudes by the characteristic value. A summary of the Weibull parameters and the rate of whipping events is given in Table 2 for all full scale test conditions which contained a sufficient number of whipping events for the Weibull analysis. Where multiple data runs existed for the same test conditions, these runs were combined to

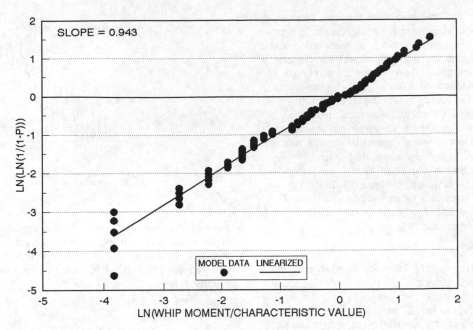

FIGURE 14 TYPICAL WEIBULL PLOTS OF MIDSHIP WHIPPING FROM CG-47 CLASS MODEL TESTS
a) VERTICAL BENDING, SEA STATE 6, 20 KNOTS, HEAD SEAS

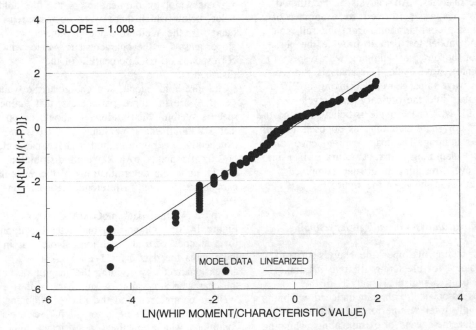

FIGURE 14 TYPICAL WEIBULL PLOTS OF MIDSHIP WHIPPING FROM CG-47 CLASS MODEL TESTS
b) VERTICAL BENDING, HURRICANE CAMILLE, 5 KNOTS, HEAD SEAS

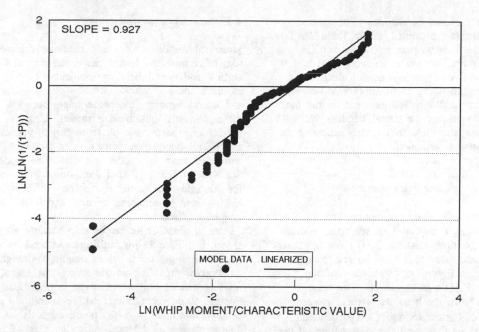

FIGURE 14 TYPICAL WEIBULL PLOTS OF MIDSHIP WHIPPING FROM CG-47 CLASS MODEL TESTS
c) LATERAL BENDING, SEA STATE 6, 20 KNOTS, HEAD SEAS

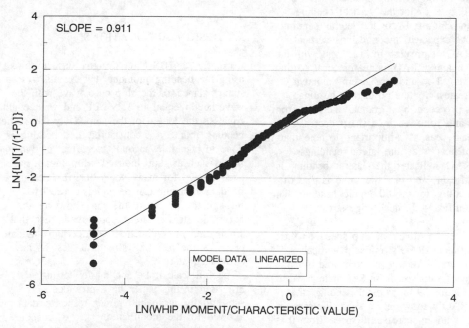

FIGURE 14 TYPICAL WEIBULL PLOTS OF MIDSHIP WHIPPING FROM CG-47 CLASS MODEL TESTS
d) LATERAL BENDING, HURRICANE CAMILLE, 5 KNOTS, HEAD SEAS

increase the data populations. The most significant fact presented in the Table 2 is that the Weibull shape parameters for vertical whipping are, in most cases, very close to 1.0, representative of an exponential distribution. For lateral bending there appears to be more scatter in the Weibull slopes, but for the more severe conditions the lateral bending Weibull slopes are also close to 1.0, also indicating an exponential distribution.

4.4 Response amplitude operators

To address the basic assumption that the RAO's are constant irrespective of sea state, midship vertical bending moment RAO's for Sea States 5, 6, 7, and Hurricane Camille are plotted together in Figure 15, as are the midship lateral bending RAO's in Figure 16. The close agreement of the head sea vertical moment RAO's of Figure 15, regardless of sea state, clearly show that the linear assumption of the RAO's may be considered validated. The same conclusion may be drawn for the 60 degree lateral moment RAO's as shown in Figure 16.

The influence of ship heading on the midship vertical and lateral bending moment RAO's is shown in Figure 17 for the vertical bending moment and Figure 18 for the lateral bending moment. As expected, the midships vertical moment RAO is greatest in head seas with the 15 degree heading RAO is nearly equal to that of head seas. The midships lateral moment RAO is greatest for the 60 degree heading, which was preceded by a gradual increase from head seas and followed by a sharp decrease toward beam seas. It is important to note that in the case of head seas, the lateral bending moment RAO indicates that lateral bending is not insignificant, and that the same is true of vertical bending for the 60 degree heading. This may be significant for deck edge stress calculations.

To examine the effects of ship speed on the bending moment RAO's, midship vertical and lateral moment RAO's were compared for a wide range of ship speeds in Sea State 6. As is shown in Figure 19 for vertical bending, the head sea RAO's show only a slight increase in magnitude with an increase in speed from 0 to 25 knots. For lateral bending, the 60 degree heading RAO's of Figure 20 also use only a slight increase in the RAO magnitude for a ship speed increase from 0 to 15 knots.

4.5 Phase angle determination

Structural responses of a ship structure are the sum of vertical and lateral responses due to vertical and lateral bending moments. To calculate these combined stresses due to vertical and lateral bending, the phase angle between these moments must be determined. Cross correlation spectra and phase angles of vertical and lateral bending responses were therefore computed from the model test data. The model Sea State 5, 5 and 10 knot data runs were used for this study due to the consistency of the wave heading and to eliminate vertical and lateral whipping effects.

Plots of phase angle versus ship heading are shown in Figure 21 for midships. In head seas this phase angle for midships bending is roughly 45 degrees, providing an indication that lateral bending cannot be ignored, even in head seas. For the 60 degree heading, where lateral bending is a maximum, the phase angle is approximately 30 degrees, indicating that vertical bending cannot necessarily be ignored. The conclusion to be drawn here is that maximum combined stresses (particularly at the deck edge) may occur for a heading other than head seas.

5.0 DESIGN APPLICATIONS

Currently, the US Navy designs surface ships using the bending moment that results from a static balance of the ship on a wave with the wave height equal to $1.1\sqrt{LBP}$ and wavelength equal to the length of the ship. Only the vertical bending moment is considered and nothing is done to take into account the effect of dynamic forces such as slam induced whipping. It has been known for many years that this design bending moment can be and is exceeded throughout the life of the ship, however the allowable stresses are proportioned such that when used in conjunction with certain standard practices the resulting structural designs have been adequate.

The full scale trials and model testing discussed in this paper have provided substantially more hull girder response data and allowed for the modification of the procedures for calculating extreme bending moments. The data that were obtained also gave indications that in addition to the vertical bending moments, the ships were also experiencing

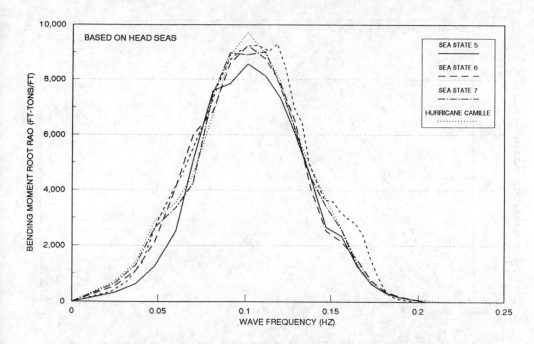

FIGURE 15 CG-47 CLASS MODEL BASED VERTICAL BENDING MOMENT ROOT RAO'S AS A
FUNCTION OF SEA STATE

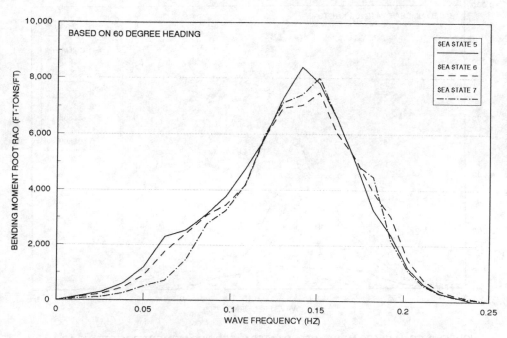

FIGURE 16 CG-47 CLASS MODEL BASED LATERAL BENDING MOMENT ROOT RAO'S AS A
FUNCTION OF SEA STATE

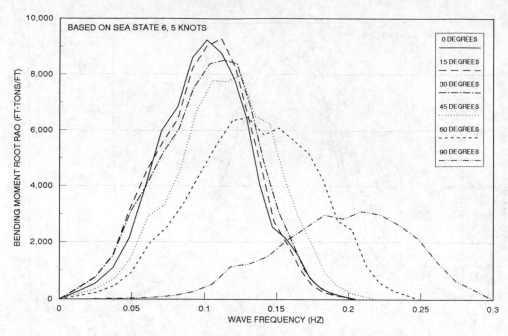

FIGURE 17 CG-47 CLASS MODEL BASED VERTICAL BENDING MOMENT ROOT RAO'S AS A
FUNCTION OF SHIP HEADING

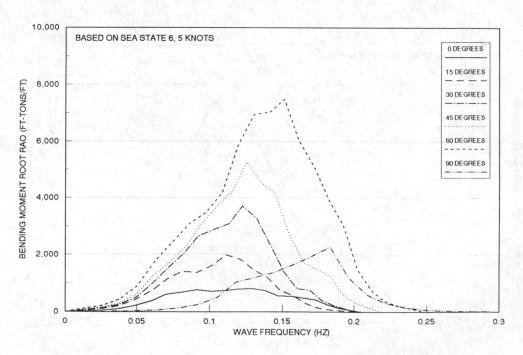

FIGURE 18 CG-47 CLASS MODEL BASED LATERAL BENDING MOMENT ROOT RAO'S AS A
FUNCTION OF SHIP HEADING

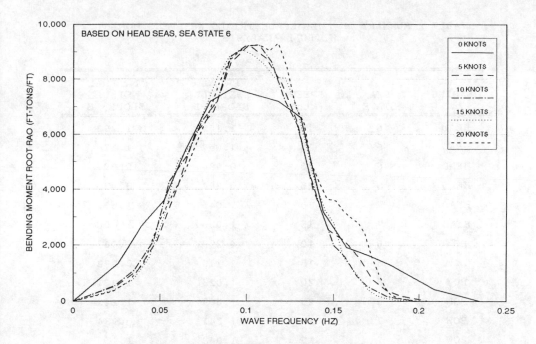

FIGURE 19 CG-47 CLASS MODEL BASED VERTICAL BENDING MOMENT RAO'S AS A
FUNCTION OF SHIP SPEED

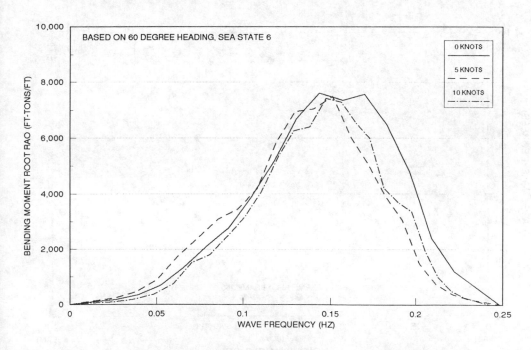

FIGURE 20 CG-47 CLASS MODEL BASED LATERAL BENDING MOMENT ROOT RAO'S AS A
FUNCTION OF SHIP SPEED

TABLE 2 SUMMARY OF FULL SCALE TRIALS WHIPPING MOMENT
WEIBULL ANALYSIS

HEADING	SEA STATE	SPEED (kts)	VERTICAL WEIBULL SLOPE, β_v	LATERAL WEIBULL SLOPE, β_l
HEAD	6	10	1.155	1.293
HEAD	6	15	1.057	1.124
HEAD	6	20	0.986	0.997
HEAD	6	25	0.984	0.976
HEAD	7	10	1.045	1.092
HEAD	7	15	1.036	1.008
BOW	6	10	4.299	1.967
BOW	6	15	2.110	0.893
BOW	6	20	1.026	1.351
BOW	6	25	1.069	1.242
BOW	7	10	1.241	0.984
BOW	7	15	1.048	1.159
BOW	7	20	1.057	0.867

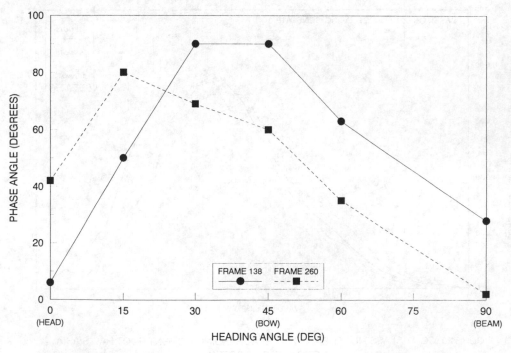

FIGURE 21 PHASE ANGLE BETWEEN VERTICAL AND LATERAL BENDING MOMENTS

186

significant lateral bending moments. These lateral bending moments were not in phase with the vertical, however there is still a lateral component that acts in conjunction with the extreme vertical bending moment. This combination of moments increases the maximum stresses in the outer edges of the hull girder. This could have an impact on the calculation of the fatigue life as well as looking at the extreme stress for yielding and buckling.

Currently the US Navy is conducting an R&D program to develop a new, reliability based design criteria. The objective of this program is to develop probabilistic design methodologies so as to quantify the risk of failure of surface ship hulls under environmental loadings; and to define an acceptable risk of failure under predetermined operational conditions. This program is investigating loads and strength determination as well as reliability based design methodologies. The full scale trials and model testing described in this paper acted as a stepping stone for the loads determination portion of this R&D program.

6.0 CONCLUSIONS/RECOMMENDATIONS

The efforts completed under this program represent a first step in a long and arduous process on collecting and interpreting hydrodynamic loads data. Much more needs to be accomplished before we can realize our goal of reliability based design. However, based upon our initial findings, the following conclusions have been reached:

1. A trapezoidal distribution with a constant maximum value between 40 and 60 percent of ship length provides the best fit for both vertical and lateral whipping moment distributions.

2. Lateral loads can be a major contributor to overall loading of a ships structure, even in head seas.

3. The application of Weibull analysis techniques provides the designer with a rigorous method to determine lifetime whipping moments.

4. Based upon Weibull analysis of model and full scale test data, vertical whipping can be best represented by an exponential distribution. Although more scatter is present in the slopes of the lateral whipping data, the more severe conditions appear to follow an exponential distribution as well.

5. For this hull form, the onset of whipping will occur at approximately 220 degrees into the ordinary wave cycle. Occurring well into the sag cycle, it is believed that this phasing is the result of bow flare slamming as opposed to hull bottom slamming.

6. Our technical findings suggest to us the need for the following future R&D efforts:

a) Before major revisions to existing structural design criteria can be made, it will be necessary for the design community to understand the effects that varying hull form characteristics (e.g., length, beam) have on longitudinal distribution. To aid in this assessment, it is recommended that the locations for primary bending strain gages be standardized. In keeping with recent trials where significant numbers of strain gages were installed, we recommend that all future trials have strain gages installed at 0.2, 0.3, 0.5, 0.7 and 0.8 along the length of the ship. Furthermore, for measurements of lateral bending, it is recommended that for each longitudinal location, gages be placed at the deck edge, both port and starboard. Data from this type of gage installation will allow analysts to develop vertical and lateral trials based RAO's, establish the longitudinal distribution of whipping moments, determine resultant combination as a structural response and, when used in conjunction with ship operational requirements, the ability to determine the distribution of lifetime wave induced loads.

b) Concurrent with our analysis of model test data, an assessment of the state of the art in predicting hydrodynamic loads was performed. Results of that assessment indicated that most frequency domain codes could predict bending moments RAO's with a fair degree of accuracy (NAVSEA 1992). However, the most advanced non-linear time domain computer codes were limited in their ability to predict responses in extreme conditions (Sikora et al 1992). Accordingly, it is felt that, at this time, the advanced non-linear technology is not sufficiently mature enough to support development of probabilistic design methods. Additional enhancements and validation work is required if these newer methods are to be utilized.

ACKNOWLEDGEMENTS

The authors would like to thank the officers and crew members of USS MONTEREY, whose dedication and cooperation resulted in the collection of an invaluable set of full scale loads data. Similar thanks are offered to Captain MacKenzie, of the Aegis Program Office, for providing us the opportunity to use a ship of the CG 47 Class.

In addition to the those involved with the full scale trials effort, the authors would like to thank Mr. Jerry Sikora, Mr. James Rodd and Mr. Robert Michaelson of DTMB's Structural Structures and Protection Department for their part in performing numerous finite element analyses and designing the pvc model. Finally, the authors would like to thank Mr. Jim Kuny, Mr. Tom Brady, Mr. Lewis Motter, Mr. Harry Jones and Mr. Rich Bishop of DTMB for performing the extensive series of model tests and its subsequent analysis.

REFERENCES

AEGIS program office, informal communication, January 1991.

Bishop R.C., Jones H.D., Thomas W.L., Private Communication, March 1992.

NAVSEA Technical Note No. 051-55W-TN-0094, "Comparative analysis of analytic predictions for hydrodynamic loads in a seaway", February 1992.

NAVSEA Technical Report 100-55Y-TR-0005, "Reliability based design criteria for surface ship structures, project management plan", June 1992.

Richardson, William M., "A probability based load estimation technique for ship structure design and technology evaluation", Naval Engineers Journal, May 1987.

Rodd James L., Michaelson Robert W., Informal communication, July 1992.

Sikora J. P., Informal communication, 1992.

Sikora Jerome P., Dalzell John F., Private communication, October 1992.

Sikora Jerome, Hess Paul, Synder Brian, DTRC report SSPD-91-173-43, Private communication, April 1991.

Slamming and whipping of ships

Hydroelasticity in Marine Technology, Faltinsen et al. (eds) © 1994 Balkema, Rotterdam, ISBN 90 5410 387 6

Wave-induced springing and whipping of high-speed vessels

P. Friis-Hansen, J. Juncher Jensen & P. Terndrup Pedersen
Department of Ocean Engineering, The Technical University of Denmark, Lyngby, Denmark

ABSTRACT: Novel materials used for high-speed vessels normally have lower modulus of elasticity than the conventionally used steels. Therefore, for large fast ships the lowest natural frequencies of the global hull modes can be relatively low compared to the frequency of wave encounter.

It is the purpose of the present paper to investigate the effect of hull flexibility on the wave-induced structural loading on high-speed ships.

Especially it is the purpose to determine whether there is an upper size of GRP and aluminium mono-hulls caused by hydroelastically induced vibrations.

A theory is described and results presented for wave-induced ship hull vibrations in stationary stochastic seaways. The calculation is performed within the framework of a non-linear, quadratic strip theory formulated in the frequency domain, so that the excitation of springing is caused partly by linear resonance and partly by non-linear excitation. The importance of springing on both extreme value predictions and fatigue damage accumulation is investigated.

Besides this continuous excitation also transient loads, so-called whipping vibrations, due to impact slamming is considered. Here special emphasis is given to the combination of the continuous wave-induced response and the high-frequency slamming induced response.

The wave-induced response becomes non-Gaussian in stationary stochastic seaways because of the non-linearities. In the present approach the statistical properties of the hull girder response are described by the first four statistical moments through a Hermite series approximation to the probability density function.

1 INTRODUCTION

The past years have seen a rapidly growing interest in fast ships for both cargo and passager transportation. The various designs considered have included catamarans, SES, SWATH etc. but recently attention has been focused on fast mono-hull displacement ships, especially through the detailed design study made by Sumitomo Heavy Industries, [1]. That study addresses many of the problems encountered in a mono-hull design (resistance and propulsion, sea-keeping, and longitudinal strength). From the experimental results obtained using a model with an elastic backbone, a slight springing vibration is observed. It was concluded in [1] that this could be of importance for the fatigue strength of the hull, but not for the extreme values. Comparisons between model experiments and calculations also indicated that conventional low speed strip theories yield reasonable results even at Froude numbers up to 0.6. Strip theory formulations specially developed for fast ships have been applied in other interesting studies [2], [3] but applied to very simplied hull forms.

In the present paper a parameter study is performed to determine the relative importance of hull flexibility and ship length on the springing and whipping response of fast mono-hull vessels.

The calculations are done within the framework of a non-linear strip theory approach. Based on the obtained results some conclusions are made on the importance of hull flexibility on the stress response - especially with respect to the use of different construction materials (steel, aluminium or GRP) for the hull beam.

2 DYNAMIC HULL CHARACTERISTICS OF FAST SHIPS

Springing and whipping responses are strongly influenced by the dynamic elastic behaviour of the hull girder.

Springing refers to the continuous vertical hull vibration resulting from the unsteady wave-induced hydrodynamic pressure field acting on the hull. There are two main excitation sources for springing. One is excitation of the lowest hull natural frequency by the harmonic components in the high frequency part of the wave energy spectrum. This part of the springing response is linear with the wave heights. The other excitation source for springing is the higher order non-linear components of the wave-induced hydrodynamic forces which can excite hull girder vibration frequencies. In both cases it is the two-node vibration mode which dominates the hull girder response.

The global whipping response considered in this paper is caused by the pressure impulse generated when a transverse section of the hull re-enters the water after bottom emergence. The midship vertical whipping bending moments resulting from these impacts will also be determined by the normal mode approach. Again the two node vibration mode dominates the response, hence, only a few modes need to be calculated in order to determine the global whipping response amidship.

In the following we shall show how the two-noded hull girder natural frequency can be expected to vary with ship length and construction material for ships built according to classification rules.

For mono-hull high speed light craft with length L less than 100 m, breadth B, and block coefficient C_B the tentative rules from Det Norske Veritas dated January 1993 show that the sum of the still-water and the wave-induced vertical hogging bending moment amidship can be expressed as

$$M_{tot, hog} = 2.39 \cdot 10^{-3} \, \rho \, g \, L^3 \, B \, C_B \qquad (2.1)$$

where ρ is the mass density of water and g the acceleration of gravity. Similarly, the total sagging bending moment can be expressed as

$$M_{tot, sag} = 1.59 \cdot 10^{-3} \, \rho \, g \, L^3 \, B \, (C_B + 0.7) \qquad (2.2)$$

provided that the maximum service speed in knots is larger than $3.0\sqrt{L}$. It is seen that for realistic block coefficients the vertical sagging bending moment is largest.

A characteristic value for the required vertical cross-sectional moment of inertia I_v can be found from the following expression:

$$I_v \approx \frac{M_{tot, sag}}{\sigma_a} \frac{D}{2} \qquad (2.3)$$

where σ_a is the maximum allowable hull girder stress and D is the depth of the hull.

For a free-free prismatic Bernoulli-Euler beam with mass M, bending stiffness EI_v, and length L the natural frequencies in rad/s can be determined from:

$$\Omega_i = \alpha_i^2 \sqrt{\frac{EI}{ML^3}} \quad ; \quad i = 2, 3, \ldots \qquad (2.4)$$

where α_i are determined as roots from the transcendental equation

$$\cos \alpha_i \, \cos h\alpha_i = 1$$

yielding $\alpha_2 = \alpha_{min} = 4.730$.

In order to transform expression (2.4) into an empirical formula for calculation of vertical hull girder vibration frequencies, it is necessary to introduce a multiplication factor which quantifies the effect of the non-homogeneous distribution of mass and bending stiffness, the shear deflections and rotatory inertia. Based on often applied empirical expressions for two-noded vertical vibration for conventional ships, we here use a multiplication factor for these effects equals 0.78.

The total apparent mass M of the vibrating hull will be taken as

$$M = \rho \, C_B B L T \left(1.2 + \frac{B}{3T} \right) \qquad (2.5)$$

where T is the draught and the term in brackets models the inertia effect of the hydrodynamic forces.

By combining Eqs. (2.2) - (2.5) the natural frequency Ω_2 is obtained for the two-noded vertical vibration mode as

$$\Omega_2 = 0.492 \sqrt{\frac{g \, E D \, (C_B + 0.7)}{\sigma_a \, C_B L T \left(1.2 + \frac{B}{3T} \right)}} \qquad (2.6)$$

For the fast mono-hull ships we shall consider here we will assume that $B/T = 4.5$, $D/T = 3.0$ and $C_B = 0.5$. This leads to

$$\Omega_2 = 0.803 \sqrt{\frac{E g}{\sigma_a L}} \qquad (2.7)$$

For ship with a steel hull where Young's modulus $E = 2.1 \cdot 10^{11}$ Pa and $\sigma_a \approx 200 \cdot 10^6$ Pa we get

$$\Omega_{2 \, steel} \approx 25 \sqrt{\frac{g}{L}} \quad (\, rad/s \,) \qquad (2.8a)$$

With a hull made of high strength aluminum with $E = 7 \cdot 10^{10}$ Pa and $\sigma_a \approx 200 \cdot 10^6$ Pa the two noded vibration mode has the approximate frequency

$$\Omega_{2 \, al} \approx 15 \sqrt{\frac{g}{L}} \quad (\, rad/s \,) \qquad (2.8b)$$

Finally for a GRP hull with $E = 1.1 \cdot 10^{11}$ Pa and $\sigma_a = 70 \cdot 10^6$ Pa we get

$$\Omega_{2 \, grp} \approx 10 \sqrt{\frac{g}{L}} \quad (\, rad/s \,) \qquad (2.8c)$$

Thus, in the following we shall assume that the variation with ship length of the lowest natural frequencies for vertical hull vibration can be characterized by the expression (2.8a)-(2.8c) for steel, aluminium, and GRP hulls, respectively.

3 SPRINGING ANALYSIS OF FAST SHIPS

The calculations are performed within the framework of a non-linear, quadratic strip theory formulated in the frequency domain. Included are non-linear effects due to changes in added mass, hydrodynamic damping and water line breadth with sectional immersion in waves. This section is limited to continuous excitations from the waves. Transient, so-called whipping vibrations due to slamming loads, are considered in the next section.

The theoretical background was developed 15 years ago [4], [5]. Recent extensions are given [6] from which some of the advantages and limitations of the method should be quoted:

- The method is formulated in the frequency domain. This implies that the correct frequency dependence of the added mass and damping is used, also in a stochastic seaway.

- The full quadratic transfer functions are determined for a ship sailing in obligue seaways. Thus the Newman approximation is not needed in the stochastic analysis.

- Bow flare slamming is partly included.

- The formulation is based on a semi-empirical generalization of the linear strip theory approach. Therefore the linear part of the solution is identical to that obtained from linear strip theory. The linear theory can be either the Gerritsma-Beukelman or the Salvesen-Tuck-Faltinsen strip theory.

- Using a two-dimensional formulation implies an extremely fast calculation of the quadratic transfer functions. This makes the procedure useful also in design studies.

- The procedure is limited to the vertical responses (heave, pitch, vertical bending moments and vertical shear forces). Linear asymmetric responses can, however, be included.

- Hull vibrations (springing) are included with the hull modelled as a non-prismatic Timoshenko beam.

- Statistical analysis in stationary, stochastic seaways can be done using either a Charlier series representation with the full joint distribution of the response and its time derivative or an approximate Hermite series approach.

- Comparisons of the quadratic sectional loads in regular sea with model tests show reasonable agreement.

- Comparisons of the statistical predictions in moderate, short term stochastic seaways with full scale measurements show that the method is able to predict the difference between the hogging and sagging wave-induced bending moments with good accuracy.

- Comparisons with computer-expensive time-simulation procedures show reasonable agreements in regular and irregular seas.

- Fatigue damage predictions in stationary stochastic seaways is included. It is found that the non-linear contributions to the fatigue damage are dominated by terms proportional to the kurtosis and to the skewness squared.

In the following a short outline of the quadratic strip theory is given. For further details, see [4]-[6].

3.1 *Equations of motions*

The lower modes of ship hull vibrations can generally be determined quite accurately by modelling the hull as a non-prismatic Timoshenko beam. The equation of motions in the vertical plane then becomes:

$$\frac{\partial}{\partial x}\left[EI\left(1+\eta\frac{\partial}{\partial t}\right)\frac{\partial\varphi}{\partial x}\right]+$$

$$\mu GA\left(1+\eta\frac{\partial}{\partial t}\right)\left(\frac{\partial w}{\partial x}-\varphi\right)=m_s r^2\frac{\partial^2\varphi}{\partial t^2}$$

$$\hspace{8cm}(3.1)$$

$$\frac{\partial}{\partial x}\left[\mu GA\left(1+\eta\frac{\partial}{\partial t}\right)\left(\frac{\partial w}{\partial x}-\varphi\right)\right]=$$

$$m_s\frac{\partial^2 w}{\partial t^2}-F(x,t)$$

in which $EI(x)$ and $\mu GA(x)$ are the vertical bending and shear rigidities, respectively, $\varphi(t,x)$ is the slope due to bending, $w(t,x)$ is the total deflection including rigid body displacement, η is an internal damping coefficient, and x is a longitudinal coordinate in a xyz-coordinate system fixed with regard to the undisturbed ship so that the z-axis is in the vertical direction. Finally, $m_s(x)$ is the hull mass per unit length, $m_s r^2(x)$ is the mass moment of inertia about the horizontal y-axis, and $F(x,t)$ is the external force per unit length which is nonlinear in w.

The boundary conditions to Eqs. (3.1) express that the bending moments and shear forces are zero at the ends of the ship.

The solution to Eqs. (3.1) and the associated boundary conditions is approximated as a series in the form:

$$\varphi(x,t) = \sum_{i=0}^{N} u_i(t)\, \alpha_i(x)$$

$$w(x,t) = \sum_{i=0}^{N} u_i(t)\, v_i(x)$$

(3.2)

where $u_i(t)$ are coefficients to be determined and where $\{v_i(x),\, \alpha_i(x)\}$ are the eigenfunctions to the homogeneous, self-adjoining part of Eqs. (3.1).

If we substitute Eqs. (3.2) into the differential equations (3.1), and make use of orthonormality relations, we obtain the following set of equations of motion:

$$\ddot{u}_j + \eta\, \Omega_j^2\, \dot{u}_j + \Omega_j^2\, u_j = \int_0^L v_j\, F\left(x, t, \sum_{i=0}^{N} u_i v_i\right) dx$$

(3.3)

$$j = 0,1,2,...,N$$

where over-dot denotes differention with respect to time t.

3.2 Forcing function

Calculation of the hydrodynamic forces will be based on the time derivative of the momentum of the added mass of water surrounding the hull. In addition to forces due to a change in momentum of the added mass of water, a damping term and a restoring term, both dependent on the relative motion are included. Thus the force per unit length of the hull acting at position x is taken in the form:

$$F(x,t) = -\left[\frac{D}{Dt}\left\{ m(\tilde{z},x)\, \frac{D\tilde{z}}{Dt} \right\} + \right.$$

$$\left. N(\tilde{z},x)\, \frac{D\tilde{z}}{Dt} + \int_{-T}^{-\tilde{z}} B(z,x)\, \frac{\partial p}{\partial z}\bigg|_{z+w} dz \right]$$

(3.4)

where the difference between the absolute displacement of the ship in the vertical direction, $w(x,t)$, and the surface of the ocean, $h(x,t)$, corrected to account for the Smith effect, is denoted by $\tilde{z}(x,t)$. Furthermore, m is the added mass per unit length, and N is the damping. The operator D/Dt is the total derivative with respect to time t

$$\frac{D}{Dt} \equiv \frac{\partial}{\partial t} - V\, \frac{\partial}{\partial x}$$

(3.5)

where V is the forward speed of the ship. The breadth of the ship is denoted by $B(z,x)$ and the draft by $T(x)$. Finally, p is the Froude-Krylov fluid pressure.

If we neglect the $\tilde{z}$ dependence in m, N, and B, the forcing function Eq. (3.4) corresponds to a linear strip theory. However, here we shall evaluate $F(x,t)$ by a perturbational method, taking linear and quadratic terms in the relative displacement $\tilde{z}$ into account, and hence in the displacement of the hull w and the wave surface elevation h. To do this we evaluate the waterline breadth B, the added mass m, and the damping coefficient N around $\tilde{z} = 0$:

$$B(\tilde{z},x) \simeq B(0,x) + \tilde{z}\, \frac{\partial B}{\partial \tilde{z}}\bigg|_{\tilde{z}=0} \equiv$$

$$B_0(x) + \tilde{z}(x,t)B_1(x)$$

(3.6)

$$m(\tilde{z},x) \simeq m(0,x) + \tilde{z}\, \frac{\partial m}{\partial \tilde{z}}\bigg|_{\tilde{z}=0} \equiv$$

$$m_0(x) + \tilde{z}(x,t)m_1(x)$$

(3.7)

$$N(\tilde{z},x) \simeq N(0,x) + \tilde{z}\, \frac{\partial N}{\partial \tilde{z}}\bigg|_{\tilde{z}=0} \equiv$$

$$N_0(x) + \tilde{z}(x,t)N_1(x)$$

(3.8)

The wave surface elevation h and the pressure p are also expressed as sums of a linear term and a quadratic term so that we have, for instance

$$h(x,t) = h^{(1)} + h^{(2)}$$

or

$$h(x,t) = \sum_{i=1}^{n} a_i \cos \psi_i +$$

$$\frac{1}{4} \sum_{i=1}^{n} \sum_{j=1}^{n} a_i a_j \Big[(k_i + k_j) \cos(\psi_i + \psi_j) -$$

$$|k_i - k_j| \cos(\psi_i - \psi_j) \Big]$$

(3.9)

where a_i are wave amplitudes, and

$$\psi_i = -k_i(x - Vt\, \cos\phi) - \omega_i t + \theta_i$$

Here k_i denotes the wave number in the x-direction, and the wave frequency $\omega_i = \sqrt{gk_i}$. Finally, the ships heading angle is ϕ, measured such that head seas correspond to 180 degrees.

Similarly, the deflections of the ship hull are expressed as a sum of a linear part and a quadratic part

$$w = w^{(1)} + w^{(2)} \tag{3.10}$$

These assumptions lead to

$$F(x,\bar{z},t) = F^{(1)} + F^{(2)} \tag{3.11}$$

where $F^{(1)}$ contains linear terms in the displacement $w^{(1)}$ and the wave surface elevation $h^{(1)}$, and $F^{(2)}$ terms which are quadratic in these quantities as well as terms linear in $w^{(2)}$ and $h^{(2)}$.

Substitution of Eq. (3.11) into Eq. (3.3) yields two sets of linear, non-homogeneous differential equations governing the linear, $u_j^{(1)}$, and quadratic, $u_j^{(2)}$, components of the deflection coefficients u_j in Eq. (3.2). Then the sectional forces and moments are determined in the form of the linear and the full quadratic transfer functions by integration of the inertia and hydrodynamic forces per unit length.

3.3 Short term statistics

Only stochastic responses in long-crested (unidirectional) seas are considered.

To model a stationary stochastic seaway, the wave amplitude a_i and the phase lag θ_i are chosen such that $\xi_j = a_j \cos\theta_j$ and $\xi_{j+n} = a_j \sin\theta_j$ are joint normally distributed with θ_j uniformly distributed in $[0,2\pi]$, and such that the half mean-squared amplitudes $\frac{1}{2}a_j^2$ are equal to the wave energy in the associated range of fre-quencies.

The theory assumes a uniform discretization of the encounter frequency $\bar{\omega}$, that is $\bar{\omega}_i - \bar{\omega}_{i-1} = \Delta\bar{\omega}$ for $i = 2,3,...,n$. The independent variables ξ_j and ξ_{j+n} $(j = 1,2,..,n)$ will then have zero mean and a variance

$$V_j = V_{j+n} = \bar{S}(\bar{\omega})\Delta\bar{\omega} \tag{3.12}$$

where $\bar{S}(\bar{\omega})$ is the wave spectrum formulated in terms of the frequency of encounter.

Now, on the basis of this stochastic description of the first-order wave elevation, an expression for instance for the wave-induced midship bending moment $M(x,t)$ at time $t = 0$ can be written in the form:

$$\frac{M(x,0)}{\lambda} = \sum_{j=1}^{2n} \lambda_j \chi_j + \varepsilon \sum_{j=1}^{2n} \sum_{k=1}^{2n} \Lambda_{jk} \chi_j \chi_k \tag{3.13}$$

Here we have introduced as independent stochastic variables

$$\chi_j = \frac{\xi_j}{\sqrt{V_j}}$$

The dependent variable $M(x,0)$ is normalized by the standard deviation λ of the first-order contribution to the wave-induced bending moment. The non-linearity parameter ε are defined so that

$$\sum_{j=1}^{2n} \sum_{k=1}^{2n} \Lambda_{jk}\Lambda_{jk} = 1 \tag{3.14}$$

From Eq. (3.13) the spectral density of the response can be calculated. However, often the stochastic properties are reasonable well described by the first four statistical moments of the response. These are the mean value μ, the standard deviation σ, the skewness κ_3, and the kurtosis κ_4 given by

$$\mu = K_1 = \lambda\varepsilon \sum_{i=1}^{2n} \Lambda_{ii}$$

$$\sigma^2 = K_2 = \lambda^2(1 + 2\varepsilon^2)$$

$$\kappa_3\sigma^3 = K_3 =$$
$$\lambda^3 \left(6\varepsilon \sum_{i=1}^{2n}\sum_{j=1}^{2n} \lambda_i \Lambda_{ij} \lambda_j + \right.$$
$$\left. 8\varepsilon^3 \sum_{i=1}^{2n}\sum_{j=1}^{2n}\sum_{k=1}^{2n} \Lambda_{ij} \Lambda_{jk} \Lambda_{ki} \right) \tag{3.15}$$

$$(\kappa_4 - 3)\sigma^4 = K_4 =$$
$$\lambda^4\, 48\,\varepsilon^2 \left(\sum_{i=1}^{2n}\sum_{j=1}^{2n}\sum_{k=1}^{2n} \lambda_i \Lambda_{ij} \Lambda_{jk} \lambda_k + \right.$$
$$\left. \varepsilon^2 \sum_{i=1}^{2n}\sum_{j=1}^{2n}\sum_{k=1}^{2n}\sum_{l=1}^{2n} \Lambda_{ij} \Lambda_{jk} \Lambda_{kl} \Lambda_{li} \right)$$

where the expressions for the cumulants K_i follow from Eq. (3.13).

3.4 Fatigue damage

Springing can be of more significance for the fatigue damage of the hull than for the extreme loads. This is so partly because of the springing contribution increases the mean upcrossing rate ν_0 and partly because of the non-linearities usually lead to a non-Gaussian response with a kurtosis κ_4 greater than 3.

For a symmetric, slightly non-Gaussian stress response the fatigue damage per cycle d can be estimated based on the Palmgren-Miner rule by, see [7]:

$$d = \frac{\Gamma\left(1 - \frac{m}{2}\right)}{a\left(2\sqrt{2}\,\sigma\right)^m} \left(1 + \frac{1}{24}\left(m^2 + m\right)\left(\kappa_4 - 3\right)\right) \qquad (3.16)$$

where $\Gamma(\bullet)$ is the Gamma function, a and m the scale and slope parameters in the S-N curve and σ and κ_4 the standard deviation and kurtosis of the stress response. A consistent analysis of non-symmetric responses, [8], shows that the fatigue damage also could depend on the skewness squared. However, due to lack of experimental proof of this, Eq. (3.16) will be used here. Eq. (3.16) assumes a narrow-band spectral density. Empirical corrections to account for finite spectral bandwidth are performed using the well known rainflow correction factor γ due to Wirsching and Light [9]. Thus, the accumulated damage D during a time periode τ in a stationary sea state becomes

$$D = \tau \gamma \nu_0 d \qquad (3.17)$$

3.5 Numerical results.

The importance of the springing contribution relative to the wave-induced loads will be illustrated by a numerical example. The body plan of a fast naval ship with smooth sectional contours without knuckle lines near the still water line is used as the parent hull form.

Two investigations are performed. In the first the lines are scaled uniformly in all three dimensions to obtain a ship with length $L = 100$ m between the perpendiculars. The variation in the springing contribution with the two-noded natural frequency Ω_2 is then determined for values of Ω_2 ranging from resonance with the peak spectral response period to infinity as for a rigid hull.

The second set of calculations examine the variation in springing with length of the ship. Here only the three natural frequencies given by Eqs. (2.8a-c) will be used, as they approximately represent the frequencies to be expected for ships built of steel, aluminium, and GRP materials, respectively.

In all calculations the same stationary seaway is used, defined by a Pierson-Moskowitz wave spectrum with a significant wave height $H_s = 2$ m and a zero crossing period $T_z = 5$ s. Furthermore, only head sea is considered with two values of the forward speed $V = 15.43$ m/s (30 kn) and $V = 20.57$ m/s (40 kn). This should represent reasonable rough operational conditions for fast ships operating in short sea voyages.

Regarding the numerical calculations Lewis form transformations have been used for the two-dimensional added mass and damping terms. The step

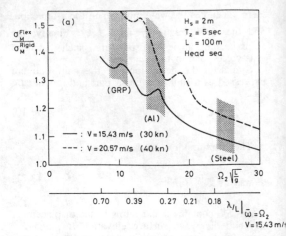

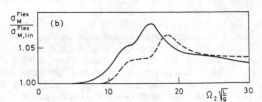

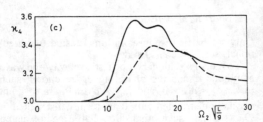

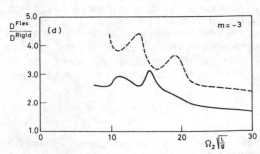

Figure 1. *Relative importance of springing on the midship bending moment for a ship a length $L = 100$ m sailing head sea in a stationary seaway with $H_s = 2$ m and $T_z = 5$ s as function of the two-noded natural frequency Ω_2. (a) Influence of springing on the standard deviation. (b) Influence of non-linear effects on springing. (c) Kurtosis κ_4. (d) Influence of springing on the fatigue damage.*

196

length $\Delta\bar{\omega}$ in encounter frequency was chosen to be 0.05 rad/s but actually a larger step could have been used. Of course, application of a conventional strip theory formulation, Eq. (3.4) can be questioned at the high Froude numbers in the present calculations. However, it is believed that the trends obtained for the importance of springing and whipping will be quite accurate.

Fig. 1a-c show the results for a ship of length $L =$ 100 m. In Fig. 1a the ratio between the standard deviation σ_M of the midship vertical bending moment calculated with an assumed hull flexibility and that of the rigid hull is given. It is clearly seen that the importance of springing increases with decreasing two-node frequency Ω_2. The wavy curves are partly due to coincidence of the two-noded frequency with wave cancelation frequencies or peaks in the transfer function for the bending moment and partly due to non-linear effects.

Consider for instance $V = 15.43$ m/s. Here the maximum peak value in the transfer function for the midship bending moment occurs at a wave length to ship length ratio $\lambda/L = 0.85$. A second peak is found for $\lambda/L = 0.29$ and in-between these two values, a minimum in the transfer function appears at $\lambda/L = 0.34$. Although somewhat averaged out in the stationary stochastic analysis the effect is clearly reflected in Fig. 1a.

The non-linear effects tend to enhance the wavy behaviour of the dynamic amplification given in Fig. 1a. In Fig. 1b, the standard deviation σ_M of the flexible hull is depicted normalized by the same standard deviation, but calculated using only the linear part of the theory. The lowest values of Ω_2 used in the analysis correspond to $\lambda/L \approx 0.7$ and, therefore, resonance occurs directly with the linear part of the transfer function. Hence, the non-linear contributions become completely negligible. For values of Ω_2 between $10\sqrt{g/L}$ and $20\sqrt{g/L}$, the quadratic terms (which are proportional to the square of the transfer function at encounter frequencies approximately equal to half the two-noded frequency) get significant contributions from the peak in the transfer function. Comparing Fig. 1a and Fig. 1b it is seen that in this range for Ω_2, the non-linear effects are responsible for a non-negligible part of the dynamic amplification due to springing. The results for the two different forward speeds are quite similar apart from a shift in the frequency axis due to the Doppler effect.

Fig. 1a-b can be used to estimate the extreme-values of the wave bending moment in the stationary sea state, using as a first approximation the assumption of a Gaussian response. In this case the extreme-values are directly proportional to the standard deviation σ_M. The non-Gaussian behaviour can also be included, [4]-[6], using the values of the skewness κ_3 and kurtosis κ_4. For the present cases the skewness is very small, less than 0.01, indicating a nearly symmetric probability distribution of the bending moment. The kurtosis κ_4 is shown in Fig. 1c. As discussed above the non-linearities, and hence the deviation from a Gaussian response, are most significant for Ω_2 between $10\sqrt{g/L}$ and $20\sqrt{g/L}$, although, the non-linearities still exist for a rigid hull ($\Omega_2 \to \infty$) with the values $\sigma_M^{Rigid}/\sigma_{M,\,lin}^{Rigid} = 1.015,\ 1.012$ and $\kappa_4 = 3.236,\ 3.145$ for $V = 15.43$ m/s and $V = 20.57$ m/s, respectively.

In the relative low sea state applied in the present analysis - a realistic daily operational condition - the extreme-values are small. However, fatigue damage accumulation can be of greater significance. As seen from Eqs. (3.16) and (3.17) the dynamic amplification due to springing of the fatigue damage depends on the exponent m in the S-N curve and on the mean crossing rate v_0. The exponent is usually taken to $m = -3$ and results for the ratio between the damage accumulation D, Eq. (3.17), calculated with and without hull flexible included are shown in Fig. 1d. Comparing this figure with Fig. 1a, it is seen that for the lower values of Ω_2 the ratio D^{Flex}/D^{Rigid} is approximately equal to $\left(\sigma_M^{Flex}/\sigma_{M,\,lin}^{Rigid}\right)^3$, whereas D^{Flex}/D^{Rigid} is much larger than $\left(\sigma_M^{Flex}/\sigma_{M,\,lin}^{Rigid}\right)^3$ for the more stiffer hull beams. This is because of the time variation of the springing contribution increases with Ω_2 and thereby also increases the total mean upcrossing rate.

The main conclusion which can be deduced from Fig. 1a-d is that if the two-noded frequency Ω_2 becomes less that about $20\sqrt{g/L}$ then significant dynamic amplifications can occur due to springing in a sea state with $H_s = 2$ m and $T_z = 5$ s. Thus, if materials like GRP or aluminium are used as construction materials for the hull beam concern should be given to the importance of springing, especially with regard to the possibility of fatigue damage.

For sea states with higher zero crossing periods the relative importance of springing decreases as the wave energy spectrum moves away from the two-noded frequency.

The next investigation is concerned with the variation of springing with length L of the vessel. Results are given in Fig. 2a-b for $V = 15.43$ m/s and in Fig. 3a-b for $V = 20.57$ m/s. As in Fig. 1a-b, the ratio between the standard deviation of the midship bending moment with and without hull flexibility included is given together with, Fig. 2b and Fig. 3b, the relative magnitude of the non-linearities. Results are shown for the three values of the two-noded frequency, defined by Eqs. (2.8a-c), representing hull beams made of GRP, aluminium and steel.

From Fig. 2a and Fig. 3a, the relative springing contribution is seen to be rather insensitive to the length of the ship. This is because both the two-noded fre- quency Ω_2 and the encounter frequency correspond- ing to the maximum value of the transfer function decreases with increasing length L. Only for $L \leq 75$m and $\Omega_2 = 15\sqrt{g/L}$ a significant increase in the spring- ing contribution is seen due to non-linear effects, see Fig. 2b and Fig. 3b. In accordance with Fig. 1b the non-linarities are largest for this value of Ω_2.

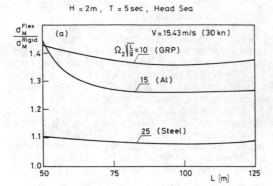

$H = 2m$, $T = 5 sec$, Head Sea

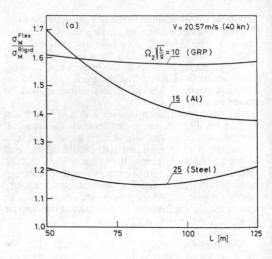

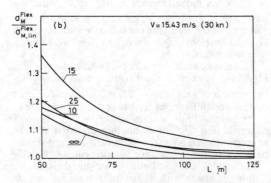

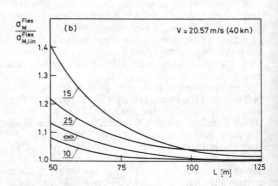

Figure 2. *Relative importance of springing in the midship bending moment for a ship sailing head sea in a stationary sea way with $H_s = 2$ m and $T_z = 5$ s as function of ship length L. V = 15.43 m/s.*
(a) Influence of springing on the standard deviation.
(b) Influence of non-linear effects on springing.

Figure 3. *Relative importance of springing in the midship bending moment for a ship sailing head sea in a stationary sea way with $H_s = 2$ m and $T_z = 5$ s as function of ship length L. V = 20.57 m/s.*
(a) Influence of springing on the standard deviation.
(b) Influence of non-linear effects on springing.

The conclusions which can be drawn from the present analysis are that springing can be of importance for fast ships made of GRP or aluminium and that the dynamic amplification due to springing does not change significantly with the length of the vessel. However, the present study only consider one hull form and one stationary sea state. For a specific design, a more thorough analysis should be made taking due account of the operational profile of the ship.

4 WHIPPING ANALYSIS DUE TO IMPACT SLAMMING

This section relates to analysis of ship hull global extreme loads and stresses arising from the combined effects of slamming- and wave-induced loads. Slamming loads are significant in many types of ocean-going vessels, e.g. those with fine form, low draft, and high speed. The calculation of slamming effects (stresses) requires the consideration of hull flexibility. The maximum slamming loads do not typically occur when the wave induced loads are the largest, and such phasing needs to be considered in the calculation of combined load effects. Another characteristic to be mentioned is the marked non-linearity of slam loads with respect to the wave height, resulting in the hull girder response being significantly different for the hogging and sagging parts of the wave cycle.

Previous investigations of load effect combination in the context of slamming, e.g. those of Ochi and Motter [10] and Ferro and Mansour [11], have all assumed the slamming impact to follow a Poisson pulse process. However, the slamming process is not exactly a Poisson process, because slamming impacts tend to occur in clusters, thereby violating the assumption of mutual independence of slam impacts inherent in a Poisson assumption of impact arrival times. When slamming plays a significant role in design, this non-Poisson character of the slamming process may affect the results significantly. Hence in this study, clustering is accounted for by applying a recently developed procedure, [12].

Time domain based methods for obtaining combined slamming and wave induced response have been used before, e.g. Kaplan [13], Bishop et al. [14]. The procedure applied here is, however, fundamentally different in comparison to existing methods in its more exact and appealing probabilistic treatment of non-linearities.

Moreover, the procedure only requires deterministic calculations of the response for regular sinusoidal waves of selected amplitude and frequency. This implies that the more elaborate calculations of the wave-induced bending moment presented in Section 3 can be used. The only restriction is that the method requires that heave and pitch motions are sufficiently accurately described by linear theory. This is usually so, and consequently almost no practical limitations are imposed on the applicability of the method.

Altogeter the assumptions embedded in the model are:

1. The ship motions are sufficiently described by a linear theory, and not influenced by the slamming impacts.

2. The spectrum of the relative motion is narrow-banded.

3. The dynamic transients are small and evolve slowly, so that the structure responds directly to the local wave sinusoid without significant effects of transients from previous waves.

These assumptions are justifiable for almost all ships.

4.1 Joint distribution of amplitude and frequency

To facilitate the basic idea of [12] the following overview is given:

1. Construct an envelope process for the relative motion at the bow section.

2. Formulate the so-called Slepian model process for the envelope process when the envelope

process exceeds the sectional draft at the bow section (i.e. when slamming will occur).

3. Construct a second-order Taylor expansion to the Slepian model process and find the point of maximum envelope excursion. This point corresponds to the maximum slamming response within a cluster of slamming impacts.

4. Obtain the joint probability distribution of the wave amplitude and the frequency for the waves that give the maximum slamming response.

5. Make a deterministic analysis of the slamming- and wave-induced response due to regular sinusoidal waves of selected amplitude and frequency.

6. Obtain the force statistic by weighing the calculated response by the probability densities of the various pairs of wave amplitude and frequency.

7. Finally, obtain the extreme-value distribution on the basis of the theory for first-passage time distributions in Poisson pulse processes. The mean interarrival times of the pulses are approximated by use of the upcrossing rate of the envelope process, modified for so-called "empty" envelope excursions.

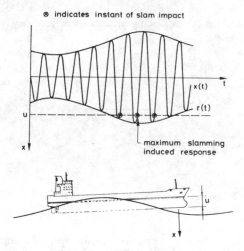

Figure 4 *Narrow-band Gaussian process with associated envelope.*

Given the envelope process of relative motion at the bow section, it is in particular of interest to describe the value of the envelope process for which the slamming-induced response reaches its maximum value within a cluster, see Fig. 4. This point in time is the point of maximum downcrossing velocity at level u of the process $X(t)$, since slam impact severity is a function of this velocity. The phase process is slowly varying, which leads to the occurrence of maximum slamming-induced response at approximately the same point in time as when the envelope excursion reaches its maximum value after an upcrossing of level u.

Hence, the key to the probabilistic model which provides for the combination of low frequency wave-induced hull girder response and the high-frequency slamming response is the calculation of the joint distribution of wave amplitude and frequency. Such calculation is necessary so that the statistics of the combined hull girder response can be obtained by appropriately weighing and summing the response to individual wave sinusoids of given amplitude and frequency.

The above envelope process may be accomplished by a Slepian regression technique. From the technique, it is possible to find approximations for both the maximum value of the envelope process, and the duration of the excursion conditional on an upcrossing of level u. Furthermore, the joint density of wave amplitude and frequency that give rise to the local maximum slamming response within a cluster of slamming impacts may also be found.

The main results are the maximum value of the excursion of the envelope process r_{max} above level u,

$$r_{max} = u + \frac{\xi^2 u}{2(u^2 - \zeta^2 + \varepsilon u \zeta)} \qquad (4.1)$$

and the corresponding instantaneous frequency

$$\omega_{max} = \left(\frac{\zeta}{u} + \frac{1}{\gamma}\right)\sqrt{\mu_2} -$$

$$\left(\frac{\zeta}{u} - \frac{\varepsilon}{2}\right)\frac{2\xi^2 \mu_2}{u^2 - \zeta^2 + \varepsilon u \zeta} \qquad (4.2)$$

in which

$$\gamma = \frac{\lambda_0 \sqrt{\mu_2}}{\lambda_1} \quad ; \quad \varepsilon = \frac{\mu_3}{\mu_2}\sqrt{\mu_2} \qquad (4.3)$$

μ_n is the n'th normalized central spectral moment, that is the spectral moment with respect to λ_1/λ_0:

$$\frac{\lambda_2}{\lambda_0} = \mu_2 + \left(\frac{\lambda_1}{\lambda_0}\right)^2 \qquad (4.4)$$

$$\frac{\lambda_3}{\lambda_0} = \mu_3 + 3\left(\frac{\lambda_1}{\lambda_0}\right)\mu_2 + \left(\frac{\lambda_1}{\lambda_0}\right)^3 \qquad (4.5)$$

ξ is Rayleigh distributed and ζ is standard normally distributed. λ_n is the n'th spectral moment of the response spectrum $S_r(\omega)$ for the process of relative motion at the bow:

$$\lambda_n = \int_0^\infty \omega^n S_r(\omega) \, d\omega \qquad (4.6)$$

4.2 Determination of slamming impact force

The slamming impact force on a vessel will be considered to arise from two sources:

- Bottom impact slamming, giving rise to an initial short duration high pressure transient.

- Momentum transfer slamming, resulting in a longer duration low pressure transient.

The duration of the bottom impact transient is of the order of 1/100 second, while that of the momentum transfer transient, when water is thrown up the vessel sides, is of the order of 1 second.

Theories for the treatment of bottom impact are semi-empirical, while those for the treatment of momentum transfer slamming are more theoretically founded. The relative contributions of the bottom impact and momentum transfer slamming in typical vessels is a matter of controversy, and their correct use (including possible overlap) is in fact not clearly understood. Experimental verifications of the theories are few, but one such correlation is provided by Belik, et al. [15], in which it is shown that the transient responses of momentum slamming is of comparable magnitude to bottom slamming.

In the present analysis, the slamming impact pressure is calculated on the basis of the Ochi and Motter [10] method. Momentum slamming is modelled through the quadratic transfer functions for the wave-induced bending moment.

In the Ochi and Motter method, the maximum impact pressure p is calculated from an expression of the following type

$$p(\dot{r}) = k \dot{r}^2 \qquad (4.7)$$

where k is a dimensional constant that depends on the section shape below 1/10th of the local design draft and $\dot{r}$ is the relative velocity at re-entry.

4.3 Hull girder response

A time-varying slamming impact force, $p(x,t)$, delivered to the forward bottom of the ship produces a vibratory whipping stress. It has been observed in many full-scale trials that although high-frequency accelerations and whipping stresses are excited by slamming impact, only the fundamental modes are generally appreciable, since the higher-mode vibrations die out quickly because of strong structural damping characteristics.

The vertical elastic, transient deflection $w_t(x,t)$ of the hull during a slam is determined by Eq. (3.1) with $F(x,t)$ replaced by the time-varying slamming impact pressure $p(x,t)$. The solution can be written

$$w_t(x,t) = \sum_{j=2}^{N} u_j(t)\, v_j(x) =$$

(4.8)

$$\sum_{j=2}^{N} \frac{v_j(x)}{\gamma_j} \int_0^t e^{-\beta_j(t-\tau)} \sin\gamma_j(t-\tau) \int_0^L p(x,\tau)\, v_j(x)\, dx\, d\tau$$

where

$$\beta_j = \frac{1}{2}\eta\,\Omega_j^2 \quad ; \quad \gamma_j = \sqrt{\Omega_j^2 - \beta_j^2} \qquad (4.9)$$

The vertical bending moment distribution is then calculated as

$$M(x,t) = M_{rigid\ body,\ quadratic}(x,t) +$$

(4.10)

$$\sum_{j=2}^{N} \ddot{u}_j \int_0^x \left[m_s r^2(\zeta)\alpha_j(\zeta) - (x-\zeta)m_s v_j(\zeta) \right] d\zeta$$

provided the slamming pressure is acting forward of section x considered. In Eqs. (4.8) ad (4.10) $\alpha_j(x)$, $v_j(x)$ are the same eigenfunctions as used in Eq. (3.2).

4.4. Extreme-value distribution of response

In the narrow-band slamming analysis presented here, all forces and motions induced on the ship by a regular sinusoidal wave $h(t)$ of the form

$$h(t) = a\cos(\omega t + \varepsilon) \qquad (4.11)$$

are calculated.

The wave process $h(t)$ in Eq. (4.11) is required to be stationary. This implies that whatever be the choice of bivariate density $f(a,\omega)$ for the wave amplitude a and frequency ω, the phase ε should be uniformly distributed in the interval $[0,2\pi]$. For this reason, it is necessary to calculate the corresponding maximum (minimum) response over the entire phase interval $[0,2\pi]$.

The calculated response includes both the low-frequency wave-induced, and the high-frequency slamming-induced components. The maximum (minimum) hull girder response at a location is denoted $M(a,\omega)$. Then response statistics are obtained by weighing $M(a,\omega)$ by the joint probability density $f(a,\omega)$ of various doublets of (a,ω) that occur in a random wave history. The joint probability density to be used was described in Section 4.1.

The response statistics (n'th order statistical moments) can be expressed as the integral

$$E[M^n] = \int_{all\ a} \int_{all\ \omega} [M(a,\omega)]^n f(a,\omega)\, d\omega\, da \qquad (4.12)$$

In practice, this integral might conveniently be evaluated by either simulation or numerical quadrature techniques. From the first four moments an approximation to the distribution function $F_M(m)$ of the response can be obtained by a Hermite transformation model, Winterstein [7], [16].

The distribution function of the maximum combined structural response M during the period of time T - including slamming- and wave-induced load effects, accounting for both slam and non-slam periods - may be expressed as:

$$F_{max\ M}(m) =$$

$$P\left[\max_{T} M \le m\right] =$$

$$P\left[\max_{T_{slam}} M \le m \cap \max_{T \backslash T_{slam}} M \le m\right] = \qquad (4.13)$$

$$1 - P\left[\max_{T_{slam}} M \ge m \cup \max_{T \backslash T_{slam}} M \ge m\right] \ge$$

$$1 - P\left[\max_{T_{slam}} M \ge m\right] - P\left[\max_{T \backslash T_{slam}} M \ge m\right]$$

in which T_{slam} is the random part of the lifetime T of which the ship is in a slamming condition, and $T \backslash T_{slam}$ the remaining time. The error in the last approximation in Eq. (4.13) is less than the smallest of the two probabilities. If the maximum value of the wave induced response cannot occur within a slamming period, then the equality sign is valid.

Turkstra's rule [17], which often provides a good estimation for the maximal effect of a linear combination of independent processes, will be used in the present case. According to this rule, only the points in time where one of the processes is at its maximum value are considered, with the companion response taken at its point in time value.

When applying Turkstra's rule to the present extreme value analysis where the wave-induced and slamming processes are dependent, a series system with two components must be considered. The two components represent the maximum response when slamming is present and the maximum response when slamming is not present, respectively.

Assuming that the point process of maximal slamming impacts may be regarded as a Poisson pulse process with intensity equal to the upcrossing rate of level u of the envelope process and the distribution function of the pulses given by $F_M(m)$, the extreme value distribution of the maximum combined slamming and wave-induced response during the time period T is

$$F_{\max M}^{slam}(m) = \exp\left[-\left(1 - F_M(m)\right)\nu T\right] \qquad (4.14)$$

where ν is the mean rate of "qualified" envelope excursions of level u, that is, excursions that are not empty.

Envelope excursions above level u are characterized as being empty if during the entire excursion the process itself stays below the level u. The long-term fraction of qualified envelope excursions above level u may be approximated using the theory developed by Ditlevsen and Lindgren [18].

By using Eq. (4.14) and the traditional extreme-value distribution for the pure wave-induced response, a lower bound on the extreme value distribution of the response during the lifetime is obtained from Eq. (4.13).

4.5 Numerical results

The importance of the whipping contribution to the wave-induced loads will be illustrated by a numerical example. The parent hull form used in Section 3.5 is analyzed for the three sets of frequencies given by Eqs. (2.8a-c). The stationary seaway and the selected ship speeds are the same as in Section 3.5.

Concerning the numerical calculations, the combined wave- and slamming-induced bending moment at midship is calculated from Eq. (4.10) by use of the first five eigenmodes. The result of the analysis is depicted in Figs. 5 and 6 in terms of the ratio of the expected value of the peak of the combined bending moment to the peak of the wave-induced moment. Fig. 5 is for vessel length 50m and Fig. 6 for 100 m.

It follows from Figs. 5 and 6 that the relative importance of the whipping-induced response is significantly dependent on vessel length. Moreover, it is seen that the whipping-induced response increases approximately linearly with hull girder natural frequency. This trend is to be expected because of the slamming impact duration is very

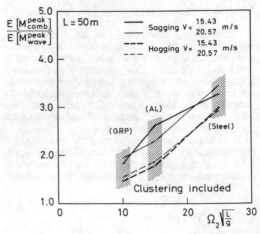

Figure 5. *Relative increase in mean peak response due to whipping (L=50m).*

fundamental period of the hull girder. Although Fig. 5 shows a considerable increase in the mean peak response due to whipping, some questions remains to be answered with respect to the validity of the expression for the slamming-impact-pressure at such high Froude numbers as is the case for the 50 m vessel. Nevertheless, a more refined slamming-impact-pressure model is not expected to change the observed trends only (perhaps) the relative

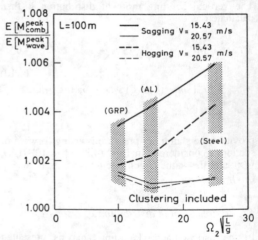

Figure 6. *Relative increase in mean peak response due to whipping (L=100m).*

short compared to both the wave period and the magnitudes.

The effect of clustering may for the present example be approximately quantified by comparing the rate of qualified envelope excursion ν to the rate of process upcrossings ν_p about level u (the sectional draft at the bow section). The upcrossing rates were found to be $v = 0.125$ and $v_n = 0.190$ for the vessel of length 50 m, and $v = 2.00 \cdot 10^{-3}$ and $v_p = 2.02 \cdot 10^{-3}$ for the vessel of length 100 m. Hence, clustering is only important for the 50 m vessel.

For other vessels or sea states effect of clustering may be different. It is not straightforward to quantify the influence of the clustering effect on the extreme-value distribution since the effect is dependent on the considered fractile. However, by arguing that the statistics of the slamming induced response for *any* wave within a cluster of slamming impacts is not significantly different from the statistics of the wave resulting in maximum slamming induced response, then the clustering effect correspond to a dirrect reduction of the considered period of time T.

5 CONCLUSION

The effect of hull flexibility on the wave-induced structural loading of high-speed mono-hull vessels has been studied.

It is shown that for ships with comparable longitudinal strength the lowest natural frequencies of GRP and aluminium hulls are only about 40 per cent and 60 per cent, respectively, of the same hull girder frequencies for steel hulls.

The effect of this difference in flexibility on springing, i.e. continuous wave-induced hull girder vibratory responses, has been investigated for parametrically scaled ships sailing in a relatively frequent moderate sea state.

The results show that when the construction material is GRP or aluminium then significant dynamic stress amplification can occur due to springing. Therefore, when these materials are used as construction materials for the hull girder concern must be given to the increase in fatigue damage due to springing. The analysis also shows that the stress amplification due to springing is insensitive to variation of the size of the vessel.

Also whipping has been considered, i.e. the effect of hull girder flexibility on midship bending stresses caused by transient slamming forces. For this analysis a new probabilistic procedure has been developed which takes into account the clustering effect of slamming incidents. The analysis results indicate significant effects of ship length and hull girder flexibility. However, since only one moderate sea state has been considered then general conclusions are difficult to make at this stage.

ACKNOWLEDGEMENT

This work was financially supported by Nordisk Industrifond and the Danish Technical Research Council.

REFERENCES

[1] Naonosuke et al., 1993. R&D of a Displacement-Type High-Speed Ship (Part 1-4). Proc. FAST'93. Yokohama, Japan. pp. 317-359.

[2] Wu, M.K., Hermundstad, O.A. and Moan T., 1993. Hydroelastic Analysis of Ship Hulls at High Forward Speed. Proc. FAST'93. Yokohama, Japan. pp. 699-710.

[3] Tønnesen, R., Vada, T. and Nestegård, A., 1993. A Comparison of an Extended Strip Theory with a Three-Dimensional Theory for Computation of Response and Loads. Proc. FAST'93. Yokohama, Japan. pp. 671-677.

[4] Jensen, J. Juncher and Pedersen, P. Terndrup, 1979. Wave-induced Bending Moments in Ships - a Quadratic Theory. Trans. RINA. Vol. 121. pp. 151-165.

[5] Jensen, J. Juncher and Pedersen, P. Terndrup, 1981. Bending Moments and Shear Forces in Ships Sailing in Irregular Waves. Journal of Ship Research. Vol. 24. No. 4. pp. 243-251.

[6] Jensen, J. Juncher and Dogliani, M., 1993. Wave-induced Ship Hull Vibrations in Stochastic Seaways. Accepted for publication in Marine Structures.

[7] Winterstein, S.R., 1985. Non-normal Response and Fatigue Damage. Journal of Engineering Mechanics. ASCE. Vol. 111. No. 10. pp. 1291-1295.

[8] Jensen, J. Juncher, 1991. Fatigue Analysis of Ship Hulls under Non-Gaussian Wave Loads. Marine Structures. Vol. 4. pp. 279-294.

[9] Wirsching, P.H. and Light, M.C., 1980. Fatigue Under Wide Band Random Stresses. Journal of the Structural Division. ASCE. Vol. 106. No. ST7. July 1980.

[10] Ochi, M.K. and Motter, L.E., 1973. Predictions of Slamming Characteristics and Hull Responses for Ship Design. Trans. SNAME, Vol. 81. pp. 144-190.

[11] Ferro, G. and Mansour, A.E., 1985. Probabilistic Analysis of Combined Slamming and Wave Induced Responses. Journal of Ship Research,.Vol. 29. No. 3. September. pp. 170-188.

[12] Friis Hansen, P., 1994. On Combination of Slamming and Wave Induced Responses. To appear in the Journal of Ship Research, June 1994.

[13] Kaplan, P. and Sargent, T.P., 1972. Further Studies of Computer Simulation of Slamming and other Wave Induced Structural Loadings on Ships in Waves. Ship Structure Committee Report 231.

[14] Bishop, R.E.D., Price, W.G. and Tam, P.K.Y., 1978. On the Dynamics of Slamming. Trans. RINA, Vol. 120. pp. 259-280.

[15] Belik, Ö., Bishop, R.E.D. and Price, W.G., 1988. Influence of Bottom and Flare Slamming on Structural Responses. Trans. RINA. Vol. 130. pp.261-275.

[16] Winterstein, S.R., 1988. Nonlinear Vibration Models for Extremes and Fatigue. Journal of Engineering Mechanics Division. ASCE. Vol. 114. No. 10. pp. 1772-1790.

[17] Turkstra, C.J., 1970. Theory of Structural Safety. SM Study No. 2, Solid Mechanics Division. University of Waterloo. Ontario.

[18] Ditlevsen, O. and Lindgren, G., 1988. Empty Envelope Excursions in Stationary Gaussian Processes. Journal of Sound and Vibration. Vol. 122. No. 3. pp. 571- 587.

Hydroelasticity in Marine Technology, Faltinsen et al. (eds) © 1994 Balkema, Rotterdam, ISBN 90 5410 387 6

Slamming loads on wetdecks of multihull vessels

J. Kvålsvold & O. M. Faltinsen
The Norwegian Institute of Technology, Trondheim, Norway

ABSTRACT: Slamming against the wetdeck of a multihull vessel in head sea waves is studied analytically and numerically. The theoretical slamming model is a two-dimensional, asymptotic method valid for small local angles between the undisturbed water surface and the wetdeck. The disturbance of the water surface as well as the local hydroelastic effects in the slamming area are accounted for. The elastic deflections of the wetdeck are expressed in terms of "dry" normal modes. The structural formulation accounts for the shear deformations and the rotatory inertia effects in the wetdeck. The findings of this work indicate that the slamming loads on the wetdeck are significantly influenced by the elasticity of the wetdeck structure.

1 INTRODUCTION

If we consider a multihull vessel advancing at high forward speed in a head sea wave system, slamming against the wetdeck in the bow of the vessel may occur. By wetdeck we are referring to the structural part connecting the two side hulls of the vessel. Wetdeck slamming causes large hydrodynamic forces on the structure and will introduce local as well as global hydroelastic effects of the multihull vessel. By hydroelastic effects we mean that the hydrodynamic pressure acting on the wetdeck is a function of the structural deformations of the wetdeck. Only the local hydroelastic effects are considered in this paper.

Slamming has been widely studied in the literature through the last decades. For instance, slamming against rigid two-dimensional bodies has been studied by Wagner (1932), Cointe and Armand (1987) and Zhao and Faltinsen (1992,1993). Slamming against rigid wetdecks has been studied by for instance Kaplan and Malakhoff (1978) and Kaplan (1987,1991). Kaplan (1992) reported that wetdeck slamming could cause a hydrodynamic loading in the order of the weight of the vessel or even larger. This may lead to severe local as well as global damages of the hull structure. Zhao and Faltinsen (1992) reported that the global

heave and pitch motions of a catamaran were influenced by wetdeck slamming. They did not account for any local or global elastic effects of the catamaran.

A slamming model where the local hydroelastic effects are accounted for, are not so frequently appearing in the literature. Anyway, the authors are aware of the work by Meyerhoff (1965) who studied slamming against elastic wedges penetrating an initially calm free water surface. Hydroelastic modelling of slamming has also been studied by Kvålsvold and Faltinsen (1993).

In this study, the work by Kvålsvold and Faltinsen (1993) is improved by modelling the wetdeck as a Timoshenko beam with rotatory springs at the beam ends. This means that the shear deformations and the rotatory inertia effects of the modelled beam are properly accounted for. It is shown that these effects influence the structural response and the hydrodynamic loads on the wetdeck during slamming.

The hydrodynamic formulation is based on an extension of Wagner's (1932) two-dimensional theory. It is assumed that the crest of an incident head sea regular wave system hits the wetdeck so that the structural deformations and the fluid flow are symmetric about a vertical plane through the line of initial impact between the wetdeck and the waves. A generalization of the method to include unsymmetry will be reported in the near future.

To properly describe the slamming loads on the wetdeck and to ensure the stability of the time integration of the hydroelastic response, this study demonstrates the importance of integrating analytically the hydrodynamic pressure loads on the wetdeck. This implies that one should be careful using numerical methods based upon for instance the boundary element method for the fluid loading and the finite element method for the structural deformations, to study slamming. The reason is the rapid variations of the hydrodynamic slamming pressure in the water impact region as a function of both time and space.

2 THE STRUCTURAL FORMULATION

A detail of the wetdeck structure of a multihull vessel is shown in Figure 1. The righthanded xyz-coordinate system is a local coordinate system moving with the forward speed U of the vessel. The x-axis is parallell to the longitudinal stiffeners and is pointing towards the stern of the multihull vessel. The y-axis is parallel to the transverse stiffeners and is pointing towards the starboard side. The z-axis is pointing upwards. The origin of the local coordinate system is located at the line of initial impact between the free surface of the waves and the wetdeck structure. It is assumed that the crest of a regular wave system hits the wetdeck mid between two of the transverse stiffeners. This is only one of many possible locations along the wetdeck where the initial water impact may take place. This method will in the near future be generalized to account for an arbitrary location of the initial water impact.

The analysed part of the wetdeck between two of the transverse stiffeners is modelled as a Timoshenko beam with length L_B corresponding to the distance between the transverse stiffeners. This means that the deflections of the beam are totally dominated by those of the longitudinal stiffeners. Local deformations of the plate field between two of the longitudinal stiffeners is a three-dimensional effect which is not covered by this two-dimensional analysis. This will be discussed in a separate section. The transverse stiffeners are assumed to be much stiffer than the longitudinal stiffeners, so that the vertical deflections at the beam ends are disregarded. Rotatory springs are introduced at the beam ends to account for the restoring mo-

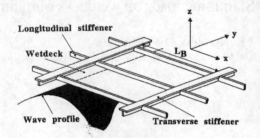

Figure 1: A detail of the wetdeck structure of a multihull vessel. The xyz coordinate system is the local coordinate system defined more clearly in Figure 2.

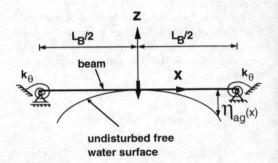

Figure 2: The definitions of the beam and the local coordinate system. k_θ is the spring stiffness of the rotatory springs and $\eta_{ag}(x)$ is the airgap between the undisturbed free surface of the waves and the wetdeck at the moment of initial impact.

ment of the transverse stiffeners as well as the part of the wetdeck structure outside the modelled beam. k_θ is the spring stiffness that is related to the restoring moment M by $M = -k_\theta\theta$. θ is the rotation angle at the beam end. No axial force effects of the modelled beam are considered. This approach does not properly account for the inertia effects of the wetdeck structure outside the modelled beam. It is believed that such effects are important and will influence the hydroelastic response. It is also assumed that the beam is horizontal at the moment of initial water impact, so that both the beam deflections and the fluid flow are symmetric about the yz-plane. The local coordinate system together with the definitions of the beam are shown Figure 2. $\eta_{ag}(x)$ is the airgap between the undisturbed free surface of the waves and the wetdeck at the moment of initial impact.

2.1 The governing beam equations of motions

Let $w(x,t)$ be the beam deflection which is to be interpreted as the difference between the actual vertical position of the wetdeck and the vertical position due to the rigid ship motions. $w(x,t)$ is assumed to be small compared to the beam length L_B. The force and the moment equilibrium of a beam element are satisfied through the following governing coupled beam equations of motions:

$$M_B\ddot{w} \; + \; GA_s\left(\frac{\partial \beta}{\partial x} - \frac{\partial^2 w}{\partial x^2}\right)$$

$$+ \; M_B\dot{V}_g(t) = p(x,w,t) \qquad (1)$$

$$M_B r^2 \ddot{\beta} \; + \; GA_s\left(\beta - \frac{\partial w}{\partial x}\right)$$

$$- \; EI\frac{\partial^2 \beta}{\partial x^2} = 0 \qquad (2)$$

M_B is the mass per unit length of the longitudinal stiffener together with the flange and divided by the width of the flange, G is the shear modulus, A_s is the shear area divided by the width of the flange and r is the mass radius of gyration of the beam element. The width of the flange is equal to the distance between the longitudinal stiffeners. Further, E is the Youngs modulus and I is the area moment of inertia of the beam cross section and divided by the width of the flange. An effective flange of 70 % of the distance between the longitudinal stiffeners is used when calculating I. $\dot{V}_g(t)$ is the local vertical accelerations due to the heave and the pitch motions of the vessel. $p(x,w,t)$ is the hydrodynamic impact pressure which is a function of time, space and the beam deflections. Dot stands for the time derivative and t is the time variable. $\beta(x,t)$ is the slope of the deflection curve when the shear deformations are neglected. The total rotation angle $\frac{\partial w(x,t)}{\partial x}$ of a beam element is related to the shear angle $\gamma_s(x,t)$ as well as to $\beta(x,t)$ through the relation:

$$\frac{\partial w(x,t)}{\partial x} = \gamma_s(x,t) + \beta(x,t) \qquad (3)$$

A normal mode approach is followed in this study. This means that the solutions of $w(x,t)$ and $\beta(x,t)$ are expressed in terms of the beam's "dry" normal modes:

$$w(x,t) = \sum_{n=1}^{\infty} a_n(t)\psi_n(x) \approx \sum_{n=1}^{N_{eig}} a_n(t)\psi_n(x) \qquad (4)$$

$$\beta(x,t) = \sum_{n=1}^{\infty} a_n(t)\phi_n(x) \approx \sum_{n=1}^{N_{eig}} a_n(t)\phi_n(x) \qquad (5)$$

where $a_n(t)$ is the principal coordinate of vibration mode n. $\psi_n(x)$ and $\phi_n(x)$ are the eigenfunctions of vibration mode n. $\phi_n(x)$ is related to $\psi_n(x)$ through the coupled equations of motions (1) and (2) with $\dot{V}_g(t)$ and $p(x,w,t)$ set equal to zero. N_{eig} is the number of eigenfunctions used in the normal mode approach. By "dry" normal modes we mean that the effect of the surrounding water is not accounted for when $\psi_n(x)$ and $\phi_n(x)$ are evaluated. The four beam end boundary conditions needed to determine the eigenfunction $\psi_n(x)$ and the eigenfrequency ω_n of vibration mode n are:

$$w = 0 \quad \text{for} \quad x = \frac{\pm L_B}{2} \qquad (6)$$

$$\frac{k_\theta}{EI}\beta + \frac{\partial \beta}{\partial x} = 0 \quad \text{for} \quad x = \frac{L_B}{2} \qquad (7)$$

$$-\frac{k_\theta}{EI}\beta + \frac{\partial \beta}{\partial x} = 0 \quad \text{for} \quad x = -\frac{L_B}{2} \qquad (8)$$

The beam end boundary conditions can physically be interpreted as no vertical deflections and continuity of the bending moment at the beam ends.

2.2 The beam eigenvalue problem

To obtain an expression of $\psi_n(x)$ and $\phi_n(x)$ one needs to solve the eigenvalue problem characterized by equations (1) and (2) together with the beam end boundary conditions expressed by equations (6) to (8). In the eigenvalue analysis, $p(x,w,t)$ and $\dot{V}_g(t)$ are set equal to zero. The solutions of $w(x,t)$ and $\beta(x,t)$ are written as:

$$w(x,t) = \psi_n(x)e^{i\omega_n t} = C_{\psi_n}e^{D_n x}e^{i\omega_n t} \qquad (9)$$

$$\beta(x,t) = \phi_n(x)e^{i\omega_n t} = C_{\phi_n}e^{D_n x}e^{i\omega_n t} \qquad (10)$$

Here, C_{ψ_n} and C_{ϕ_n} are constants. D_n is interpreted as the wave number which is not yet known and i is the imaginary unit. By substituting equations (9) and (10) into equations (1) and (2) two equations are obtained, from which C_{ψ_n} and C_{ϕ_n} are solved. By setting the coefficient determinant of the two by two equation system equal to zero, it follows that:

$$D_n^4 + D_n^2(K_n^2 + r^2 a_n^4) + K_n^2 r^2 a_n^4 - a_n^4 = 0 \quad (11)$$

Here

$$K_n^2 = \frac{M_B \omega_n^2}{GA_s} \quad (12)$$

$$a_n^4 = \frac{M_B \omega_n^2}{EI} \quad (13)$$

Equation (11) is the relation between the wave number D_n and the eigenfrequency ω_n and is therefore interpreted as the dispersion relation. The four solutions $D_n^1, D_n^2, D_n^3, D_n^4$ of D_n can be complex or real:

$$D_n^{1,2} = \pm i p_n \quad (14)$$

$$D_n^{3,4} = \begin{cases} \pm q_n & \text{for} \quad \omega_n < \sqrt{\frac{GA_s}{M_B r^2}} \\ \pm i q_n & \text{for} \quad \omega_n > \sqrt{\frac{GA_s}{M_B r^2}} \end{cases} \quad (15)$$

The wave numbers p_n and q_n are:

$$p_n = \sqrt{\frac{K_n^2 + r^2 a_n^4}{2} + \sqrt{\left(\frac{K_n^2 - r^2 a_n^4}{2}\right)^2 + a_n^4}} \quad (16)$$

and

$$q_n = \sqrt{-\frac{K_n^2 + r^2 a_n^4}{2} + \sqrt{\left(\frac{K_n^2 - r^2 a_n^4}{2}\right)^2 + a_n^4}} \quad (17)$$

for $\omega_n < \sqrt{\frac{GA_s}{M_B r^2}}$. For $\omega_n > \sqrt{\frac{GA_s}{M_B r^2}}$ one has:

$$q_n = \sqrt{\frac{K_n^2 + r^2 a_n^4}{2} - \sqrt{\left(\frac{K_n^2 - r^2 a_n^4}{2}\right)^2 + a_n^4}} \quad (18)$$

The wave numbers described by equations (14) and (15) together with the assumptions made in equations (9) and (10) lead to the following solutions of $\psi_n(x)$ and $\phi_n(x)$ for $\omega_n < \sqrt{\frac{GA_s}{M_B r^2}}$:

$$\psi_n(x) = A_n \sin(p_n x) + B_n \cos(p_n x)$$
$$+ C_n \sinh(q_n x) + D_n \cosh(q_n x) \quad (19)$$

$$\phi_n(x) = E_n \sin(p_n x) + F_n \cos(p_n x)$$
$$+ G_n \sinh(q_n x) + H_n \cosh(q_n x) \quad (20)$$

For $\omega_n > \sqrt{\frac{GA_s}{M_B r^2}}$:

$$\psi_n(x) = A_n \sin(p_n x) + B_n \cos(p_n x)$$
$$+ C_n \sin(q_n x) + D_n \cos(q_n x) \quad (21)$$

$$\phi_n(x) = E_n \sin(p_n x) + F_n \cos(p_n x)$$
$$+ G_n \sin(q_n x) + H_n \cos(q_n x) \quad (22)$$

The coefficients E_n, F_n, G_n, H_n are related to the coefficients A_n, B_n, C_n, D_n through the coupled beam equations of motions (1) and (2) with $p(x, w, t)$ and $\dot{V}_g(t)$ set equal to zero. The coefficients A_n, B_n, C_n, D_n together with the eigenfrequency ω_n are determined in order to satisfy the four beam end boundary conditions in equations (6) to (8). Non-trivial solutions of the four coefficients are achieved when the coefficient determinant is set equal to zero. There is an infinite number of ω_n that causes the coefficient determinant to be zero. This determines the eigenfrequencies ω_n for $n = 1, \infty$. To each ω_n, three of the four coefficients A_n, B_n, C_n, D_n are uniquely determined, while the fourth may be chosen arbitrarily. Here, the fourth coefficient is selected in order to normalize the eigenfunction $\psi_n(x)$ so that the sum of the coefficients A_n, B_n, C_n, D_n is equal to 1.

2.3 *The governing modal beam equation of motions*

The governing modal beam equation of motions will be derived. The normal mode formulation of $w(x, t)$ and $\beta(x, t)$ described by equations (4) and (5) are substituted into equations (1) and (2) which are multiplied by $\psi_m(x)$ and $\phi_m(x)$, respectively. The arising equations are added together and integrated over the length of beam. By using the orthogonality conditions of the eigenfunctions, the governing modal beam equation of motions of vibration mode m becomes:

$$M_{mm}\ddot{a}_m(t) + C_{mm}a_m(t)$$

$$+ \; M_B\dot{V}_g(t)\int_{-\frac{L_B}{2}}^{\frac{L_B}{2}} \psi_m(x)dx$$

$$= \int_{-c(t)}^{c(t)} p(x,w,t)\psi_m(x)dx \qquad (23)$$

M_{mm} and C_{mm} denote the generalized mass and restoring coefficients, respectively. $2c(t)$ is an approximation of the wetted length of the beam that will be discussed later in the text. M_{mn} and C_{mn} are expressed as:

$$M_{mn} = M_B\int_{-\frac{L_B}{2}}^{\frac{L_B}{2}} \left[\psi_m(x)\psi_n(x)\right.$$

$$\left. +r^2\phi_m(x)\phi_n(x)\right]dx \qquad (24)$$

$$C_{mn} = \int_{-\frac{L_B}{2}}^{\frac{L_B}{2}} \left[-EI\phi_n''(x)\phi_m(x)\right.$$

$$+ \; GA_s\Big(\psi_m'(x) - \phi_m(x)\Big)$$

$$\times\Big(\psi_n'(x) - \phi_n(x)\Big)\bigg]dx \qquad (25)$$

The primes stand for the derivative with respect to x. Due to the orthogonality conditions of the beam eigenfunctions, M_{mn} and C_{mn} are zero for $m \neq n$. It also follows from the orthogonality conditions that $C_{mm} = \omega_m^2 M_{mm}$

2.4 *The elementary beam formulation*

No shear deformations and no rotatory inertia effects are considered in the elementary beam formulation. This means that K_n in equation (12) and r in equation (2) become zero. The wave numbers then become equal, i.e. $p_n = q_n = a_n = \frac{M_B\omega_n^2}{EI}$. Additionally, the four roots of equation (11) are always two complex and two real roots. Consequently, the eigenfunction $\psi_n(x)$ is expressed by equation (19) for all eigenfrequencies. M_{mm} and C_{mm} are expressed as:

$$M_{mn} = M_B\int_{-\frac{L_B}{2}}^{\frac{L_B}{2}} \psi_m(x)\psi_n(x)dx \qquad (26)$$

$$C_{mn} = \int_{-\frac{L_B}{2}}^{\frac{L_B}{2}} EI\phi_n''''(x)\phi_m(x)dx \qquad (27)$$

As for the Timoshenko beam formulation, M_{mn} and C_{mn} are zero for $m \neq n$ and $C_{mm} = \omega_m^2 M_{mm}$ due to the orthogonality conditions.

3 THE HYDRODYNAMIC BOUNDARY VALUE PROBLEM

One needs to solve a hydrodynamic boundary value problem (HBVP) to obtain an expression for the hydrodynamic pressure p in equation (23). Some additional assumptions are made before defining the HBVP. First the fluid accelerations are assumed to be much greater than the acceleration of gravity g. Next, the vertical velocities due to the waves are neglected in the impact region. This is reasonable as long as the line of initial contact between the free surface of the waves and the wetdeck is located along a wave crest. The disturbances of the free surface of the waves due to the side hulls of the vessel are not considered when solving the HBVP. The wetdeck is assumed to be horizontal in the impact region at the moment of initial water impact. This means that both the structural deformations as well as the fluid flow are symmetric about the yz-plane. Further, no airpocket is allowed to be trapped between the free water surface and the wetdeck. This will be discussed in a separate section. Cavitation is not considered but could actually occur for high impact velocities some time after the initial stage of the fluid flow.

A simplified two-dimensional HBVP (see Figure 3) often referred to as the outer solution, is set up to express the fluid flow in the impact region. Assuming incompressible fluid and irrotational flow, there exists a velocity potential ϕ which satisfies the Laplace equation in the fluid domain. The boundary conditions are $\frac{\partial\phi}{\partial z} = V(t) + \dot{w}(x,t) \equiv V_e(x,t)$ on the wetted length of the beam and $\phi = 0$ on the free water surface. $V(t)$ is the relative vertical velocity between the rigid wetdeck and the free surface of the waves in the impact region. Since the vertical velocities due to the waves

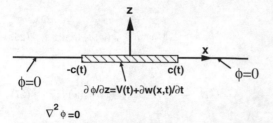

$\phi=0$ $\phi=0$

$\nabla^2\phi=0$

Figure 3: The hydrodynamic slamming model used in this study.

are neglected in this study, the only contribution to $V(t)$ comes from the global heave and the pitch motions of the multihull vessel. $V_e(x,t)$ is defined as the effective velocity. $2c(t)$ is an approximation of the wetted length of the beam. $c(t)$ is unknown and part of the solution of the impact problem. How to obtain an expression of $c(t)$ is discussed in a separate section. The boundary conditions are transferred to $z=0$. Close to the edges of the flow ($|x|=c(t)$) the outer solution breaks down due to a singular hydrodynamic pressure. In this region an inner solution has to be defined. The outer and the inner solutions can be matched in the same way as shown by Cointe and Armand (1987) for impact of a rigid and blunt body. Here, only the outer solution is considered. The reason is that the details of the inner flow do not influence the solution of the hydroelastic problem.

3.1 The solution procedure of the HBVP

The HBVP is solved analytically and the hydro-dynamic loading due to wetdeck slamming is expressed in terms of analytical functions in this study. The fluid flow in the impact region is expressed in terms of a vortex distribution $\gamma(x,t)$ on the wetted length of the beam. The vortex density is solved from the integral equation (see Newman 1977, page 180):

$$V_e(x,t) = -\frac{1}{2\pi}\fint_{-c(t)}^{c(t)}\frac{\gamma(\xi,t)}{\xi-x}d\xi \qquad (28)$$

$\fint$ denotes the principal value integral. Similarly as Newman (1977) did for the lifting problem, the vertical velocities on the wetted length of the beam are expressed in terms of a Fourier cosine series as:

$$V_e(\theta) = A_0 + \sum_{k=1}^{\infty}A_k\cos(k\theta)$$

$$\approx A_0 + \sum_{k=1}^{N_{fcom}}A_k\cos(k\theta) \qquad (29)$$

N_{fcom} is the number of Fourier components used to represent the vertical velocities on the wetted length of the beam and θ is the transformed co-ordinate on the wetted length of the beam. θ is related to x through:

$$x = c(t)\cos\theta \qquad (30)$$

The Fourier coefficients are for $\omega_n < \sqrt{\frac{GA_s}{M_B r^2}}$:

$$A_0(t) = V(t) + \sum_{n=1}^{\infty}\dot{a}_n(t)\Big[B_nJ_0(p_nc(t))$$
$$+ D_nI_0(q_nc(t))\Big] \qquad (31)$$

$$A_{2k}(t) = 2\sum_{n=1}^{\infty}\dot{a}_n(t)\Big[(-1)^kB_nJ_{2k}(p_nc(t))$$
$$+ D_nI_{2k}(q_nc(t))\Big] \qquad (32)$$

and for $\omega_n > \sqrt{\frac{GA_s}{M_B r^2}}$:

$$A_0(t) = V(t) + \sum_{n=1}^{\infty}\dot{a}_n(t)\Big[B_nJ_0(p_nc(t))$$
$$+ D_nJ_0(q_nc(t))\Big] \qquad (33)$$

$$A_{2k}(t) = 2\sum_{n=1}^{\infty}\dot{a}_n(t)\Big[(-1)^kB_nJ_{2k}(p_nc(t))$$
$$+ (-1)^kD_nJ_{2k}(q_nc(t))\Big] \qquad (34)$$

J_k are Bessel functions of the first kind and I_k are modified Bessel functions of the first kind. The Fourier coefficients $A_{2k-1}(t) = 0$ are equal to zero due to the assumed symmetry properties of the structural deformations and the fluid flow in the water impact region. In Newman (1977) (see pages 180-188) the integral equation (28) is solved for the vortex density and expressed in terms of the transformed coordinate θ:

$$\gamma(\theta) = -2A_0(t)\frac{\cos\theta}{\sin\theta} + 2\sum_{k=1}^{\infty}A_{2k}(t)\sin(2k\theta) \quad (35)$$

This leads to the velocity potential:

$$\phi(\theta) = A_0(t)c(t)\sin\theta - c(t)\sum_{k=1}^{\infty}A_{2k}(t)$$

$$\times\frac{1}{2}\left[\frac{\sin(2k-1)\theta}{2k-1} - \frac{\sin(2k+1)\theta}{2k+1}\right] \quad (36)$$

The hydrodynamic pressure p can be approximated by the "$-\rho\frac{\partial\phi}{\partial t}$" term in Bernoulli's equation. The total modal hydrodynamic force described by the right hand side of equation (23) can, after some algebraic manipulations, be rewritten as:

$$\int_{-c(t)}^{c(t)} p(x,w,t)\psi_m(x)dx = -\sum_{n=1}^{\infty}B_{mn}(t)\dot{a}_n(t)$$

$$-\sum_{n=1}^{\infty}A_{mn}(t)\ddot{a}_n(t)$$

$$+\ F_{exc,m}(t) \quad (37)$$

The detailed expressions of the coefficients $B_{mn}(t)$, $A_{mn}(t)$ and $F_{exc,m}(t)$ are shown in Appendix A. The total hydrodynamic slamming force $F_{tot}(t)$ is reproduced in the appendix as well. Equation (37) needs some explanation. First, all the terms which are explicitly proportional to $\dot{a}_n(t)$ are collected. The force term is called the modal hydrodynamic damping force $F_{dam,m}(t)$ and the coefficient is the modal hydrodynamic damping coefficient. Next, all the terms which are explicitly proportional to $\ddot{a}_n(t)$ are collected. The force term is called the modal hydrodynamic added mass force $F_{add,m}(t)$ and the coefficient is the modal hydrodynamic added mass coefficient. The remaining terms are called the modal excitation force $F_{exc,m}(t)$. Actually, $F_{exc,m}(t)$ is the modal hydrodynamic force caused by the hydrodynamic pressure part $-\rho V(t)\frac{c(t)}{\sqrt{c^2(t)-x^2}}\frac{dc(t)}{dt} - \rho\frac{dV(t)}{dt}\sqrt{c^2(t)-x^2}$.

One should note that $c(t)$ is influenced by $a_n(t)$. The same expression also applies for the total hydrodynamic force acting on the beam in the non-vibratory case, but then $c(t)$ depends only on $V(t)$, t and the curvature of the incident waves. We know that our definition of the hydrodynamic forces may be misleading, since $A_{mn}(t)$, $B_{mn}(t)$ and $F_{exc,m}(t)$ are implicitly functions of $a_n(t)$ through their dependency on $c(t)$. The only intention for doing this, is to be able to move as much as possible of the total modal hydrodynamic force to the left hand side of equation (23). This

improves the accuracy of the hydrodynamic loading as well as the stability of the time integration of the differential equations.

3.2 The wetted length of the beam

How to obtain an approximation of half the wetted length $c(t)$ will now be discussed. $c(t)$ is solved by generalizing the integral equation that Wagner used in his theory (Wagner 1932). In a slamming problem where the local angles between the undisturbed free water surface and the wetdeck are small, it is important to account for the effect of the pileup water when estimating the wetted length of the beam. This is taken care of in this study. $c(t)$ is the solution x of the integral equation:

$$\eta_{ag}(x) = \int_0^t \left[\frac{\partial\phi(x,\tau)}{\partial z} - \left(V(\tau) + \dot{w}(x,\tau)\right)\right]d\tau$$

$$(38)$$

where $\eta_{ag}(x)$ is the air gap between the undisturbed free water surface and the wetdeck at the moment of initial water impact. Here, $\eta_{ag}(x) = \zeta_a(1 - \cos\nu x)$, where $\nu = \frac{\omega_o^2}{g}$ is the wave number and ζ_a is the wave amplitude. ω_o is the circular frequency of oscillations of the waves. $\frac{\partial\phi}{\partial z}$ is the vertical fluid velocity <u>outside</u> the wetted part of the beam along $z = 0$. $\frac{\partial\phi}{\partial z}$ is written in terms of the transformed coordinate χ outside the wetted length of the beam as:

$$\frac{\partial\phi(\chi,t)}{\partial z} = -A_0(t)\frac{\sin^2\chi}{\cos\chi(1+\cos\chi)}$$

$$+\frac{\sin^2\chi}{(1+\cos\chi)^2}\sum_{k=1}^{\infty}A_{2k}(t)\tan^{(2k-2)}\left(\frac{\chi}{2}\right)(39)$$

χ is related to x through:

$$c(t) = x\sin\chi \quad (40)$$

$\frac{\partial\phi}{\partial z}$ has a square root singularity at $|x| = c(t)$. The singular term is the first term on the right hand side of equation (39). Anyway, the singularity is integrable and the integral described by equation (38) is finite. The wetted length $2c(t)$ plays an important role for the hydroelastic response of the wetdeck. This means that the numerical evaluation of the singular term of the integrand in equation (38) has to be carried out carefully. Here, this is taken care of by identifying the singular part of the integrand and integrate it analytically over each time step. The remaining part is assumed to be constant over the time step. This means:

$$\int_0^t \frac{\partial \phi(x,\tau)}{\partial z} d\tau = \int_0^x \frac{\partial \phi(x,c)}{\partial z} \frac{dt}{dc} dc$$

$$= \int_0^x \frac{g_A(x,c)}{\sqrt{x^2 - c^2}} dc + \int_0^x g_B(x,c) dc$$

$$\approx \sum_{j=1}^{N_{step}} g_A(x, c_{j-\frac{1}{2}}) \int_{c_{j-1}}^{c_j} \frac{dc}{\sqrt{x^2 - c^2}}$$

$$+ \sum_{j=1}^{N_{step}} g_B(x, c_{j-\frac{1}{2}}) \cdot (c_j - c_{j-1}) \qquad (41)$$

c_j is the value of $c(t)$ at the previous time step j and N_{step} is the actual time step in the numerical solution procedure. $\frac{g_A(x,c)}{\sqrt{x^2-c^2}} \frac{dc}{dt}$ and $g_B(x,c)\frac{dc}{dt}$ are identical to the first and the second term on the right hand side of equation (39), respectively.

3.3 *Alternative solution procedures of the HBVP*

The HBVP described by Figure 3 has been solved by two other approaches. Each of those approaches will now be discussed.

The first approach is a simplified analytical method where the vertical velocity on the wetted length of the beam is approximated by its mean value in space over the wetted length of the beam. Then the solution of the velocity potential becomes the classical solution for a flat plate with transverse velocity in infinite fluid (See for instance Newman 1977 page 122). The width of the flat plate is equal to the wetted length of the beam. The total modal hydrodynamic force on the beam is decomposed into different force terms in the same way as described by equation (37).

The second approach is a pure numerical method based on Green's second identity. This formulation breaks down with zero plate thickness. The geometry of the flat plate is therefore modelled as a diamond with height to length ratio 0.01. This means that the square root singular behaviour of the hydrodynamic pressure near $x = \pm c(t)$ is satisfied in an approximate way. The numerical method predicts the slamming force on a rigid flat plate with relative error 1%. The following difficulties occurred for the hydroelastic part: a) Decomposition of the total modal hydrodynamic force on the beam as illustrated by equation (37) and b) Convergence problems. Each of

these problems will now be discussed. It turned out to be difficult to decompose the total modal hydrodynamic force on the beam into excitation, damping and added mass terms as illustrated by equation (37). In particular we did not manage to identify the part of the total modal hydrodynamic force explicitly proportional to $\dot{a}_n(t)$. In that way we were not able to move the main parts of the total modal hydrodynamic force to the left hand side of equation (23). Two reasons are proposed to cause the convergence problems. As a consequence of a), the right hand sides were large compared to the left hand sides in the set of differential equations. This may lead to numerical instabilities in the time integration. The structural and the hydrodynamic parts were not solved simultaneously in time in this method. It is believed that a simultaneous solution of the principal coordinates together with the unknowns at the boundary elements on the flat plate (the wetted length) could improve the numerical stability of the integration in time. Our use of a boundary element method was unsuccessful, but we will not state that it is impossible to use a direct numerical solution of the HBVP shown in Figure (3). In general, one should be careful when using a numerical method based on the boundary element method for the fluid loading combined with the finite element method for the structural deformations, to study slamming. The reason is the rapid variations of the hydrodynamic slamming pressure in the water impact region as a function of both time and space.

4   VERIFICATION

In order to verify the computer code and the mathematical formulation of the local hydroelastic analysis, different methods may be used. Verification methods based on conservation of energy, mass and momentum should ideally be used, but are not suitable to use in this case. This is due to the inadequate modelling of the jet flow near the edges ($x = \pm c(t)$) of the wetted length of the beam. In order to conserve energy, mass as well as momentum during the time simulation, the jet flow has to be accounted for in a similar way as Zhao and Faltinsen (1993) did. Anyway; convergence tests by varying the time step, the number of eigenfunctions in the modal representation of the beam deflections as well as the number of Fourier components used to represent the effective vertical velocity on the wetted length of the beam, have successfully been carried out.

The 30 first eigenfrequencies and the corresponding eigenfunctions of the Timoshenko beam used in this study, have been compared with similar calculations by a separate computer program PUSFEA (1993) based on the finite element method. The agreement of the eigenfrequencies and the eigenfunctions between the two approaches were satisfactory. However, the computer program PUSFEA detects an additional eigenfrequency and a corresponding eigenfunction for $\omega = \sqrt{\frac{GA_s}{M_B r^2}}$. By following the approach presented in this work, the eigenfrequencies are evaluated by requiring a coefficient determinant to be zero. Limited numerical studies show that the determinant is close to zero for $\omega = \sqrt{\frac{GA_s}{M_B r^2}}$. However, it turned out to be difficult, for that particular eigenfrequency, to evaluate coefficients A_n, B_n, C_n, D_n that satisfied the four beam end boundary conditions described by equations (6) to (8). Therefore, we believe that $\omega = \sqrt{\frac{GA_s}{M_B r^2}}$ is not an eigenfrequency. The corresponding eigenfunction is not among the eigenfunctions used to express the beam deflections in this study. Anyhow, in a normal mode approach the solution of the governing equation is expressed in terms of a convergent series of the eigenfunctions corresponding to the lowest eigenfrequencies. $\omega = \sqrt{\frac{GA_s}{M_B r^2}}$ corresponds to an eigenfrequency between the vibration modes 16 and 17. It is believed that the possible lack of this particular eigenfunction in the normal mode approach will not significantly influence the solution of the governing equations of motions.

5 RESULTS

A local hydroelastic slamming analysis is carried out. This means that we are considering an elastic beam that is forced with constant vertical velocity V through the crest of a regular wave system. Then $\dot{V}_g(t)$ in equation (1) and $\frac{dV(t)}{dt}$ in equation (63) are zero. The velocity V is due to the global heave and pitch motions of the multihull vessel and will in principle be influenced by the local slamming analysis. The interaction effects between the local slamming forces and the global rigid ship motions are discussed in Kvålsvold and Faltinsen (1993). Aluminium is selected as a basis for the wetdeck material. The set of N_{eig} second order differential equations described by equation (23) is simulated in time by a fourth order Runge-Kutta integrator. Linear structural damping is in-

Table 1: The input data for the time simulation.

Description	Unit	Value
Youngs modulus E	$[N/m^2]$	$70 \cdot 10^9$
Shear modulus G	$[N/m^2]$	$27 \cdot 10^9$
Area moment of inertia I	$[m^4/m]$	0.000011
Shear area A_s	$[m^2/m]$	0.0012
Mass of the beam M_B	$[kg/m^2]$	36.6
Beam length L_B	$[m]$	1.5
Radius of gyration r	$[m]$	0.02856
Eigenmodes N_{eig}	[-]	40
Time step dt	[s]	0.000002
Fourier components N_{fcom}	[-]	20
Wave period T_w	[s]	5.7
Wave amplitude ζ_a	$[m]$	2.0
Constant fall velocity V	$[m/s]$	-6.0

troduced in the numerical solution and added to each vibration mode m along the diagonal of the damping matrix. The structural damping for each vibration mode is chosen as 1 % of $2M_{mm}\omega_m$. M_{mm} is the modal mass defined in equation (24). Since the wetted length $2c(t)$ is a priori unknown in this slamming problem, the solution of the hydroelastic response is iterated at each time step. Three iterations are used in this study. In the initial phase of the water impact an approximate acoustic formulation is followed by requiring that the total hydrodynamic force described by equation (63) has to be below the total force based upon an integration of the one-dimensional acoustic pressure $p_{ac}(t)$ over the wetted length of the beam. $p_{ac}(t)$ is expressed as $p_{ac}(t) = -\rho \overline{V}_e(t) c_e$ where $\overline{V}_e(t)$ is the mean value, in space over the wetted length of the beam, of the effective vertical velocities $V_e(x,t)$. c_e is the speed of sound in water. The rotatory spring stiffness is $k_\theta = \frac{3.5EI}{L_B}$ (see Figure 2) in the presented results. The input data for the time simulations are shown in Table 1.

The total hydrodynamic force acting on the beam is shown as a function of time in Figure 4. 40 eigenfunctions are used in the normal mode approach. The simulations are continued until the whole beam is wet. Results from both a rigid wetdeck and an elastic wetdeck are reproduced. This analysis covers the hydroelastic response in the first 0.0014 [s] after the initial water impact. Results from two different structural formulations are shown for the elastic case. The first is the elementary beam formulation (no shear deformations and rotatory inertia) and the second is the Timoshenko beam formulation. Following the rigid wet-

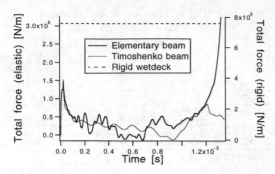

Figure 4: The total hydrodynamic force on the beam as a function of time. Results from both a rigid (right axis) and an elastic (left axis) wetdeck structure are shown. For the elastic wetdeck, results from two different structural formulations are reproduced. The input data for the time simulation are presented in Table 1.

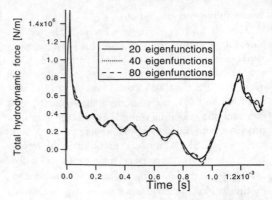

Figure 5: The total hydrodynamic force on the Timoshenko beam as a function of time. Results for different number of eigenfunctions N_{eig} used to describe the beam deflections are reproduced. The input data for the time simulations are presented in Table 1.

deck approach, the total hydrodynamic force on the wetdeck is large and almost constant in time. By taking into account the flexibility of the wetdeck, the total hydrodynamic force is significantly reduced. This means that the first and the second terms on the right hand side of equation (37) become important when evaluating the total modal hydrodynamic force on the beam. One should note that the force component $F_{exc,m}(t)$ in equation (37) contributes to the reduction of the total hydrodynamic force as well. This is discussed later in the text. Initially, the total hydrodynamic force is equal to zero and it rapidly increases to a local maximum value. The simplified acoustic formulation is followed in that time range. Later, the impact force is reduced again. The negative force in certain time intervals indicates the possibility of cavitation. It is also clear from the figure that the shear and the rotatory inertia effects influence the hydrodynamic loading. The flexibility of the elementary beam is less than the flexibility for the Timoshenko beam. This means that the eigenfrequencies of the elementary beam are larger than the corresponding eigenfrequencies for the Timoshenko beam. This is a possible explanation to the rapid oscillations in time of the total hydrodynamic force when the elementary beam formulation is used. It has been found that the linear structural damping introduced in the time simulations does not significantly influence the hydroelastic response. As the spray root of the jet moves towards the beam ends the total hydrodynamic

force due to the elementary beam formulation increases. This part of the force curve is believed to be unrealistic since the inertia effects of the wetdeck structure outside the modelled beam will become important. Such effects are not accounted for in this formulation and will be studied in the near future. Overall, Figure 4 shows that the total hydrodynamic force due to wetdeck slamming is significantly reduced when the flexibility of the wetdeck structure is accounted for.

The convergence of the hydroelastic response of the Timoshenko beam with the number of eigenfunctions N_{eig} used to express the beam deflections will now be discussed. Figure 5 reproduces the total hydrodynamic force on the beam as a function of time. Results for $N_{eig} = 20, 40, 80$ are shown. A large number of eigenfunctions is needed to achieve convergence of the total hydrodynamic force. Actually, if the hydrodynamic loading had been a delta pulse in time and space, all the vibration modes would have been triggered. One should note that the convergence of the eigenfunction series of the elementary beam formulation is better than for the Timoshenko beam. A possible explanation is that the eigenfrequencies of the vibrations modes for the elementary beam formulation are higher than the corresponding eigenfrequencies for the Timoshenko beam. The convergence at the initial stage of water impact is better than at a later stage of the impact. A possible explanation is the use of the "dry" eigenfunctions, which means that the added mass effect of the surround-

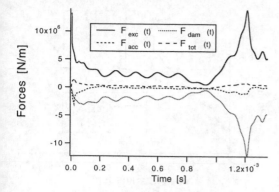

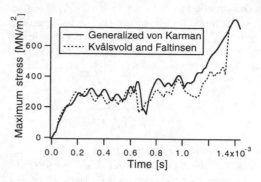

Figure 6: The different force components of the total hydrodynamic force on the beam as a function of time. Input data for the time simulation are presented in Table 1. $F_{exc}(t)$, $F_{dam}(t)$ and $F_{add}(t)$ are identical with $F_{exc,m}(t)$, $F_{dam,m}(t)$ and $F_{add,m}(t)$, respectively, when $\psi_m(x) = 1.0$ (see equation (37) together with the followed discussion of the interpretation of the force decomposition).

Figure 7: The maximum von Mises stress along the Timoshenko beam as a function of time. The stresses due to two different methods of calculating the wetted length of the beam are presented. These are the generalized von Karman approach and the method described by equation (38). The input data for the time simulation are presented in Table 1.

ing water is not considered when the eigenfunctions are evaluated. The "dry" eigenfunctions are very similar to the exact eigenfunctions at the initial stage of the water impact. At the later stage of the water impact, the effect of the surrounding water becomes important for the shape of the eigenfunctions. This implies that the "dry" eigenfunctions deviate from the correct eigenfunctions and consequently, the convergence of the hydroelastic response as a function of the number of eigenfunctions N_{eig} is slower.

We will now focus our attention on the Timoshenko beam formulation and decompose the corresponding total hydrodynamic force on the beam into local excitation force $F_{exc}(t)$, local hydrodynamic damping force $F_{dam}(t)$ as well as local hydrodynamic added mass force $F_{add}(t)$. These force components are identical with $F_{exc,m}(t)$, $F_{dam,m}(t)$ and $F_{add,m}(t)$, respectively, when $\psi_m(x) = 1.0$ (see equation (37) together with the followed discussion of the interpretation of the force decomposition). Figure 6 reveals each of these force components as a function of time. One should note that $F_{exc}(t)$ will for the non-vibratory case, be equal to the total hydrodynamic force on the beam. This is discussed in connection with equation (37). By comparing $F_{exc}(t)$ with the non-vibratory total hydrodynamic force shown in Figure 4, it can be seen that this part of the total hydrodynamic force is, in a particular time interval, reduced with approx-

imately a factor of 3. At the initial and at the later stage of the water impact, $F_{exc}(t)$ may even be larger than the total hydrodynamic force for the non-vibratory case. However, the total hydrodynamic force on the elastic wetdeck is reduced and the reduction is mainly due to a reduction of $\frac{dc(t)}{dt}$ which again is due to the deformations of the beam. As the edge of the flow moves towards the beam ends, the boundary conditions at the beam ends restrict the beam from deflecting. This leads to an increase of $\frac{dc(t)}{dt}$ and subsequently an increase of the total hydrodynamic force. The curves show that the local maxima of the total hydrodynamic force correspond to local maxima of $F_{exc}(t)$ as well as local minima of $F_{dam}(t)$. These minima again correspond to local maxima of the time derivative $2\frac{dc(t)}{dt}$ of the wetted length. This can be seen from equations (46) and (48) in Appendix A. In that way $\frac{dc(t)}{dt}$ and in particular the reduction of $\frac{dc(t)}{dt}$ becomes important in a hydroelastic wetdeck slamming analysis.

The influence on the hydroelastic response by using two different methods of calculating the wetted length of the beam has been investigated. One method has already been described and the other method is a generalization of the von Karman method (von Karman 1929). In the generalized von Karman method, half the wetted length $c(t)$ is the solution x of the integral equation (38) when $\frac{\partial \phi(x,\tau)}{\partial z}$ and $\dot{w}(x,\tau)$ are equal to zero. This means that neither the effect of the pileup water nor the elasticity effects of the wetdeck structure are accounted for when evaluating the wetted length of the beam. We will now discuss the effect the different formulations of estimating the wetted length of the beam have on the stress level. Let σ_{vm} denote the von Mises stress defined as:

$$\sigma_{vm} = \sqrt{\sigma_b^2 + 3\tau_s^2} \qquad (42)$$

where σ_b and τ_s are the maximum normal bending stress and the maximum shear stress, respectively, in the beam cross section area. Here, it is used that:

$$\sigma_b = -E z_{na} \frac{\partial \beta}{\partial x} \qquad (43)$$

$$\tau_s = G \gamma_s \qquad (44)$$

where z_{na} is the distance from the neutral axis in the cross section area to the point where the maximum bending stress occurs. $z_{na} = 0.12$ [m] in this work. γ_s is the shear angle. Figure 7 reproduces the maximum von Mises stress along the beam as a function of time due to the two methods. One should note that the location $x_{stress}(t)$ along the beam where the maximum von Mises stress occurs is a function of time as well. This is discussed later in the text. As seen from the figure, the generalized von Karman method overestimates the stresses in the wetdeck structure. This is also the case for the hydrodynamic loading on the wetdeck. A third approach to evaluate the wetted length of the beam is to neglect the effect of the pileup water but to include the elasticity effects of the wetdeck when solving the integral equation (38). In that approach, $\frac{dc(t)}{dt}$ tended to zero and the solution broke down.

The location $x_{stress}(t)$ along the beam where the maximum stress occurs, is shown in Figure 8 as a function of time. The wetted length is calculated by the method described by equation (38). Initially, the maximum stress occurs at the midpoint of the beam and later $x_{stress}(t)$ moves toward the beam ends. In some time intervals the maximum stress even occurs at the beam ends. The figure also includes half the wetted length $c(t)$ as a function of time. As seen from the figure, the maximum stress occurs in close vicinity to $c(t)$ in the initial and at the later stage of the water impact.

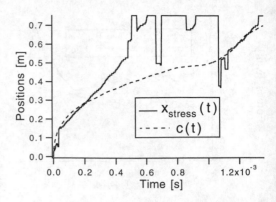

Figure 8: The x-position $x_{stress}(t)$ along the beam there the maximum von Mises stress occurs as a function of time. $2c(t)$ is an approximation of the wetted length of the beam. The input data for the time simulation are presented in Table 1.

6 DISCUSSION

This study is based upon a two-dimensional approach, both when it comes to the structural and the hydrodynamic formulations. It is expected that three-dimensional effects will influence the results and probably reduce the total hydrodynamic forces in the initial stage of water impact. In principle, we know how to generalize this two-dimensional formulation to three dimensions. However, studies are needed to examine the accuracy of the solution of the hydroelastic response as well as the stability of the time integration of the differential equations. Since we have to rely on a numerical method for solving the three-dimensional hydrodynamic flow, difficulties are anticipated of a similar nature as we discussed earlier for the two-dimensional flow.

In the initial stage of water impact an approximate acoustic formulation is followed. Limited numerical studies have shown that the absolute maximum von Mises stress which occurs at the later stage of the water impact is not very sensitive to the initial acoustic formulation.

We do not allow for an air pocket to be trapped between the free surface of the waves and the wetdeck structure in this study. If we for a moment consider a wedge of small deadrise angle moving toward an initially calm free water surface, the creation of an airpocket is partly connected to the large pressure gradients in the airflow near the knuckle of the wedge. If we are located away from

216

the bow end of the multihull vessel, the wetdeck structure does not have such sharp corners. This is at least an indication that an airpocket does not necessarily have to be trapped, but this will be studied in the future. It should be noted that some part of the wetdeck between the stiffeners may be initially buckled due to fabrication or permanent deformations caused by wetdeck slamming. This may cause an aircushion to be created. There are also designers of high-speed multihull vessels that utilize this phenomena to reduce the hydrodynamic loading on the wetdeck.

This study assumes that the crest of a regular wave system hits mid between two of the transverse stiffeners in the wetdeck. In addition, the inertia effects of the wetdeck structure outside the modelled beam are not accounted for. This formulation is now being generalized to account for an arbitrary relative position between the wetdeck and the undisturbed free water surface at the moment of first water impact. This is done by modelling the wetdeck structure as a system of Timoshenko beams. In that way the inertia effects of the wetdeck are more properly accounted for as well. The HBVP expressing the fluid flow in the impact region will differ from what is used here, but can by a linear coordinate transformation be transformed to the HBVP shown in Figure 3. In addition, when the wetted length is larger than the length of the beam, it is not convenient to express the vertical velocities along the wetted length of the beams in terms of a Fourier series as shown in equation (29).

This study is limited to describe the hydroelastic response of the wetdeck at an early stage of water impact, while the wetted length is limited to one beam. The vertical beam vibration velocities $\dot{w}(x,t)$ on the wetted part of the beam were positive in our case study. This will cause a reduction of the magnitude of the effective vertical velocity $V_e(x,t) = V(t) + \dot{w}(x,t)$ on the beam and subsequently a reduction of the total hydrodynamic force on the beam. However, the water impact is not over and we do not know before examining the hydroelastic response of a set of several Timoshenko beams, if the magnitude of the effective vertical velocity $V_e(x,t)$ will increase at a much later stage of the impact so that the hydrodynamic loads and the stresses increase.

7 CONCLUSIONS

The hydroelastic response due to slamming against the wetdeck of a multihull vessel in a head sea regular wave system is investigated. It is assumed that the crest of a regular wave system hits the wetdeck mid between two transverse stiffeners. Shear deformations and rotatory inertia effects are accounted for in the modelling of the wetdeck structure. The hydrodynamic load model is a generalization of Wagner's asymptotic theory (Wagner 1932). This theory is valid for small angles between the undisturbed free water surface and the wetdeck structure.

This study is limited to the early stage of the water impact, when the wetted area is limited between two transverse stiffeners in the wetdeck. At that early stage of the water impact, this work indicates that the hydrodynamic loading is reduced when the elasticity of the wetdeck structure is accounted for. However, further studies to account for a larger impact area are needed before any final conclusions about the slamming load level can be drawn. It is demonstrated that the shear deformations as well as the rotatory inertia effects influence the hydroelastic response of the wetdeck. The wetted area as a function of time is important for the hydrodynamic loads and the maximum stress. It is demonstrated that the effects of the pileup water and the elasticity of the beam are important for the evaluation of the wetted area.

To properly describe the slamming loads on the wetdeck and to ensure the stability of the time integration of the hydroelastic response, it is important to express the local fluid loading in terms of analytical functions. This implies that one should be careful when using a pure numerical method based upon for instance the boundary element method for the fluid loading and the finite element method for the structural deformations, to study slamming. The reason is the rapid variations of the hydrodynamic slamming pressure in the water impact region as a function of both time and space.

8 ACKNOWLEDGEMENT

This work is part of a dr.ing study of one of the authors (J. Kvålsvold) and has received financial support by The Research Council of Norway (NFR). The computer time is supported by the Norwegian Supercomputing Committee (TRU).

REFERENCES

Cointe, R. and Armand, J.L. 1987. Hydrodynamic impact analysis of a cylinder. *J. Offshore Mechanics and Arctic Engineering* 109: 237-243.

Kaplan, P. and Malakhoff, A. 1978. Surface effect ship loads: lessons learned and their implications for other advanced marine vehicles. *Proc. SNAME spring meeting/STAR Symp.*, Pittsburg, PA: SNAME.

Kaplan, P. 1987. Analysis and prediction of flat bottom slamming impact of advanced marine vehicles in waves. *Int. Shipbuilding Progress* 34: 44-53.

Kaplan, P. 1991. Structural loads on advanced marine vehicles, including effects of slamming. *Proc. First Int. Conf. on Fast Sea Transportation FAST'91*: 781-795. Trondheim: Tapir.

Kaplan, P. 1992. Advanced marine vehicles structural loads - present state of the art. *Proc. HPMV'92 Conference and exhibit*: S&T1-S&T12. USA: ASNE, Flagship section.

Kvålsvold, J. and Faltinsen, O. 1993. Hydroelastic modelling of slamming against the wetdeck of a catamaran. *Proc. Sec. Int. Conf. on Fast Sea Transportation FAST'93*: 681-697. Japan: Sos. Nav. Arc.

Meyerhoff, W.K. 1965. Die Berechnung hydroelastischer Stösse. *Schiffstechnik*: 12, 60.

Newman, J.N. 1977. *Marine Hydrodynamics*. Cambridge: The MIT Press.

PUSFEA 1993. PUSFEA - A Package of Utility Fortran Subroutines for the Finite Element Analysis. By Wu M., Dept. of Marine Structures, The Norwegian Institute of Technology (unpublished).

von Karman (1929). The impact of seaplane floats during landing. NACA, Technical Note 321, Washington.

Wagner, H. 1932 *Über Stoss und Gleitvergänge an der Oberfläche von Flüssigkeiten. Zeitschr. f. Angewendte Mathematik und Mechanik* 12: 192-235.

Zhao, R. and Faltinsen, O. 1992. Slamming loads on high-speed vessels. *Proc. of the 19th ONR Conference*, Korea. Washington DC: National Academy Press.

Zhao, R. and Faltinsen, O. 1993. Water entry of two-dimensional bodies. *J. Fluid Mech.* 246: 593-612.

APPENDIX A

The coefficients $B_{mn}(t)$, $A_{mn}(t)$ and $F_{exc,m}(t)$ in equation (37) change depending on whether ω_m and/or ω_n are less or greater than $\sqrt{\frac{GA_s}{M_B r^2}}$. It may be shown that:

$$F_{exc,m}(t) = -\rho V(t)\frac{dc(t)}{dt}\left(I_m^{(6)}(t) + I_m^{(7)}(t) \right)$$

$$-\rho\frac{dV(t)}{dt}c(t)\left(I_m^{(5)}(t) + I_m^{(8)}(t) \right) \qquad (45)$$

For $\omega_n < \sqrt{\frac{GA_s}{M_B r^2}}$ $B_{mn}(t)$ and $A_{mn}(t)$ are written as:

$$B_{mn}(t) = \rho\frac{dc(t)}{dt}\left\{ \left[B_n J_0(p_n c(t)) + D_n I_0(q_n c(t)) \right] \right.$$

$$\times \left[I_m^{(6)}(t) + I_m^{(7)}(t) \right]$$

$$+ \left[-B_n p_n c(t) J_1(p_n c(t)) + D_n q_n c(t) I_1(q_n c(t)) \right]$$

$$\times \left[I_m^{(5)}(t) + I_m^{(8)}(t) \right]$$

$$+ \sum_{k=1}^{\infty}\left[\left(B_n (-1)^k \right. \right.$$

$$\times\frac{p_n c(t) J_{2k+1}(p_n c(t)) - 4k J_{2k}(p_n c(t))}{2k - 1}$$

$$-D_n\frac{q_n c(t) I_{2k+1}(q_n c(t)) + 4k I_{2k}(q_n c(t))}{2k - 1}\right)$$

$$\times \left(I_{km}^{(1)}(t) + I_{km}^{(3)}(t) \right)$$

$$+ \left(-B_n (-1)^k\frac{p_n c(t) J_{2k+1}(p_n c(t))}{2k + 1} \right.$$

$$\left. +D_n\frac{q_n c(t) I_{2k+1}(q_n c(t))}{2k + 1}\right)$$

$$\left.\left.\times \left(I_{km}^{(2)}(t) + I_{km}^{(4)}(t) \right) \right]\right\} \qquad (46)$$

$$A_{mn}(t) = \rho c(t)\left\{ \left[B_n J_0(p_n c(t)) + D_n I_0(q_n c(t)) \right] \right.$$

$$\times \left[I_m^{(5)}(t) + I_m^{(8)}(t) \right]$$

$$- \sum_{k=1}^{\infty}\left[(-1)^k B_n J_{2k}(p_n c(t)) + D_n I_{2k}(q_n c(t)) \right]$$

$$\times \left[\frac{I_{km}^{(1)}(t) + I_{km}^{(3)}(t)}{2k - 1} - \frac{I_{km}^{(2)}(t) + I_{km}^{(4)}(t)}{2k + 1} \right] \right\} \quad (47)$$

Furthermore, for $\omega_n > \sqrt{\frac{GA_s}{M_B r^2}}$, $B_{mn}(t)$ and $A_{mn}(t)$ are written as::

$$B_{mn}(t) = \rho \frac{dc(t)}{dt} \left\{ \left[B_n J_0(p_n c(t)) + D_n J_0(q_n c(t)) \right] \right.$$

$$\times \left[I_m^{(6)}(t) + I_m^{(7)}(t) \right]$$

$$+ \left[-B_n p_n c(t) J_1(p_n c(t)) - D_n q_n c(t) J_1(q_n c(t)) \right]$$

$$\times \left[I_m^{(5)}(t) + I_m^{(8)}(t) \right]$$

$$+ \sum_{k=1}^{\infty} \left[\left(B_n (-1)^k \right. \right.$$

$$\times \frac{p_n c(t) J_{2k+1}(p_n c(t)) - 4k J_{2k}(p_n c(t))}{2k - 1}$$

$$+ D_n (-1)^k$$

$$\times \frac{q_n c(t) J_{2k+1}(q_n c(t)) - 4k J_{2k}(q_n c(t))}{2k - 1} \right)$$

$$\times \left(I_{km}^{(1)}(t) + I_{km}^{(3)}(t) \right)$$

$$+ \left(-B_n (-1)^k \frac{p_n c(t) J_{2k+1}(p_n c(t))}{2k + 1} \right.$$

$$\times - D_n (-1)^k \frac{q_n c(t) J_{2k+1}(q_n c(t))}{2k + 1} \right)$$

$$\left. \left. \times \left(I_{km}^{(2)}(t) + I_{km}^{(4)}(t) \right) \right] \right\} \quad (48)$$

$$A_{mn}(t) = \rho c(t) \left\{ \left[B_n J_0(p_n c(t)) + D_n J_0(q_n c(t)) \right] \right.$$

$$\times \left[I_m^{(5)}(t) + I_m^{(8)}(t) \right]$$

$$- \sum_{k=1}^{\infty} (-1)^k \left[B_n J_{2k}(p_n c(t)) + D_n J_{2k}(q_n c(t)) \right]$$

$$\left. \times \left[\frac{I_{km}^{(1)}(t) + I_{km}^{(3)}(t)}{2k - 1} - \frac{I_{km}^{(2)}(t) + I_{km}^{(4)}(t)}{2k + 1} \right] \right\} \quad (49)$$

In equations (45) to (49) the following notations are implicit:

$$I_m^{(5)}(t) = \frac{B_m \pi}{p_m} J_1(p_m c(t)) \quad (50)$$

$$I_m^{(6)}(t) = B_m \pi c(t) J_0(p_m c(t)) \quad (51)$$

$$I_{km}^{(1)}(t) = \frac{B_m \pi c(t)}{2} (-1)^{k+1} \quad (52)$$

$$\times \left[J_{2k-2}(p_m c(t)) + J_{2k}(p_m c(t)) \right] \quad (53)$$

$$I_{km}^{(2)}(t) = \frac{B_m \pi c(t)}{2} (-1)^k \quad (54)$$

$$\times \left[J_{2k}(p_m c(t)) + J_{2k+2}(p_m c(t)) \right] \quad (54)$$

For $\omega_m < \sqrt{\frac{GA_s}{M_B r^2}}$:

$$I_m^{(7)}(t) = D_m \pi c(t) I_0(q_m c(t)) \quad (55)$$

$$I_m^{(8)}(t) = \frac{D_m \pi}{q_m} I_1(q_m c(t)) \quad (56)$$

$$I_{km}^{(3)}(t) = \frac{D_m \pi c(t)}{2}$$

$$\times \left[I_{2k-2}(q_m c(t)) - I_{2k}(q_m c(t)) \right] \quad (57)$$

$$I_{km}^{(4)}(t) = \frac{D_m \pi c(t)}{2}$$

$$\times \left[I_{2k}(q_m c(t)) - I_{2k+2}(q_m c(t)) \right] \quad (58)$$

For $\omega_m > \sqrt{\frac{GA_s}{M_B r^2}}$:

$$I_m^{(7)}(t) = D_m \pi c(t) J_0(q_m c(t)) \quad (59)$$

$$I_m^{(8)}(t) = \frac{D_m \pi}{q_m} J_1(q_m c(t)) \quad (60)$$

$$I_{km}^{(3)}(t) = \frac{D_m \pi c(t)}{2} (-1)^{k+1}$$

$$\times \left[J_{2k-2}(q_m c(t)) + J_{2k}(q_m c(t)) \right] \quad (61)$$

$$I_{km}^{(4)}(t) = \frac{D_m \pi c(t)}{2} (-1)^k$$

$$\times \left[J_{2k}(q_m c(t)) + J_{2k+2}(q_m c(t)) \right] \quad (62)$$

The total hydrodynamic force $F_{tot}(t)$ on the beam is expressed as:

$$F_{tot}(t) = -\rho V(t)\frac{dc(t)}{dt}c(t)\pi - \rho\frac{dV(t)}{dt}c^2(t)\frac{\pi}{2}$$

$$- \rho\frac{dc(t)}{dt}c(t)\pi \sum_{n=1}^{N_{eig}} \dot{a}_n(t)\left\{ B_n J_0(p_n c(t)) \right.$$

$$+ D_n\Big(I_0(q_n c(t)) + \big[J_0(q_n c(t)) - I_0(q_n c(t))\big]$$

$$\left. \times H(\omega_n - \sqrt{\frac{GA_s}{M_B r^2}})\Big)\right\}$$

$$- \rho c^2(t)\frac{\pi}{2}\sum_{n=1}^{N_{eig}} \ddot{a}_n(t)\left\{ \vphantom{\sum} \right.$$

$$B_n\Big(J_0(p_n c(t)) + J_2(p_n c(t))\Big)$$

$$+ D_n\Big(I_0(q_n c(t)) - I_2(q_n c(t))\Big)$$

$$+ D_n\Big[J_0(q_n c(t)) + J_2(q_n c(t))$$

$$-I_0(q_n c(t)) + I_2(q_n c(t))\Big]$$

$$\left. \times H(\omega_n - \sqrt{\frac{GA_s}{M_B r^2}})\right\} \qquad (63)$$

$H(x)$ is the Heaviside function defined by:

$$H(x) = \begin{cases} 0 & \text{for } x < 0 \\ 1 & \text{for } x \geq 0 \end{cases} \qquad (64)$$

Hydroelasticity in Marine Technology, Faltinsen et al. (eds) © 1994 Balkema, Rotterdam, ISBN 90 5410 387 6

Structural response analysis of cylinders under water impact

T. Shibue, A. Ito & E. Nakayama
Ishikawajima-Harima Heavy Industries Co. Ltd, Yokohama, Japan

ABSTRACT: A series of transient structural response analysis is carried out for drop tests of two dimensional cylinders on water surface to reproduce the time histories of strain under water impact pressure. Measured and calculated strain values show good correlations, showing this method can be an useful tool. Then the effects of maximum impact pressure values on the maximum strain values are evaluated with this method. The measured time histories of pressure are divided into two parts. Each of them are applied to the structural model to estimate the maximum strain values. The strain values obtained from only maximum impact pressure part show less effects on the maximum strain values than that without maximum impact pressure part. This means that not only maximum impact pressure part but the remaining part has major effects on the maximum strain values on the simple local structures. The mass effects on the water impact loads are also discussed.

1. INTRODUCTION

A way to estimate the response of ship structures under the water impact loads caused by the slumming and sloshing is introduced. Many research works (Von Karman 1929, Verhagen 1967, Chuang 1969, Miyamoto 1984) has been carried out for the estimation of water impact load since Karman's works on hydroplane. The design of ship structures under water impact loads had been based on the impact loads estimated by these works. The maximum impact pressure had been thought to be a major index of impact loads in design stage.

The estimation of impact pressure for the objects with arbitrary configurations is attained recently by the use of numerical methods such as marker and cell method(Arai 1993). The effects of elasticity of a hull girder are included into the estimation of ship motion by strip method(Yamamoto 1978). The transient response of a hull girder under wave environment becomes available with this method from the view of longitudinal strength of a ship.

On the other hand, it is necessary to estimate the strain values actually initiate within structures, for the purpose of strength evaluation of local structures under water impact loads. The relations between eigen frequencies of structures and the time duration of impact pressure are discussed by the use of Bagnold model into the plate structure (Takagi 1972).

This report focuses on the transient response of local structures, and present a method to estimate the magnitude of strain under water impact loads. Then, the effects of maximum pressure values on the maximum strain values are examined and the effects of factors on the magnitude of maximum strain values are discussed.

First, a series of drop tests are carried out with thick and thin cylinder in two dimensional water way. The time histories of both water impact pressure and strain are measured on the surface of a thick cylinder at a time. For the thin cylinder, only the time histories of strain are measured. The maximum impact pressure and the maximum strain values initiated on the cylinder surfaces are discussed.

Then, the measured strain are compared with the numerically estimated strain under the measured time histories of impact pressure for the thick cylinder model to ensure the validity of the numerical estimation method. After the confirmation of the validity of the numerical estimation method, the effects of the peak pressure part on the maximum strain values are examined with the same method. The transient analysis are carried out with two different time histories of impact pressure, one is with only peak pressure part and the other is without peak pressure part to see the quantitative effects on the maximum strain values.

The effects of the maximum impact pressure values on the maximum strain values are also evaluated by the transient numerical analysis with two different pressure time histories, one is without a peak

221

pressure data, the other is with a peak pressure data of twice as large as that of measured. The results are compared with the maximum strain values to evaluate the effects of peak pressure data on the maximum strain values.

Finally, the measured time histories of pressure are applied to the thin cylinder model and the calculated strain values are compared with those of the measured values to evaluate the difference of the water impact pressure between the thick and thin cylinder models, i.e. between heavy and light cylinder models.

2. DROP TESTS OF CYLINDERS ON WATER SURFACE

2.1 *Drop tests overview*

The drop tests are carried out as follows to measure time histories of both strain and pressure along the circumference of cylinder models. The drop test apparatus is shown in figure 1. A cylinder model with length of 600mm and outer diameter of 312mm, falls down on two dimensional water way with depth of 1,800mm. A couple of vertical flat plates are settled in the water way to assure two dimensional flow around the cylinder model. The cylinder model is hung beneath a rigid beam by a couple of strings.

The rigid beam is guided by two blocks at both ends among two vertical guide rails to keep the beam horizontal during its fall. The cylinder model is lifted up by an electric magnet to a certain height at tests. The cylinder model with a rigid beam falls on water surface by switching off the electric magnet. As the rigid beam collides with a pair of dumper at the roots of vertical guide rails, just before the cylinder collides with water surface, the cylinder model collides with water surface without effects of the rigid beam. Latex sheets are used to cover both ends of the cylinder model for the purpose of water tightening.

Two kinds of cylinder models with same scantlings except their thicknesses, are prepared as shown in figure 2. One is the thick model with 5.1mm thickness which is made of mild steel(JIS STPG370), the other is the thin model with 1.0mm thickness which is made of mild steel(JIS SPCC).

The semi-conductor type strain gages (Kyowa: KFG-2-120-D16-11 L5 M3S) are attached on the inner surface of both models as shown in figure 2. The diaphragm type pressure gages (Druck:PDCR-81, Kyowa:PGM-5KC & PGM-2KC) are also attached through the holes on thick cylinder model. The signals from these sensors are recorded by a personal computer through dynamic strain meters and analogue/digital converters with maximum conversion rate of 200kHz. The maximum response

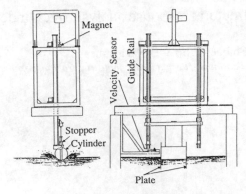

Figure 1 Drop test setup

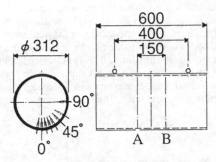

Figure 2 Cylinder model for drop tests

frequency of the pressure gages are 100kHz and that of the strain gages are 5kHz. The drop heights emploied are 0.5m, 1.0m and 1.5m.

2.2 *Drop tests with a thick cylinder model*

The thick cylinder model is equipped with both pressure and strain gages. Mass of the cylinder is 23.8kg. The normalized values of measured maximum pressure are compared with the values obtained from approximating equations by Wagner and Chuang as shown in figure 3. According to this figure, measured maximum pressure can be estimated fairly well by the estimating equation by Chuang. The angle of 0 degree means the first contact point with water surface. The maximum pressure at 0 degree varies according to the drop heights. The variations may caused by the differences of impact angles between the diaphragm surface and water surface. The results of the drop test with drop height of 1.0m shows the best fit with the Chuang's estimations.

Measured time histories of pressure and strain for the angles of 0, 10, 20 and 30 degrees, in case of 1.0m drop height are shown in figure 4. The

222

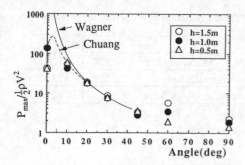

Figure 3 Normalized maximum impact pressures

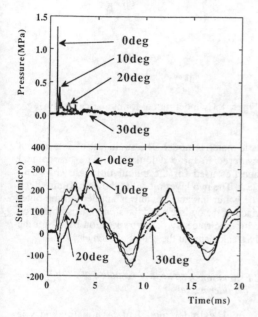

Figure 4 Measured pressures and strains
(Thick cylinder test)

$$f_i = \frac{1}{2\pi}\sqrt{\frac{E}{\rho}\cdot\frac{I}{Ar^4}\cdot\frac{i^2(i^2-1)^2}{i^2+1+v}} \qquad (1)$$

where:
  f_i :eigen frequency for wave number i
  ρ:Mass density of cylinder
  E :Young's modulus
  v:Poisson's ratio
  r:Radius of cylinder
  A:Sectional area
  I:2nd moment of inertia of section

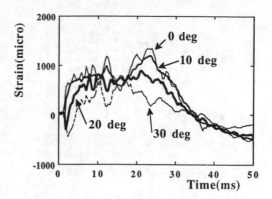

Figure 5 Measured strains (Thin cylinder test)

maximum pressure at 0 degree appears about 0.05milliseconds after the contact, whereas the maximum strain at 0 degree appears about 4milliseconds after the contact. The first peak of the strain values gives the maximum values. The pressure at 10 degrees shows apparent time delay of about 0.3milliseconds compared with that of 0 degree, whereas the strain shows no time delay between these angles. The vibration period of the cylinder is about 8milliseconds according to the measured results, which coincides with the estimated flexural vibration period of two dimensional cylinder estimated with equation(1) (Timoshenko 1955), which is 7.5milliseconds.

2.3 Drop tests with a thin cylinder model

The thin cylinder model is equipped with only strain gages, because it is difficult to set pressure gages on the cylinder shell of 1mm thickness. Mass of the cylinder is 5.0 kg. Measured time histories of strain for the angles of 0, 10, 20 and 30 degrees, in case of 1.0m drop height are shown in figure 5. The maximum strain at 0 degree appears about 22milliseconds after the contact. The duration is about 5 times longer than that of thick cylinder model.

The maximum strain values are shown in figure 6 compared with those of the thick cylinder model. The maximum strain values of the thin cylinder model exceed the elastic limit, whereas those of the thick cylinder model do not exceed it.

The maximum strain ratio between thick and thin cylinder models are shown in figure 7. These ratio is about 4 between 0 degree and 20 degrees. The maximum strain ratio of about 4 implies the difference of stiffness, mass and material nonlinearities between two models.

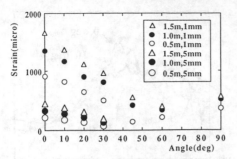

Figure 6 Measured maximum strains

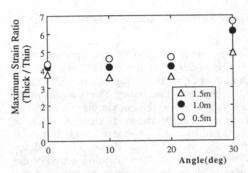

Figure 7 Ratio of maximum strains between thick and thin tests

Table 1 Mechanical properties of cylinders

	SPCC	STPG370
Young's Modulas(MPa)	206,000	206,000
Yield Stress(MPa)	157	215
Tangent Modulus(MPa)	306	621

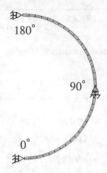

Figure 8 Cylinder model for numerical analysis

3. NUMERICAL ANALYSIS

3.1 *Numerical simulation overview*

The objects of numerical simulations are the drop tests with drop height of 1.0m for both thick and thin cylinder models. The aim of numerical simulations is to ensure the capability of the analytical tool to reproduce transient strain response of the two dimensional cylinder under water impact pressure, and to validate the effect of peak part of the water impact pressure on the maximum strain values on the surface of cylinder models.

DYNA3D is used as a analytical tool to estimate the transient strain response.

The model of the two dimensional cylinder is expressed by a slice of cylinder with length of 10mm, and is applied the continuous boundary conditions at the both ends. The lowest point of the model is identified as 0 degree and highest as 180 degrees. Right left symmetricity is also introduced and the vertical support is applied at the point of 90 degrees. The models are divided into 36 meshes in circumferential direction, 2 or 3 meshes in thickness direction and 1 mesh in longitudinal direction. A 8 nodes solid element is used along with the elastic plastic material model. The strain rate effects are not considered in these calculations. The mechanical properties used for the calculations are shown in table 1. The two layer model is shown in figure 8.

The water impact pressure is applied on the outer surface of the cylinder and the strain are estimated on the inner surface by the extrapolation based on the values calculated at the center of each elements.

3.2 *Estimation of structural response of a thick cylinder model*

The thick cylinder model of 5.1mm thickness is expressed by two layers model, i.e. the shell thickness is divided into two layers. The measured time histories of pressure are applied to the cylinder model. The calculated strain time histories are compared with the measured values at the points of 0, 10, 20 and 30 degrees as shown in figure 9. The maximum strain values are less than 600 micro strain. This means the responses are fully elastic.

The calculated maximum strain values are compared with the measured values in table 2. The estimation errors at 0 and 10 degrees are considerably large. In spite of these error, calculated strain time histories show the same tendencies and good agreement with measured values on the whole. These results show that this estimation method can be used for simple structures to reproduce the time histories of strain under water impact pressure.

Then, we divided the measured time histories of

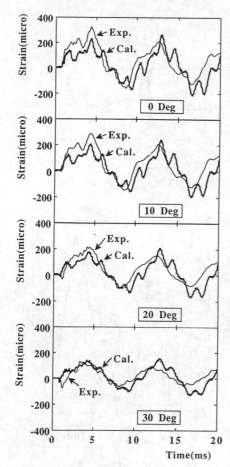

Figure 9 Calculated and measured strains
(Thick cylinder test)

Table 2 Maximum strain values
(Thick cylinder test)

Ang.	Exp.	Cal.	Exp./Cal.
0	322	230	1.40
10	286	208	1.38
20	215	179	1.20
30	131	139	0.94

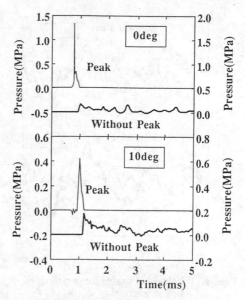

Figure 10 Devided pressure models

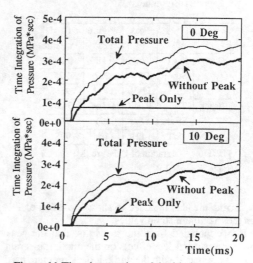

Figure 11 Time integration of devided pressure
models

pressure into two parts, one is with peak part only, and the other is without peak part, as shown in figure 10. These time histories of pressure are applied to the same model to see the effects of peak part of the water impact pressure on time histories of strain. To evaluate the pressure energy, we introduced the value of TIP(Time integration of pressure). TIP values are shown in figure 11 for both 0 and 10 degrees. The calculated time histories of strain are compared with the calculated values with the full time history of pressure as shown in figure 12. It is clear that the time histories of pressure with peak part has less effects on maximum strain values than that without peak part.

Table 3 shows the maximum strain values within the first half wave of the strain for two divided pressure models and full pressure model. The TIP values at the time when the maximum strain

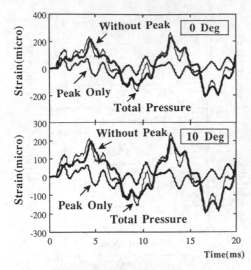

Figure 12 Calculated strains for two devided
pressure models

Table 3 Maximum strain values for two different
pressure models

		Peak	Without Peak	Total Pressure
0deg	TIP	0.066	0.182	0.233
	Strain	75.5	207	225
	S/T	1144.	1137.	966.
10deg	TIP	0.047	0.164	0.204
	Strain	66.8	191	203
	S/T	1421.	1165.	995.

TIP:Time Integration of Pressure (MPa* ms)
S/T: Strain/T.I.P.

occurred, are also included to show the effect of TIP on the maximum strain values.

It is concluded from these results, that the peak pressure part has less effects on the maximum strain values compared with the without peak pressure part, and maximum strain values are proportional to TIP values. The impulse period of peak pressure is about 0.08 milliseconds, and is much shorter than 7.5milliseconds which is the eigen period of the two dimensional vibration of cylinder.

Finally, two kinds of modified time history of pressure are introduced. The measured peak data of the time history of pressure is intentionally modified to its twice value and zero as shown in figure 13. As pressure data is recorded at every 40 microseconds, maximum amount of pressure energy comes to 40 milliseconds multiplied by maximum pressure (1.3MPa), i.e. 5.2×10^{-5}(MPa x second). Two sets

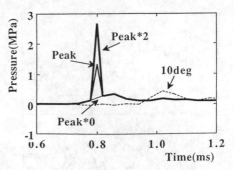

Figure 13 Modified pressure models

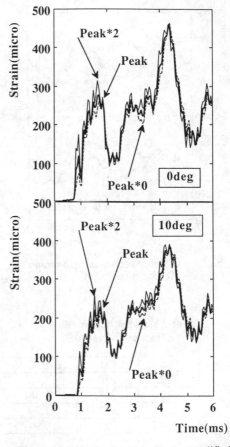

Figure 14 Calculated strains for two modified
pressure models

of modified pressure are applied to the thick cylinder model to see the effect of the peak value of pressure to the maximum strain value.

The calculated time histories of strain are shown in figure 14. There are two peaks of strain, first peak occurs at 1.7milliseconds and the maximum peak

226

Table 4 Maximum strain values for two modified pressure models

		Peak*2	Peak	Peak*0
T.I.P.		0.256	0.233	0.211
Strain Values	1.70ms	313	277	261
	4.35ms	459	459	459
S/T	1.70ms	1223	1189	1237
	4.35ms	1793	1970	2175

T.I.P.:Time Integration of Pressure (MPa x ms)

S/T: Strain/T.I.P.

occurs at 4.35milliseconds after the time origin. Calculated peak values are shown in table 4 with the TIP values up to these times. Certain effect of the peak value of pressure is observed at the first peak of strain, but no effect is observed at the second and the maximum peak.

3.3 Estimation of structural response of a thin cylinder model

The thin cylinder model of 1.0 mm thickness is expressed by three layers model, i.e. the shell thickness is divided into three layers. The measured time histories of pressure are applied to this model. The calculated time histories of strain are compared with the measured values at the points of 0, 10, 20 and 30 degrees as shown in figure 15. The calculated maximum strain values are greater than 15,000 micro strain. This means the responses are fully plastic, in spite of the fact that the measured strain values are less than 1,500 micro strain. The difference in strain values varies with the angle, and the maximum difference comes to 10 times.

4. CONSIDERATIONS

The differences in measured strains between thick and thin cylinders are possibly initiated from the difference of stiffness, the difference of water impact load and the nonlinearities of materials. The strain is proportional to the inverse of square of thickness, which means the strain initiates on the thin cylinder is 26 times as large as that of thick cylinder, so far as water impact loads are the same and the elasticity is assumed. In actual, the strain on thin cylinder is about 4 times larger than that of thick cylinder even within the area of elastic range.

The remaining factor is the difference in water impact load. According to the above considerations, the load actually applied to the thin cylinder must be

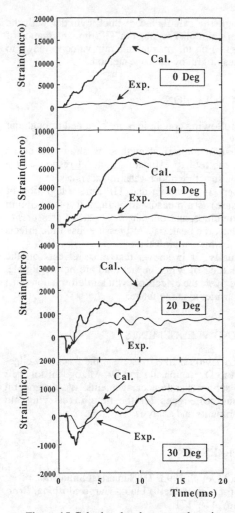

Figure 15 Calculated and measured strains
(Thin cylinder test)

about one sixth(i.e. 4/26) to that applied to the thick cylinder. This difference may caused by the difference in mass. The ratio of mass between two cylinder models are about one fifth(i.e.5.0/23.8). This tendency is also confirmed by the comparison of strain values of thin cylinder tests and its transient calculation.

The effectiveness of the numerical tool for the estimation of transient strain response of simple structure under water impact load is confirmed through the comparison of measured and calculated strain for the drop test of a thick cylinder model.

The effect of peak part of water impact pressure on the maximum strain value is shown to be not so large compared with the effect of the remaining part, through the calculation of thick cylinder with the numerical tool. The effect of TIP(time integration of pressure) on the maximum strain value is shown to be meaningful by the same method.

5. CONCLUSIONS

The following conclusions are obtained through our study.

First, a numerical method is shown to be an efficient tool to estimate structural response of a simple structure under water impact pressure.

Then, it is shown that TIP(time integration of pressure) is a measure in estimation of maximum strain values, and only the peak part of pressure, as much as the peak data of pressure, has little effects on maximum strain values.

Finally, it is shown that mass effects on the magnitude of water impact load are not negligible. More over, the effects of cylinder deformation must be considered in the future.

ACKNOWLEDGEMENTS

Authors would like to express their acknowledgement to Dr. Yamasaki and Dr. Mizoguchi for their kind supports and encouragements. We are grateful to our colleagues for the help in carrying out experiments and analysis.

REFERENCES

Arai, M. et al. 1993. 3D Numerical Simulation of Impact Load due to Liquid Cargo Sloshing. *Proc. of S.N.A.J.* 171

Chuang, S.L. 1969. Impact Pressure Distribution on Wedge Shaped Hull Bottoms of High Speed Craft. *NSRDC* 2953

Karman, V. 1929. The Impact on Sea Plane Floats during Landing. *NACA* TN321

Miyamoto, T. & Tanizawa, K. 1984. A Study of the Impact Load on Ship Bow. *Proc. of S.N.A.J.* 156

Takagi, M. & Togano, Y. 1972. On the Impulsive Pressure Acting on Elastic Structures Induced by Compression of Fluid. *Proc. of S.N.A.J.* 132

Timoshenko, S. 1955. Vibration Problems in Engineering. Tokyo Tosyo

Verhagen, J.H.G. 1967. The Impact of a Flat Plate on a Water Surface. *J. of Ship Research* 11.

Yamamoto, Y. et al.1978. Motion and Longitudinal Strength of a Ship in Head Sea and the Effects of Non-Linearities. *Proc. of S.N.A.J.* 132

Springing of ships

Hydroelastic analysis of a SWATH in waves

W.G. Price & P. Temarel
Department of Ship Science, University of Southampton, UK

A.J. Keane
Department of Engineering Science, University of Oxford, UK

ABSTRACT: This paper presents results for the steady state responses of a SWATH in waves. The issues arising from idealising the structure of the SWATH from detailed structural information are addressed. The dynamic *in vacuo* analysis of the refined finite element model of the structure is investigated with particular reference to modelling difficulties and use of mode shapes. Comparisons are made between simplified and refined finite element models. Stress distributions in the structure are obtained for the SWATH travelling at various speeds and heading angles in irregular seas described by wave spectra (frequency domain analysis). The application of the three-dimensional hydroelasticity theory as a design tool is discussed.

1 INTRODUCTION

Two-dimensional hydroelasticity theory for flexible beamlike hulls was developed in the seventies and applied successfully to investigate the steady state and transient (e.g. slamming) loads and stresses of a variety of merchant and naval ships at seas (Bishop & Price 1979). This idealisation is not suitable, however, for structures which are not beamlike such as multi-hull vessels, jack-up rigs, catamarans, semi-submersibles etc. Thus in the eighties, a three-dimensional hydroelasticity theory was developed to describe the behaviour of fixed, floating or submerged structures of arbitrary shape subjected to internal or external excitation (Bishop, Price & Wu 1986). This theory has been used to investigate the dynamic behaviour of SWATHs in waves (Bishop, Price & Temarel 1986; Price, Temarel & Wu 1987), the problem of jack-up transportation (Fu, Price & Temarel 1987), the behaviour of a dry dock (Lundgren, Price & Wu 1989), a thin cylindrical shell immersed below the free water surface (Ergin, Price, Randall & Temarel 1992), etc. Studies incorporating the responses to transient excitation are also in progress (Aksu, Price & Temarel 1991a,b) following the earlier work on slamming by Belik, Bishop & Price (1988).

Estimation of fluid loading using traditional hydrodynamic approaches, with particular reference to multi-hull vessels, has been evolving since the mid seventies (Lee & Curphey 1977; McCreight 1987; Wu & Price 1987). Following this approach, however, difficulties arise in specifying the dynamic force boundary conditions for the subsequent structural analysis needed to asses the adequacy of the design. Due to the complexity of this problem simplified static boundary conditions have to be introduced leading to a static or quasi-static prediction of structural responses (Reilly, Shin & Kotte 1988). Such an approximation, undoubtedly, results in creating some uncertainties. Unlike these traditional hydrodynamic approaches, the unified hydroelasticity theory can be used to predict the seakeeping performance as well as the structural responses of SWATHs and other multi-hull vessels in waves without introducing such approximations. The assumptions involved in this theory are those inherent to potential flow theory. This theory has been modified to allow for fluid viscous damping effects based on a Morison type approximation (Price & Wu 1989).

The hydroelastic approach is not only more rigorous theoretically but also provides more flexibility regarding its use at different stages of the design of SWATHs (Keane, Temarel, Wu & Wu 1991). These features are also illustrated in this paper by the presentation of selective results from the analysis of a vessel similar in type to the T-AGOS 19 SWATH.

2 THEORETICAL BACKGROUND

Linear hydroelasticity theory was developed to predict the dynamic responses of flexible structures in a seaway. This is carried out in two stages. First the *dry* analysis in which the free vibrations of the structure in vacuo, in the absence of external forces and structural damping, is examined. Subsequently, in the *wet* analysis external forces due to fluid actions, mechanical excitation, etc are applied and allowance is made for the structural damping. The principles of the analysis are the same for the two and three dimensional hydroelasticity theories. The idealisation of the structure and the evaluation of the fluid-structure interaction, however, employ different methods. In the former the structure is idealised as a Timoshenko beam whilst a strip theory is used to determine the hydrodynamic coefficients and wave excitation associated with rigid body motions and distortions (Bishop & Price 1979). On the other hand for the the the three-dimensional theory, a finite element idealisation of the structure and a panel element discretisation of the wet surface of the hull is employed. In this case a source whose strength is determined from boundary conditions is situated at the centre of each panel and use is made of suitable Green's functions (Bishop, Price & Wu 1986).

The three-dimensional hydroelasticity theory has been extensively described in the literature and only its main features are summarised here. The free undamped vibrations of a structure in vacuo are governed by

$$\mathbf{M}\,\ddot{\mathbf{U}} + \mathbf{K}\,\mathbf{U} = 0 \qquad (1)$$

where $\mathbf{M}$ and $\mathbf{K}$ are the structural mass and stiffness matrices and $\mathbf{U}$ the nodal displacement vector. The solution of eq.(1) provides the natural frequencies ω_r and corresponding principal mode shapes of the structure, as well as, dynamic characteristics such as generalised masses, modal stresses etc. It can be shown that eq.(1), with the addition of structural damping and external forces, can be expressed in terms of the corresponding principal coordinates as (Bishop, Price & Wu 1986)

$$\mathbf{a}\,\ddot{\mathbf{p}}(t) + \mathbf{b}\,\dot{\mathbf{p}}(t) + \mathbf{c}\,\mathbf{p}(t) = \mathbf{Z}(t) . \qquad (2)$$

In this equation the $n \times n$ diagonal matrices $\mathbf{a}$, $\mathbf{b}$ $(b_{rr} = 2\,\nu_r\,\omega_r\,a_{rr})$ and $\mathbf{c}$ $(c_{rr} = \omega_r^2\,a_{rr})$ represent the generalised mass, generalised structural damping (with rth modal damping factor ν_r) and generalised stiffness of the structure,

respectively. $\mathbf{p}$ denotes the $n \times 1$ principal coordinate vector. n is the number of principal mode shapes admitted to the analysis, consisting of six rigid body modes with zero natural frequencies $(\omega_r = 0 , r = 1,...,6)$ corresponding to surge, sway, heave, roll, pitch and yaw and $n-6$ distortional modes obtained from the dry hull analysis. The generalised excitation in regular sinusoidal waves is represented by the $n \times 1$ vector $\mathbf{Z}$ whose elements are expressed as

$$Z_r(t) = \left\{ \Xi_{or} + \Xi_{dr} + \sum_{k=1}^{n} p_k (\omega_e^2 A_{rk} - i\,\omega_e B_{rk} - C_{rk}) \right\} e^{i\omega_e t} . \qquad (3)$$

Here the $n \times 1$ vectors Ξ_o and Ξ_d are the generalised excitation due to incident waves (Froude-Krylov) and diffracted waves and depend on wave amplitude a and frequency ω, forward speed U and heading angle χ. The $n \times n$ matrices $\mathbf{A}$ and $\mathbf{B}$ represent the encounter frequency $[\omega_e = \omega - (U\,\omega^2/g)\cos\chi]$ dependent generalised added mass and hydrodynamic damping and $\mathbf{C}$ is a generalised fluid restoring matrix.

Assuming the principal coordinates to vary with time as $\mathbf{p}(t) = \mathbf{p}\,e^{i\omega_e t}$, we find that eq.(2) becomes

$$\left[-\omega_e^2(\mathbf{a}+\mathbf{A}) + i\,\omega_e(\mathbf{b}+\mathbf{B}) + (\mathbf{c}+\mathbf{C}) \right]\mathbf{p} = \Xi_o + \Xi_d , \qquad (4)$$

providing the solution for the complex principal coordinate amplitudes $\mathbf{p}$ in regular sinusoidal waves.

The corresponding responses of the structure can be obtained using modal superposition. For example, the stress component σ_x anywhere in the structure is

$$\sigma_x(x,y,z,t) = e^{i\omega_e t} \sum_{r=7}^{n} \sigma_{x_r}(x,y,z)p_r \qquad (5)$$

where the summation commences from $r=7$ as the modal stress σ_{x_r} are zero for $r = 1,...,6$.

3 FINITE ELEMENT MODELLING

In order to establish the *dry* or *in vacuo* dynamic characteristics of non-beamlike vessels such as SWATHs, semi-submersibles, dry docks and jack-up rigs, idealisation of the structure can, currently, be only achieved by Finite Element Analysis. In some previous applications of hydroelasticity theory a simplified idealisation of the SWATH structure using a relatively small number of finite elements and degrees of freedom was employed. Nevertheless, these idealisations encapsulated the global dynamic behaviour of the structure well, were capable of predicting stresses arising from dynamic wave loading anywhere in the structure and identified regions where large loads and stresses arise as a result of global dynamic loading (Price, Temarel & Wu 1987; Keane, Temarel, Wu & Wu 1991). Such idealisations, being less time consuming and not requiring detailed information about the distribution of structural material and weights, are very useful during preliminary design stages. On the other hand for the main model under discussion in this paper, a more detailed structural idealisation and weight distribution is adopted which is more suitable for the evaluation of the final design and the assessment of the effects of dynamic loads.

3.1 Refined Model

The model adopted uses plate and shell elements to account for all the principal steel-work and stiffeners. Remaining items, including those not contributing to the structure, are represented by lumped and distributed masses and beams. Individual stiffeners, for example, are subsumed in plate elements with equivalent properties of stiffened plating. Minor bulkheads, on the other hand, do not contribute to structural strength and, thus, their mass is smeared into the structural weight.

Forty five percent of the full load displacement weight of the SWATH is modelled as the principal structure which defines the overall structural rigidity and includes smeared mass contributions accounting for minor bulkheads, local stiffening, welding and mill tolerance. Lumped masses corresponding to individual items of equipment account for 14% of the total weight and beam elements modelling masts, stacks, tetrapods and some large items of equipment correspond to 3% of the total weight. The main fuel tanks, corresponding to 17% of the total weight, are modelled by distributing their mass locally. The remaining 21% of the total weight, corresponding to design and build margins and various systems and outfits which cannot be identified individually, is represented as mass smeared over the model so that the correct centres of mass for these items are attained.

For simplicity the structure, but not the lumped and distributed masses, has been modelled with port-starboard symmetry. In all, the model consists of approximately 4700 (including 343 lumped mass and 393 beam) elements with 34000 degrees of freedom. A view of the model is shown in Fig.1. The bulk of the structure, thus, is modelled using a mixture of isoparametric flat plate(with three and four nodes) and semi-loof curved shell (with six and eight nodes) elements. Triangular (i.e. three or six noded) elements have only been used where required by changes of shape and/or mesh density.

Initial calculations using only six and eight noded semi-loof elements, which simplifies data preparation and maximizes consistency, resulted

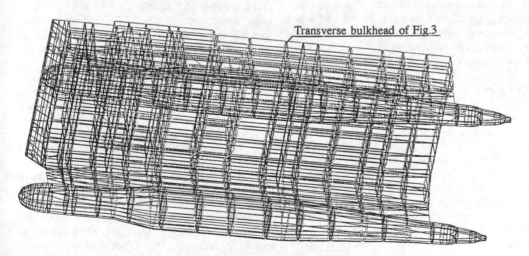

Fig.1 A three-dimensional view of the refined finite element model.

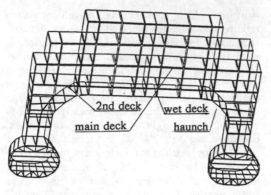

Fig.2 A typical cross section of the refined finite element model.

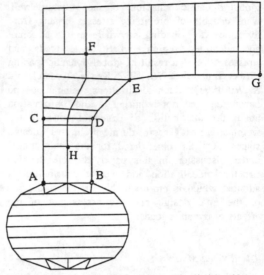

Fig.3 A typical bulkhead (see Fig.1) identifying various nodes of interest.

not only in longer run times but, more importantly, the presence of *local modes* in the solution to the eigen value problem. Such modes have significant amplitudes over very small, localised regions of the structure. They usually arise from lumped masses on, relatively, weakly supported plating which is a consequence of not modelling all the detailed stiffening supporting such equipment represented by the lumped masses. They can also occur due to instabilities in the semi-loof elements, particularly where small semi-loof elements are grouped together. These local modes have low natural frequencies and, thus, swamp the global modes required for use in the hydroelasticity analysis. To reduce this difficulty three and four noded isoparametric flat plate elements can be used as they do not allow for local bending mechanisms. However, they may, strictly, only be used for flat regions and they require their transverse degrees of freedom to be supported by other elements to prevent elements rotating about each other. These have only been used for bulkheads, upper decks and the parallel middle body of the hull.

Thirty five longitudinal divisions were used to model the geometry of the SWATH. These longitudinal divisions reflect the frame spacing of the hull design as well as changes in the geometry. Between these positions, typically 102 elements were used to model the shell plating, main and second decks, platforms and longitudinal bulkheads, as shown in Fig.2. The complete superstructure shell plating and upper decks were modelled using 136 isoparametric flat plate elements. Transverse bulkheads, as shown in Figures 2 and 3 are, typically, modelled using 96 quadrilateral and 24 triangular isoparametric flat

plate elements, the latter restricted to the lower hull where rapid changes of geometry occur. Stiffeners, as mentioned previously, are incorporated into plates with equivalent properties in the principal direction of stiffening. Large openings in bulkheads, decks and platforms which are reinforced by deep frames are modelled as plates with equivalent stiffness and (reduced) mass properties.

Lumped mass elements, with no associated stiffnesses, used to model individual large items of equipment are attached with suitable arms to existing nodes of the finite element model. Some large items are represented by several lumped masses whilst maintaining the correct centre of mass. Items such as masts and tetrapods are suitably represented by beam elements. In addition, beams were used for large and heavy non-structural items to avoid heavy lumped mass elements or items such as stacks which require excessively long arms of attachment. Some beam elements were also used to support isoparametric flat plate elements where no other support is available.

It is well known that stiffened plate structures do not have the ability to transmit full loads across the complete width of plating. Thus, midway between stiffeners, especially when the stiffening is widely spaced, the effectiveness of plating is reduced. This effect, known as *shear lag*, is allowed for in the model by reducing the value of Young's modulus in certain regions, such as main and wet decks, strut and haunch lying in between transverse bulkheads and the superstructure.

3.2 Simplified model

This model, also referred to as *stick* model, uses thick plate elements for the deck and struts and beam elements with offset for the pontoons or lower hulls. In addition, further beam elements were used in the region of the struts and deck to simulate the behaviour of the model more realistically. The model, shown in Fig.4, comprises 344 elements and has port-starboard symmetry. Various models were tested using different combinations of element types as suggested by Stirling, Jones & Clarke (1988). Allowance for shear lag effects was also considered, although it is difficult to apply to this

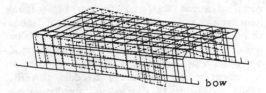

Fig.4 The simplified finite element model with the broken lines denoting the mode shape corresponding to 1.319 Hz.

simplified idealisation. Comparisons of the simplified and refined models are discussed in the dry analysis section.

4 DRY ANALYSIS

The PAFEC finite element package was used in idealising the SWATH and obtaining its *dry* or *in*

Table 1. Natural frequencies and generalised masses (normalised to unit maximum displacement) of the first 50 distortion modes of the refined model.

Mode number	Natural Frequency (Hz)	Generalised Mass (tonnes)	Mode type *	Mode number	Natural Frequency (Hz)	Generalised Mass (tonnes)	Mode type *
49	0.765	0.156	L	74	2.596	1.33	L
50	0.786	0.114	L	75	2.636	2.09	L
51	0.894	722.0	G	76	2.640	1.94	L
52	0.901	0.095	L	77	2.740	1.85	L
53	1.007	0.084	L	78	2.780	3.72	G
54	1.136	0.065	L	79	2.848	2.40	G
55	1.230	0.087	L	80	2.861	0.016	L
56	1.422	0.036	L	81	2.886	0.018	L
57	1.422	0.036	L	82	2.975	5.22	L
58	1.721	0.021	L	83	2.979	1.08	L
59	1.799	81.5	G	84	2.981	2.73	L
60	1.837	2.32	L	85	2.984	0.707	G
61	1.840	2.13	L	86	2.990	1.58	L
62	1.843	2.31	L	87	2.991	2.96	G
63	1.844	2.27	G	88	2.993	1.40	G
64	1.856	4.44	L	89	3.000	0.623	L
65	1.934	1.74	L	90	3.006	0.997	L
66	1.949	1.73	L	91	3.014	0.057	L
67	2.163	2.99	L	92	3.014	0.057	L
68	2.305	3.35	L	93	3.096	0.014	L
69	2.480	2.72	L	94	3.099	1.83	G
70	2.531	4.83	L	95	3.108	0.017	L
71	2.548	1.68	L	96	3.117	0.670	L
72	2.551	1.51	L	97	3.130	0.828	G
73	2.593	1.47	L	98	3.206	0.762	L

* L: local   G: global

vacuo dynamic characteristics such as natural frequencies and corresponding principal mode shapes and modal stresses. The eigen value problem is solved using Guyan's reduction method by which the problem is reduced to one of equivalent master degrees of freedom (PAFEC 1986). 200 master degrees of freedom were selected and the natural frequencies and generalised masses corresponding to the first 50 distortion modes are shown in Table 1. It should be noted that the solution includes 48 rigid body modes, comprising of 6 true rigid body modes and 42 mechanisms as a result of lumped mass attachments to isoparametric flat plate elements, with natural frequencies below 0.0018 Hz. Theoretically, these natural frequencies should be zero and the small values obtained is an indication of the numerical accuracy of the process.

Since the use of semi-loof elements in conjunction with lumped masses was not, altogether, eliminated, Table 1 still contains *local* modes. For a more efficient and less time consuming wet analysis, a relatively small number of global distortion modes, namely 10, is included. This necessitates a distinction between *global* and *local* modes. It is easily seen from Table 1 that modes 51 and 59 are, undoubtedly, global. The remaining modes were identified examining, in detail, the corresponding mode shapes. Although some mode shapes are easily identifiable as local or global, this process is, in general, difficult and time consuming. It should be noted, however, that the first two global modes are bound to provide the dominant contributions to the responses of the SWATH in waves. Classification of the global

modes in terms of their symmetry or not, about a longitudinal centre plane of symmetry, is also made by observation. The resultant global distortion modes are shown in Table 2. A view of the first global distortion mode shape is shown in Fig.5.

The natural frequencies for the simplified model are shown in Table 3, together with the effects of shear lag. A comparison of Tables 2 and 3 illustrates the influence of modelling. Similarities as well as variations are observed with the overall trends showing good agreement. An example of a mode shape is shown in Fig.4, denoted by broken lines.

5 WET ANALYSIS

5.1 Hydrodynamic properties

The wet surface of the SWATH was discretised by three or four noded panels. The discretisation of the surface of the SWATH below the water line used in the finite element analysis of the refined model was adopted for convenience, although such a one to one correspondence is not necessary. The resulting model contained 573 panels for one half of the SWATH. Note that for computing efficiency the structure was treated as port-starboard symmetric, namely using only the port/starboard half of the mode shapes obtained in the dry analysis, as the resultant non-symmetry was very small indeed. The underwater port-starboard symmetric or antisymmetric global distortions were used to evaluate the generalised

Table 2. *Dry* natural and *wet* resonance frequencies of the refined model - global distortion modes only.

Modal index r	Mode type *	Natural frequency ω_r (Hz)	Resonance frequency (Hz)
7	A	0.894	0.649
8	S	1.799	1.078
9	S	1.844	1.840
10	A	2.780	1.873
11	A	2.848	2.819
12	S	2.984	2.978
13	A	2.991	2.984
14	A	2.993	2.992
15	A	3.099	3.089
16	A	3.206	3.118

*   S: symmetric   A: antisymmetric

Table 3. *Dry* natural frequencies ω_r of the simplified model.

Modal index r	without shear lag		with shear lag	
	ω_r (Hz)	Mode type *	ω_r (Hz)	Mode type *
7	1.319	A	1.319	A
8	1.503	S	1.503	S
9	1.566	S	1.566	S
10	3.270	A	2.737	A
11	3.519	A	3.270	A
12	3.687	A	3.344	S
13	3.918	A	3.687	A
14	3.981	A	3.918	A
15	4.032	A	3.981	A
16	4.044	A	4.032	A

*   S: symmetric   A: antisymmetric

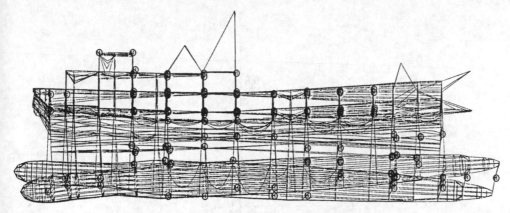

Fig.5 A view of the mode shape corresponding to 0.894 Hz of the refined finite element model.

added mass, hydrodynamic damping and restoring coefficients as well as the generalised wave excitation as described in section 2. These properties were also calculated for the rigid body modes, including the cross coupling effects between bodily motions and distortions. These calculations were performed for the refined model only over a frequency range between 0.02 and 3.34 Hz containing all the resonances which arise from the modes included in the analysis. A selection of the calculated diagonal elements of the of the non-dimensional generalised added mass and hydrodynamic damping coefficients are shown in Fig.6. The trends observed in these coefficients are similar to those previously derived from the application of hydroelasticity theory (Bishop, Price & Wu 1986; Keane, Price, Temarel, Wu & Wu 1988). It was found that the magnitudes of the coefficients corresponding to distortion modes $r \geq 9$ are much smaller than those with modal index $r = 7$ or 8 .

5.2 Principal coordinates and resonances

The principal coordinates were evaluated from eq.(4) using the information obtained in the previous sections. Structural damping was allowed for assuming modal damping factors used in previous applications, i.e. $v_7 = 0.0052$, $v_8 = 0.0054$, $v_9 = 0.0055$, $v_{10} = 0.0059$ etc (Bishop, Price & Temarel 1986). Additional heave and pitch damping contributions due to viscous effects were included empirically. Examples of principal coordinates obtained for the SWATH travelling at 3 knots in beam sinusoidal waves of unit amplitude are shown in Fig.7. It can be seen

that, particularly in the low frequency range ($\omega_e < 2$ rad/s or 0.32 Hz), the amplitudes of the principal coordinates p_8 and p_9 are larger than those of p_7 . However, it should be remembered that during the evaluation of the responses using modal summation the amplitude as well as the phase of the principal coordinate together with the magnitudes of modal characteristics, such as distortion, stress etc, determine the magnitude of the response and, thus, the contribution from each principal mode. In addition these amplitudes vary with heading angle as well as forward speed.

The resonance (wave encounter) frequencies were identified from the behaviour of the principal coordinates and are shown in Table 2 for the distortion modes.

6 RESPONSES AT SEA

In the refined finite element model of the SWATH stresses are evaluated for each flat plate or curved shell element at each distortion mode. Thus direct stresses σ_x , σ_y and σ_z as well as shear stresses τ_{xy} , τ_{yz} and τ_{zx} can be obtained anywhere on the hull using eq.(5).

Presentation of these results, however, poses a problem. Accordingly it was decided to use the von Mises stress as an effective stress illustrating the combination of direct and shear stress and also being relevant to the condition of yield, namely

$$\sigma_e = \sqrt{\frac{\left[(\sigma_x - \sigma_y)^2 + (\sigma_y - \sigma_z)^2 + (\sigma_z - \sigma_x)^2\right]}{2} + 3(\tau_{xy}^2 + \tau_{yz}^2 + \tau_{zx}^2)}.$$

(6)

Results were required for a set of operating

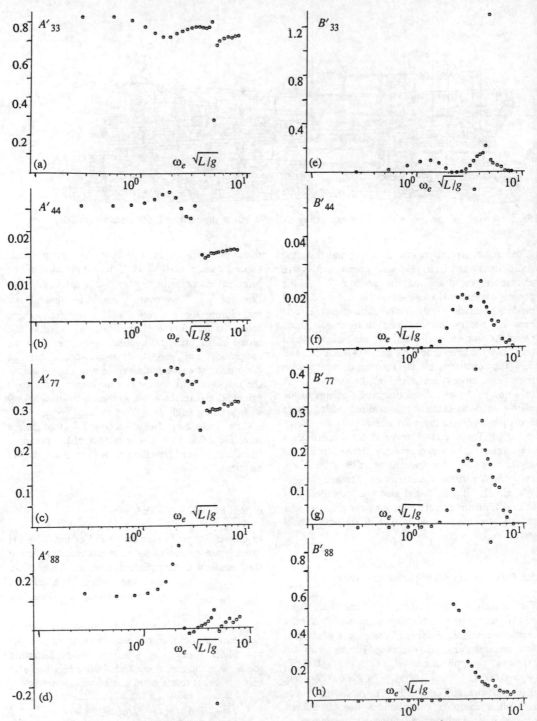

Fig.6 A selection of non-dimensional generalised (a-d) added mass ($A'_{rr} = A_{rr}/\rho \nabla l^2$) and (e-h) hydrodynamic damping ($B'_{rr} = B_{rr}/\rho \nabla l^2 \sqrt{g/L}$) corresponding to the refined model. Note that ρ denotes density, g acceleration due to gravity, ∇ displacement volume and $l = 1$ m except for $r = 4,5,6$ where $l = L$ the length of the SWATH.

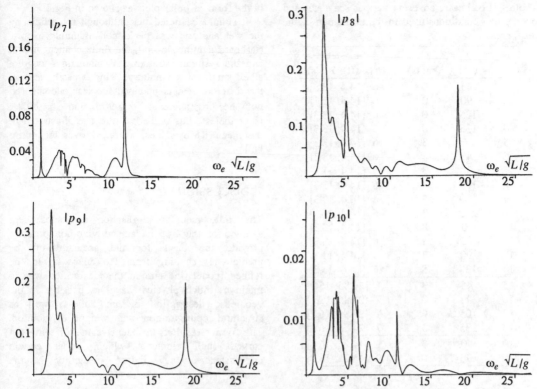

Fig.7 Principal coordinate amplitudes $|p_r|$ r=7,8,9,10 for the refined model travelling at 3 knots in beam sinusoidal waves of unit amplitude.

conditions identified by sea state, forward speed and heading angle as illustrated in Table 4. The seaways were modelled by ISSC wave energy spectra defined by significant wave height $h_{1/3}$ and characteristic wave period T_1. The range of the values chosen is shown in Table 4. A frequency domain analysis was employed and the von Mises stress response spectrum $\Phi_{\sigma_e \sigma_e}$ at a particular position on the structure can, accordingly, be calculated from the expression

$$\Phi_{\sigma_e \sigma_e}(\omega_e,x,y,z)=|H_{\sigma_e \sigma_e}(\omega_e,x,y,z)|^2 \Phi_{\zeta\zeta}(\omega_e) \quad .(7)$$

Here $\Phi_{\zeta\zeta}$ is the wave spectrum and $H_{\sigma_e \sigma_e}$ denotes the Response Amplitude Operator associated with the von Mises stress at a particular position (x,y,z) calculated according to eqs.(5) and (6) with the SWATH travelling at forward speed U and heading angle χ in sinusoidal waves of unit amplitude and frequency ω.

Statistical properties of the response spectrum can thus be evaluated. For example, the RMS von Mises stress values at various positions on the transverse bulkhead shown in Fig.3 are given in Table 5 for various load cases. The longitudinal position of this bulkhead is shown in Fig.1. Examining Table 5 and Fig.3 we see that the largest stresses occur at the connection between strut and pontoon, outboard and inboard respectively. Large stresses also occur at the haunch, i.e. the inboard connection between strut and deck box structure. Furthermore, the information in Table 5 shows that beam seas excite the largest RMS values of the von Mises stress. Increase of forward speed results in increased RMS values. It can also be seen that results for the quartering seas ($\chi = 45^o$) are larger than the corresponding ones for bow seas ($\chi = 135^o$). Furthermore, head and following seas, in that order, produce the smallest RMS values. These comments are also verified by the last column in Table 5 which contains the spatial average of all the RMS von Mises stresses. This latter quantity is very useful in identifying critical operational conditions rather than stress levels on the hull. These variations can best be represented

Table 4. Load cases covering various sea states and operational conditions used in the evaluation of the responses.

Load Case	χ (^{o})	U (knots)	Sea state [$h_{1/3}$ (ft); T_1 (s)]
1	45	0	6 [14.4; 13.3]
2	90	0	6 [14.4; 11.1]
3	105	0	6 [14.4; 13.3]
4	135	0	6 [14.4; 13.3]
5	0	3	6 [14.4; 13.3]
6	45	3	6 [14.4; 13.3]
7	90	3	6 [14.4; 11.1]
8	135	3	6 [14.4; 13.3]
9	180	3	6 [14.4; 13.3]
10	270	3	6 [14.4; 11.1]
11	0	8	6 [14.4; 13.3]
12	45	8	6 [14.4; 13.3]
13	90	8	6 [14.4; 11.1]
14	135	8	6 [14.4; 13.3]
15	45	3	7 [24.85; 16.7]
16	90	0	7 [24.85; 11.1]
17	105	0	7 [24.85; 16.7]
18	135	3	7 [24.85; 16.7]
19	0	3	9 [44.25; 16.7]
20	90	0	9 [44.25; 16.7]

in the form of polar plots as shown in Fig.8.

It should be noted that, although for efficiency the wet analysis was performed assuming a port-starboard symmetric hull, the finite element model and the resultant stresses still maintain a degree, albeit small, of asymmetry. This is easily seen in the different results obtained for seas encountered with heading angles of $\chi = 90^o$ and $\chi = 270^o$. Nevertheless, this is a local effect as illustrated in the mean RMS values, averaged over the entire hull.

7 DISCUSSION

The finite element idealisation and subsequent solution of the free undamped vibration problem provide the basis for the accuracy of the predictions resulting from the analysis. In the refined model, the aim of the idealisation was to retain as much physical detail as possible in the geometric description of the hull as well as structural configuration and weight distribution. The former resulted in extensive use of semi-loof curved shell elements which, under certain circumstances, may induce instability in the

Table 5. Spatial variations of the RMS von Mises stress (MPa) within a transverse bulkhead as a function of sea state and operation conditions. The mean RMS von Mises stress (MPa) in the last column is an average over the entire hull.

Load case	RMS von Mises stresses at various nodes (see Fig.3)								Mean RMS von Mises stress
	A	B	C	D	E	F	G	H	
1	6.57	5.94	2.28	4.54	2.83	2.33	1.00	0.53	21.52
2	17.39	14.58	5.95	11.81	7.70	6.18	2.71	1.23	57.92
3	16.90	14.00	5.77	11.44	7.52	6.02	2.66	1.17	56.88
4	9.20	7.67	3.14	6.23	4.10	3.28	1.46	0.66	31.27
5	1.28	0.96	0.44	0.84	0.59	0.47	0.22	0.08	4.66
6	7.59	8.00	2.60	5.46	3.08	2.59	1.06	0.80	22.52
7	17.54	14.70	6.00	11.91	7.77	6.23	2.74	1.24	58.48
8	8.46	7.26	2.89	5.77	3.74	3.00	1.32	0.63	28.29
9	1.04	0.78	0.36	0.69	0.48	0.38	0.18	0.06	3.80
10	14.99	11.20	5.14	9.82	7.05	5.61	2.68	0.93	58.48
11	1.03	0.77	0.35	0.68	0.47	0.38	0.17	0.06	3.75
12	9.57	11.25	5.64	7.11	3.67	3.14	1.23	1.18	25.53
13	17.97	15.05	6.15	12.20	7.96	6.39	2.81	1.26	60.05
14	8.67	7.96	2.96	6.01	3.73	3.02	1.29	0.73	27.47
15	15.02	20.39	5.00	11.77	5.06	4.58	1.56	2.23	29.03
16	21.79	21.06	7.40	15.28	9.23	7.52	3.20	2.03	67.93
17	21.10	20.23	7.16	14.76	8.99	7.30	3.13	1.95	66.51
18	12.00	13.33	4.04	8.75	4.78	3.99	1.63	1.39	34.01
19	3.17	2.39	1.09	2.09	1.45	1.16	0.53	0.18	11.56
20	38.79	37.50	13.18	27.22	16.43	17.09	5.69	3.61	120.95

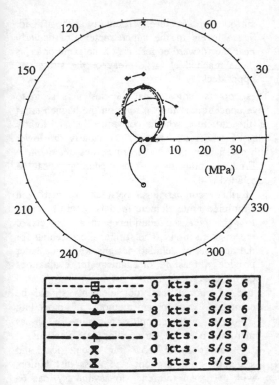

Fig.8 Variation of the RMS von Mises stress at point A (see Fig. 3) with speed, heading angle ($\chi = 180°$ head seas) and sea state.

solution. These factors also contribute to the extensive use of lumped masses attached to various nodes of the structure, resulting in *local* modes. This, however, was not thought to have affected the final results as the two fundamental, and dominant, modes were identified correctly. In principle, any mode shape, whether *local* or *global*, can be incorporated into the wet analysis at the expense of computing time and space. In practice, however, it may be preferable to compromise some of the geometrical and structural features in order to obtain a more workable model. This includes neglecting the small asymmetry in this SWATH at all stages of the analysis. Inclusion of the shear lag effects poses another difficulty. Its effect on the *dry* hull characteristics is not known as no refined model was constructed excluding these effects. We believe that, in principle, the level of detail included in the refined model is right and tractable. Assessing the accuracy of the solution of such a large model is rather difficult. The process of reduction, although widely used, remains another imponderable.

The process of determining the mass and flexibility properties of a simplified model is by no means unique or clear cut. Various models using different idealisations were tested. The scheme followed during the development of this model was to achieve equivalence of structural and mass properties. Dynamic equivalence is bound to require further adjustments. This is believed to be the way forward. The aim of this exercise was to verify the modelling as well as to gain experience with various degrees of refinement. The results shown are encouraging, although further work is necessary. One of the by-products of such an analysis was assessment of the shear lag effects. Incorporation of such effects in a simplified model is even more difficult. The results obtained are interesting in themselves, illustrating the selectivity of such effects, though no definite conclusions can be drawn from this limited evidence.

The results of the wet analysis are broadly in agreement with previous experience (Bishop, Price & Wu 1986). The hydrodynamic coefficients, for example, display the jumps and peaks associated with the formation of standing waves between the hulls. The principal coordinates display peaks at the various resonance frequencies shown in Table 2, depending on the degree of cross coupling. They also show peaks in the low frequency range due to standing wave effects and ship-wave matching. For this model the first two principal distortion modes provide the dominant contribution to the responses. However, the possibility of higher modes contributing to dynamic influences such as slamming, fatigue, etc should always be born in mind.

Results for the dynamic response of the structure, presented in section 5.2, relate to the assessment of structural strength. The analysis also produces results for the seakeeping qualities of the SWATH. These, however, are not discussed here as the emphasis of the modelling has been in favour of structural behaviour. The post processing of the stresses allows for variations of direct and shear stresses to be observed throughout the structure on a modal as well as structural level. Although stress components are important, they do not easily identify the complex stressing occurring in a region of the structure; nor do they reflect the actual stress magnitudes experienced by the material. Principal stresses may be considered as a better alternative for such a representation. In this respect, the von Mises stress, based on principal stresses, is a viable criterion in complex yield conditions. In this paper it is used as a convenient measure of the stress field anywhere in the structure.

The spatial as well as frequency (or time) dependence of the stresses caused us to use statistical properties resulting from dynamic wave loads while assessing the strength of the structure. A preferable way of illustrating the stress variations in the SWATH is by colour contour plots (Keane, Temarel, Wu & Wu 1991). Here typical variations with forward speed, heading angle and sea state are illustrated in tabular form at various positions within a transverse bulkhead. Overall, it was observed that the magnitudes of the RMS von Mises stress are realistic and only rising to levels greater than 100 MPa in a few localised regions of the structure for sea states 6 and 7. In the case of sea state 9 the beam seas condition results in high stresses. It must be noted, however, that the characteristic wave period chosen is larger than the one recommended by the ISSC.

8 CONCLUSIONS & RECOMMENDATIONS

This investigation has shown that hydroelasticity analysis is capable of describing the dynamics of a complex structure, such as a SWATH, in any degree of detail as required at different design stages. The hydroelastic approach, moreover, has the advantage of working directly from first principles and the theory allows the evaluation of loadings, responses and the strength of the structure to be undertaken in the same set of assumptions.

For a structure representative of the SWATH T-AGOS 19 it is shown that using this analysis there exist areas of the structure which appear to have stress levels of particular interest:

- the lower hulls at the strut aft ends are highly stressed. This may be due to the rather narrow transverse bulkheads in the region which support the considerable overhang of the lower hulls as well as the attached propellers.
- the lower hulls at the strut forward ends are significantly, but not as severely as at the strut aft ends, stressed again, presumably, due to the narrow transverse bulkheads and the overhang.
- the main box structure outboard of the longitudinal bulkheads, situated 24 ft either side of the centre line, are lightly stressed even in extreme load cases.
- the upper deck between the longitudinal bulkheads, situated 24 ft either side of the centre line, is more highly stressed than the

surrounding structure, particularly at the aft end.
- stress levels in the superstructure, particularly near the forward edges, are of the same order as those reached at large areas of the struts and upper deck.

These points indicate that the designers seem to have concentrated their efforts on the haunch areas of the structure which, consequently, are lightly stressed whilst other areas, particularly the lower hull overhangs, have had less rigorous treatment.

From the point of view of modelling, it can be concluded that:

- further comparisons between refined and simplified finite element models should be made to gain increased confidence in all aspects of the discretisation. The former are essential for the analysis of detailed designs whilst the latter provide a feasible mechanism for evaluations during concept design.
- use of curved semi-loof elements should be compared with other methods of modelling structures with complex curvature, such as increased use of flat triangular plate elements.
- the question of how far geometrical and structural configurations and mass distribution, with particular reference to asymmetry, can be compromised without affecting accuracy of the results is significant and should be addressed.
- the effects due to shear lag and the way it is allowed for, in either refined or simplified models, should be further investigated.
- modelling large items of equipment as lumped masses without corresponding compensatory increases in local stiffness needs further investigation, with particular reference to the occurrence of local modes.
- consideration should be given to the use of a dynamic mode refinement approach which avoids a detailed finite element model but allows for detailed stress distribution with greatly reduced computational effort (Wu, Xia & Du 1991).

The fluid-structure interaction is attained using three-dimensional potential flow analysis and discretising the wet surface with panels each containing a source at its centre. Such an analysis, within the constraints of its inherent assumptions, is very useful in the prediction of the steady state dynamic behaviour at sea. In this respect, nevertheless, the following topics require further investigation:

- transient excitation and response due to slamming of the underdeck of the box structure as well as the emergence of the lower hulls.

Initial studies of this problem with an application on a rectangular barge have been successful (Aksu, Price & Temarel 1991a,b).
- allowance for the effects of non-linear fluid forces. A time domain simulation using a Morison type approximation has already been incorporated into the three-dimensional hydroelasticity theory (Price & Wu 1989).
- a hybrid analysis, using a three-dimensional structural model and a two-dimensional fluid analysis (Wu & Price 1987). Such an approach may be thought of more practical use in the design process of SWATHs and other multi-hull vessels.

REFERENCES

Aksu, S., Price, W.G. & Temarel, P. 1991a. A comparison of two-dimensional and three-dimensional hydroelasticity theories including the effect of slamming. *Proc. Inst. Mech. Engrs.* 205:3-15.

Aksu, S., Price, W.G. & Temarel, P. 1991b. A three-dimensional theory of ship slamming in irregular oblique seaways. In C.S.Smith & R.S.Dow (eds), *Advances in Marine Structures* 2:208-229. London:Elsevier.

Belik, O., Bishop, R.E.D. & Price, W.G. 1988. Influence of bottom and flare slamming on structural responses. *Trans. RINA* 130:261-275.

Bishop, R.E.D. & Price, W.G. 1979. *Hydroelasticity of ships*. Cambridge University Press.

Bishop, R.E.D., Price, W.G. & Temarel, P. 1986. On the hydroelastic response of a SWATH to regular oblique waves. In C.S.Smith & J.D.Clarke (eds), *Advances in Marine Structures* :89-110, London: Elsevier.

Bishop, R.E.D., Price, W.G. & Wu, Yousheng 1986. A general linear hydroelasticity theory of floating structures moving in a seaway. *Phil. Trans. Roy. Soc.* London A316:375-426.

Ergin, A., Price, W.G., Randall, R. & Temarel, P. 1992. Dynamic characteristics of a submerged, flexible cylinder vibrating in finite water depths. *J. Ship Research* 36:154-167.

Fu, Y., Price, W.G. & Temarel, P. 1987. The 'dry and wet' towage of a jack-up in regular and irregular waves. *Trans. RINA* 129:147-159.

Keane, A.J., Price, W.G., Temarel, P., Wu, X.J. & Wu, Yongshu 1988. Seakeeping and structural responses of SWATH ships in waves. *Proc. 2nd Int. Conf. on SWATH Ships and Advanced Multi-Hulled Vessels*. London:RINA.

Keane, A.J., Temarel, P., Wu, X.J. & Wu, Yongshu 1991. Hydroelasticity of non-beamlike structures. *Phil. Trans. Roy. Soc. London* A334:339-355.

Lee, C.M. & Curphey, R.M. 1977. Prediction of motion, stability and wave load of small waterplane twin hull ships. *Trans. SNAME* 85:94-130.

Lundgren, J., Price, W.G. & Wu, Yongshu 1989. A hydroelastic investigation into the behaviour of a floating 'dry' dock in waves. *Trans. RINA* 131:213-231.

McCreight, K.K. 1987. Assessing the seaworthiness of SWATH ships. *Trans. SNAME* 95:189-214

PAFEC 1986. *Pafec data preparation user manual*.

Price, W.G., Temarel, P. & Wu, Yongshu 1987. Responses of a SWATH travelling in irregular seas. *J. Underwater Techn.* 13:2-10.

Price, W.G. & Wu, Yongshu 1989. The influence of non-linear fluid forces in the time domain responses of flexible SWATH ships excited by a seaway. *Proc. Int. Conf. OMAE* 2:125-135. The Netherlands.

Reilly, E.T., Shin, Y.S. & Kotte, E.H. 1988. A prediction of structural load and response of a SWATH ship in waves. *Naval Engineers Journal* 251-264.

Stirling, A.J., Jones, G.L. & Clarke, J.D. 1988. Development of a SWATH structural design procedure for Royal Naval vessels. *Proc. 2nd Int. Conf. on SWATH Ships and Advanced Multi-hulled Vessels*. London:RINA.

Wu, X.J. & Price, W.G. 1986. A multiple Green's function expression for the hydrodynamic analysis of multi-hull structures. *Applied Ocean Research* 9:58-66.

Wu, Yousheng, Xia, Jinzhu & Du, Shuangxing 1991. Two engineering approaches to hydroelastic analysis of slender ships. In W.G.Price, P.Temarel & A.J.Keane (eds), *Dynamics of Marine Vehicles and Structures in Waves*: 157-165. London:Elsevier.

Hydroelasticity in Marine Technology, Faltinsen et al. (eds) © 1994 Balkema, Rotterdam, ISBN 90 5410 387 6

Hydroelastic response analysis of a high speed monohull

Ole Andreas Hermundstad, MingKang Wu & Torgeir Moan
The Norwegian Institute of Technology, Trondheim, Norway

ABSTRACT: A method for hydroelastic analysis of high speed ships is presented. The method is based on the use of "dry" modes, and the generalized hydrodynamic boundary value problem is solved by the method presented by Faltinsen and Zhao (1991a,b). The standard deviation of the midship vertical bending moment is calculated for a 120 meter monohull with different combinations of stiffness, speed and hull damping. The results indicate that hydroelastic effects can influence the fatigue performance of high speed vessels, and for the investigated ship, the internal hull damping is of little importance for the response in the first and second vertical bending modes.The damping in these modes was found to be dominated by hydrodynamic damping due to the forward speed. Only steady-state forces and short term responses in head sea have been considered.

1 INTRODUCTION

In recent years we have witnessed a trend in the high speed ship market towards larger vessels. Catamarans up to 144 meters (Svensen & Valsgård 1993) and monohulls of 220 meters length (Jullumstrø, Leppänen & Sirviö 1993) are being designed with speed ranges between 35 and 40 knots.

For such large ships, the hull structure makes up the dominant part of the ship weight, and much effort is made to reduce the mass of the hull. These efforts may include the use of materials such as aluminum, high strength steel and for moderate ship lengths, fiber reinforced plastics.

Focusing on the minimization of the mass/strength ratio however, may result in a design which is globally quite flexible, since the stiffness of the materials does not necessarily increase as the strength increases.

High global flexibility may also result from the use of sandwich structures. While sandwich plates have a highly competitive mass/stiffness ratio for local loadings, this merit is not present for the global stiffness of the hull girder.

The reduction in the hull structural weight leads to a higher cargo capacity, i.e. the total mass remains the same, while the stiffness is reduced. Hence, a decrease in the ship's natural frequencies results.

Due to the high speed of these vessels, the frequency of encounter is quite large, and resonant encounter (springing) may occur for relatively large waves. But even for ships with a very high speed and a low global stiffness, the waves that cause resonance in the 2-node vertical vibration mode will generally be much shorter than the ship length. Intuitively, we would therefore not expect steady-state hydroelastic phenomena to be very important with respect to the ultimate strength of the ship.

The more moderate responses however, could be influenced by hydroelasticity, and to evaluate the fatigue performance of large high speed ships, a hydroelastic analysis may prove necessary. Some attention will be paid to this question in the present paper.

In the cases where dynamic amplification of structural vibrations turns out to be important, the question of damping will be of interest. The total damping will consist of the internal hull damping as well as the hydrodynamic damping.

Hull damping increases for higher modes of vibration, and there seems to be a general agreement that for high modes, this is the dominant part of the total damping. However, for the few lowest modes the picture is blurred by the hydrodynamic damping caused by the forward speed of the ship. The relative importance of these two damping contributions will be investigated in the paper.

The matemathical model adopted in this

paper is based on modal technique using so-called "dry" modes. These modes are the eigenmodes of the ship when no account is made for the surrounding fluid. The hydrodynamic boundary value problem is then solved for the generalized body boundary conditions that evolve from the use of m modes of vibration instead of the traditional 6 modes, known from conventional seakeeping analysis.

For monohulls, the structure is usually modelled as a Timoshenko beam, and strip theory is used for the hydrodynamic calculations; see e.g. Bishop and Price (1979). Catamarans and other non-beamlike structures are often modelled by the finite-element method, while a three-dimensional singularity-distribution method is applied to determine the fluid actions; see e.g. Bishop, Price and Wu (1986) and Wu (1984).

However, a strip theory assumes implicitly that the forward speed is zero or moderate, which means that its applicability for high-speed vessels should be questioned (Faltinsen & Zhao 1991b).

A complete 3D method on the other hand, requires substantional computer time, and such methods are therefore not very practical for routine calculations (Faltinsen & Svensen 1990).

The authors' approach is a generalization of the method presented by Faltinsen and Zhao (1991a,b). Their method was intended for sea-keeping and resistance calculations of rigid high speed vessels, and it has been implemented in the FASTSEA (1991) computer program. A two-dimensional Laplace equation is solved for each section of the ship, but as opposed to strip theory, a three-dimensional free surface condition with forward speed is used. In this way, the important wave-systems generated by a high speed vessel is properly incorporated. The method has been extended to include the flexibility of the ship (Wu, Hermundstad & Moan 1993). It is applicable to both monohulls and catamarans. An outline of the problem formulation and solution method is included in the next section of this paper.

The original FASTSEA program has been validated at various levels for both monohulls (Faltinsen & Zhao 1991a,b) and catamarans (Ohkusu & Faltinsen 1991), (Faltinsen et al. 1992). The generalized hydroelastic version however, has not yet been verified by experiments.

Results are presented for a 120 meter monohull with two different stiffnesses; one relatively flexible and one moderately stiff. Three different speeds and three levels of hull damping are investigated. Althogh the method is applicable for all

wave headings, only head sea results are presented in this paper. As a response quantity, we have selected the vertical bending moment (VBM) at L/2 from the bow. Only steady-state conditions are investigated; i.e. no slamming loads and other transient effects are included.

2. THEORETICAL BACKGROUND

2.1 *The equations of motion*

Consider a flexible ship hull moving in incident waves on deep water. Let (x,y,z) be a right-handed coordinate system, fixed with respect to the mean oscillatory position of the ship. The origin is in the plane of the undisturbed free surface with the positive z-direction vertically upwards through the center of gravity, and the ship has a forward speed U in the negative x-direction; see Figure 1. The fluid loads cause deflections of the ship hull, which

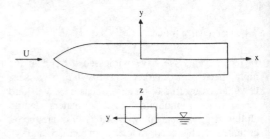

Figure 1. Coordinate system.

include both rigid body motions and structural deformations. We will assume that the global deflections are small and that the fluid is ideal with irrotational flow. Hence, the problem can be formulated in terms of linear elastic structural theory and potential flow theory.

For the discretised ship hull, the equations of motion can be written

$$m\ddot{u} + c\dot{u} + ku = f(\ddot{u}, \dot{u}, u, t) \qquad (1)$$

where **u**=u(x,y,z,t) is the displacement vector; **m**, **c** and **k** are the structural mass-, damping- and stiffness matrices, respectively; **f** is the vector of external fluid forces. Overdots denote differentiation with respect to time t.

For the global vibration, **u** may be approximated adequately by an aggregate of the m lowest eigenmodes $\mathbf{u}_r(x,y,z)$

$$u(x,y,z,t) = \sum_{r=1}^{m} u_r(x,y,z)p_r(t) \qquad (2)$$

in which $p_r(t)$, r=1,2,...,m, are the principal coordinates. The first 6 modes are the rigid body modes: surge, sway, heave, roll, pitch and yaw. The remaining are "dry" flexible modes, i.e. they are determined for the structure without taking the fluid actions into account.

Inserting Eq.(2) into Eq.(1) and then premultiplying Eq.(1) by $\mathbf{u}_k$, we get

$$\sum_{r=1}^{m} [M_{kr}\ddot{p}_r(t)+C_{kr}\dot{p}_r(t)+K_{kr}p_r(t)] = F_k \qquad (3)$$
$$k=1,2,...,m$$

where $M_{kr}=\mathbf{u}_k^T\mathbf{m}\mathbf{u}_r$, $C_{kr}=\mathbf{u}_k^T\mathbf{c}\mathbf{u}_r$ and $K_{kr}=\mathbf{u}_k^T\mathbf{k}\mathbf{u}_r$. $F_k=\mathbf{u}_k^T\mathbf{f}$ is the generalized fluid force in mode k.

2.2 The internal structural forces

The structural mass matrix, **m**, the structural stiffness matrix, **k,** and the eigenmodes, $\mathbf{u}_k$, are obtained from a finite element analysis of the ship hull, using the program PUSFEA (1993).

Since the dry flexible modes are orthogonal with respect to both **m** and **k** , the generalized matrices, **M** and **K**, will be diagonal. Hence, the dry modes can be normalized so that $M_{kk}=1$ and $K_{kk}=\omega_k^2$ for k>6. The 6 rigid body modes are not generally orthogonal, so **M** is not diagonal for k and r less than 7. The entries K_{kr}, in the generalized structural stiffness matrix are all zero for k and r less than 7.

Establishing the generalized hull damping matrix, **C**, requires detailed knowledge of the damping characteristics of the ship. Since this is generally not available, some assumptions must be made. It is reasonable to assume that the off-diagonal terms in **C** are zero (Newland, 1989). The remaining problem is then to estimate the damping in each mode k, which can be expressed by for example the logarithmic decrement, δ_k. Several formulas for δ_k exist, but these are based on measurements that show a large variation in the damping level (Betts, Bishop & Price, 1977).

Due to these uncertainties about the damping

level, one may simplify the problem further and choose a Rayleigh damping model: $\mathbf{C}=\alpha_1\mathbf{M}+\alpha_2\mathbf{K}$. For steel ships, correlations between calculations and measurements have shown that damping proportional to stiffness is the best representation of damping over a wide frequency range (Jensen & Madsen, 1977), (Hylarides, 1974). For a beam model, $\alpha_2=0.0014$ has been suggested (Catley & Norris, 1976). The hull damping may be larger for aluminum vessels, and significantly larger for FRP sandwich structures. Results for stiffness proportional damping are presented in this paper, for three different values of α_2.

2.3 The external fluid forces

The external fluid forces are found by solving a boundary value problem for the unknown velocity potential Φ. To obtain a set of equations that can be solved in an efficient manner, some assumptions regarding the hull form and the fluid motions must be introduced.

Using the slenderness parameter ε, so that L is O(1) and B is O(ε) (L and B is the ship length and beam, respectively), we assume that the Froude number (based on the ship length) is O(1). The frequency of encounter ω_e, is assumed to be O($\varepsilon^{-1/2}$). Since a ship hull is generally slender, it is reasonable to assume that the x-component n_1, of the vector normal to the ship's wetted surface is O(ε). The two other components are O(1). For a flow variable f, caused by the body in some region near it, $\partial f/\partial x=O(f\varepsilon^{-1/2})$, $\partial f/\partial y=O(f\varepsilon^{-1})$ and $\partial f/\partial z=O(f\varepsilon^{-1})$ is assumed. Further, the steady and the unsteady motions of the ship and the fluid are assumed to be small. The above assumptions are the same as those made by Faltinsen and Zhao (1991b).

The total velocity potential $\Phi(x,y,z,t)$, can be decomposed into the steady potential ϕ_s, the unsteady potential ϕ_u and the potential caused by the forward speed

$$\Phi(x,y,z,t)=\phi_s(x,y,z)+\phi_u(x,y,z,t)+Ux \qquad (4)$$

It can be shown that with the above assumptions, the steady potential and the unsteady potential can be found separately. We will only be concerned with the unsteady potential in this paper.

For a sinusoidal wave excitation with encounter frequency ω_e, the displacement vector **u** of the ship hull and the unsteady potential ϕ_u take the oscillatory forms,

$$u(x,y,z,t) = \sum_{r=1}^{m} p_r u_r(x,y,z)e^{i\omega t} \qquad (5)$$

and

$$\phi_u(x,y,z,t)=\phi_I(x,y,z)e^{i\omega t}+\phi_d(x,y,z)e^{i\omega t}$$
$$+\sum_{r=1}^{m} p_r \phi_r(x,y,z)e^{i\omega t} \qquad (6)$$

where p_r, r=1,2,...,m, are the amplitudes of the principal coordinates, $\phi_I(x,y,z)$ is the incident wave potential, $\phi_d(x,y,z)$ is the diffraction potential and $\phi_r(x,y,z)$ is the radiation potential due to the ship motion in mode r.

It can be shown that both the diffraction potential and the radiation potentials must satisfy the two-dimensional Laplace equation

$$\frac{\partial^2 \phi}{\partial y^2}+\frac{\partial^2 \phi}{\partial z^2}=0 \qquad (7)$$

subject to the following boundary conditions on the free surface, z=0

$$\frac{\partial \zeta}{\partial t}+U\frac{\partial \zeta}{\partial x}-\frac{\partial \phi}{\partial z}=0 \qquad (8)$$

$$g\zeta+\frac{\partial \phi}{\partial t}+U\frac{\partial \phi}{\partial x}=0 \qquad (9)$$

in a region near the ship. Here g is the acceleration of gravity, and ζ is the unsteady surface elevation.

For the diffraction potential, the body boundary condition becomes

$$\frac{\partial \phi_d}{\partial N}=-\frac{\partial \phi_I}{\partial n} \qquad (10)$$

on the mean wetted body surface S. The normal vectors are positive into the fluid domain and evaluated for the ship in its mean position. N is the projection of the normal vector **n**, in the transverse plane.

For the radiation problem, the body boundary condition takes the form

$$\frac{\partial \phi_r}{\partial N}=i\omega_e u_r \cdot n-m_r \qquad (11)$$

for all r=1,....,m.

The m_r-terms can be written

$$m_r=U\mathbf{i}\cdot\Delta\mathbf{n}_r \qquad (12)$$

The expression $\Delta\mathbf{n}_r$ represents the change in the normal vector when the body deforms in the rth mode. Introducing the two tangential vectors α and β, so that $\alpha\times\beta=\mathbf{n}$, we can express $\Delta\mathbf{n}_r$ as

$$\Delta\mathbf{n}_r=[\alpha\times(\beta\cdot\nabla)-\beta\times(\alpha\cdot\nabla)]u_r \qquad (13)$$

(Wu, Hermundstad & Moan, 1993).

Far away from the ship, the complete solution of the problem must satisfy a radiation condition. The solution of the inner problem stated above, must therefore be matched with the solution of an outer problem that satisfies the radiation condition (Faltinsen & Zhao, 1991a,b). The form of the outer solution and the matching process is not influenced by the generalization from 6 to m modes.

Faltinsen and Zhao (1991a,b) have presented a method to solve the problem described by Eqs.(7)-(11) for the 6 rigid body modes. The same solution scheme can be used for the general m-mode problem.

The numerical solution is found by starting at the bow and then use the free surface boundary conditions Eqs.(8) and (9) to step the solution ϕ, on z=0 from section to section in the x-direction. For each section, the velocity potential can be found by a two-dimensional analysis. In this way, the interaction between hull sections is accounted for as an upstream effect. Starting conditions are obtained by setting $\phi=0$ and $\zeta=0$ on z=0 at the bow. Hence, it is assumed that there are no waves propagating ahead of the ship. This assumption, together with the fact that only the divergent wave systems are accounted for (Ohkusu & Faltinsen 1991), makes the method questionable for Froude numbers lower than about 0.4.

Transom stern effects are included by assuming that the flow leaves the transom stern tangentially in the downstream direction, so that there is atmospheric pressure at the entire transom stern.

When the velocity potential is obtained, the fluid pressure acting on the ship hull is written by Bernoulli's equation

$$p = -\rho(\frac{\partial \Phi}{\partial t}+\frac{1}{2}\nabla\Phi\cdot\nabla\Phi+gz-\frac{1}{2}U^2) \qquad (14)$$

where ρ is the water density.

The kth generalized unsteady fluid force is obtained by integrating the pressure over the mean wetted surface S

$$F_k(t) = -\iint_S u_k \cdot n p \, dS$$

$$= \rho \iint_S u_k \cdot n [\frac{\partial \phi_u}{\partial t} + U \frac{\partial \phi_u}{\partial x} + g w] dS \quad (15)$$

where w is the vertical displacement of a point on the hull surface.

The generalized Eq.(3) for the unsteady principal coordinate deflections of the ship hull now takes the form

$$\sum_{r=1}^{m} [-\omega_e^2 (M_{kr} + A_{kr}) + i\omega_e (C_{kr} + B_{kr}) + K_{kr} + R_{kr}] p_r = F_k^f + F_k^d \quad (16)$$

The generalized added mass, damping and restoring coefficients can be written

$$A_{kr} = \frac{\rho}{\omega_e^2} Re\{\iint_S u_k \cdot n [i\omega_e \phi_r + U \frac{\partial \phi_r}{\partial x}] dS\} \quad (17)$$

$$B_{kr} = -\frac{\rho}{\omega_e} Im\{\iint_S u_k \cdot n [i\omega_e \phi_r + U \frac{\partial \phi_r}{\partial x}] dS\} \quad (18)$$

and

$$R_{kr} = -\rho g \iint_S u_k \, n w_r \, dS \quad (19)$$

respectively, for k,r=1,.....,m. Since the center of gravity is located a vertical distance z_g away from the origin of the coordinate axes, the roll and pitch restoring coefficients must be modified by subtracting the term $z_g M_{33}$. The motions in the flexible modes are much smaller than in the rigid body modes, and modification due to gravity is therefore not included for other restoring terms.

The terms on the right hand side of Eq(16) are the generalized Froude-Krylov forces

$$F_k^f = \rho \iint_S u_k \cdot n_s (i\omega_e \phi_I + U \frac{\partial \phi_I}{\partial x}) dS \quad (20)$$

and the generalized diffraction forces

$$F_k^d = \rho \iint_S u_k \cdot n_s (i\omega_e \phi_d + U \frac{\partial \phi_d}{\partial x}) dS \quad (21)$$

Eq (16) can be expressed in matrix form

$$H^{-1}(\omega_e) p = F^f(\omega_e) + F^d(\omega_e) \quad (22)$$

and the unsteady principal coordinate deflections are found as

$$p = H(\omega_e)[F^f(\omega_e) + F^d(\omega_e)] \quad (23)$$

where

$$H(\omega_e) = [-\omega_e^2 (M+A) + i\omega_e (C+B) + K + R]^{-1} \quad (24)$$

is the frequency-response function. The displacement at any point in the ship hull is obtained from Eq(2).

2.4 Response in irregular waves

As illustrated by Wu, Hermundstad and Moan (1993), flexible hull forms travelling at high speeds in regular waves may display quite large modal deformations due to hydroelastic effects. But since these effects are generally significant only at relatively high frequencies of encounter, the energy content of the corresponding waves may be rather small. Hence, to evaluate the importance of hydroelasticity with respect to design, we need to study the response in irregular waves.

A long term statistical analysis considering all possible sea states in the ship's operational area would be an important part of a design process, especially with respect to fatigue. In the present investigation however, we will only study the response in various short term sea states.

Each sea state will be represented by an ISSC wave spectrum

$$S(\omega) = \frac{A}{\omega^5} \exp(-\frac{B}{\omega^4}) \quad (25)$$

where the parameters A and B can be written

$$A = 0.11 H_s^2 (\frac{T_1}{2\pi})^{-4} \quad (26)$$

and

$$B = 0.44(\frac{T_1}{2\pi})^{-4} \qquad (27)$$

Here, H_s is the significant wave height, and T_1 is the mean (or characteristic) wave period, $T_1=2\pi m_0/m_1$. m_0 and m_1 are spectral moments.

3. RESULTS AND DISCUSSION

3.1 *Description of the vessel*

Calculations have been made for a 120 meter high speed ship with the same hull form as "model 1" in Blok and Beukelman (1984); see Figure 2. The ship's draft is 3 meter and the beam is 12 meter. We have assumed a simplified internal hull structural arrangement, and a longitudinal mass distribution. The longitudinal distribution of mass, cross section area, cross section shear area, second moment of area and moment of inertia about the y-axis as well as the height of the centroid above the mean water level, can be found in Figure 4 a)-f). A Young's modulus and a Poisson ratio of 200GPa and 0.3, respectively, have been used in the eigenvalue analyses.

By varying the thickness of the hull structural members, two different stiffnesses were obtained. Hence, ship 1 has a dry natural frequency of 9.71 rad/s in its 2-node vibration mode, while the corresponding frequency for ship 2 is 12.7 rad/s. The total mass of the two ships are the same.

Talvia and Wiefelspütt (1991) present wet natural frequencies for the 2-node mode for different 60 meter long high speed monohulls. These range from 5.7 to about 10 rad/s. The corresponding dry frequencies will then most probably lie between 9 and 20 rad/s. For a 220 meter high speed ship, Jullumstrø, Leppänen and Sirviö (1993) report that the dry 2-node natural

Table 1. The 8 different cases that have been analysed.

Case	Stiffness (Ship no.)	Hull damping (α_2, [s])	Speed (U, [knots])
case111	ship 1	0.0007 (low)	30.0 (low)
case121	ship 1	0.0014 (med)	30.0 (low)
case131	ship 1	0.0028 (high)	30.0 (low)
case122	ship 1	0.0014 (med)	37.5 (med)
case123	ship 1	0.0014 (med)	45.0 (high)
case113	ship 1	0.0007 (low)	45.0 (high)
case133	ship 1	0.0028 (high)	45.0 (high)
case222	ship 2	0.0014 (med)	37.5 (med)

Table 2. Calculated dry eigenfrequencies for the first 4 symmetric flexible modes. Non-dimensional frequency: $\bar{\omega}=\omega(L/g)^{1/2}$.

Mode no.	Frequency [rad/s]		Non-dim. frequency	
	Ship 1	Ship 2	Ship 1	Ship 2
7	9.71	12.7	34.0	44.4
8	23.5	30.8	82.2	108
9	41.7	54.5	146	191
10	63.2	82.7	221	289

frequency is 8.29 rad/s. Bishop and Price (1979) have found the corresponding frequency for a 107 meter destroyer to be 13.5 rad/s. Based on these reports, we may conclude that our ship no. 1 is quite flexible, while ship no. 2 is moderately stiff.

For the stiffness proportional internal hull damping, $C=\alpha_2 K$, 3 different values of α_2 have been tested. These correspond to damping ratios (percentage of critical damping) for the first vibration mode of 0.34%, 0.68% and 1.36% in the flexible ship case. The corresponding values for the stiffer ship are 0.44%, 0.89% and 1.78%.

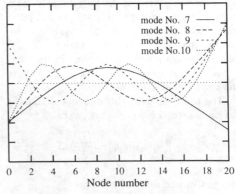

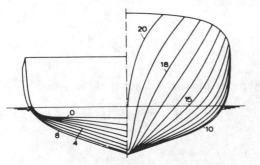

Figure 2. Body plan of the 120 meter monohull (Blok & Beukelman 1984).

Figure 3. Shape of the first 4 symmetric modes. Node number 20 is at the bow.

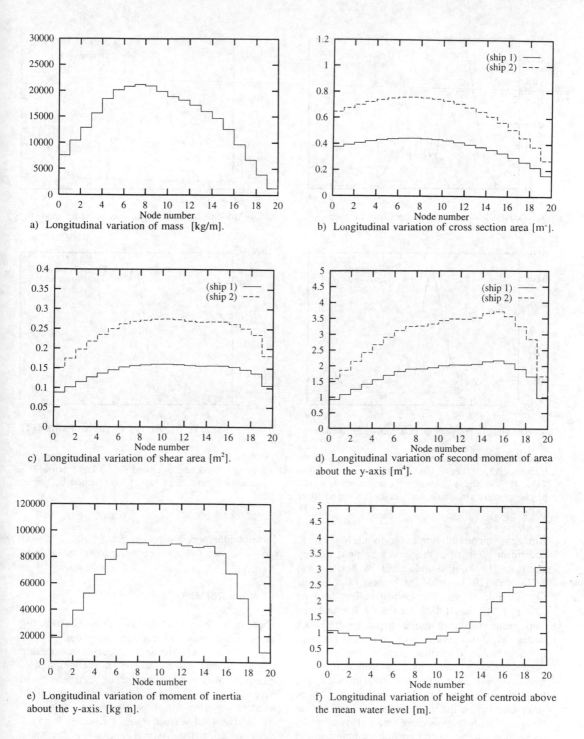

a) Longitudinal variation of mass [kg/m].

b) Longitudinal variation of cross section area [m²].

c) Longitudinal variation of shear area [m²].

d) Longitudinal variation of second moment of area about the y-axis [m⁴].

e) Longitudinal variation of moment of inertia about the y-axis. [kg m].

f) Longitudinal variation of height of centroid above the mean water level [m].

Figure 4. Longitudinal variation of mass, cross section area, cross section shear area, second moment of area and moment of inertia about the y-axis; and height of centroid above the mean water level for the 120 meter monohull. Ship 1 and ship 2 have natural frequencies in the 2-node vibration mode of 9.71 and 12.7 rad/s, respectively. Node no. 20 is at the bow.

251

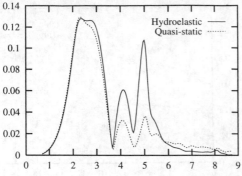

a) Modulus of the principal coordinate for mode 7.

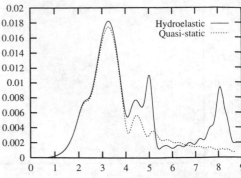

b) Modulus of the principal coordinate for mode 8.

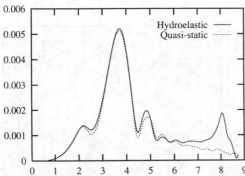

c) Modulus of the principal coordinate for mode 9.

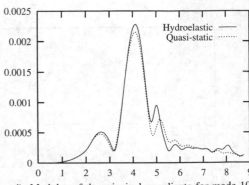

d) Modulus of the principal coordinate for mode 10.

Figure 5. Modulus of the principal coordinates p_k, for modes 7, 8, 9 and 10, plotted as functions of non-dimensional *wave* frequency: $\hat{\omega}_0=\omega_0(L/g)^{1/2}$. The flexible ship with medium hull damping is used, and the speed is 37.5 knots. (case122 in Table 1). A modulus value of 1.0 corresponds to a deflection amplitude at the bow of 1.01, 1.14, 1.22 and 1.27 meters per meter wave amplitude, for modes 7, 8, 9 and 10, respectively.

Among the empirical expressions for the logarithmic decrement δ_k, listed in the survey by Betts, Bishop and Price (1977), a formula given by Aertssen and de Lembre (1971) yields the highest value for the first flexible mode. The corresponding damping ratio is 1.13% and 1.48% for the flexible and stiff ship, respectively. Formulas given by the other authors yield damping ratios that are less than half of these values. All the formulas are based on measurements, and the contribution from hydrodynamic damping is therefore included. The hydrodynamic term should therefore have been subtracted before the values from the formulas are compared with our damping ratios. However, the magnitude of the hydrodynamic damping is not known, and it may have been insignificant for the ships for which the measurements were made.

Three different speeds have been used. The corresponding Froude numbers are 0.45, 0.56 and 0.67. The various cases are summarized in Table 1.

3.2 *Numerical models*

The ship has been modelled as a Timoshenko beam, using 20 elements with constant properties.

In the hydrodynamic calculations, the ship was subdivided into 80 sections in the longitudinal direction. Such a high number of sections proved necessary in order to represent the relatively small incident wave-lengths in the high-frequency range.

The total wave frequency range is from 0.2 rad/s to 2.5 rad/s. To make sure that the sharp resonance peaks for the flexible modes are properly included, the calculations were carried out for 200 equally spaced frequencies within this range. The

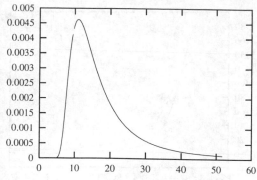

a) Non-dimensional wave spectrum: $\hat{S}=S(g/H_s^4L)^{1/2}$.

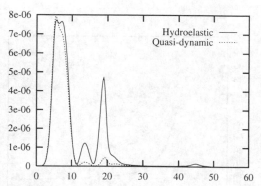

b) Non-dimensional VBM transfer function (modulus squared): $(|\hat{H}_{VBM}|)^2=(|H_{VBM}|/\rho g L^3)^2$.

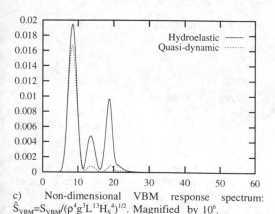

c) Non-dimensional VBM response spectrum: $\hat{S}_{VBM}=S_{VBM}/(\rho^4 g^3 L^{13}H_s^4)^{1/2}$. Magnified by 10^6.

Figure 6. Non-dimensional wave spectrum, VBM transfer function (squared) and VBM response spectrum, plotted as functions of non-dimensional encounter frequency: $\hat{\omega}=\omega(L/g)^{1/2}$. Mean wave period: 4.5 sec. Speed: 37.5 knots. Medium hull damping. Flexible ship (case122).

difference obtained by using 300 frequencies was negligable.

In addition to the 6 rigid body modes, 4 flexible modes have been used. Since head sea was considered, only flexible modes with deformation in the vertical plane (symmetric modes) were included; see Figure 3. For one of the cases (case122), the analysis was also carried out with 10 symmetric flexible modes.

3.3 Dry eigenvalue analysis

The dry eigenvalue analyses were carried out by using the FEM-program PUSFEA (1993). In Figure 3, the shape of the first 4 symmetric flexible modes are presented. The mode shapes are similar for the two vessels. The corresponding dry frequencies are presented in Table 2.

3.4 Typical results

As an example of typical results, the modulus of the generalized coordinates p_k, is plotted in Figure 5a-d for modes 7 to 10. The plots refer to the flexible ship with medium speed and medium hull damping. A non-dimensional wave spectrum for $T_1=4.5$ seconds is depicted in Figure 6a. Figure 6b contains the square of the modulus of the VBM transfer function. The resulting response spectrum is shown in Figure 6c.

The continous lines in the figures represent the current hydroelastic theory, as described in the previous section of this paper. The dotted lines result from a quasi-dynamic analysis where the fluid excitation forces are calculated only for the rigid body modes, and the dynamic amplification due to the response in the flexible modes is not included. Hence, the hydroelastic effects are not properly accounted for in the quasi-dynamic analysis, and it therefore represents a more conventional approach in ship design. When we discuss the influence of hydroelasticity, we will refer to the difference between these two methods.

Figure 5 shows that there are peaks in the transfer functions due to resonance in modes 7 and 8. They occur at a non-dimensional wave frequency of approximately 5 and 8, respectively. Modes 9 and 10 do not exhibit resonant behavior in the calculated frequency range. The resonance peaks are absent in the results from the quasi-dynamic analysis. The less sharp peaks are due to ship-wave matching and coupling with the rigid body modes heave and pitch.

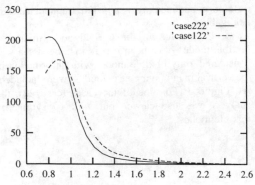

a) Stiff (case222) and flexible (case122) ship.
Medium hull damping. Speed: 37.5 knots.

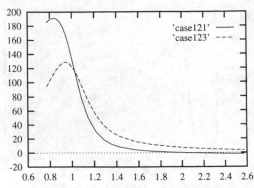

b) Low (case121) and high (case123) speed.
Medium hull damping. Flexible ship.

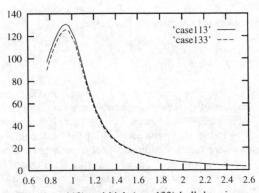

c) Low (case113) and high (case133) hull damping.
Flexible ship. Speed: 45 knots.

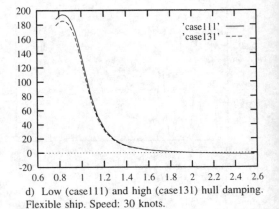

d) Low (case111) and high (case131) hull damping.
Flexible ship. Speed: 30 knots.

Figure 7. Percentage increase in the standard deviation for VBM at L/2 due to hydroelastic effects. Plotted as function of non-dimensional mean wave period: $\bar{T}_1 = T_1(g/L)^{1/2}$.

These effects are properly accounted for in the quasi-dynamic approach.

In Figure 6b, there is a large contribution from hydroelastic effects at a non-dimensional encounter frequency of approximately 19. This corresponds to a wave frequency of 5, for which there was a large response in modes 7 and 8. There is also a small peak at $\bar{\omega}_e \approx 44$. The corresponding wave frequency is 8, which is the resonance frequency of mode 8. But since the wave energy is relatively low at these frequencies, this last peak is not present in the response spectrum.

Figure 6c shows that there is a significant difference in the response spectra calculated by the hydroelastic and the quasi-dynamic methods.

3.5 Parametric study

In the example above, it was seen that hydroelastic effects were relatively large for a mean wave period of 4.5 seconds. We would expect the importance of hydroelasticity to decrease for larger wave periods. In order to investigate the influence of hydroelastic effects in different sea states, we have calculated the standard deviation of the vertical bending moment at L/2 both by use of the current hydroelastic theory, and by the quasi-dynamic method. The two different standard deviations are denoted σ_{HE} and σ_{QD}, respectively.

In Figure 7a-d we have plotted the percentage increase in the VBM standard deviation due to hydroelastic effects, $(\sigma_{HE}/\sigma_{QD}-1)100\%$, as a function of non-dimensional mean wave period, $\bar{T}_1$. The magnitude of the significant wave height can vary a great deal for the same value of T_1, but the T_1-range used in these plots would typically correspond to H_s-values in the range between 0.5 and 3.5 meters. For comparison, two different cases are plotted in each figure.

It is immediately evident that the hydroelastic effects are large for the lowest sea states, while their significance decreases rapidly for higher mean wave periods. However, the curves decay for the very lowest T_1-values. The reason for this phenomenon is that the peak of the encounter wave spectrum is moved to a frequency which is higher than the resonance frequency for the first flexible mode. Studying Figure 6b, we see that this is an area where hydroelastic effects are small. For the stiffer ship and for the cases with the lowest speed, this phenomenon happens for even smaller wave periods.

In Figure 7a, the continous line represents data for the stiffer ship (ship 2), while the punctured line is for the more flexible ship (ship 1). In both cases, a medium speed and a medium hull damping have been used. We see that a 20-25% reduction in the ship's 2-node natural frequency results in significantly larger hydroelastic effects for all but the lowest T_1-values.

Figure 7b shows results for the flexible ship for two different speeds; U=30 knots and U=45 knots. The punctured curve corresponds to the higher speed. A medium hull damping is used in both cases. The 50% increase in ship speed more than doubles the hydroelastic effects on the VBM for non-dimensional wave periods higher than approximately 1.3. It is seen from the same figure that for the highest periods, the calculated VBM standard deviation for the 30-knot ship is reduced when hydroelasticity is accounted for. This happens because the magnitude of the VBM transfer function is slightly reduced for the lowest frequencies in the hydroelastic calculations (Figure 6b).

The influence of hull damping has been investigated in Figures 7c and d. The study has been carried out for two different speeds. Since the hydrodynamic damping increases with forward speed, it would be expected that the influence of hull damping would be larger for the lower speed. This proves to be the case, but it is evident from the two figures that hull damping has a very small influence on the results, even for the lower speed. There is a noticable influence only for the lowest wave periods, where modes higher than no. 7 are

important. The difference between the hull damping in the two cases is a factor of 4, as can be seen from Table 1.

When the number of flexible modes was increased from 4 to 10, there was a small reduction in the influence of hydroelasticity. The peak value for case122 in Figure 7a was reduced from approximately 170 to 150. The influence of the additional modes were rapidly diminishing for higher mean wave periods, and for $\bar{T}_1$-values larger than 1.1-1.2, their effect was negligable. Hence, 4 flexible modes were considered as sufficient for the purpose of the present study.

In Figure 8, we have plotted the non-dimensional VBM standard deviation as a function of the mean wave period for 4 of the above cases.

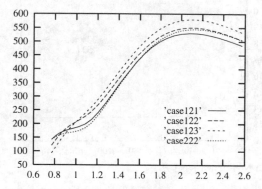

Figure 8. Non-dimensional VBM standard deviation $(\hat{\sigma}_{HE}=\sigma_{HE}/\rho g L^3 H_s)$ for 4 of the analysed cases. Plotted as a function of non-dimensional mean wave period: $\bar{T}_1=T_1(g/L)^{1/2}$. Magnified by 10^6.

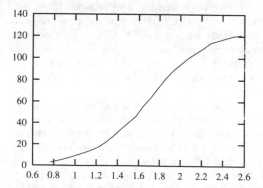

Figure 9. Most probable largest VBM [MNm] in a 6 hrs sea state with mean wave period: $\bar{T}_1=T_1(g/L)^{1/2}$. Plotted as a function of $\bar{T}_1$. Case122.

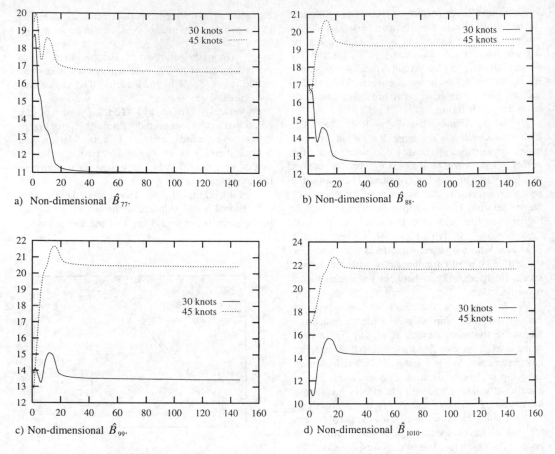

Figure 10. Non-dimensional generalized hydrodynamic damping coefficients: $\hat{B}_{kk}=B_{kk}/(\rho^2 gL^9)^{1/2}$. Plotted as functions of non-dimensional frequency: $\hat{\omega}=\omega(L/g)^{1/2}$. Flexible ship. Medium hull damping.

3.6 *The stress level*

We have seen that hydroelastic effects can increase the calculated vertical bending moment significantly for sea states with low and moderate mean wave periods. In order to evaluate the importance of hydroelasticity for the monohull, we need to estimate the typical stress levels in these sea states.

The proper way of doing this would be to select a scatter diagram for a representative sea area, and then perform a long term statistical analysis. Such an analysis should take into account the fact that the vessel will not operate at all, or it will operate at a reduced speed, for the high sea states. Typical criteria for speed reduction would be maximum values of hull accelerations or maximum frequency of slamming.

As a first approach, however, we can calculate the expected value of H_s for each T_1, based on the selected scatter diagram.

An operational area between Korea and Japan has been selected. This area was also used by Faltinsen et al. (1992), and the scatter diagram was taken from area E02S in the weather atlas presented by Takaiski, Matsumoto and Ohmatsu (1980).

Multiplying the values of the curve for case 122 in Figure 8, by the expected H_s for each wave period, we can find the most probable vertical bending moment in a 6 hour sea state as a function of wave period; see Figure 9. The most probable VBM is found from the relation

$$VBM_N = \sigma_{HE}\sqrt{2\ln N} \qquad (28)$$

where N is the number of cycles during the 6 hours.

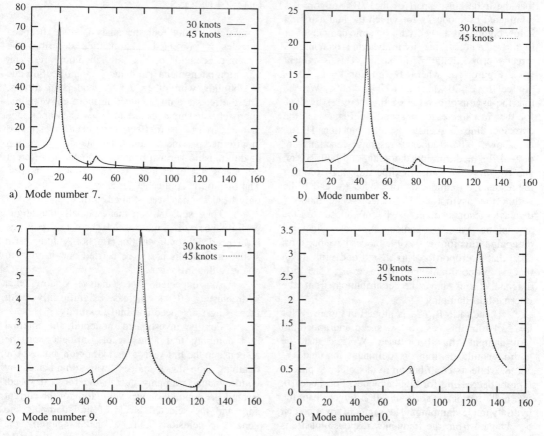

a) Mode number 7.

b) Mode number 8.

c) Mode number 9.

d) Mode number 10.

Figure 11. Non-dimensional modulus of frequency-response functions: $\hat{H}_{kk}=H_{kk}\rho g L^4$. Plotted for non-dimensional frequency: $\hat{\omega}=\omega(L/g)^{1/2}$. Magnified by 10^4. Flexible ship. Medium hull damping.

From the DNV Light Craft Rules (DNV 1993), we find a design VBM of approximately 300 MNm. There are however, some uncertainties connected to this value. The Light Craft Rules are generally valid only for vessel lengths up to 100 meter (Svensen & Valsgård 1993), and their applicability to the current vessel is therefore questionable. Large deviations between values obtained by calculations and those obtained from the Rules have been reported for 50-120 meter catamarans (Faltinsen et al. 1992) and for a 220 meter monohull (Jullumstrø, Leppänen & Sirviö 1993).

Figure 9 shows that the most probable largest VBM in the different sea states is much lower than the DNV design moment. However, sea states where the most probable largest stresses are in the order of 10-20% of the expected maximum long term stresses during the ship's lifetime

contribute significantly to the fatigue damage of the ship (Soares & Moan 1991) . If we, in the absence of a better alternative, take the DNV design moment as an estimate of the expected maximum long term moment, we see from Figure 9 that $\bar{T}_1$-values between 1.4 and 1.7 would produce stresses that are important with respect to fatigue. It is apparent from Figure 7 that there can be a 5-30% influence of hydroelasticity for these sea states, depending on which of the different cases we study. Hence, the fatigue performance of the ship can be influenced by hydroelastic effects.

3.7 *Some comments on damping*

It was discovered in the parametric study that the hull damping played an insignificant role for all but the very lowest mean wave periods. The reason is that the hydroelastic part of the VBM response is dominated by mode 7, for which the hull damping is low compared to the hydrodynamic damping. For high frequencies, the hydrodynamic damping is directly proportional to the speed. This is evident from Figure 10, where $B_{kk}(\omega)$ for $k=7,...,10$ is plotted for U=30 knots and U=45 knots. We see that the asymptotic value of B_{kk} increases by 50% for the 50% increase in speed. In Figure 11, the corresponding frequency-response functions $H_{kk}(\omega)$ are shown. The frequency-response functions are plotted for medium hull damping, and both $B_{kk}(\omega)$ and $H_{kk}(\omega)$ relate to the flexible ship. It is seen that the resonance frequencies for modes 8, 9 and 10 lie within the asymptotic B_{kk}-range, while mode 7 displays resonance at a $B_{77}(\omega)$ close to the asymptotic value. Hence, the total diagonal damping terms for the flexible modes have one term (B_{kk}) that is proportional to U, and one term (C_{kk}) that is proportional (equal) to $\alpha_2\omega_k^2$. The off-diagonal terms have no contribution from the internal hull damping.

The ratio B_{kk}/C_{kk} is plotted in Figure 12 for $k=7,...,10$. In this case, a low speed and a medium hull damping has been used. We see that the hydrodynamic damping is dominant for modes 7 and 8, while its contribution to the total damping is rapidly decreasing for higher modes. For mode 10, the hull damping is about 50% larger than the hydrodynamic damping. The same information can be obtained from the frequency-response functions in Figure 11. It is seen that for mode 7, the peak value of H_{kk} is inversely proportional to U, while the dependency on U decreases for the higher modes.

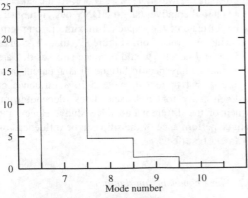

Figure 12. Relative difference between hydrodynamic damping and internal hull damping: B_{kk}/C_{kk} for the first 4 symmetric flexible modes. Speed: 30 knots. Medium hull damping: $\alpha_2=0.0014$. Flexible Ship.

4 CONCLUSIONS

A method for hydroelastic analysis of high speed vehicles is presented. The method describes the dynamic behavior of the ship, including resonant vibration phenomena and dynamic amplification due to coupling with other flexible and rigid modes. These effects are not included in more conventional approaches in ship response analysis. The importance of these hydroelastic effects have been investigated in a limited parametric study. The standard deviation of the midship vertical bending moment in different sea states has been used as the response quantity. The method is applicable for all wave headings, but only head sea has been studied in this paper.

The results indicate that especially the fatigue performance of high speed vehicles can be influenced by hydroelastic effects. A long term statistical analysis has to be carried out in order to quantify this influence.

As expected, the analyses show that hydroelastic effects increase significantly with increasing ship speed and hull flexibility.

For the investigated monohull, the internal hull damping had a negligable influence on the response in the first and second vibration modes. The damping in these modes was dominated by hydrodynamic damping due to the forward speed. Even if the domination was very large for the current ship and the selected levels of hull damping, no general conclusions can be made regarding the importance of hull damping in high speed vessels. Other high speed ships may display a much higher dependency on hull damping. This may particularly be the case for FRP sandwich vessels.

This paper is part of an on-going study of hydroelastic effects in high speed marine vehicles, and it is by no means conclusive. In addition to the calculation of long term statistical responses, more response quantities should be investigated, and different wave headings should be included. Analyses should be made of other high speed monohulls as well as catamarans. An important future task is to verify the method by experiments.

5. ACKNOWLEDGEMENT

The authors wish to thank professor Odd Faltinsen at the Dept. of Marine Hydrodynamics, Norwegian Institute of Technology, for his helpful advice and comments. The original version of the FASTSEA computer program was developed as a joint project between MARINTEK, Det Norske Veritas, The

Norwegian Institute of Technology and The Royal Norwegian Navy. Computer facilities have partly been provided by MARINTEK.

REFERENCES

Aertssen, G. & de Lembre, R. 1971. A Survey of Vibration Damping Factors found from Slamming Experiments on Four Ships. *Trans. NECIES* 87: 83-86.

Betts, C.V., R.E.D. Bishop & W.G. Price 1977. A Survey of Internal Hull Damping. *Trans. RINA* 119: 125-142.

Bishop, R.E.D. & W.G. Price 1979. *Hydroelasticity of ships*. Cambridge, U.K.: Cambridge University Press.

Bishop, R.E.D., W.G. Price & Y. Wu 1986. A general linear hydroelastic theory of floating structures moving in a seaway. *Phil. Trans. R. Soc. Lond.* A 316: 375-426.

Blok, J.J. & W. Beukelman 1984. The High-Speed Displacement Ship Systematic Series Hull Forms - Seakeeping Characteristics. *SNAME Trans.* 92: 125-150.

Catley, D. & C. Norris 1976. Theoretical Prediction of the Vertical Dynamic Response of Ship Structures Using Finite Elements and Correlation with Ship Mobility Measurements. *Proc. 11th Symp. Nav. Hydrodyn.*: VI.23-VI38. London, U.K.: Dept. Mech. Eng., Univ. College London.

DNV 1993. *Tentative rules for classification of high speed and light craft.* Høvik, Norway: Det Norske Veritas Classification.

Faltinsen, O., J.R. Hoff, J. Kvålsvold & R. Zhao 1992. Global loads on high speed catamarans. *Proc. Practical Design of Ships and Mobile Units PRADS'92*: 1.360-1.373. Newcastle,U.K. : Elsevier Applied Science

Faltinsen, O. & T. Svensen 1990. Incorporation of seakeeping theories in CAD. *CFD and CAD in Ship Design*: 147-164. Elsevier Sc. Publ.

Faltinsen, O. & R. Zhao 1991a. Numerical prediction of ship motions at high forward speed. *Phil. Trans. R. Soc. Lond.* A 334: 241-257.

Faltinsen, O. & R. Zhao 1991b. Flow predictions around high-speed ships in waves. *Matemathical approaches in hydrodynamics SIAM*.

FASTSEA 1991. *FASTSEA User's Manual.* Techn. Rep. no. 91-2012. Høvik, Norway: Det Norske Veritas Research.

Hylarides, S. 1974. Damping in Propeller-Generated Ship Vibrations. Rep. no. 468: Netherlands Ship Model Basin.

Jensen J.J., & N.F. Madsen 1977. A review of ship hull vibration. Part II: Modeling physical phenomena. *Shock and Vibration Digest* 9 (5): 13-22.

Jullumstrø, E., J. Leppänen & J. Sirviö 1993. Performance and Behaviour of the Large Slender Monohull. *Proc. Sec. Int. Conf. on Fast Sea Transportation FAST'93*: 1477-1488. Yokohama, Japan: Soc. Nav. Arch. Japan.

Newland, D.E. 1989. *Mechanical vibration analysis and computation.* Burnt Mill, U.K.: Longman.

Ohkusu, M. & O. Faltinsen 1991. Prediction of Radiation Forces on a Catamaran at High Froude number. *Proc. 18th Symp. Nav. Hydrodyn.*:5-19. Washington D.C., USA: National Academy Press.

PUSFEA 1993. *PUSFEA - A Package of Utility Fortran Subroutines for Finite Element Analysis.* M. Wu, Rep. Dept. of Marine Structures: The Norwegian Institute of Technology.

Soares, C.G. & T. Moan 1991. Model Uncertainty in the Long-term Distribution of Wave-induced Bending Moments For Fatigue Design of Ship Structures. *Marine Structures* 4: 295-315.

Svensen, T.E. & S. Valsgård 1993. Design Philosophy and Design Procedures for Large High Speed Craft. *Proc. Sec. Int. Conf. on Fast Sea Transportation FAST'93*: 1597-1612. Yokohama, Japan: Soc. Nav. Arch. Japan.

Takaiski, Y., T. Matsumoto & S. Ohmatsu 1980. Winds and Waves of the North Pacific Ocean 1964-1973. Statistical Diagrams and Tables. Tokyo, Japan: Ship Research Institute.

Talvia, J. & R. Wiefelspütt 1991. Offshore Measurements on Large Scale Model and Investigation of Structural Response Aspects of Slamming Loads for High Speed Monohulls. *Proc. First Int. Conf. on Fast Sea Transportation FAST'91*: 797-809. Trondheim, Norway: Tapir.

Wu, M., O.A. Hermundstad & T. Moan 1993. Hydroelastic Analysis of Ship Hulls at High Forward Speed. *Proc. Sec. Int. Conf. on Fast Sea Transportation FAST'93*: 699-710. Yokohama, Japan: Soc. Nav. Arch. Japan.

Wu, Y. 1984. *Hydroelasticity of floating bodies.* Ph.D thesis. Dept. Mech. Eng. Brunel University, Uxbridge, U.K.

Hydroelasticity in Marine Technology, Faltinsen et al. (eds) © 1994 Balkema, Rotterdam, ISBN 90 5410 387 6

Hydrodynamic added mass of a floating vibrating structure

A. Nestegård
Det Norske Veritas Research AS, Høvik, Norway

M. Mejlænder-Larsen
Det Norske Veritas Classification AS, Høvik, Norway

ABSTRACT: A floating structure may be excited to vibrate in its elastic modes due to disturbances from the engine, propeller, wave impact, etc. The eigen-frequencies for a floating structure are different from the "dry" structural eigen-frequencies of the structure vibrating in air. To account for the effect of the surrounding fluid, the solution of the fluid problem must be incorporated into the structural equations. A symmetric boundary integral equation method for the fluid flow is coupled with a finite element method for the structure displacements. Numerical results for eigenfrequencies and mode shapes of different partly submerged three-dimensional structures are presented.

1 INTRODUCTION

A floating structure (ship, offshore platform, etc.) may be excited to vibrate in its elastic modes due to disturbances from engine, propeller, wave impact, etc. The eigen-frequencies for a partly or fully submerged structure are different from the "dry" structural eigen-frequencies of the structure vibrating in air. This is due to the influence of the water pressure causing an added mass effect. In order to account for this effect, the solution of the fluid problem must be incorporated into the structural equations.

The most commonly used numerical methods for determining the fluid pressure field are the finite element method (FEM) (Armand & Orsero 1979, Hakala 1984) and the boundary integral equation method (BIEM) (Vorus & Hylarides 1981, Chung 1987). The use of finite elements requires the introduction of pressure nodes in the fluid domain which may be infinite. Hence the size of the coupled problem becomes rather large. So called infinite elements may be utilized to reduce the number of nodes. The resulting assembled finite element matrices are symmetric and can easily be merged with the structural matrices and handled by standard symmetric eigenvalue solvers.

Boundary integral equations provide an economical way of satisfying the fluid flow equations by reducing the dimensionality of the problem. Pressure nodes are needed only on the wetted surface of the structure. The resulting matrices however are nonsymmetric and can not be used in standard finite element codes which utilize symmetric matrices and corresponding equation solvers. Zienkiewicz (1977) and Mathews (1986) have shown how to obtain symmetric matrices from a boundary integral equation formulation for fluid-structure interaction problems. In this study symmetric matrices are derived by utilizing a variational formulation together with the boundary integral equation method.

The fluid-structure coupling is effected through matching the normal surface velocity defined in the elastic equilibrium relationship to that defined by the governing equations of the fluid field. In the coupling of finite element - boundary integral formulations, isoparametric elements are used to approximate the fluid pressure and the normal surface velocity in a manner consistent with the structural finite element discretization. The fluid pressure is eliminated from the problem and appears implicitly through a generalized added mass matrix which is merged with the structural mass matrix. This "virtual mass" matrix is then used in the eigenvalue problem to determine the natural frequencies for the coupled hydroelastic problem. For relatively simple geometries, like slender ships or vertical piles, the analysis of hydroelastic response may be treated in terms of predefined modal shape functions. This idea is pursued by Newman (1994). For more complex structures these mode shapes must be precalculated by finite element methods and would then naturally be the "dry" eigen-modes of the structures. For such geometries the general methodology suggested in this study would be preferred.

In Section 2 the fluid problem is formulated in terms of the fluid presure and the integral equation for the fluid pressure is set up. In Section 3 a discrete relationship between the pressure and its normal derivative on the wetted surface of the structure is derived. In Section 4 a variational functional is utilized to provide a symmetric relationship between the pressure and the normal displacements of the vibrating structure. In Section 5 the structural and fluid equations are coupled to give the hydroelastic eigenvalue problem. Numerical results for a vibrating spherical steel shell, torsion and bending of a steel rudder and vertical bending of a cruise vessel are presented.

2 FORMULATION OF THE FLUID PROBLEM

Assuming an ideal fluid and irrotational flow (viscous effects are neglected), potential theory can be applied. Also effects like surface tension and cavitation are ignored. The fluid can be assumed to be incompressible provided that a characteristic structural wavelength λ_s are much smaller than the acoustic wavelength in the water $\lambda_w = c/f$ where c is the speed of sound in the fluid and f is the frequency of vibration. Within potential theory, the fluid velocity is written as

$$\mathbf{v} = \nabla\Phi \qquad (2.1)$$

where

$$\Phi = \text{Re}[\phi(\mathbf{x})e^{i\omega t}] \qquad (2.2)$$

is the velocity potential. The motion is assumed to be time-harmonic with angular frequency of vibration ω.

Within linear theory the fluid pressure P is related to the velocity potential by

$$P = \text{Re}[pe^{i\omega t}] = -\rho\frac{\partial\Phi}{\partial t} \qquad (2.3)$$

or

$$p(\mathbf{x}) = -i\rho\omega\phi(\mathbf{x}) \qquad (2.4)$$

where ρ is the fluid density. In the fluid domain Ω the governing equation for ϕ is the Laplace's equation,

$$\nabla^2\phi = 0 \ , \ \mathbf{x}\epsilon\Omega \qquad (2.5)$$

If compressibility effects are included, the governing equation would be the Helmholtz equation. On the wetted surface Σ of the vibrating structure the fluid normal velocity must equal the normal velocity v_n of the surface. Hence,

$$\frac{\partial\Phi}{\partial n} = v_n \ , \ \mathbf{x}\epsilon\Sigma \qquad (2.6)$$

The normal velocity of the surface of a vibrating structure can be written

$$v_n = \text{Re}[i\omega u_n e^{i\omega t}] \qquad (2.7)$$

where u_n is the normal displacement of the surface of the structure.

We will assume that the structure vibrates at relatively high frequencies so that the effect of surface waves can be neglected. This is a

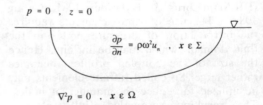

Figure 1. A partly submerged structure. The wetted part of the structure is denoted by Σ. The fluid domain is denoted by Ω.

good approximation in the frequency range of propeller induced vibrations of a ship, but is questionable at low frequencies corresponding to rigid body motions or the first fundamental mode of vibration of the hull girder of large ships. In the limit of very high frequencies ($\omega \rightarrow \infty$) the free surface condition for ϕ can be approximated by

$$\phi = 0 \ , \ z = 0 \qquad (2.8)$$

This means that the effect of gravity is neglected so that generation of surface waves is prohibited. Since the pressure p within linear theory is proportional to the potential ϕ, the boundary value problem for ϕ can easily be transformed to a boundary value problem for p. Laplace's equation (2.5) and the homogeneous free surface condition (2.8) also hold for the fluid pressure p

$$\nabla^2 p = 0 \ , \ \mathbf{x} \epsilon \Omega \qquad (2.9)$$

$$p = 0 \ , \ z=0 \qquad (2.10)$$

Combining (2.3), (2.6) and (2.7) the boundary condition for p on the wetted surface Σ becomes

$$\frac{\partial p}{\partial n} = \rho \omega^2 u_n \ , \ \mathbf{x} \epsilon \Sigma \qquad (2.11)$$

The displacement in the direction normal to the surface of the structure, u_n, which appears in (2.11) is unknown and must be found as part of the solution of the coupled problem. The fluid depth is assumed to be infinite.

A direct relationship between the fluid pressure p and its normal derivative $\partial p / \partial n$ on the wetted surface can be obtained by applying Green's second identity to p and a Green function G. Requiring G to satisfy the same free surface condition (2.8) as p the integration domain is reduced to the wetted surface Σ of the structure. The Green function is given by

$$G(\mathbf{x},\xi) = \frac{1}{R} - \frac{1}{R'} \qquad (2.12)$$

where

$$\begin{pmatrix} R' \\ R \end{pmatrix} = [(x-\xi)^2+(y-\eta)^2+(z\pm\varsigma)^2]^{\frac{1}{2}} \qquad (2.13)$$

and $\mathbf{x} = (x,y,z)^T$, $\xi = (\xi,\eta,\varsigma)^T$. Superscript T means transpose. $R = |\mathbf{x} - \xi|$ is the distance from $\mathbf{x}$ to the source point ξ and $R' = |\mathbf{x} - \xi'|$ the distance to the image source $\xi' = (\xi,\eta,-\varsigma)^T$. Assuming a smooth surface Σ, the following integral equation is obtained,

$$- 2\pi p(\mathbf{x}) + \int_{\Sigma} p(\xi) \frac{\partial G(\mathbf{x},\xi)}{\partial n_\xi} d\sigma_\xi$$

$$= \int_{\Sigma} G(\mathbf{x},\xi) \frac{\partial p(\xi)}{\partial n_\xi} d\sigma_\xi \qquad (2.14)$$

where $\mathbf{x} \epsilon \Sigma$ and $\mathbf{n}_\xi$ is the outward normal at ξ. For non-smooth surfaces the factor 2π in (2.14) is replaced by the solid angle at $\mathbf{x}$.

3 DISCRETIZATION

The discretization of the composite integral formulation is carried out using the isoparametric concept from finite elements. The same interpolation functions are used to model both the surface of the structure as well as the fluid pressure and its normal derivative which is proportional to the normal displacement of the surface. This concept allows the integration of the interpolated variables over a 3D surface to be carried out within a standard basis square in the (s,t) coordinate space.

Using isoparametric elements to discretize the surface, the global Cartesian coordinates $\xi = (\xi,\eta,\varsigma)$ on the surface are related to nodal coordinates $(\xi_i,\eta_i,\varsigma_i)$ by

$$\begin{Bmatrix} \xi(s,t) \\ \eta(s,t) \\ \varsigma(s,t) \end{Bmatrix} = \sum_{i=1}^{n_{en}} N_i(s,t) \begin{Bmatrix} \xi_i \\ \eta_i \\ \varsigma_i \end{Bmatrix} \qquad (3.1)$$

where n_{en} is the number of nodes per element. For example the shape functions $N_i(s,t)$ for the bilinear ($n_{en}=4$) elements are given by

$$N_i(s,t) = \frac{1}{4}(1\pm s)(1\pm t) \ , \ i=1,4 \qquad (3.2)$$

Using the isoparametric element concept, p

263

and $\partial p/\partial n$ are approximated over each element by the same relationship used to interpolate the surface geometry. This means that on element no. e,

$$p^e = \sum_{i=1}^{n_{en}} N_i p_i^e \qquad (3.3)$$

$$\left(\frac{\partial p}{\partial n}\right)^e = \sum_{i=1}^{n_{en}} N_i \left(\frac{\partial p}{\partial n}\right)_i^e \qquad (3.4)$$

Discretization of the wetted surface of the structure makes it possible to approximate the integral equation (2.14) by a system of simultaneous equations. The surface integrals in (2.14) can be written as a sum of integrals over each element,

$$-2\pi p(\mathbf{x}) + \sum_{e=1}^{n_{el}} \int_{\Sigma^e} p^e(\xi)\frac{\partial G(\mathbf{x},\xi)}{\partial n}d\sigma_\xi$$

$$= \sum_{e=1}^{n_{el}} \int_{\Sigma^e} \left(\frac{\partial p}{\partial n}\right)^e G(\mathbf{x},\xi)\, d\sigma_\xi \quad (3.5)$$

where Σ^e is the surface of element e and n_{el} is the number of elements on the wetted surface. Introducing the expressions (3.3) and (3.4),

$$-2\pi p(\mathbf{x}) + \sum_{e=1}^{n_{el}}\sum_{i=1}^{n_{en}} D_i^e(\mathbf{x})\, p_i^e$$

$$= \sum_{e=1}^{n_{el}}\sum_{i=1}^{n_{en}} S_i^e(\mathbf{x}) \left(\frac{\partial p}{\partial n}\right)_i^e \quad (3.6)$$

where

$$S_i^e(\mathbf{x}) = \int_{\Sigma^e} N_i(\xi)G(\mathbf{x},\xi)\, d\sigma_\xi \qquad (3.7)$$

$$D_i^e(\mathbf{x}) = \int_{\Sigma^e} N_i(\xi)\, (\mathbf{n}\cdot\nabla_\xi)\, G(\mathbf{x},\xi)\, d\sigma_\xi \quad (3.8)$$

Assembling the contributions from each element and satisfying the equation (3.6) for each of the nodes of the wetted surface, the algebraic system of equations can be represented as

$$\mathbf{Ap} = \mathbf{Bq} \qquad (3.9)$$

where

$$\mathbf{p} = \left\{p_1, p_2, \ldots\ldots, p_M\right\}^T \qquad (3.10)$$

$$\mathbf{q} = \left\{\left(\frac{\partial p}{\partial n}\right)_1, \left(\frac{\partial p}{\partial n}\right)_2, \ldots\ldots, \left(\frac{\partial p}{\partial n}\right)_M\right\}^T \quad (3.11)$$

are the nodal values of the pressure and normal derivative of pressure on the wetted surface. M is the number of nodes on the wetted surface. The matrices $\mathbf{A}$ and $\mathbf{B}$ are defined by equation (3.6). The integrals over each element in (3.6) are evaluated using quadrature formulaes. It is necessary, however, to consider separately the case in which a node is one of the nodes of the element to be integrated. Such elements are denoted as singular elements. For nonsingular elements the integrand varies smoothly over the surface and Gaussian quadrature formulas are used. The numerical evaluation of the integrals S_i^e and D_i^e in the case of bilinear elements (3.2) are described in Appendix A.

4 SYMMETRIC RELATIONSHIPS

The matrices $\mathbf{A}$ and $\mathbf{B}$ are nonsymmetric and therefore the relationship (3.9) is not suitable for use together with the symmetric structural matrices in the coupled hydroelastic problem. It turns out that it is possible to obtain a symmetric pressure - displacement relationship making use of a variational formulation together with the relationship (3.9) (Zienkiewicz 1977), (Mathews 1986). A variational functional for the boundary value problem (2.9) and (2.10) is given by

$$\Pi(p) = \tfrac{1}{2}\int_\Sigma p\frac{\partial p}{\partial n}d\sigma - \rho\omega^2\int_\Sigma pu_n d\sigma \quad (4.1)$$

The boundary value problem for p is obtained by requiring that the variation of Π w.r.t. p vanishes. Introducing the expressions (3.3) and (3.4), the contribution from element e to the discrete functional $\hat{\Pi}$ is

$$\hat{\Pi}^e = \frac{1}{2} \int_{\Sigma^e} N_i p_i^e N_j \left(\frac{\partial p}{\partial n} \right)_j^e d\sigma$$

$$- \rho \omega^2 \int_{\Sigma^e} N_i p_i^e u_n d\sigma \qquad (4.2)$$

Assembling the elements, the functional becomes

$$\hat{\Pi} = \frac{1}{2} \mathbf{p}^T \mathbf{C} \mathbf{q} - \mathbf{f}^T \mathbf{p} \qquad (4.3)$$

where the contributions to $\mathbf{C}$ and $\mathbf{f}$ from each element are

$$C_{ij}^e = \int_{\Sigma^e} N_i N_j d\sigma \qquad (4.4)$$

$$f_i^e = \rho \omega^2 \int_{\Sigma^e} N_i u_n d\sigma \qquad (4.5)$$

From the discretized integral equation (3.9) we can eliminate the normal derivative of the pressure at the nodes, $\mathbf{q}$,

$$\mathbf{q} = \mathbf{B}^{-1} \mathbf{A} \mathbf{p} \qquad (4.6)$$

Substituting into the expression for the discrete functional (4.3),

$$\hat{\Pi}(\mathbf{p}) = \frac{1}{2} \mathbf{p}^T \mathbf{C} [\mathbf{B}^{-1} \mathbf{A}] \mathbf{p} - \mathbf{f}^T \mathbf{p} \qquad (4.7)$$

To obtain a symmetric relationship the functional $\hat{\Pi}$ is minimized by requiring that the variation of $\hat{\Pi}$ with respect to $\mathbf{p}$ vanishes. Hence, we obtain,

$$\delta\hat{\Pi}(\mathbf{p}) = \mathbf{p}^T \mathbf{K}_f \delta\mathbf{p} - \mathbf{f}^T \delta\mathbf{p} = 0 \qquad (4.8)$$

where the symmetric "fluid stiffness" matrix is defined by

$$\mathbf{K}_f = \frac{1}{2} \mathbf{C}(\mathbf{B}^{-1} \mathbf{A}) + \frac{1}{2} [\mathbf{C}(\mathbf{B}^{-1} \mathbf{A})]^T \qquad (4.9)$$

Equation (4.8) should hold for arbitrary variations $\delta\mathbf{p}$, therefore

$$\mathbf{p}^T \mathbf{K}_f = \mathbf{f}^T \qquad (4.10)$$

or by taking the transpose and utilizing the symmetry of $\mathbf{K}_f$,

$$\mathbf{K}_f \mathbf{p} = \mathbf{f} \qquad (4.11)$$

Equation (4.11) provides a discrete sym-

metric relation between the pressure at the nodes and the normal displacements on the wetted surface.

5 COUPLING OF STRUCTURAL AND FLUID EQUATIONS

The finite element equations for the dynamic fluid-structure system, assuming free vibrations with no structural damping are,

$$(-\omega^2 \mathbf{M}_s + \mathbf{K}_s)\mathbf{u} = \mathbf{g} \qquad (5.1)$$

where $\mathbf{M}_s$ and $\mathbf{K}_s$ are the global structural mass and stiffness matrices respectively. $\mathbf{u}$ are the nodal displacements (each node has three displacement components) and $\mathbf{g}$ represents the interaction forces generated by the fluid pressure acting on the wetted surface. The interaction force vector $\mathbf{g}$ can be defined through a coupling matrix $\mathbf{L}$ and the nodal pressure distribution $\mathbf{p}$ where

$$\mathbf{g} = \mathbf{L}\mathbf{p} \qquad (5.2)$$

The contribution from each element to the coupling matrix is

$$L_{3(i-1)+k, j}^e = \int_{\Sigma^e} N_i n_k N_j d\sigma, \ k=1,2,3 \ (5.3)$$

Although nonsquare, the assembled matrix $\mathbf{L}$ is similar in form to a consistent structural mass matrix.

Enforcing continuity of displacement on the wetted surface, one may show that the r.h.s. of (4.11) can be expressed in terms of the nodal displacements $\mathbf{u}$ by

$$\mathbf{f} = \rho \omega^2 \mathbf{L}^T \mathbf{u} \qquad (5.4)$$

The fluid pressure can now be related to the structural displacements by using the relationship (4.11),

$$\mathbf{p} = \mathbf{K}_f^{-1} \mathbf{f} = \rho \omega^2 \mathbf{K}_f^{-1} \mathbf{L}^T \mathbf{u} \qquad (5.5)$$

Thus, a symmetric expression for the fluid interaction forces in terms of the nodal displacements is given by

$$\mathbf{g} = \rho \omega^2 (\mathbf{L} \mathbf{K}_f^{-1} \mathbf{L}^T) \mathbf{u} \qquad (5.6)$$

Substituting this expression into equation (5.1) gives the following equations of motions for free vibrations of the coupled system,

$$[-\omega^2(\mathbf{M}_s + \mathbf{M}_a) + \mathbf{K}_s]\mathbf{u} = 0 \qquad (5.7)$$

where

$$\mathbf{M}_a = \rho\mathbf{L}\mathbf{K}_f^{-1}\mathbf{L}^T \qquad (5.8)$$

is the generalized "added mass" matrix. This matrix is independent of mode and frequency. It is only a function of the wetted geometry and its discretization. Thus the solution to the coupled system is given in terms of the structural displacements. The eigen-frequencies ω_k and the corresponding eigenmodes $\mathbf{u}_k$ are found from the solution of the eigenvalue problem

$$|\mathbf{K}_s - \omega^2(\mathbf{M}_s + \mathbf{M}_a)| = 0 \qquad (5.9)$$

For the more general moderate frequency problem including surface waves, the "added mass" matrix would be a function of the frequency of vibration, $\mathbf{M}_a(\omega)$. The wave damping represented by a frequency dependent generalized damping matrix $\mathbf{B}_a(\omega)$ should then be added to the structural damping matrix. In this general case the eigenvalue problem is non-linear and a better way to identify the natural frequencies would be to perform a forced analysis with time-harmonic forcing on the right hand side of (5.7). For a "hyper-flexible" structure, the hydrostatic contribution must be included by adding a generalized restoring matrix $\mathbf{K}_a$.

The method described above can be generalized to the case of hydroelastic response of flexible structures caused by incoming surface waves. A typical application is *springing* of large ships. The fluid pressure must then also include the effect of the diffracted waves resulting in a generalized pressure excitation force on the right hand side of (5.7)

6 NUMERICAL RESULTS

Modules for calculation of the generalized "added mass" matrix $\mathbf{M}_a$ have been implemented in the general purpose linear structural analysis program SESTRA in the SESAM suite. The analysis features of SES-

TRA include static and dynamic analysis using the super-element technique. Dynamic analyses can be performed using Master-Slave or Component Mode Synthesis reduction techniques. Free vibrations are solved by utilizing the Householder, Subspace Iteration or Lanczos eigenvalue solvers. SESTRA can be used for arbitrary large structural systems using out-of-core equation solvers.

At present 3-noded linear and 4-noded bilinear elements have been used for the representation of the fluid pressure and the wetted surface of the structure. Symmetry is exploited so that only one half or a quarter of a symmetric structure needs to be discretized. Also, the structure can be partly or fully submerged.

During the development of the modules for calculation of the matrices $\mathbf{A},\mathbf{B},\mathbf{C},\mathbf{L}$ and $\mathbf{M}_a$, several numerical tests were done. The wetted surface of the structure can be approximated in terms of the entries of the $\mathbf{C}$ matrix as follows

$$S = \sum_{i=1}^{M} \sum_{j=1}^{M} C_{ij} \qquad (6.1)$$

The displaced volume of the structure can be approximated by

$$V = \sum_{i=1}^{M} \sum_{j=1}^{M} L_{3(i-1)+1,j} x_j \qquad (6.2)$$

Two different expressions can be derived for the high frequency limit of rigid body added mass A_{kl}^{∞},

$$A_{kl}^{\infty} = \mathbf{n}_k^T \mathbf{C}\mathbf{A}^{-1}\mathbf{B}\mathbf{n}_l \qquad (6.3)$$

and

$$A_{kl}^{\infty} = \sum_{i=1}^{M} \sum_{i=1}^{M} M_{a3(i-1)+k,3(j-1)+l} \qquad (6.4)$$

These formulas were used to check the numerical results for geometries where exact values of the quantities above exist. For a partly (half) submerged sphere the values are $S = 2\pi r^2$, $V = 2/3\pi r^3$ and $A_{kl}^{\infty} = 1/3\rho\pi r^3$. Convergence tests have been performed on a half submerged sphere by calculating the displaced volume and the vertical infinite frequency added mass from the formulas (6.2) and (6.4) above. The results are presented in Table 1. Liu (1988) has investigated the accuracy obtained by using dif-

Table 1: Convergence of displaced volume V and infinite frequency added mass A_{33}^{∞} for increasing number of bilinear 4-noded elements on wetted surface.

n_{el}	V	$A_{33}^{\infty}/\rho V$
16	2.082	0.53
64	2.093	0.52
576	2.094	0.51
Exact	2.094	0.50

Table 2: Eigen-frequencies [Hz] of a spherical steel shell in air and submerged in water.

Mode	In vacuo		Submerged	
	Numerical	Theory	Numerical	Theory
1	604	603	273	293
2	713	707	337	361

Figure 2a: Finite element model of spherical steel shell. 216 4-noded elements.

ferent boundary elements. His conclusion was that compared to the linear or bilinear boundary elements, the higher order elements (8- and 9-noded) are advantageous both w.r.t. accuracy and computational efficiency.

Three cases will be considered. These are a totally submerged spherical steel shell, a steel rudder and a cruise vessel.

6.1 Submerged spherical shell

The effect of added mass on the vibrations of a spherical steel shell was investigated. The geometry and material data were chosen as: Radius of shell r = 1 m, thickness t = 0.01 m, density $\rho = 2700$ kg/m^3, Youngs modulus E = $2.1 \cdot 10^{11}$ N/m^2, Poisson ratio $\nu = 0.30$. The finite element model (using 4-noded shell elements) of the spherical shell is shown in Figure 2a. The number of elements are 216. The eigen-frequencies of the shell were found both in air and submerged in water. The numerical values for the eigen-frequencies are compared with theoretical values from Junger and Feit (1972). The comparison is shown in Table 2. It is seen that the calculated eigen-frequencies for vibrations in vacuo agree very well with the theoretical values, while there is some discrepancy for the case of a submerged shell. It should be noted that the theoretical values for the submerged case is based on an analysis including the effect of compressibility of the fluid.

The mode shapes for the two first modes are shown in Figures 2b and 2c.

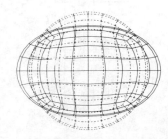

Figure 2b: First mode of vibration for spherical shell.

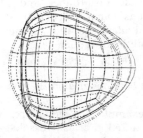

Figure 2c: Second mode of vibration for spherical shell.

6.2 Rudder

The next example is an analysis of the natural frequencies of the steel rudder / rudderhorn of a 300.000 TDW tanker. The background for such an analysis is to investigate whether the natural frequencies in both full load and ballast condition occur outside the range of bladepassing frequencies in the full speed range. The superelement technique is used with the aft part of the ship as superelement no. 1 and the rudder including the rudder-stock as Superelement no. 2. The rudder-stock is assumed clamped at the tank-top 23 m above base line. Two loading conditions have been investigated; full load condition with a fully submerged rudder (aft draught 22.0m), and a ballast conditions with an aft draught of approximately 11.0 m. The rudder is in this condition submerged to approximately 1 m above the lowest pintle.

A hidden line plot of the total model is shown in Figure 3a. The internal structure of the rudder is shown in Figure 3b. The free vibration analysis of the rudder connected to the rudder stock has been made using a finite element model of the rudder, rudder-stock and rudderhorn including parts of the aftship. 3-noded and 4-noded shell elements and 2-noded beam elements are used throughout the model. The total degrees of freedom in the model is 3732.

The lowest natural frequencies of the rudder / rudderhorn in both loading conditions are shown in Table 3. It can be seen that the effect of added mass is most pronounced for the torsion ($n=1$) and the coupled transverse bending / torsion modes ($n=2, n=4$) as can be expected from physical reasoning. The longitudinal bending modes ($n=3, n=5$) cause a smaller fluid disturbance and the added mass effect is minor. The natural frequencies in the ballast condition lie between the "dry" and full load frequencies.

6.3 Vertical bending of a cruise vessel

The third example is a global free vibration analysis of a 260 m cruise vessel. Finite element models of the vessel are shown in Figures 4a and 4b. The hull girder of the ship has been constructed by using 4-noded shell

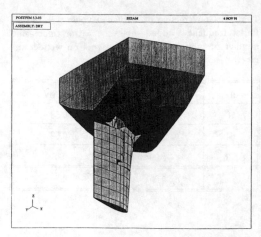

Figure 3a: Panel model of rudder / rudderhorn.

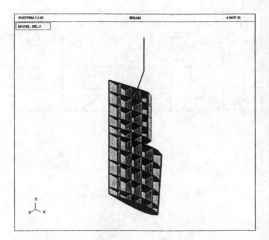

Figure 3b: Internal structure of the rudder.

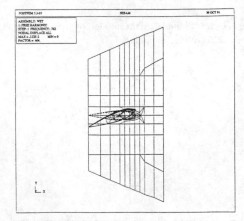

Figure 3c: Pure torsion of the rudder stock. $f_1 = 0.8\ Hz$.

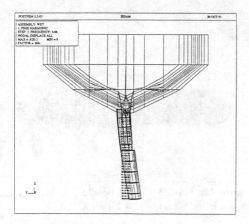

Figure 3d: First coupled transverse bending / torsion of rudder. $f_2 = 3.7\ Hz$.

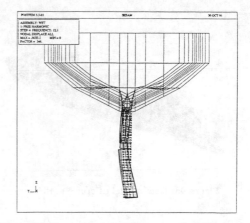

Figure 3f: Second coupled transverse bending / torsion of rudder. $f_4 = 12.1\ Hz$

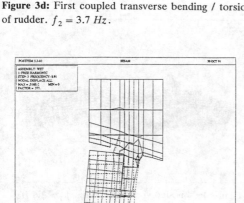

Figure 3e: First longitudinal bending of rudder. $f_3 = 8.9\ Hz$.

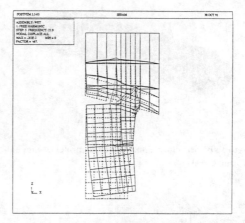

Figure 3g: Second longitudinal bending of rudder. $f_5 = 21.9\ Hz$

Table 3: Natural frequencies of rudder/rudderhorn [Hz]

Vibration mode	"Dry"	Full load	Bal-last
Pure torsion of the rudder stock	1.7	0.8	1.0
First coupl. transv. bending/ torsion of rudder	7.6	3.7	4.1
First longitudinal bending	9.8	8.9	9.5
Second coupled transv. bending/torsion of rudder	21.7	12.1	14.6
Second longitudinal bending	27.1	21.9	24.1

Table 4: Natural frequencies of a cruise vessel [Hz]

Mode	Comments to modes	Freq. (Hz)
1	2-noded vertical bending	1.4
2	3-noded vertical bending, including local modes in the aft ship	2.1
3	4-noded vertical bending	2.6
4	5-noded vertical bending	3.5
5	6-noded vertical bending, including local modes in the aft ship	4.4

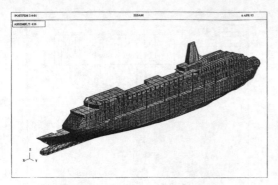

Figure 4a: Panel model of cruise vessel.

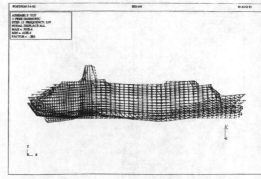

Figure 4d: 3-noded vertical bending, including local modes in the aft ship. $f_2 = 2.1\ Hz$

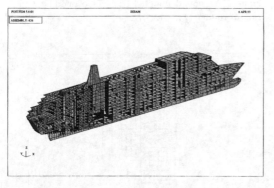

Figure 4b: Structural finite element model of one half of cruise vessel

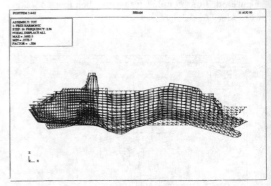

Figure 4e: 4-noded vertical bending. $f_3 = 2.6\ Hz$

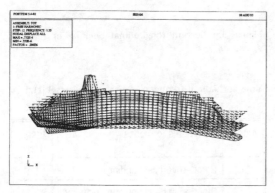

Figure 4c: 2-noded vertical bending. $f_1 = 1.4\ Hz$

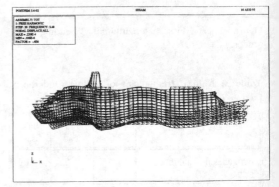

Figure 4f: 5-noded vertical bending. $f_4 = 3.5\ Hz$

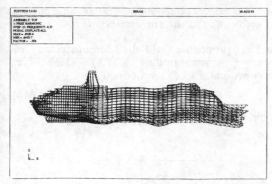

Figure 4g: 6-noded vertical bending, including local modes in the aft ship. $f_5 = 4.4\ Hz$

elements and 2-noded beam elements. The vessel model consists of 4 first level super-elements with a total number of 26616 degrees of freedom. The natural frequencies and the mode shapes including the effect of added mass have been calculated using the Householder iteration method. With increasing frequency, the modes become more coupled and complex, making the separation of the different mode shapes difficult. The calculated natural frequencies of the vertical bending modes are shown in Table 4. The mode shapes for the first 6 modes are presented in Figures 4c - 4h. The lowest frequencies of the hull girder bending modes appear in a range where the high frequency approximation may be questionable.

7 CONCLUSIONS

A symmetric formulation of the coupled fluid-structure interaction problem for vibrating floating structures has been derived. A boundary integral equation formulation for the fluid flow has been coupled to a finite element formulation for the structure. The coupling has been successfully implemented in a general purpose structural analysis program, SESTRA. Numerical results have been presented for three test cases of practical interest.

Convergence tests have shown that the coupling may be better effected by utilizing higher-order fluid boundary elements. An extension to acount for generation of surface waves is proposed, likewise the inclusion of the effect of compressibility of the fluid for elasto-acoustic analyses.

8 ACKNOWLEDGEMENTS

Jens Bloch Helmers, DNV Research developed parts of the computer program for the fluid part. Brita Carlin, DNV Sesam implemented the hydrodynamics modules in the structural analysis program SESTRA. Torgeir Vada and Rune Tønnessen of DNV Research both contributed to this study. Their help is greatly appreciated.

REFERENCES

Armand, J-L. and Orsero, P. (1979) A Method for Evaluating the Hydrodynamic Added Mass in Ship Hull Vibrations. *SNAME Transactions,* Vol.87, pp. 99-120.

Vorus, W.S. and Hylarides, S. (1981) Hydrodynamic Added-Mass Matrix of Vibrating Ship Based on a Distribution of Hull Sources. *SNAME Transactions*, Vol. 89, pp. 397-416.

Zienkiewicz, O.C., Kelly, D.W. and Bettess, P. (1977) The Coupling of The Finite Element Method and Boundary Solution Procedures. *Int. J. Num. Methods Eng.* 11, pp. 355-375.

Mathews, I.C. (1986) Numerical techniques for three-dimensional steady-state fluid-structure interaction. *J.Acoust.Soc.Am.,* 79(5), pp. 1317-1325.

Junger, M.C. and Feit, D. (1972) *Sound, Structures and Their Interaction.* MIT Press.

Liu, Y.H. 1988 Analysis of Fluid-Structure Interaction by using Higher Order Boundary Elements in Potential Problems and its Application in Coupling Vibrations of Bending and Torsion of Ships. Ph.D. Thesis, Shanghai Jiao Tong University.

Newman, J.N. (1994) Wave Effects on Deformable Bodies. To be published in *Applied Ocean Research.*

Hakala, M. (1984) Some Aspects on the Numerical Modelling of Added Mass in Ship Vibration". *International Symposium on Ship Vibration.* Paper No. 15, Genova, Italy.

Chung, K. (1987) On the Vibration Analysis of the Floating Elastic Body Using the Boundary Integral Method in Combination with Finite Element Method". Vol. 13. Report No. 10064, Korean Register of Shipping.

SESAM:SESTRA User's Manual. Det Norske Veritas Sesam (1991).

APPENDIX A.

The source and dipole integrals to be evaluated are defined by

$$S_i^e(\mathbf{x}) = \int_{\Sigma^e} \frac{N_i(\xi)}{|\mathbf{x} - \xi|} d\sigma_\xi \qquad (A.1)$$

$$D_i^e(\mathbf{x}) = \int_{\Sigma^e} N_i(\xi)(\mathbf{n} \cdot \nabla_\xi) \frac{1}{|\mathbf{x} - \xi|} d\sigma_\xi \quad (A.2)$$

where $N_i(\xi)$ are bilinear basis functions defined in (3.2). For simplicity we consider here only the $1/R$-part of the Green function (2.12). The second term $(1/R')$ is treated similarly. Transforming to parameter space (s,t), the source ane dipole integrals can be written as

$$S_i^e(\mathbf{x}) = \int\int_{-1}^{1} \frac{N_i(s,t)}{|\mathbf{x} - \xi|} J(s,t) ds dt \quad (A.3)$$

$$D_i^e(\mathbf{x}) = \int\int_{-1}^{1} N_i \frac{\mathbf{J}(s,t) \cdot (\mathbf{x} - \xi)}{|\mathbf{x} - \xi|^3} ds dt \quad (A.4)$$

where the vector and scalar Jacobians are given by

$$\mathbf{J}(s,t) = \mathbf{U}_1 \times \mathbf{U}_2 \quad , \quad J(s,t) = |\mathbf{J}| \quad (A.5)$$

$$\mathbf{U}_1 = \frac{\partial \xi}{\partial s} \quad , \quad \mathbf{U}_2 = \frac{\partial \xi}{\partial t}$$

where ξ as a function of (s,t) is given in (3.1). When $\mathbf{x} \neq \xi$, i.e. the field point is not on the element, the integrals S_i^e and D_i^e are evaluated by straightforward Gauss quadrature over the unit square in the (s,t) space. When the field point $\mathbf{x}$ is on one of the vertices ξ_j of the element Σ^e, the integrand becomes singular at $\xi = \xi_j$ and a special treatment is required. The singularity can be removed by introducing polar coordinates with origin at the singular point.

The procedure will be demonstrated for $\mathbf{x} = \xi_1$, corresponding to $(s,t) = (1,1)$. Polar coordinates (ρ,θ) are defined by

$$1 - s = \rho\cos\theta \quad , \quad 1 - t = \rho\sin\theta \quad (A.6)$$

as shown in Figure 5.

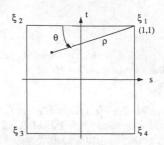

Figure 5. Definition of polar coordinates for integration of singularity.

The source integral is transformed to

$$S_i^e(\xi_1) = \int_0^{\pi/4} d\theta \int_0^{2/\cos\theta} d\rho \, f_i(\rho,\theta)$$

$$+ \int_{\pi/4}^{\pi/2} d\theta \int_0^{2/\sin\theta} d\rho \, f_i(\rho,\theta) \quad (A.7)$$

where

$$f_i(\rho,\theta) = \frac{J(\rho,\theta)N_i(\rho,\theta)}{|\xi_1 - \xi|(\rho,\theta)} \quad (A.8)$$

The singularity is removed and the integral is approximated by Gauss quadrature. A similar procedure is applied to the dipole integral.

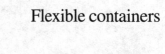

Flexible containers

Hydroelasticity in Marine Technology, Faltinsen et al. (eds) © 1994 Balkema, Rotterdam, ISBN 90 5410 387 6

Fabric as construction material for marine applications

Geir Løland & Jan Vidar Aarsnes
MARINTEK, Trondheim, Norway

ABSTRACT: The purpose of this paper is to describe the main hydrodynamic and structural problems related to the use of elastic fabric as construction material for marine applications with emphasize on the difference between the behaviour of flexible and rigid structures. The constraints on the geometrical shapes that can be obtained and the manufacturing process are also discussed. A method for calculation of the static shape of two-dimensional and axisymmetric structures are derived. Different methods for determining the dynamic tension and the necessity of performing model tests and developing special model test techniques are also discussed. A floating fabric container for transportation of water and a floating fabric bag pen for fish farming are discussed in some detail.

1 INTRODUCTION

The development of high quality elastic fabric of low weight, high strength and relative low price has increased the use of fabric as construction material for marine applications. In this context the word fabric means a sandwich structure consisting of a textile coated on both sides with a polymer. Utilizing the properties of the elastic fabric are especially important for structures that need high flexibility, high strength and low weight in order to operate as intended. Although, the most important advantage with a flexible construction material, is that one can manufacture units that requires a small storage space compared to the volume that can be enclosed in the structure.

The flexibility of a fabric structure is governed by the tension in and the elasticity of the material; the flexibility decreases with increasing tension, or in other words the pretension in the material. In spite of this the majority of such structures will be extremely flexible, and the flexibility will dramatically reduce the structural loads compared to that experienced by a rigid body of the same initial shape. Owing to the way a flexible structure responds to external loads; the structure changes shape instead of restraining deformation. This is in a sense the main reason for using flexible construction materials for marine

applications. However, the introduction of very flexible materials in a marine structure, introduces a new dimension of difficulties in the hydrodynamic and structural analysis. The structure can no longer be treated as rigid, neither can it in many cases be treated as linear. However, the most important difference between a flexible fabric structure and a rigid structure is that one can no longer decouple the hydrodynamic and the structural analysis. This means that we are dealing with a hydroelastic problem.

2 STATIC SHAPE

A fabric structure has little or no bending stiffness and the forces acting on the structure is transferred to tension in the fabric. In general, the loads on and the response of a membrane structure are non-linear due to the large deflections. We will in this section discuss the mathematical formulation of hydrostatic loads on two dimensional and axisymmetric membrane structures, and a numerical solution method for two-dimensional shapes.

2.1 *Pressure loads on a two-dimensional membrane.*

The external loads acting on a membrane or a cable are usually given by a distributed tangential load p_t

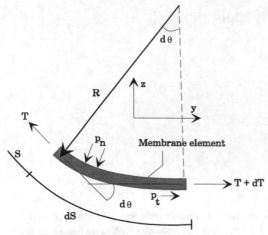

$$2 \cdot T \cdot \sin(\frac{d\Theta}{2}) - p_n(y,z) \cdot dS = 0$$

$$2 \cdot T \cdot (\frac{d\Theta}{2}) - p_n(y,z) \cdot dS \approx 0 \qquad (2)$$

$$\frac{d\Theta}{dS} = \frac{p_n(y,z)}{T}$$

Figure 1 Definition of forces acting on a infinitesimal membrane element.

Where Θ is the angle between the horizontal plane and the tangent of the membrane and $d\Theta$ is the changing in the tangent over the membrane element.

The change in the surface tangent over the element can be expressed in terms of the local radius of curvature R as $d\Theta/dS = 1/R$. This implies that equation (2) can be rewritten as:

$$\frac{1}{R} = \frac{p_n(y,z)}{T}$$

$$\updownarrow \qquad (3)$$

$$\frac{d^2 a(S)}{dS^2} \cdot [1 + (\frac{da(S)}{dS})^2]^{-3/2} = \frac{p_n(y,z)}{T}$$

and a normal component p_n, as shown in Figure 1. A detailed description of the cable and membrane theory can be found in a number of textbooks, for example Leonard (1988) or Kuznetsov (1991). For the special cases considered in the present paper, the normal component is caused by a pressure difference between an inner and an outer fluid domain. Furthermore, the tangential frictional force component p_t vanishes in the static cases, i.e no motion of the fluid. It can also be neglected in waves and steady flow since tangential friction is an order of magnitude less than the normal pressure loads. Hence, force equilibrium of an infinitesimal element in tangential direction yields:

$$dT + p_t \cdot dS = 0 \quad \Rightarrow \quad dT = 0 \qquad (1)$$

This implies that the tension T is constant in the circumferential direction, which is a very important observation.

In order to establish analytical and numerical solution procedures it is convenient to use a coordinate system following the membrane surface S, as shown in Figure 1. Force equilibrium of the infinitesimal element shown in Figure 1, provides a differential equation for the shape of the cross-section.

where S is the surface coordinate and a(S) is the deflection of the surface in a global coordinate system. Analytic solutions of the above differential equation exists only for a few special cases, for example constant pressure or hydrostatic pressure on an nonelastic membrane. Some of these solutions are discussed by Leonard (1988), Kuznetsov (1991) and Hawthorne (1961).

2.2 Pressure loads on an axially symmetrical membrane.

A detailed description of the derivation of the axi-symmetric equilibrium equation for membrane structures is given by Leonard (1988). The variables describing an axially symmetric structure in a cylindrical co-ordinate system (r,θ,z) are independent of the angle θ, and the shape of the body is given by r(z) as shown in Figure 2.

The force equilibrium equation is derived by assuming constant pressure δp over a small element with length ds along the meridian and length $r(z)d\theta$ in the circumferential direction. The external pressure force is balanced by the tension in the membrane, hence:

276

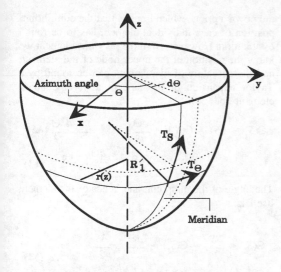

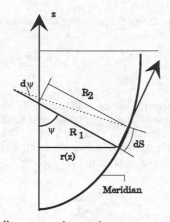

Figure 2 Axially symmetric membrane structure.

$$\delta p(z) \cdot dS \cdot r(z) \cdot d\Theta = T_S \cdot r(z) \cdot d\Theta \cdot d\psi + T_\theta \cdot ds \cdot n_r \cdot d\Theta$$

$$\updownarrow \qquad (4)$$

$$\frac{d\psi}{ds} = \frac{p(z) \cdot r(z) - T_\theta \cdot n_r}{T_S \cdot r(z)}$$

where n_r is the component of the surface normal vector in the r-direction. T_S and T_θ are the tension in the direction along the meridian and in the tangential direction (hoop stress) respectively and ψ is the angle between the horizontal plane (z=constant) and the surface tangent. Furthermore, $dS/d\psi$ is the same as the local radius of curvature R_1

and $r(z)/n_r = R_2$ is the same as the distance from the surface to the z-axis along the surface normal. The previous equation can thus be rewritten as:

$$\frac{T_S(z)}{R_1(z)} + \frac{T_\theta(z)}{R_2(z)} + \delta p(z) = 0 \qquad (5)$$

In addition to the above equation, vertical force equilibrium has to be satisfied for each horizontal cross-section. This implies that the vertical component of the meridian tension is equal to the external pressure load in the vertical direction F_V, on the portion of the structure below the cross-section. Hence;

$$T_S(z) = \frac{F_V(z)}{2\pi \cdot r(z) \cdot \sin(\psi)} \qquad (6)$$

Equation (5) and (6) give us means to find the membrane tension for a given geometry or to find the resulting geometry for a prescribed relationship between the membrane tensions.

The example shown in Figure 2 will now be investigated in some detail. Let us consider a case where the pressure difference between the inside and the outside fluid is give as:

$$\delta p(z) = g \cdot z \cdot (\rho_i - \rho_o) + g \cdot p_i \cdot Z_0 \qquad (7)$$

where Z_0 is the difference between the inside and outside water elevation and g is the acceleration of gravity. The vertical pressure load on a small circumferential element is thus given as:

$$dF_V = \delta p(z) \cdot 2\pi \cdot r(z) \cdot dr \qquad (8)$$

In the case of a spheroid one have that $R_1 = R_2 = R$, and by integrating the above equation to a certain angle ψ one get the following expression for the meridian tension T_S :

$$T_S(\psi) = [-gR \cdot \cos^2(\psi) \cdot (2R \cdot \cos(\psi) \cdot (\rho_i - \rho_o) \\ + 3\rho_i \cdot Z_0) + gR(2R \cdot (\rho_i - \rho_o) \\ + 3\rho_i \cdot Z_0)] \cdot \frac{1}{6 \cdot \sin^2(\psi)} \qquad (9)$$

and the corresponding hoop stress T_θ is:

$$T_\theta(\psi) = g \cdot R \cdot [R \cos(\psi) \cdot (\rho_i - \rho_o) \\ + g \cdot \rho_i \cdot Z_0] - T_s(\psi) \qquad (10)$$

The tensions in the upper end of the membrane are:

$$\lim_{\psi \to \pi/2}(T_s) = \frac{gR}{6}(2R \cdot (\rho_i - \rho_o) + 3\rho_i \cdot Z_0) \qquad (11)$$

$$\lim_{\psi \to \pi/2}(T_\theta) = \frac{gR}{6}(-2R \cdot (\rho_i - \rho_o) + 3\rho_i \cdot Z_0) \qquad (12)$$

It is important to note that the meridian tension is positive unless the outside fluid density is larger than the inside fluid density. The hoop stress can be negative for certain combinations of density differences and internal pressure. Negative stress means buckling of the membrane and should be avoided.

2.3 Numerical approximation for two-dimensional static shape.

The static shape of a two-dimensional membrane is governed by the non-linear first order differential equation given in equation (2). As already mentioned, analytical solutions is only available for a few cases. So, in order to obtain solutions for a general case including hydrostatic pressure, internal pressure and membrane elasticity one has to solve the governing differential equation numerically. This can be done by a discretization of the spatial coordinate, i.e the distance along the membrane. This means that we divide the membrane into a number of elements with length δl. The numerical integration of the governing differential equation is started at the top of the membrane at s=0.0, with initial estimates for the different parameters governing the static shape of the structure. Based on these initial estimates we calculate the shape of the membrane, then we estimate new values for the unknown parameters until the correct solution is obtained. This method will be referred to as a shooting method.

The discretization model is described in some detail below. It is assumed that the pressure is constant over each element and equal to the pressure at the midpoint of the element. The position of the midpoint of the element and thus the pressure is

unknown a priori, which implies that the equilibrium position of each individual element has to be found by iteration. However, from the previous element we know the position of the upper node of the element, and the curvature of the element can be found by a Taylor expansion of the tangent about the upper element node, S_1.

$$\Theta(S_1 + \Delta S) = \Theta(S_1) + \frac{d\Theta}{dS}\Big|_{S_1} \cdot \Delta S + \frac{1}{2} \cdot \frac{d^2\Theta}{dS^2} \cdot (\Delta S)^2 \\ + \cdots + \approx \Theta(S_1) + \frac{d\Theta}{dS}\Big|_{S_1} \cdot \Delta S \qquad (13)$$

The angle of the midpoint, and hence of the element itself is:

$$\Theta(S_1 + \frac{\Delta S}{2}) \approx \Theta(S_1) + \frac{d\Theta}{dS}\Big|_{S_1} \cdot \frac{\Delta S}{2} \\ = \Theta(S_1) + \frac{1}{2} \cdot \frac{d\Theta}{dS}\Big|_{S_1} \cdot \Delta S \qquad (14)$$

Introducing the governing differential equation gives:

$$\Theta(S_1 + \frac{\Delta S}{2}) \approx \Theta(S_1) + \frac{1}{2} \cdot \frac{\delta p}{T}\Big|_j \cdot \Delta S_j = \Theta_j \qquad (15)$$

Where the pressure difference δp is referred to the midpoint of the jth element, which also the curvature is related to. Θ_j is the angle of the jth element relative to the global coordinate system. The coordinate of the lower node is thus given directly as:

$$y_2 \approx y_1 + \cos(\Theta_j) \cdot \Delta S \\ z_2 \approx z_1 - \sin(\Theta_j) \cdot \Delta S \qquad (16)$$

The numerical shooting method can be unstable and convergence thus difficult to achieve for certain configurations. Stability problems increase with increasing number of unknown parameters and decreasing pressure difference. Standard routines for solution of a set of non-linear equations are used in the computer code.

The mathematical formulation of the iteration process depends on the actual problem in question. In general, the iteration process can be formulate as a search for the solution of a set of non-linear equations; one equation for each unknown parameter. We will show the mathematical formulation for a floating closed membrane with inside fluid density less than the outside fluid, as shown in Figure 3.

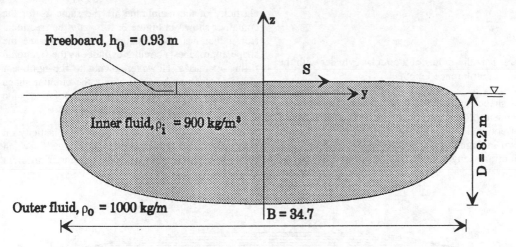

Figure 3 Cross-sectional shape of a floating fabric container.

Two-dimensional floating container.

The geometrical shape of the cross-section is in this case governed by the following three unknown parameters; the freeboard h_0, the tension T and the internal pressure p_0, which together with the density of the inner fluid ρ_i and the outer fluid ρ_0, the elasticity of the membrane Et and the initial length of the membrane O shall describe a cross-section of a given area A. E and t is the modules of elasticity and the thickness of the membrane respectively. Furthermore, the cross-section shall be closed and symmetrical about z axis. Hence, concentrating on one half of the cross-section one can formulate 3 nonlinear equations in the three unknown parameters as:

$$F_1(T,\ h_0,\ p_0,\ \rho_i,\ \rho_0,\ O,\ Et) = 0$$
$$F_2(T,\ h_0,\ p_0,\ \rho_i,\ \rho_0,\ O,\ Et) = \pi \qquad (17)$$
$$F_3(T,\ h_0,\ p_0,\ \rho_i,\ \rho_0,\ O,\ Et) = A$$

where

F_1 is the y-position of the bottom of the cross-section. This equation is given by the symmetry of the solution.

F_2 is the tangent of the surface at the bottom of the cross-section, which is equal to π. This equation is also given by the symmetry of the solution.

F_3 is the area A enclosed by the membrane.

The pressure or more correct, the difference in pressure between the inside and the outside of the container is in this case given as:

$$p = p_0 + \rho_i \cdot g \cdot (h_0 - z) \qquad\qquad z \geq 0$$

$$p = p_0 + g \cdot (\rho_i \cdot h_0 + z(\rho_0 - \rho_i)) \qquad z \leq 0$$

The numerical integration of the governing differential equation is started at the top of the container at (y=0.0, z=h_0) with initial estimates for h_0, T and p_0.

Figure 3 shows the static cross-sectional shape of a floating fabric container with initial membrane length O= 76.3 m, membrane elasticity Et= 1000 kN/m, outside water density ρ_0=1000 kg/m³ and inside fluid density ρ_i=900 kg/m³. This give membrane tension T= 19.04 kN/m, freeboard h_0= 0.93 m and internal over pressure p_0= 0.22 N/m². The container has in this case 60 % fractional filling and the cross-sectional area is 278 m², the draft is D = 8.2 m and the width is B = 34.7 m.

The fractional filling, f, is defined as the ratio between the actual filling and the maximum possible filling referred to the initial circumference length, i.e

279

$$f = \frac{A}{\pi R^2}$$

where R is the radius of a circular cylinder with the same circumference length.

The shape and the fabric tension are strongly dependent of the fractional filling. The influence of the fractional filling on the cross sectional shape is shown in Figure 4 for f=0.5, 0.6, 0.7, 0.8 and 0.9. The initial length, the density difference and the

elasticity of the membrane are the same as for the container shown in Figure 3. It is only the fractional filling that has been changed. As we can see the shape approaches a circular cylinder as the fractional filling increases. However, owing to the nonlinear effect of the membrane elasticity the circular shape will never be reached. The elongation of the membrane increases with increasing tension.

The nonlinear effect of the membrane elasticity on the static shape is clearly seen in Figure 5, which shows the cross-sectional shape for three different

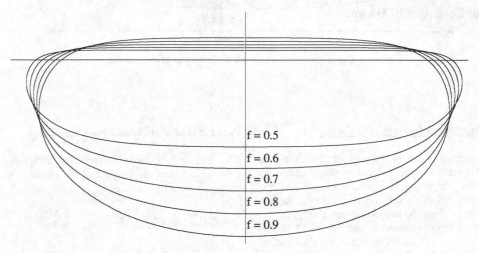

Figure 4 Static shape as a function of the fractional filling for a floating container with initial circumferential length 0=76.3 m, Elasticity Et = 1000 kN/m and a density difference of 100 kg/m³.

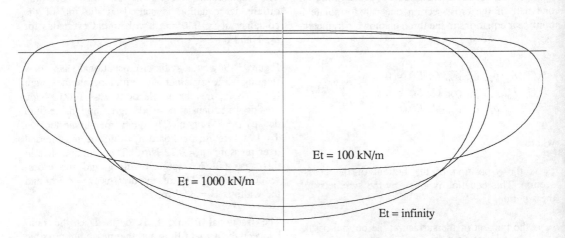

Figure 5 Static shape as a function of membrane elasticity for a floating container with initial circumferential length 0=76.3 m, a density difference of 100 kg/m³ and three different fabric elasticities.

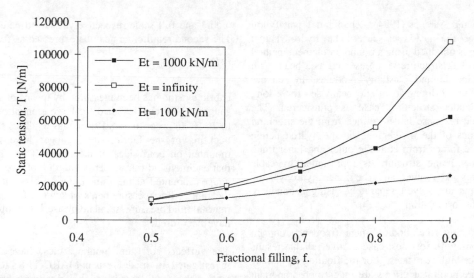

Figure 6 Static tension as a function of fractional filling for a floating container with initial circumferential length 0=76.3 m, a density difference of 100 kg/m³ and three different elasticities.

membrane elasticities; an nonelastic membrane and two membranes with elasticity Et = 1000 kN/m and Et = 100 kN/m. The initial membrane length and the density difference are the same as for the container shown in Figure 3, and the fractional filling is f=0.8.

The static fabric tension shown in Figure 6 is seen to increase rapidly with the fractional filling for nonelastic or stiff membranes. For example, the static tension in an nonelastic membrane is about 3 times higher at 80 % than 60 % fractional filling. However, the increase in the static tension is not so dramatic for more elastic membranes. This effect is clearly shown in Figure 6 for the two elasticities Et=1000 kN/m and Et=100 kN/m. For large container sizes the strength of the fabric is the critical parameter and 60 % fractional filling has therefore been chosen for practical applications.

3 DYNAMIC WAVE LOADS ON FLOATING MEMBRANE STRUCTURES

3.1 *General*

Depending on the actual application of the floating membrane structure, dynamic loading is caused by wave induced loads, towing forces or mooring forces. A theoretical analysis of this problem is very complex, mainly due to the highly elastic behaviour of the structure.

Due to the complexity of the problem, the dynamic loading and resulting fabric tension have so fare been determined from model tests. Theoretical calculations have mainly been used to investigate possible scale effects between model and full scale.

3.2 *Numerical Calculations*

A review of available methods for numerical analysis of the interaction between a deformable structure and waves is given by Broderick and Leonard (1990). Most numerical works related to membrane structure in waves deal with a 2-dimensional problem of an air-filled membrane or a submerged membrane with fixed boundaries, see for example Fukasawa (1990) and Tanaka et.al (1992). These works are typically intended for practical applications as breakwaters. Another area which cover some of the same aspects as for our application, is the internal fluid motions in elastic containers. Within this area much work have been done, see for example Liu and Ma (1982), Schulkes (1990) and Løland (1994)

However, for our case the fabric structure is always fluid filled and floating in the free surface zone. Both these factors add complexity to the numerical solution. Based on the assumptions of small motions and small wave amplitudes and linearization, frequency domain solutions for a long and slender bag structure are given by Zhao and Triantafyllou

(1994) and Aarsnes (1994). Zhao and Triantafyllou consider the head sea case. The hydrodynamic loading is obtained using a boundary element method and the bag structure is represented as a beam with given inertia and elasticity modules. In Aarsnes (1994) the 2-dimensional (i.e beam sea for a long and slender structure) case is considered. The dynamic response is determined from the structural properties of the membrane and the hydrodynamic pressure forces from both the outer and inner fluid. The membrane structure is represented by cable elements. The outer and the inner hydrodynamic problems are solved separately, using a boundary element formulation.

For floating rigid bodies linear frequency domain solutions will, for most cases, give reliable results even for high sea states. For the floating membrane structure nonlinear effects are much more important and will have a significant influence on the response for a typical wave condition (see results from model tests presented below). This implies that the linear solutions can not be used to determine the wave loading for a design wave or for a design sea state condition.

In general, calculation of the nonlinear dynamic interaction between waves and an elastic deformable membrane structure requires a time domain analysis which includes the most important nonlinear effects. For the case of a floating rigid body nonlinear numerical methods have been developed. However, to our knowledge, a general nonlinear solution for elastic deformable structure with an interior fluid is not yet developed.

3.3 Model Tests

To perform reliable model tests two aspects are important, namely the physical modelling and the measuring technique applied to measure the main parameters such as fabric tension, pressure etc.

In model tests for dynamic loading in waves, Froude scaling (similarity between inertial and gravitational forces) have to be used. In general, the most important scaling requirement which should be satisfied for an elastic model are:

*   Geometrical similarity;
*   Elastic similarity.

The first requirement imply that all length scales $\lambda = L_P/L_M$, where subscripts $_M$ and $_P$ indicate the model and full scale respectively, have to be equal. The second requirement can be expressed as follow:

$$(Et)_M = (Et)_P/\lambda^2$$

Fabric tension has been measured by means of strain gauges which are designed for large elongation, up to 20 %. In order to avoid erroneous readings due to buckling of the fabric, strain gauges have to be mounted on both sides of the fabric. This implies that each tension transducer is build up of two strain gauges mounted at the same point on both sides of the fabric. For measurements of towing forces and internal fluid pressure, standard measuring equipment have been used.

The effects of the nonlinearities have been investigated in model tests in MARINTEKs towing tank with a 1:15 scaled model of a fabric container, intended for transport of water. The length of the full scale container was 150 m and the enclosed volume was 25.000 m^3. During the tests, fabric tension and towing performance were measured. An example of measured fabric tension in head sea is shown in Figure 7. The figure shows the time history of the longitudinal fabric tension at the upper side at the midships position. The tension is shown for two different regular wave heights H=0.8 m and H=6.0 m, both with wave period T= 9.5 s. The second wave condition corresponds to a typical wave steepness. For the lowest wave height H=0.8 m one can see that the variation in the fabric tension is relatively regular, but the absolute value of the maxima is seen to be significantly larger than the minima. This shows that nonlinear effects are of some importance even for low wave heights. For the highest wave height H=6 m one can see that the variation in the measured fabric tension is irregular and completely governed by the nonlinear effects. The maximum value of the fabric tension is about 14 kN/m for H=0.8 m and about 35 kN/m for H=6.0 m. This implies that the tension is only 2.5 times higher while the waves is 8 times higher. The energy spectrum for the time series of the fabric tension are shown in Figure 8. For the lowest wave height H=0.8 m almost all energy is concentrated at the wave frequency $f_W = 1/T$. For the highest wave H=6.0 m the energy is mainly distributed at three different frequencies, $0.5 f_W$, f_W, and around $2 f_W$. As one can see the nonlinearities in the response are observed even for low wave heights and will be increasingly important for increasing wave height. Results from a linear method will therefore give limited information.

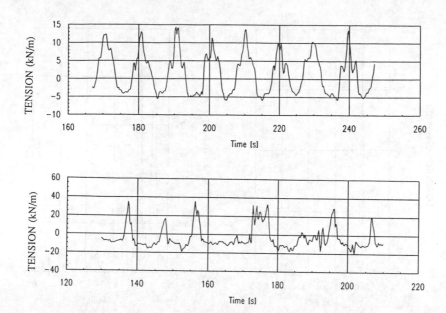

Figure 7 Measured longitudinal tension in regular head sea waves with wave period 9.6 sec. A: Wave height 0.8 m. B: Wave height 6.0 m.

4 APPLICATIONS

We will in this section discuss the use of fabric as construction material for marine structures. Both existing structures and possible new applications are discussed.

4.1 General

There is a large difference between the behaviour of a flexible membrane structure and a rigid structure when exposed to waves. An elastic and deformable fabric structure follows the wave motion and the waves will propagate almost undisturbed through the structure. This behaviour will dramatically reduce the structural loads compared to the shear forces and the bending moment experienced by a conventional rigid structure. The required structural strength of a fabric structure is therefore only a small fraction of the required strength of an equal size rigid structure.

Another large advantage of fabric structures is the low weight. For example, the total weight of a 25.000 m³ fabric container for transportation of water will be about 30 tons including towing equipment. This is only about 1 % of the steel weight of a ship or a barge of the same size. The low weight makes the fabric container easy to transport and handle,

which are important factors for some applications. The ratio between the production costs for a steel barge and a fabric container will be of the same order of magnitude as the weight ratio.

A fabric structure will automatically take its natural shape depending on the surface area and the loading. This means that it is only the circumferential length as a function of the longitudinal position that can be used to determine the shape of a long and slender container. Optimization of the geometrical shape with respect to towing resistance and course stability is therefore not possible for a fabric container in the same way as for rigid hulls.

In order to avoid stress concentration in the fabric it is important that the structure is manufactured with sufficient accuracy in agreement with the intended geometrical shape. A fabric structure is assembled from prefabricated fabric sheets which are welded together, and even small inaccuracies in the assembly process will introduce local stress concentrations. Another important requirement which has to be satisfied in order to avoid severe stress concentration, is that the elasticity has to be uniform in each cross-section. Furthermore, it is especially important to avoid stress concentrations in highly loaded parts of the structure.

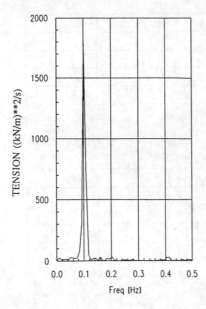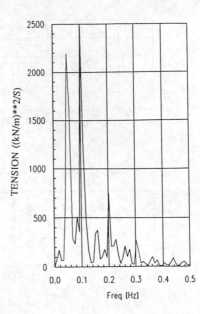

Figure 8 Energy spectra of the measured time series of the longitudinal tension shown in Figure 7. A: Wave height H = 0.8 m, B: Wave height H = 6.0 m.

The fabric structure should not be exposed to high tear loading due to the relatively low tear strength of the fabric. Fabric structures should not be used in situations where there are physical contact between the fabric structure and other rigid or sharp edged structures.

4.2 *Floating fabric containers*

The concept of producing floating containers for transportation of water and oil was first described by Hawthorne (1961) and the product was named Dracone. The Dracone structure is made of rubber and it is long and slender. Owing to the development of high quality coated fabric, one can now manufacture such structures in this material. The fabric material is more collapsible and has lower weight than comparable rubber material.

MARINTEK has since 1989 been involved in the design and documentation of floating fabric containers both for transportation of fresh water and for temporary storage and transportation of oil. At present, containers up to 30.000 m³ have been designed and larger sizes have been considered. The oil container is intended to be used for oil spill recovery, either by direct transfer of oil from a damaged cargo tanker or from an oil skimmer unit.

The idea for this concept is to have a low weight unit that can be used for temporary storage and transportation of oil, oil-water emulation or contaminated water. The low weight of the container is a great advantage compared to other concepts, owing to the easy transport from a depot to an oil spill emergency field.

The main dimensions of the floating containers are determined by the required enclosed volume (the fractional filling), the towing resistance and the sea keeping performance. Based upon these basic requirements one ends up with a shape that is relatively long and slender, consisting of a parallel part in the middle, axisymmetric stern and a relative long and slender bow. The cross-sectional shape along the parallel part and in the bow region can be determined by the two-dimensional method described previously. This can be done because the two dimensional solution is accurate in regions where the longitudinal curvature is much smaller than the transverse curvature, and hence where the pressure difference is balanced by the transverse tension. The stern region can be approximated by a part of a three-dimensional axisymmetric body.

4.3 *Floating bag pens for fish farming*

Floating flexible fabric bag pens are also used for aquaculture purposes. The common floating fish farm has so far used net pens for fish enclosure. However, with increasing focus on the environmental consequences of fish farming in open sea, floating bag pens may be a competitive concept compared to land based fish farming. Floating bag pens can also be used for operations that requires special control with the fish and the water, for example medication, sorting and transportation.

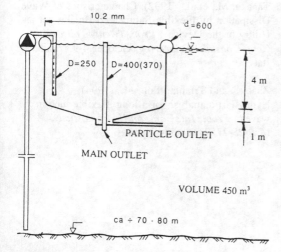

Figure 9 Floating fabric bag pen used in fish farming, from Solaas (1993).

Figure 9 shows an example of a floating bag pan for fish farming. Besides the bag pen, the cage consist of a floating collar with sufficient buoyancy, water inlet, water outlet and particle outlet. Details about the design are given in Solaas et al. (1993). The bag has an axially symmetric shape and the tension in the fabric can be found as described in equation (5), (6) and (7). Z_0 in equation (7) represents the difference in the water level between the inside and outside, and it is determined by the velocity and the losses in the outlet pipe. The water body inside the bag circulates as a large vortex in order to generate sufficient current for the fish and for the removal of particles. However, the vortex introduces a drop in the free surface elevation at the center of the cage, reducing the hydrostatic pressure, which has to be accounted for when calculating Z_0. In addition to the static shape, both dynamic wave loads, internal sloshing and static deformation in current are important parameters that have to be analyzed. Finally, the floating bag pen is not only a hydroelastic structure, it also has to satisfy the biological requirements for the fish.

4.4 *Other applications*

Floating fabric container can also be used for transportation of other low weight cargo besides water and oil, for example grain. Owing to the increased focus on the environment, energy consumption and pollution, we may see an increase in the use of low weight and low speed transportation units. The use of floating fabric containers may be a competitive concept for this type of transportation.

Another interesting concept is to use floating containers for transportation of live fish from the fishing grounds to factories on shore or directly to the marked. In this case the structure has to be submitted with additional buoyancy, because there is no density difference between the inside and outside water in this case.

Fabric sheets are also used in many environmental protection situations, for example in oil booms and other barriers for separation of contaminated water. Fabric structures may also be used for energy production through extraction of wave energy or as wave breakers to protect exposed areas from the wave actions. Finally, the coated fabric structures are also used as drift anchor and as skirts and air bags for Surface Effect Ships (SES).

5 CONCLUSION

The purpose of this paper has been to describe advantages and problems related to the use of coated fabric as construction material for marine applications. Numerical methods for determining the hydrostatic shape of two-dimensional and axially symmetric shapes has been described. A floating fabric container and a floating bag pen are discussed in some detail.

Coated fabric can be a competitive construction material for several marine applications due to its low price and high strength/weight ratio. However, there are several constrains both related to the design and the use of fabric structures. There are limitations on shapes that can be used, the structure has to be manufactured to the correct shape in order to avoid stress concentrations spots. Owing to the relative low tear strength, fabric structures cannot be used in situations where there are contact between the fabric structure and rigid objects with sharp corners.

The dynamic response of elastic membrane structure in waves is a highly nonlinear process even at moderate wave heights. This implies that linear solutions methods cannot be used to determine the wave loading in a design wave condition or for a practical situation. At present model tests results have to be used to determine the design loading in the fabric.

6 ACKNOWLEDGMENT

The financially support of this work by The Norwegian Marine Technology Research Institute (MARINTEK) and The Research Council of Norway (NFR) are gratefully acknowledged.

REFERENCES

Aarsnes, J.V., (1994). Linearized Response of a Membrane Structure in Waves. Submitted for publication.

Broderick, L.L. and Leonard, J.W., (1990). Selective review of boundary element modelling for the interaction of deformable structures with water waves. *Eng. Struct.* Vol. 12, pp 269-276

Fukasawa, T., (1990). Hydroelastic Response of a Membrane Structure in Waves, *RINA Conf. Dynamics of Marine Vehicles and Structures in Waves,* London

W. R. Hawthorne (1961). The Dracone Flexible Barge. *The Engineer* Feb. 1961.

Kuznetsov (1991). *Underconstrained Structural Systems.* Springer-Verlag New York. ISBN 0 387 97594 2

J. W. Leonard (1988). *Tension structures.* McGraw -Hill Book Company. ISBN 0 07 037226 8 E. N.

Liu, W.K., (1982). Coupling effect between Liquid Sloshing and Flexible Fluid-Filled Systems, *Nuclear Engineering and Design*, Vol 72. pp 345-357.

Løland, G. Nonlinear internal fluid-structure motion in a partly flexible fluid domain. Submitted for publication.

F. Solaas and H. Rudi (1993). Floating fish farms with bag pens. *First International Conference in Fish Farming Technology,* 9-12 August 1993. pp. 317-323. ISBN 90 5410 326 4.

R. M. S. M. Schulkes (1990). Fluid oscillation in a open, flexible container. *Journal of Engineering Mathematics* 24, pp. 237-259, 1990.

Tanaka, M., et.al., (1992). Characteristics of Wave Dissipation by Flexible Submerged Breakwater and Utility of the Device", *Proc. Twenty-Third Int Conf. Coastal Engineering*, October 4-9, Venice, Italy

Zhao, R and Triantafyllou, M., (1994): Hydroelastic analysis of a long, flexible tube in waves. *Proc. Int. Conf. on Hydroelasticity*, May 25-27, Tronheim, Norway

Hydroelasticity in Marine Technology, Faltinsen et al. (eds) © 1994 Balkema, Rotterdam, ISBN 90 5410 387 6

Hydroelastic analyses of a long flexible tube in waves

Rong Zhao
MARINTEK, Trondheim, Norway

Michael Triantafyllou
Massachusetts Institute of Technology, Cambridge, Mass., USA

ABSTRACT: A linear theory to predict motions, hydroelastic deformations and stresses of an along flexible tube in head sea waves has been presented. For the fluid inside the tube, one-dimensional approach has been applied. A strip theory has been used to estimate hydrodynamic forces from the fluid outside the tube. The hydroelastic deformation has been taken care of in the body boundary condition. Importance of the nonlinear effects has been discussed.

1 INTRODUCTION

The development of the "Dracone" barge which is a long flexible tube designed to carry and transport oil and other liquids lighter than water, was first reported by Hawthorne(1961). Full scale, experimental results and theoretical studies were presented. From a hydrostatic and hydrodynamic point of view there are three major problems which are important for a flexible tube.

1) Static stresses and shape of a flexible floating tube.

2) Stresses, motions and shapes of a flexible tube in waves.

3) Directional stability of a flexible tube under tow.

The first and third problem have been extensively studied by Hawthorne(1961). Semi-analytical solutions of static stresses and shapes of a two-dimensional flexible floating body were presented. No details about the semi-analytic solution for the shape of a floating tube with fluid densities $\rho_i/\rho_o \neq 0.5$ were reported, here ρ_i and ρ_o are the fluid densities inside and outside the tube. For the problem of directional stability of a flexible tube under tow, both full scale and analytical analyses were carried out. The problem of directional stability has been further studied by for instance Paidoussis(1968) and Paidoussis and Yu(1976).

The second problem of a flexible tube in waves has not been investigated in a rational way. This is due to the complexity of the hydroelastic analysis. A numerical analysis is necessary to solve the problem. In this paper we are going to present a new approach to predict stresses, motions and hydroelastic deformations of a long flexible tube in head sea waves. Before we carry out the dynamic analysis, the static shape and stresses of a flexible tube have been evaluated by a numerical iteration scheme. An alternative approach to predict the static shapes has been presented by Løland and Aarsnes(1994). In our analyses, the motions, hydroelastic deformations and stresses of the tube in waves are found by a linear perturbation of the static solution. For the fluid inside the tube, a one-dimensional equation of motion and one-dimensional equation of continuity have been used. Outside the tube a strip theory is assumed. The hydroelastic deformations have been taken care of in the body boundary condition. It is found that the hydroelastic deformations for the cross sections are mainly due to change of percentage filling of the fluid inside the tube. The importance of nonlinear effects has been discussed in the last section of the paper.

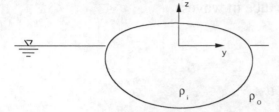

Figure 1: A two-dimensional membrane structure and a y-z coordinate system are shown. The fluid densities inside and outside the membrane are ρ_i and ρ_o.

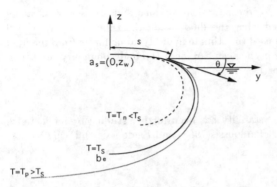

Figure 2: The y-z coordinates are dependent on the tension T one chosen. The parameters used in the iteration process (for solving eq.(1)) are illustrated.

2 STATIC SHAPE AND STRESSES

Fig.1 shows a typical cross section of a membrane structure(tube) in calm water. The fluid densities inside and outside the tube are ρ_i and ρ_o. One assumes that ρ_o is larger than ρ_i. Further one assumes the thickness of the membrane(skin of the tube) is infinitely thin. This means that we can neglect the mass of the membrane. The static shape of a 2-D cross section is dependent on the percentage filling and the densities of the fluid inside and outside the tube. The filling ratio γ is defined as $\gamma = A_0/A_{max}$, where A_0 is the area inside the two-dimensional membrane structure and A_{max} is the area for the maximum filling. Here we neglect the elastic deformation of the membrane (skin of the tube). The geometry of a two-dimensional membrane structure will be a circle when the filling ratio γ is 1.0.

The static shape of a membrane structure is given by the following equation of the condition of equilibrium,

$$\frac{d\theta}{ds} = \frac{\Delta P}{T} \qquad (1)$$

where θ and s are defined in Fig.2, ΔP is the difference of the static pressures of the fluids inside and outside the membrane, and T is the static hoop tension. The solutions of static shape and stresses of a two-dimensional floating membrane structure have been presented by Hawthorne(1961). The solution involved two sets of incomplete elliptic integrals for the portions above and below the water line. Difficulties in solving eq. (1) for the cases $\rho_i/\rho_o \neq 0.5$ were pointed out by Hawthorne(1961). No details about the numerical scheme to solve the general problem of a floating two-dimensional membrane structure were given in the paper. In this section we will present a new numerical iteration scheme to estimate static shape and stresses of a two-dimensional floating membrane structure. The same iteration scheme has also been applied to study the dynamic pressures and tensions of a floating membrane structure in head sea waves.

The iteration procedure to predict the static shape and hoop tensions is described in the following text.

We know that the static shape of a two-dimensional membrane structure is symmetric about a vertical axis. One assumes that the axis is the z-axis. One starts the solution from a point a_s with coordinate $(0,Z_w)$ (see Fig.2) and assumes the pressure inside the membrane at point a_s is P^0_{in}. For a given point a_s and pressure P^0_{in}, there exists only one static shape. The hoop tension for the static shape is assumed to be T_s. Because the static shape is symmetric about the z-axis, we only need to predict the y-z coordinates of the two-dimensional membrane structure for $y \geq 0$. It can be shown that the single value solution as function of z (for $y \geq 0$) and a convex shape are the only possible solution. If we assume the lowest point is b_e, so the angles θ are 0 and π at the points a_s and b_e(definition of θ is given in Fig.2). Then one uses eq.(1) to evaluate y-z coordinates for $\theta = 0$ to $\theta = \pi$ by one chosen value of T (we do not know the correct value of T yet). For $T \neq T_s$, the y coordinate for $\theta = \pi$ will different from 0. The next step

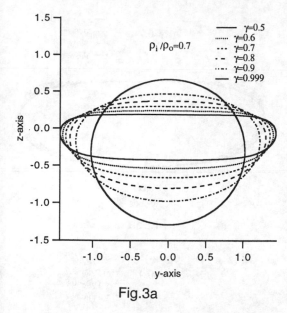

$\rho_i/\rho_o=0.7$

— $\gamma=0.5$
...... $\gamma=0.6$
· · · $\gamma=0.7$
· – · $\gamma=0.8$
·... · $\gamma=0.9$
— $\gamma=0.999$

Fig.3a

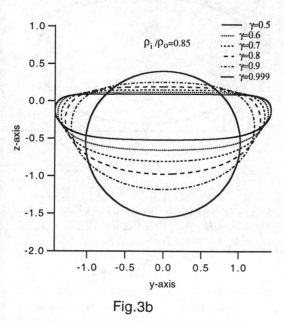

$\rho_i/\rho_o=0.85$

— $\gamma=0.5$
...... $\gamma=0.6$
· · · · $\gamma=0.7$
– – $\gamma=0.8$
·... · $\gamma=0.9$
— $\gamma=0.999$

Fig.3b

Figure 3: The static shapes are dependent on the filling ratio γ and the relative density ρ_i/ρ_o.

is to find two values of T—T_n and T_p , so the end points based on the two values are located both for y large and less than zero. Then one chooses the mean value of these two T values, and based on the new T value one finds a new end point. The y-value of the end point is then located between the two early predicted values. This new value of T will replace one of the T_n and T_p, dependent on which one has the corresponding y coordinate (of the early predicted values) which has the same sign as the new y-coordinate. Then one repeats the procedure until one finds the correct T-value. This means that T_n and T_p are the same correctly to the accuracy one pre-specified. Based on the yz-coordinates of the membrane structure we have found, one can estimate the filling ratio γ for the static shape. For a given point a_s, the filling ratio is dependent on the chosen P_{in}^o. To predict the static shape for a given filling ratio γ, we should iterate again in a similar way by choosing different values of P_{in}^0.

The numerical code that we used to predict static shape and tensions has been tested and checked in many different cases. Special care has been taken when the filling ratio is small. For example when $\gamma = 0.5$ and $\rho_i/\rho_o = 0.98$, $P_{in}^o/P_o^{max} = o(10^{-10})$ (where P_o^{max} is the maximum static pressure outside the tube) . This means that $\Delta\theta$ used in the numerical integration should be extremely small close to the point a_s. The convergence tests have been done in a systematical way. The final results have been checked with different physical quantities and relations. For instance, after the static shape has been obtained, we can calculate the tension based on the static pressures inside and outside the tube and estimate the fluid density inside the tube by evaluating the volume of the fluid inside the tube for $z > 0$ and $z < 0$. The estimated relative errors are less than 10^{-3} for the results that we have presented here.

The static shape and tensions for different filling ratios and densities ρ_i/ρ_o are presented in Fig.3 and Fig.4. From Fig.3 one may observe the static shape is strongly dependent on the filling ratio, not the fluid density ρ_i/ρ_o. This can be illustrated in Fig.5, where the graph is plotted in a such way that the highest points of each static shape are identical for different fluid densities ρ_i/ρ_o when filling ratio is constant. One may observe that the lowest points for each curves are almost the same (or the same) for different fluid densities ρ_i/ρ_o. The geometries for different fluid densities ρ_i/ρ_o, but same filling ratio γ have only small differences for large $| y |$ values. This property is very important for dynamic analyses of a membrane structure in head sea. It means that when one estimates hydrodynamic forces from fluid outside the tube, we may assume that the hydroelastic deformations of

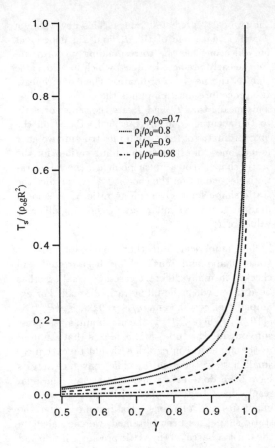

Figure 4: The hoop tensions T_s are function of filling ratio γ and the relative density ρ_i/ρ_o. R is the radius of the membrane structure when $\gamma = 1.0$, and g is the acceleration of gravity.

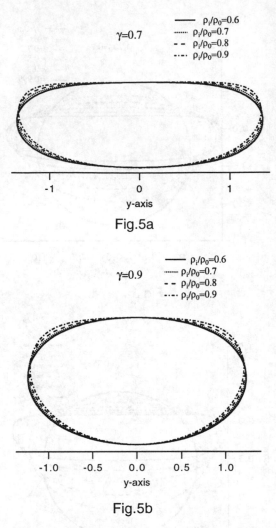

Fig.5a

Fig.5b

Figure 5: The static shapes for different relative densities ρ_i/ρ_o are plotted in a such way that the highest points of different curves are identical.

the tube are only dependent on the filling ratio. The other interesting phenomenon is for the hoop tensions of a 2-D membrane structure. Fig.6 shows the relative tensions for the different fluid densities ρ_i/ρ_o ($\rho_o = const.$). The results show the relative tensions are almost linearly dependent on the filling ratio. This relation may reduce CPU-time in the numerical analyses.

3   HYDROELASTIC RESPONSES AND DYNAMIC STRESSES IN WAVES

A new linear theory to predict hydroelastic responses and dynamic stresses of a long flexible tube in head sea waves is presented. We assume that the thickness of the skin(membrane) is in-finitely thin, so we can neglect the mass of the membrane in our analyses. The elastic deformation of the skin(membrane) in the transverse direction is neglected. To carry out the dynamic analyses we should further assume that the incident waves are linear regular waves, the wave amplitude is small compared with a characteristic cross sectional dimension(D) of the tube and the length of the tube(L) is large compared with D. Due to the slenderness approximation(i.e D/L is small), the flow inside the tube may be treated as a one-dimensional problem for the cases when the wave

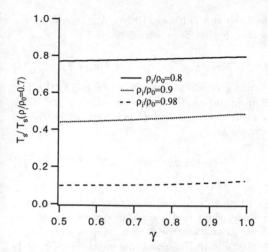

Figure 6: The relative tensions T_s for different relative densities ρ_i/ρ_o as the function of filling ratio γ ($\rho_o = const.$).

length is much larger than the cross-sectional dimension. Outside the tube a boundary element method based on two-dimensional approach (strip theory) has been applied(see Zhao and Faltinsen 1988). The hydroelastic deformation of the tube (due to change of the filling for each section as function of time) has been taken care of by the body boundary condition. The approach is valid for wave length smaller than the length of the tube. One may question if one applies the method for the wave length is same order as the length of the tube. Theoretically we should assume $D \ll \lambda \ll L$. In a practical application one may apply the theory for larger range of λ. The problem is solved in the frequency domain. For each strip (cross section) we have unknowns $P_1(x,t)$, $\eta_3(x,t)$, $A_1(x,t)$ and $V_1(x,t)$ with time dependence $e^{i\omega t}$, where P_1 is the average dynamic pressure inside the tube for each section, η_3 is the vertical motion of the rigid body (based on the static shape of 2-D sections), A_1 is the change of the filling of the fluid for each section and V_1 is the longitudinal velocity (one-dimensional) inside the tube. The four equations for solving the problem are

$$\frac{\partial P_1}{\partial x} = -\rho_i \frac{\partial V_1}{\partial t} - \rho_i g \frac{\partial z(\eta_3, A_1)}{\partial x} \qquad (2)$$

$$\frac{\partial A_1}{\partial t} + \frac{\partial V_1}{\partial x} A_0 = 0 \qquad (3)$$

$$\triangle F_3 - \triangle mg = m\ddot{\eta}_3 \qquad (4)$$

$$\begin{aligned}
P_1 &= \frac{\partial P_1}{\partial P_{A_1}} A_1 + \frac{\partial P_1}{\partial P_{\dot{A}_1}} \dot{A}_1 + \frac{\partial P_1}{\partial P_{\ddot{A}_1}} \ddot{A}_1 + \frac{\partial P_1}{\partial P_{\eta_3}} \eta_3 \\
&+ \frac{\partial P_1}{\partial P_{\dot{\eta}_3}} \dot{\eta}_3 + \frac{\partial P_1}{\partial P_{\ddot{\eta}_3}} \ddot{\eta}_3 + \frac{\partial P_1}{\partial P_{Re(F_{ex})}} e^{i\omega t} \\
&+ i \frac{\partial P_1}{\partial P_{Im(F_{ex})}} e^{i\omega t} + \frac{\partial P_1}{\partial P_{\frac{\partial^4 z}{\partial x^4}}} EI \\
&+ \frac{\partial P_1}{\partial P_{\frac{\partial^2 z}{\partial x^2}}} T \qquad (5)
\end{aligned}$$

where $\triangle F_3 = F_{ex}e^{i\omega t} - A_{33}\ddot{\eta}_3 - B_{33}\dot{\eta}_3 - C_{33}\eta_3 - A_{A_1}\ddot{A}_1 - B_{A_1}\dot{A}_1 - C_{A_1}A_1 - EI\frac{\partial^4 z}{\partial x^4} + T(x)\frac{\partial^2 z}{\partial x^2}$, $\triangle m = A_1\rho_i$ and $z = \eta_3 + \frac{\partial z}{\partial A} A_1$.

Eq.(2) is the one-dimensional equation of motion for fluid inside the tube, eq.(3) is the one-dimensional equation of continuity and eq.(4) is applied Newton's second low for each section. The last equation assumes that the pressure P_1 for each section can be expressed as function of η_3, A_1 and its derivatives, as well as F_{ex}, EI and T(x). P_{A_1}, $P_{\dot{A}_1}$, $P_{\ddot{A}_1}$... are the dynamic pressure components (outside the tube) which are proportional to A_1, $\dot{A}_1$, $\ddot{A}_1$... . Details to estimate the coefficients are given in the following subsections. Here the tube is regarded as a beam of moment of inertia I and material of modulus E. z is the average vertical motion of the fluid inside the 2-D membrane as function of time, A_0 and A_1 are the filling of the static and dynamic part, $A = A_0 + A_1$, m is the mass of each section of the static equilibrium position, T(x) is that part of the longitudinal tension which has contribution to the vertical force (A discussion of that is given by Hawthorne(1961)). The skin friction force due to fluid motion inside the tube does have contribution to the vertical force, but it has been found that the effect can be neglected. In our numerical calculations we neglect totally this term. A_{33}, B_{33} and C_{33} are two-dimensional added mass, damping and stiffness force in the vertical direction due to heave motion of the rigid body. F_{ex} is the 2-D vertical excitation force of the incident wave. A_{A_1}, B_{A_1} and C_{A_1} are two-dimensional added mass, damping and stiffness force in vertical direction due to the mode motion A_1. The hydrodynamic coefficients can be found by a two-dimensional boundary element method. This will be presented in the next subsection. The coefficients in eq.(5) can be evaluated in a similar way as

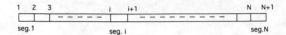

seg.1   seg. i   seg.N

Figure 7: Illustrated segments and sections in the numerical computation for solving eq. (2) to eq.(5).

one calculates static pressures and tensions. The details are presented in the subsection after next. After one obtained all the coefficients in eq.(2) to eq.(5), we can solve these four equations numerically.

As shown in Fig.7 one can divide the tube into N segments and N+1 sections. For each section we have unknown P_1, η_3, A_1 and V_1. The problem can be solved by using the finite difference method. The boundary conditions at the ends (for section 1 and N+1) are $V_1 = 0$, bending moments and vertical shear forces are equal zero. Eq.(2) and eq. (3) describe the one dimensional motion of the fluid inside the tube. For the problem there exist numerous resonances. This has been discussed by Hawthorne(1961). The damping for the problem is due to mode motion $A_1^m(A_1, \gamma, \rho_i/\rho_o)$ (change of the filling for each section). The mode motion A_1^m will generate wave which propagating away from the tube. The damping coefficient is predicted in the next subsection.

After we have solved the problem, the dynamic tension in the transverse direction can be evaluated by following equation.

$$
\begin{aligned}
T_1 = & \frac{\partial T_1}{\partial P_{A_1}} A_1 + \frac{\partial T_1}{\partial P_{\dot{A}_1}} \dot{A}_1 + \frac{\partial T_1}{\partial P_{\ddot{A}_1}} \ddot{A}_1 + \frac{\partial T_1}{\partial P_{\eta_3}} \eta_3 \\
& + \frac{\partial T_1}{\partial P_{\dot{\eta}_3}} \dot{\eta}_3 + \frac{\partial T_1}{\partial P_{\ddot{\eta}_3}} \ddot{\eta}_3 + \frac{\partial T_1}{\partial P_{Re(F_{ex})}} e^{i\omega t} \\
& + i \frac{\partial T_1}{\partial P_{Im(F_{ex})}} e^{i\omega t} + \frac{\partial T_1}{\partial P_{\frac{\partial^4 z}{\partial x^4}}} EI \\
& + \frac{\partial T_1}{\partial P_{\frac{\partial^2 z}{\partial x^2}}} T
\end{aligned}
\tag{6}
$$

The coefficients can be estimated in the same way as we predict the coefficients in eq. (5).

3.1 Calculating hydrodynamic coefficients

A_{33}, B_{33}, A_{A_1}, B_{A_1} and F_{ex} can be obtained by solving boundary value problem in a similar way

as Zhao and Faltinsen(1988). A 2-D velocity potential Φ_T has been introduced,

$$
\Phi_T = \phi_0 e^{i\omega t} + \phi_3 e^{i\omega t} + \phi_1 \eta_3 + \phi_2 A_1 \tag{7}
$$

which satisfies the Laplace equation in the fluid domain outside the membrane. Here $\phi_0 e^{i\omega t}$ is the velocity potential of incident wave which can be written as

$$
\phi_0(x, y, z) e^{i\omega t} = \frac{g}{\omega} \zeta_a e^{i\omega t - ikx + kz} \tag{8}
$$

where ζ_a, ω, t and k are the incident wave amplitude, the circular frequency of oscillation, the time variable and the wave number, $\phi_1 \eta_3$ is the potential due to the heave motion of the rigid body (based on the static shape), $\phi_2 A_1$ is the potential due to mode motion A_1^m (change of the filling for the cross-section) and $\phi_3 e^{i\omega t}$ is the diffraction potential of the incident wave. Further ϕ_i (i=1,2,3) satisfy the linear free surface condition

$$
-\omega^2 \phi_i + g \frac{\partial \phi_i}{\partial z} = 0 \quad on \quad z = 0 \tag{9}
$$

and the body boundary condition

$$
\frac{\partial \phi_i}{\partial n} = \vec{V}_i \cdot \vec{N} \quad on \quad S_b \tag{10}
$$

where S_b is the mean wetted body surface, $\vec{N} = (n_2, n_3)$ is the 2-D normal vector to the body surface with positive direction into the fluid domain. $\vec{V}_1 \eta_3$ and $\vec{V}_2 A_1$ are the velocity of the body boundary surface for heave motion and mode motion A_1^m, and $\vec{V}_3 e^{i\omega t}$ is the reverse particle velocity of the incident waves. The velocity potential can be solved by applying Green's second identity. Far from the body a multipole expansion has been used to satisfy correct radiation condition. Details about the numerical method can be found in Zhao and Faltinsen(1988). Having obtained ϕ_i the vertical wave excitation force and the added mass and damping coefficients can be evaluated as

$$
F_{ex} = \rho \int_{s_b} i\omega n_3 (\phi_0 + \phi_3) ds \tag{11}
$$

$$
A_{33} = Re(\rho \int_{s_b} i\omega n_3 \phi_1 ds)/\omega^2 \tag{12}
$$

$$
B_{33} = -Im(\rho \int_{s_b} i\omega n_3 \phi_1 ds)/\omega \tag{13}
$$

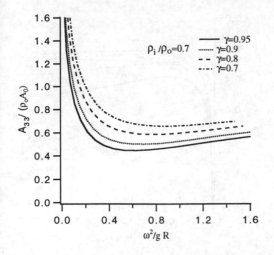

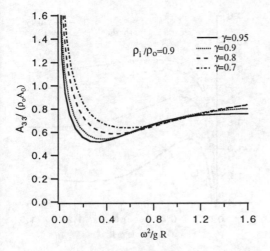

Figure 8: Added mass coefficient A_{33} as function of filling ratio γ, fluid density ρ_i/ρ_o and the oscillation frequency ω. Here R and A_0 are the radius and area of a 2-D membrane structure when the filling ratio $\gamma = 1.0$.

$$A_{A_1} = Re(\rho \int_{s_b} i\omega n_3 \phi_2 ds)/\omega^2 \qquad (14)$$

$$B_{A_1} = -Im(\rho \int_{s_b} i\omega n_3 \phi_2 ds)/\omega \qquad (15)$$

The numerical results have been carefully checked in the similar way as Zhao and Faltinsen(1988). Numerical results of added mass, damping and excitation forces are presented in Fig.8 to Fig.13. The results show that the coefficients are strongly dependent on the filling ratio

γ and the fluid density ρ_i/ρ_o. The damping coefficients B_{33} are also predicted by the principle of conservation of energy. It means the damping coefficients are calculated based on the wave amplitude far away from the body, which is generated by the forced heave motion. The damping coefficient B_{A_1} cannot be checked in a similar way, because the damping force is represented a vertical force due to the mode motion A_1^m. To check the numerical results for mode motion A_1^m, we can calculate the mode damping B_{B_1} in the following way

$$B_{B_1} = -Im(\rho \int_{s_b} i\omega \frac{\partial \phi_2}{\partial n} \phi_2 ds)/\omega \qquad (16)$$

and the damping coefficient B_{B_1} can also be checked in a similar way as B_{33}. This damping coefficient is very important for the internal surge motion of the tube. The numerical results of B_{B_1} is given in Fig.12. From the results one can find the damping for mode motion A_1^m is close to zero for some frequencies when the filling ratio is large. It means that we may have large surge motion inside the tube if the resonance frequencies of the internal motion are close to these frequencies.

3.2 Calculating coefficients in eq.(5) and. eq.(6)

The dynamic pressure P_1 and tension T_1 for each section can be estimated by eq. (5) and eq.(6). The coefficients $\frac{\partial P_1}{\partial P_{A_1}}$, $\frac{\partial P_1}{\partial P_{\dot{A}_1}}$, $\frac{\partial T_1}{\partial P_{A_1}}$, $\frac{\partial T_1}{\partial P_{\dot{A}_1}}$,... can be found approximately by a similar approach as the static problem (not the terms $P_{\frac{\partial^4 z}{\partial x^4}}$ and $P_{\frac{\partial^2 z}{\partial x^2}}$).

One assumes the total dynamic pressure for each section outside the tube is $P_D(y,z)e^{iwt}$ and can be written as

$$\begin{aligned} P_D = & P_{A_1}A_1 + P_{\dot{A}_1}\dot{A}_1 + P_{\ddot{A}_1}\ddot{A}_1 + \\ & P_{\eta_3}\eta_3 + P_{\dot{\eta}_3}\dot{\eta}_3 + P_{\ddot{\eta}_3}\ddot{\eta}_3 + \\ & P_{Re(F_{ex})}e^{i\omega t} + iP_{Im(F_{ex})}e^{i\omega t} + \\ & P_{\frac{\partial^4 z}{\partial x^4}}EI + P_{\frac{\partial^2 z}{\partial x^2}}T \qquad (17) \end{aligned}$$

For each component of dynamic pressure outside the tube, one may evaluate the corresponding dynamic pressure inside the tube and the dynamic tension. The dynamic pressure and tension we obtained are the coefficients in eq. (5) and eq.(6). The numerical method to evaluate these terms can be described as following.

Eq.(1) has been used in a similar way as the static problem. The pressure outside the tube is

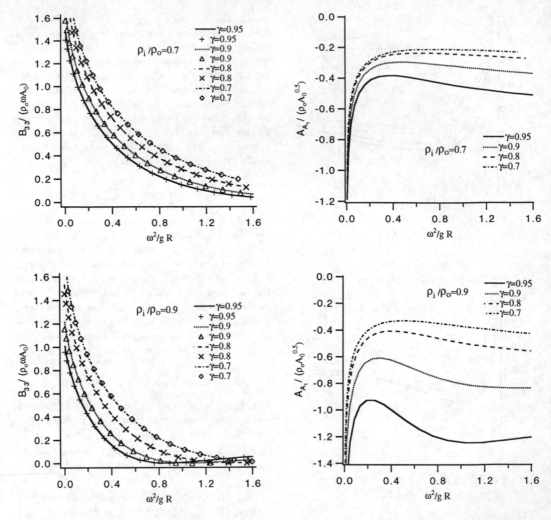

Figure 9: Damping coefficient B_{33} as function of filling ratio γ, relative fluid density ρ_i/ρ_o and the oscillation frequency ω. Here R and A_0 are the radius and area of a 2-D membrane structure when the filling ratio $\gamma = 1.0$. The results of the solid and dotted lines are based on the formula of eq.(13). The symbols $+$, $\triangle$, $\times$ and $\diamond$ are based on the wave amplitude far from the body which is generated by the forced heave motion.

Figure 10: Added mass coefficient A_{A_1} as function of filling ratio γ, relative fluid density ρ_i/ρ_o and the oscillation frequency ω. Here R and A_0 are the radius and area of a 2-D membrane structure when the filling ratio $\gamma = 1.0$.

the static pressure plus the dynamic pressure. We should keep in mind that eq. (1) is a nonlinear equation. Because we need only the linear solution, so the dynamic pressure should be much smaller than the static pressure. In practice we should multiple the dynamic pressure by a factor, so the pressure of the dynamic part is much

smaller than static part. After we found the dynamic pressure and tension based on the factor , the results should be divided by this factor. The integrated vertical force of the dynamic pressure will give contribution to a corresponding vertical acceleration of the section. We assumed the hydroelastic deformation is mainly due to the change of the filling ratio. If this is the case, the shape (for a given filling ratio) should not change significantly due to the dynamic pressure. Otherwise we should take care of the contribution in the body bound-

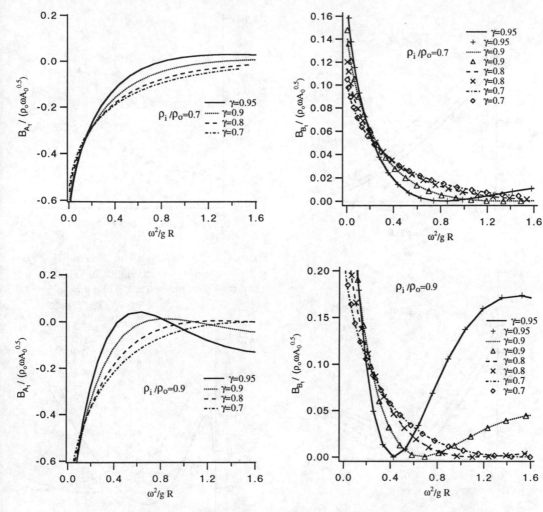

Figure 11: Damping coefficient B_{A_1} as function of filling ratio γ, relative fluid density ρ_i/ρ_o and the oscillation frequency ω. Here R and A_0 are the radius and area of a 2-D membrane structure when the filling ratio $\gamma = 1.0$.

ary condition, like we solve the problem for mode motion A_1^m. To summarize the method for solving the problem, we have the following items.

1) Integrating the dynamic pressure for each component on the body surface to find the corresponding vertical force.

2) Solving eq.(1), the fluid density inside the tube is modified (due to dynamic pressure) so the effect of vertical acceleration is taken

Figure 12: Damping coefficient B_{B_1} as function of filling ratio γ, relative fluid density ρ_i/ρ_o and the oscillation frequency ω. Here R and A_0 are the radius and area of a 2-D membrane structure when the filling ratio $\gamma = 1.0$. The results of the solid and dotted lines are based on the formula of eq.(15). The symbols $+$, $\triangle$, $\times$ and $\diamond$ are based on the wave amplitude far from the body which is generated by the mode motion A_1^m.

care of. The pressure outside the tube is the static pressure plus the dynamic pressure.

3) The coefficients in eq.(5) and eq.(6) can be evaluated based on the results one has obtained both with and without dynamic pressure.

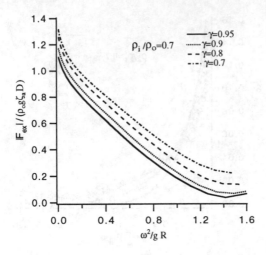

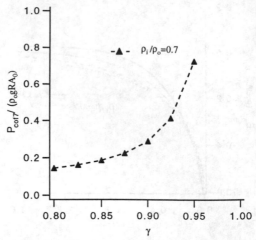

Figure 14: $\frac{\partial P_1}{\partial P_{A_1}} = P_{cof7}$ as function of filling ratio γ. Here R and A_0 are the radius and area of a 2-D membrane structure when the filling ratio $\gamma = 1.0$.

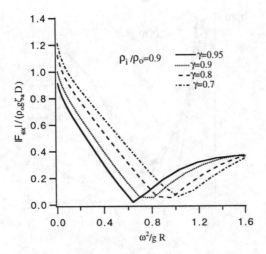

Figure 13: Excitation force F_{ex} as function of filling ratio γ, relative fluid density ρ_i/ρ_o and the oscillation frequency ω. Here R and A_0 are the radius and area of a 2-D membrane structure when the filling ratio $\gamma = 1.0$

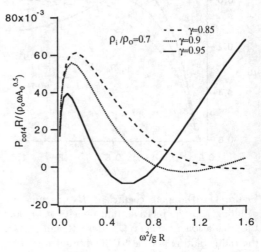

Figure 15: $\frac{\partial P_1}{\partial P_{A_1}} = P_{cof4}$ as function of oscillation frequency ω and filling ratio γ. Here R and A_0 are the radius and area of a 2-D membrane structure when the filling ratio $\gamma = 1.0$.

The coefficients $\frac{\partial P_1}{\partial P_{A_1}}$, $\frac{\partial P_1}{\partial P_{A_1}}$, $\frac{\partial T_1}{\partial P_{A_1}}$, $\frac{\partial T_1}{\partial P_{A_1}}$, ... are dependent on the filling ratio, oscillation frequencies and densities of fluid inside and outside the tube. Fig.14 to Fig.20 show the results of the coefficient $\frac{\partial P_1}{\partial P_{A_1}}$, $\frac{\partial P_1}{\partial P_{A_1}}$,... as function of filling ratio γ, fluid density ρ_i/ρ_o and the oscillation frequency ω.

3.3 Numerical results of simulations

Eq.(2) to eq.(4) have been solved by a finite differ-ence method. The hydroelastic deformations and tensions of the tube depend on many parameters. The results we can present here are limited. Fig.21 shows the maximum values of heave motion as the function of wave frequency. The length of the tube we have chosen is 20D, where D is diameter of the tube when filling ratio is 1.0. We assume also the tube has the uniform static shape for all the

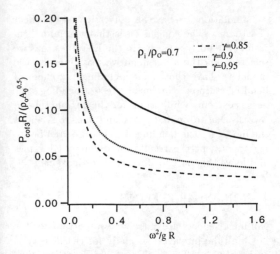

Figure 16: $\frac{\partial P_1}{\partial P_{A_1}} = P_{cof3}$ as function of oscillation frequency ω and filling ratio γ. Here R and A_0 are the radius and area of a 2-D membrane structure when the filling ratio $\gamma = 1.0$.

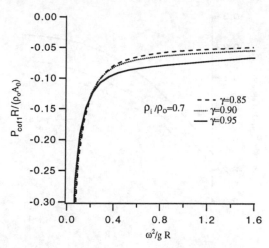

Figure 18: $\frac{\partial P_1}{\partial P_{\dot{\eta}_3}} = P_{cof1}$ as function of oscillation frequency ω and filling ratio γ. Here R and A_0 are the radius and area of a 2-D membrane structure when the filling ratio $\gamma = 1.0$.

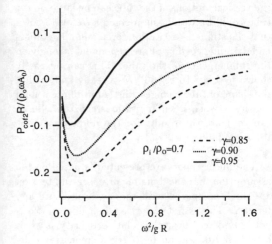

Figure 17: $\frac{\partial P_1}{\partial P_{\eta_3}} = P_{cof2}$ as function of oscillation frequency ω and filling ratio γ. Here R and A_0 are the radius and area of a 2-D membrane structure when the filling ratio $\gamma = 1.0$.

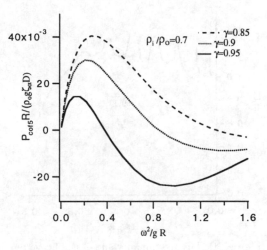

Figure 19: $\frac{\partial P_1}{\partial P_{Re(F_{ex})}} = P_{cof5}$ as function of oscillation frequency ω and filling ratio γ. Here R, D and A_0 are the radius, diameter and area of a 2-D membrane structure when the filling ratio $\gamma = 1.0$.

cross-sections. Here the maximum means the max. value of all the sections. Similar results of P_1 and T_1 have been presented in Fig.22 and Fig.23. In most cases the maximum tension and pressure occur at the region close to the end points of the tube. The peaks in the results of the dynamic pressures

and tensions are due to the resonances of the internal surge motion. The first and second peaks (lowest frequencies) can be significantly reduced if one includes the viscous damping for the internal surge motion.

297

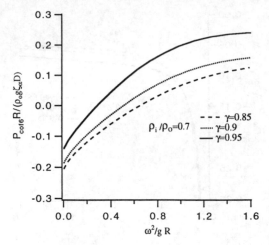

Figure 20: $\frac{\partial P_1}{\partial P_{Im(Fex)}} = P_{cof6}$ as function of oscillation frequency ω and filling ratio γ. Here R, D and A_0 are the radius, diameter and area of a 2-D membrane structure when the filling ratio $\gamma = 1.0$.

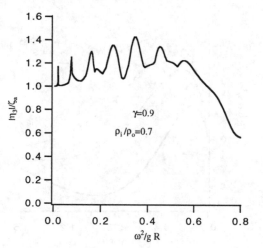

Figure 21: The maximum value of the heave motions η_3 for all the sections as function of wave number $k = \omega^2/g$. R and η_3 are the radius of a 2-D membrane structure when the filling ratio $\gamma = 1.0$ and the incident wave amplitude.

The effects of EI and T(x) have been neglected in the numerical results that we have presented here . This is correct only for the wave length is much larger than the cross-section dimensions of the tube. In practical applications the effect of EI is very important(Aarsnes(1994)). In the numeri-

cal simulations we have also investigated the effect of EI-term. The conclusion is that the numerical results are fairly sensitive to the EI-values that we selected. On the other hand we may question the model we have chosen for including the effect of bending stiffness. Nonlinear effect of bending stiffness is extremely important for the flexible tube in waves. Due to limited space and questions about the model for the bending stiffness, we are not going to present the results for different EI values and leave this part for further investigations.

4 NONLINEAR EFFECTS

Nonlinear effects are extremely important for the flexible tube in waves, specially for the density ρ_i is closed to ρ_o. This is the case for many practical applications. From Fig.3 one may observe that the intersection between fluid (outside the tube) and the tube is not wall-sided and the distance between the water line and the top of the membrane structure is usually small compared with depth of the tube. This tells us that under normal sea states, the vertical motions of the tube can be large, so the tube could be totally submerged. We also known that the stiffness, excitation force and added mass and damping coefficient are dependent strongly on the submergence of the tube. This can be partly shown by Fig.8 to Fig.13. We already known that the shape of tube changes a little for different fluid density ρ_i when the filling ratio is constant. So we can compare the results for the different ρ_i, but the same filling ratio γ.

The linear theory we presented is based on the assumption that the dynamic pressure and tension are small compared with the static pressure and tension. This is not the case when filling ratio is not closed to 1.0 or the relative density ρ_i/ρ_o is closed to 1.0. It can be partly demonstrated by Fig.4 for the hoop tension. In the case the dynamic tension is large than the static tension the approach will totally break down, because we know the total tension cannot be less than zero. What happened when tension is going to zero is not well understood yet. But from experiment results of testing a flexible bag in beam sea (Aarsnes 1994), one observed that the flexible bag may have large hydroelastic deformations when the incident wave is larger than some limit. In these cases the experiment results cannot be well repeated. It may exist instability problems. Theoretically one may have several solutions or no solution.

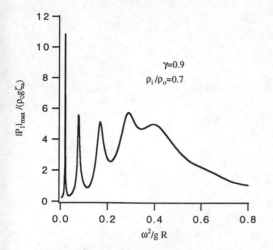

Figure 22: The maximum value of the dynamic pressures P_1 for all the sections as function of wave number $k = \omega^2/g$. R and η_3 are the radius of a 2-D membrane structure when the filling ratio $\gamma = 1.0$ and the incident wave amplitude.

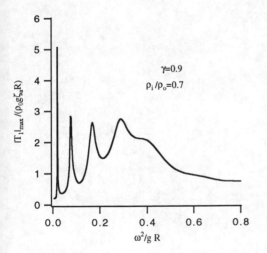

Figure 23: The maximum value of the hoop tensions T_1 for all the sections as function of wave number $k = \omega^2/g$. R and η_3 are the radius of a 2-D membrane structure when the filling ratio $\gamma = 1.0$ and the incident wave amplitude.

The other parameters which have strong nonlinear effects are the tension and pressure inside the tube. This is partly illustrated in Fig.4. For an extremely long wave one may study the problem by a quasi steady analysis. It means we may assume the tube is followed the waves, the dynamic pressure inside the tube can be found by a quasi static analysis. The results show there are strong nonlinear effect even for wave amplitudes $\zeta_a kL/2 \sim 0.1D_1$, where L and D_1 are the length and depth of the tube.

5 CONCLUSION

A linear solution for a long flexible tube in head sea waves has been presented. The hydrodynamic coefficients (used in the formulation for the linear solution) have been predicted by a boundary element method. The dynamic pressures, tensions and one-dimensional internal motion have been evaluated in an approximate way. Nonlinear effects are discussed and shown to be important. Further investigation of the effect of bending stiffness is needed.

6 ACKNOWLEDGEMENT

The work presented in this paper was done when the first author was working at Department of Ocean Engineering, MIT in the period Sep.92 — Aug.93. It is a particular pleasure to thank Professor O.M. Faltinsen and Professor D.K.P. Yue for their interesting discussions.

The financial support was provided by MARINTEK under the project HYDROELASTICITY of NTH and MARINTEK.

7 REFERENCES

Aarsnes, J.V.,Personal Communication, 1994.

Hawthorne,W.R., " The early development of the Dracone flexible Barge," Proceedings of Institution of Mechanical Engineerings, Vol.175, pp.52-83, 1961.

Løland, G. and Aarsnes, J.V., "Fabric as construction material for marine applications," Proceedings of International Conference on Hydroelasticity in Marine Technology, Trondheim, 1994.

Paidoussis,M.P.,"Stability of towed, totally submerged flexible cylinders," J. Fluid Mech., Vol.34, part 2, pp.273-297, 1968.

Paidoussis,M.P. and Yu,B.K., "Elastohydrodynamic of towed slender bodies: the effect of nose and tail shapes on stability," J.Hydronautics, Vol.10, No.4, pp.127-134,1976.

Zhao, R. and Faltinsen, O.M., "Interaction between waves and current on a two-dimensional body in the free surface," Applied Ocean Res., Vol.10, No.2 , pp.87-99, 1988.

Hydroelasticity in Marine Technology, Faltinsen et al. (eds) © 1994 Balkema, Rotterdam, ISBN 90 5410 387 6

Computation of wave induced motions on a flexible container

Dariusz E. Fathi
Division of Marine Hydrodynamics, The Norwegian Institute of Technology, Trondheim, Norway

Chang-Ho Lee & J. Nicholas Newman
Department of Ocean Engineering, Massachusetts Institute of Technology, Cambridge, Mass., USA

ABSTRACT: The linearized wave radiation and diffraction theory is applied to analyze interaction between free-surface waves and floating flexible containers. The geometry of the container is approximated as a circular cylinder with spherical caps at the ends. The motion of the container is assumed to consist of six rigid body modes and several hydroelastic mode shapes. The hydroelastic mode shape is specified by normal vectors on the panels of the discretized body surface. The panel program (WAMIT) is used for the computation of the radiation and diffraction solution. Computational results include the amplitude of the motions and exciting forces, as well as hydrodynamic coefficients for two 'surge-pressure' modes.

1. INTRODUCTION

The behavior of marine structures in waves is usually dealt with as two separate problems where the hydrodynamic and structural analysis are performed separately. There are however cases where these are so closely related that they cannot be completely separated. This is when the flexibility of the structure can not be disregarded in the computations of the fluid motions around the body. The term *hydroelasticity* appeared in 1958 (Heller & Abramson [5]) as a classification of these problems, and since then many approaches have been suggested in solving hydroelastic problems. For slender bodies, these include different two-dimensional theories, where the hydrodynamic problem is solved using strip-theory (see e.g. Bishop & Price [1]) or using more refined two-dimensional theories (Wu *et al* [12]). Another way of solving this problem is using three-dimensional theories to solve the radiation-diffraction problem, and hence to obtain more accurate computations of the hydrodynamic coefficients in the hydroelastic equations of motions.

Newman [8] has presented a method of solving three-dimensional problems including 'generalized modes' extending beyond the usual six rigid-body motion modes. The intention of this paper is to present results from calculations using the same method to solve the radiation-diffraction problem for a flexible container.

Hydroelastic problems in connection with flexible containers are wave induced loads and motions, including bending motions of the structure, and internal surging motions of the cargo, inducing large pressures inside the container. Hence, for the exterior problem, both an analysis of the bending motions, and the contributions from the outside to the 'internal-surge' problem is desirable. We have restricted ourselves to computing only the hydrodynamic coefficients of the 'internal-surge' problem, where linear analysis can be justified due to the small motion amplitudes. The thought is that such results can be connected with a separate analysis of the interior dynamics. An example of flexible containers where the above mentioned problems arise, can be found in Hawthorne [4], where the containers are made of coated fabrics and designed to be used for transportation of liquids such as water or oil. In our analysis, the geometry that is chosen is a simple 'generic' one to suggest a methodology for solving this kind of hydroelastic problem. However, it should be straightforward to use this method to solve problems for more complex geometries as well.

The hydrodynamic analysis is performed using the three-dimensional radiation-diffraction code WAMIT based on the boundary-element 'panel' method. Further details regarding this code are given in references [6, 7, 10, 11] and [8] where the last one refers to the modifications including generalized modes of motion.

2. THEORY

2.1 Generalized body modes

In addition to the six conventional rigid-body motions (surge, sway, heave, roll, pitch, yaw) denoted by $j = 1, \ldots, 6$, we define additional *generalized* modes. We denote these extended modes by $j = 7, 8, \ldots, N$. The idea is to define each generalized mode by specifying the normal velocities on the body in the form

$$n_j = \mathbf{S}_j \cdot \mathbf{n} = u_j n_x + v_j n_y + w_j n_z \qquad (1)$$

where the unit normal vector $\mathbf{n}$ points out of the fluid domain and into the body. The modes may be defined by a vector 'shape function' $\mathbf{S}_j(\mathbf{x})$, with Cartesian components (u_j, v_j, w_j), which can be any physically relevant real function of the body coordinates (x, y, z). The displacement of an arbitrary point of the body, due to motion in the j'th mode with complex amplitude ξ_j, is then represented by the product $\xi_j \mathbf{S}_j(\mathbf{x})$. Any specified distribution of normal velocity on the (global) body surface, is considered to be a 'mode' in the present work. The decomposition of complicated body motions into separate modes is justified by linear superposition, provided suitable modal functions are included.

2.2 The hydroelastic equations of motion

The coordinate system which will be used is a right-handed Cartesian coordinate system, with the z-axis pointing upwards and the x-y plane on the mean position of the free surface. The structural stiffness is assumed to be linear. Furthermore, we assume an ideal fluid and irrotational flow. The unsteady motions of the structure and the fluid are of small amplitude.

For steady-state sinusoidal motions we may write the hydroelastic equations of motion of the flexible container in the form

$$\sum_n \xi_n [-\omega^2 (A_{mn} + M_{mn}) + i\omega (B_{mn} + B^*_{mn})$$

$$+ C_{mn} + C^*_{mn}] = X_m \qquad (2)$$

where A_{mn}, B_{mn} and C_{mn} are the added mass, damping and hydrostatic restoring force matrices, respectively. Further, M_{mn} is the generalized mass matrix, and B^*_{mn} and C^*_{mn} are the generalized structural damping and stiffness matrices of

the dry hull. X_m is the generalized wave-exciting force vector, and ξ_n is the complex amplitude of mode n.

2.3 The boundary value problem

Potential theory is built on the assumption that the fluid is ideal, and that the fluid motion is irrotational. This permits the definition of the flow velocity as the gradient of the velocity potential Φ satisfying the Laplace equation

$$\nabla^2 \Phi = 0 \qquad (3)$$

in the fluid domain. The harmonic time dependence allows the definition of a complex velocity potential ϕ, related to Φ by

$$\Phi = \Re(\phi e^{i\omega t}) \qquad (4)$$

where ω is the frequency of the incident wave and t is time. In the following text, the potential will be expressed in terms of the complex velocity potential ϕ.

The velocity potential ϕ can be expressed in the form

$$\phi = \phi_R + \phi_D \qquad (5)$$

where ϕ_R is the radiation potential, and ϕ_D is the diffraction potential. The radiation potential can be expressed as

$$\phi_R = \sum_{j=1}^{N} \xi_j \phi_j \qquad (6)$$

where N is the number of motion modes, ξ_j is the complex amplitude in mode j, and ϕ_j is the corresponding unit-amplitude radiation potential.

The diffraction potential can be divided in two parts:

$$\phi_D = \phi_I + \phi_S \qquad (7)$$

where ϕ_I is the incident wave potential, and ϕ_S is the scattering component representing the disturbance of the incident wave by the fixed body. The velocity potential of the incident wave is defined by

$$\phi_I = \frac{igA}{\omega} \frac{\cosh[k(z+h)]}{\cosh kh}$$

$$\cdot exp[-ik(x \cos \beta + y \sin \beta)] \qquad (8)$$

where A is the wave amplitude, h is the water depth, β is the wave heading angle and k the wavenumber defined as the positive real root of the dispersion relation

$$\frac{\omega^2}{g} = k \tanh kh \ . \tag{9}$$

The problem of finding the velocity potential consists of finding the solution of the Laplace equation (3) with relevant boundary conditions. These are:

$$\frac{\partial \phi}{\partial z} - K\phi = 0 \quad \text{on} \quad z = 0 \tag{10}$$

$$\left. \begin{array}{l} \dfrac{\partial \phi_j}{\partial n} = i\omega n_j \\[2mm] \dfrac{\partial \phi_D}{\partial n} = 0 \end{array} \right\} \quad \text{on} \quad S_b \tag{11}$$

$$\frac{\partial \phi}{\partial z} = 0 \quad \text{on} \quad z = -h \tag{12}$$

$$\text{Radiation condition} \tag{13}$$

where $K = \omega^2/g$ is the infinite-depth wave number.

Corresponding to each mode of motion, generalized first-order pressure forces are defined in the form

$$\begin{aligned} F_i &= \int\!\!\int_{S_b} p n_i dS \\ &= -\rho \int\!\!\int_{S_b} (i\omega\phi + gz) n_i dS \end{aligned} \tag{14}$$

where p is the fluid pressure, which is evaluated in the last form of (14) from the linearized Bernoulli equation.

The contribution to (14) from the term $i\omega\phi$ involves a straightforward extension of the corresponding rigid-body analysis. After substituting the body boundary conditions for the components of the radiation potential from (11), the added mass and damping matrices are defined in the form

$$\begin{aligned} \omega^2 A_{ij} - i\omega B_{ij} &= -i\omega\rho \int\!\!\int_{S_b} \phi_j n_i dS \\ &= -\rho \int\!\!\int_{S_b} \phi_j \frac{\partial \phi_i}{\partial n} dS \end{aligned} \tag{15}$$

Similarly, the generalized wave-exciting force is:

$$\begin{aligned} X_i &= -i\omega\rho \int\!\!\int_{S_b} \phi_D n_i dS \\ &= -\rho \int\!\!\int_{S_b} \phi_D \frac{\partial \phi_i}{\partial n} dS \end{aligned} \tag{16}$$

Green's theorem can be applied to (15) to establish reciprocity, and to (16) to derive the Haskind relations between the generalized exciting force and the corresponding radiation potential.

The extended body modes can be applied directly to the equations for the pressure and exiting forces in equations (14) and (16). However, the deformation of the body geometry must be considered in evaluating the contribution to the generalized force from the hydrostatic pressure. The derivation of this force is given in Newman [8], and the result is that the generalized force can be expressed as

$$\begin{aligned} C_{ij} &= \rho g \int\!\!\int_{S_b} n_j \nabla \cdot (z\mathbf{S}_i) dS \\ &= \rho g \int\!\!\int_{S_b} n_j (w_i + zD_i) dS \end{aligned} \tag{17}$$

Here the generalized force is defined in a fixed reference frame, and only the hydrostatic pressure is considered. As a result, equation (17) includes some contributions which are normally balanced by the gravitational force due to the body mass. In the analysis of a deformable body the corresponding generalized mass force must be evaluated separately for each mode, depending on the mode shape and mass distribution. In general the hydrostatic matrix (17) is not symmetric (Newman [9]).

3. NUMERICAL EXAMPLES AND RESULTS

3.1 Geometry

The geometry used in these examples is a horizontal cylinder with spherical caps at the ends, floating with its axis in the plane of the mean water surface. The dimensions are equal to those of the 'D-type Dracones' from Hawthorne [4], with a total length of 30.48 m and a diameter of 1.46 m, giving a length to diameter ratio of approximately 20/1. The flexible container is modeled with 385 panels on a quadrant giving a total of 1540 panels on the wetted body surface. The examples include computations of the bending motion of the container, and computations of the hydrodynamic coefficients of two 'surge-pressure' modes, which can be used to represent the contribution from the outside to the 'internal-surge' problem (i.e. longitudinal sloshing of the cargo). The water depth is taken to be infinite in the following examples. Further results can be found in Fathi [2].

303

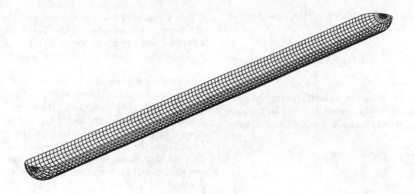

Figure 1: Geometric model discretized with 385 panels on a quadrant.

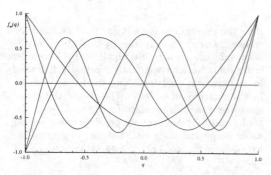

Figure 2: The first four natural bending modes for a uniform beam with free ends. The normalized coordinate q is equal to ± 1 at the ends of the body.

Table 1: The factors κ_m for the first four natural bending modes for a uniform beam with free ends.

1st Mode	2nd Mode	3rd Mode	4th Mode
$\kappa_2 = 2.3650$	$\kappa_3 = 3.9266$	$\kappa_4 = 5.4978$	$\kappa_5 = 7.0686$

3.2 Bending modes for the flexible container

To describe the bending of the container, the natural modes for a uniform beam with free ends are chosen (cf. Gran [3, Section 3.8]). A total of eight generalized modes are defined, with four modes in the vertical and transverse directions, respectively.

The natural modes for a free-free beam are given by the symmetric and anti-symmetric functions:

$$f_{2m}(q) = \frac{1}{2}\left(\frac{\cos \kappa_{2m} q}{\cos \kappa_{2m}} + \frac{\cosh \kappa_{2m} q}{\cosh \kappa_{2m}}\right) \quad (18)$$

$$f_{2m+1}(q) = \frac{1}{2}\left(\frac{\sin \kappa_{2m+1} q}{\sin \kappa_{2m+1}} + \frac{\sinh \kappa_{2m+1} q}{\sinh \kappa_{2m+1}}\right) \quad (19)$$

Here $q = 2x/L$ is the normalized horizontal coordinate varying from -1 to $+1$ over the length of the container. Equations (18) and (19) apply for $(m = 1, 2, \ldots)$. The functions $f_0 = 1$ and $f_1 = q$ correspond to the heave and pitch modes, respectively. The factors κ_m are the positive real roots of the equation

$$(-1)^m \tan \kappa_m + \tanh \kappa_m = 0 \quad (20)$$

The first four roots of this equation are shown in Table 1, and the corresponding modes are plotted in Figure 2.

The functions $f_m(q)$ are orthogonal, with the normalized values $f_m(1) = 1$, and the integrated values

$$\int_{-1}^{1} f_m(q) f_n(q) dq = \frac{1}{2}\delta_{mn} \quad (m \geq 2) \quad (21)$$

where δ_{mn} is the Kronecker delta which is equal to one when $m = n$ and zero otherwise.

The generalized mass and stiffness matrices can be evaluated considering the flexible container as a uniform slender beam, with constant mass m' per unit length and bending stiffness EI. The transverse and vertical deformation may be described as $\Re(\xi(x)e^{i\omega t})$ along the length. The structural deflection is then governed by the beam equation

$$-\omega^2 m'\xi + EI\xi^{(iv)} = F_\xi(x) \qquad (22)$$

where $\xi^{(iv)}$ denotes the fourth derivative of the deflection and $F_\xi(x)$ is the local pressure force acting on a vertical section of the flexible container, in the same direction as the deflection ξ. The appropriate boundary conditions for a free beam are

$$\xi'' = 0 \ and \ \xi''' = 0 \ at \ x = \pm L/2 \qquad (23)$$

where L is the length of the container, and the origin is at the mid-section.

Expanding the deflection ξ in an appropriate set of modes, in the form

$$\xi(x) = \sum_n \xi_n f_n(x) \qquad (24)$$

where ξ_n is the unknown complex amplitude of each mode, and using the method of weighted residuals, expressions for the mass and stiffness matrix can be found as (cf. Newman [8]):

$$M_{mn} = m' \int_{-L/2}^{L/2} f_m(x)f_n(x)dx \qquad (25)$$

$$C_{mn}^* = EI \int_{-L/2}^{L/2} f_m(x)f_n^{(iv)}(x)dx \qquad (26)$$

Using the orthogonality relation (21) the generalized mass matrix can be evaluated as

$$M_{mn} = m' \int_{-L/2}^{L/2} f_m(2x/L)f_n(2x/L)dx$$
$$= \frac{1}{4} M \, \delta_{mn} \qquad (27)$$

where $M = mL$ is the total mass.

Furthermore, the generalized stiffness matrix can be evaluated from the differential equation

$$f_m^{(iv)}(q) - \kappa_m^4 f_m(q) = 0 \qquad (28)$$

and the orthogonality relation (21) to give

$$C_{mn}^* = 4\left(\frac{EI}{L^3}\right)\kappa_m^4 \delta_{mn} \qquad (29)$$

In the results presented for the wave-induced motions, the stiffness parameter EI is set to 5000 $[kNm^2]$.

Results for the exciting forces[1] are presented in Figures 3 to 5. For high wave periods (long wavelengths), the exciting forces decrease with increasing order of the mode shapes. For shorter wave-

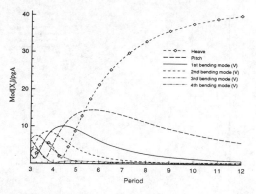

Figure 3: Exciting forces for the vertical modes in head seas.

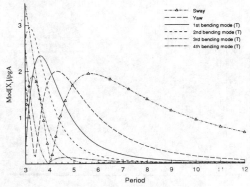

Figure 4: Exciting forces for the transverse modes at a wave heading of 30 deg.

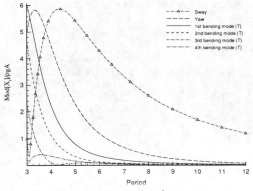

Figure 5: Exciting forces for the transverse modes at a wave heading of 60 deg.

lengths the exciting forces are relatively large when the mode shape and the longitudinal component of the incident wave are correlated.

---

[1]The length scale $L/2$ has been used to nondimensionalize the results in Figures 3-8.

305

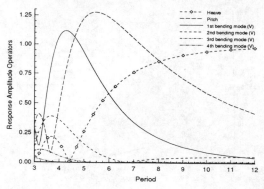

Figure 6: Amplitudes of the vertical modes in head seas.

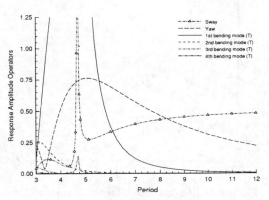

Figure 7: Amplitudes of the transverse modes at a wave heading of 30 deg.

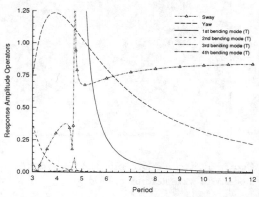

Figure 8: Amplitudes of the transverse modes at a wave heading of 60 deg.

The Response Amplitude Operators are presented in Figures 6 to 8. The amplitudes of the generalized modes are largest for the first bending mode, and decrease with increasing order. The small val-

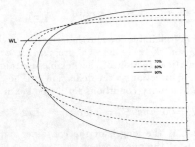

Figure 9: Cross-sectional shapes of a floating tube for various filling ratios. Cargo density $= 0.8 \times$ water density (based on data from Hawthorne [4]).

ues for the third and fourth bending modes and the rapid decrease in response amplitudes indicate that including more modes will have little, or no effect on the calculations of the total motions of the container (the amplitudes of the third and fourth bending modes are negligible for wave periods above 5 seconds). The results for the transverse bending modes show a large resonant peak at 4.7 seconds. The amplitudes are actually as large as 40 for the first bending mode, indicating that factors such as nonlinear structural stiffness, structural damping and internal damping may become important in this period range.

3.3 Hydrodynamic coefficients for two 'surge-pressure' modes

To describe the 'surge-pressure' modes (i.e. internal surging motions of the cargo), we have used two modes describing the change of cross-sectional area along the length of the body. The motivation for doing this can be seen in Figure 9 where the variation in cross-sectional shape with filling ratio of a floating container is presented[2]. The filling ratio can also be interpreted as an increase/decrease in local internal pressure, and the thought is that this longitudinal change of cross-sectional shape can be used to represent the 'internal surge' problem.

To account for the local change of area we assume that each section of the body is circular, with a varing radius. The normal velocity on the body is then described by

---

[2]The filling ratio is defined as the volume of tube occupied by liquid divided by the volume occupied when the tube has a circular cross-section.

$$u_n = 0,$$
$$v_n = f_n(q)\, r\cos\theta,$$
$$w_n = f_n(q)\, r\sin\theta \qquad (30)$$

where $f_n(q)$ represent the longitudinal variation, q is the normalized length, r is the radius of the cylinder, and θ is the angle with the y-axis in the cross-sectional plane.

Appropriate functions to describe the longitudinal variation of these modes must satisfy conservation of volume and, since the structure cannot change its cross-sectional area at the ends, $f_n(q) = 0$ at $q = \pm 1$. For this reason, we have chosen the following symmetric and anti-symmetric functions:

$$f_{2n}(q) = \sin(n\pi q) \qquad (31)$$

$$f_{2n+1}(q) = C_{2n+1}[(2n+1)\cos((2n+1)\frac{\pi}{2}q) \\ - (-1)^n \cos(\frac{\pi}{2}q)] \qquad (32)$$

where C_{2n+1} is chosen to normalize the functions so that $Max|f_{2n+1}| = 1$. Equations (31) apply for $(n = 1, 2, \ldots)$, and the functions for the two first modes are:

$$f_2(q) = \sin(\pi q) \qquad (33)$$

$$f_3(q) = \frac{1}{4}\left[3\cos(\frac{3\pi}{2}q) + \cos(\frac{\pi}{2}q)\right] \qquad (34)$$

These modes are plotted in Figure 10.

The mass and stiffness matrices have not been evaluated here. However, the stiffness is likely to be governed by the local change of pressure due

to the change of cross-sectional area (i.e. dp/dA). The generalized mass matrix for these modes, must be able to represent the internal surge flow of the fluid. The restoring coefficients can be evaluated from (17), but must in addition be modified to take care of the motion of mass inside the body.

The results for the added mass and damping coefficients are shown in Figure 11. The results indicate that the values of the coefficients decrease with increasing order. At the same time, the peak periods also decrease.

The exciting forces in head seas are shown in Figure 12. The forces decrease with increasing order, and the peak values are found at approximately 5.2 sec for the 1st mode, and 4 sec for the 2nd mode.

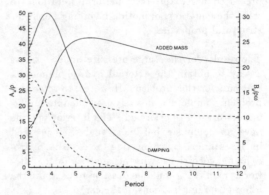

Figure 11: Added mass and damping coefficients for the 'surge-pressure' modes. The first mode is represented by solid curves, and the second with dashed curves.

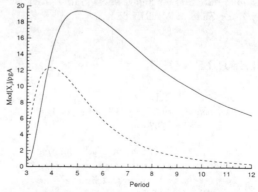

Figure 12: Exciting forces for the 'surge-pressure' modes in head seas. The first mode is represented by solid curves, and the second with dashed curves.

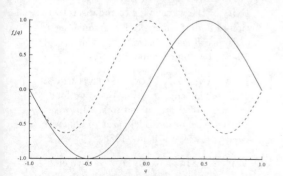

Figure 10: The longitudinal variation of the change in cross-sectional shape for the two first 'surge-pressure' modes.

4. CONCLUSIONS

This paper shows a methodology for solving the hydrodynamic coefficients and motion amplitudes for the hydroelastic motions of a flexible container. Three-dimensional linear potential theory is used to solve the radiation-diffraction problem including several hydroelastic mode shapes in addition to the conventional rigid body motions. The structural stiffness is assumed to be linear. This assumption is perhaps not always true for very elastic structures, and the results for the transverse bending modes indicate that nonlinear stiffness should be considered to obtain reasonable results for the motion amplitudes in the resonant period range. This could for instance be done by computing the hydrodynamic coefficients with the method presented here, and then perform iterations to solve the equations of motion including nonlinear structural properties.

The results for the 'surge-pressure' modes, shows a way to obtain the external hydrodynamic coefficients for this problem. However, to solve this problem completely, an analysis of the interior dynamics must be performed. The interior may be linear or nonlinear, but linear analysis is justified in the exterior. By connecting the external hydrodynamic coefficients with an analysis of the interior dynamics it should be possible to compute the internal surging motions of the cargo.

ACKNOWLEDGEMENT

Tabular data files for Figure 9 was kindly provided by Rong Zhao at MARINTEK.

REFERENCES

[1] BISHOP, R.E.D. AND PRICE, W.G. *Hydroelasticity of ships.* Cambridge University Press, 1979.

[2] FATHI, D.E. *Linear Hydrodynamic Loads due to Elastic Deformations of Large Volume Structures.* (Siv.Ing. Thesis), Division of Marine Hydrodynamics, Norwegian Institute of Technology (NTH), December 1993.

[3] GRAN, S. *A Course in Ocean Engineering.* Developments in Marine Technology, vol. 8, Elsevier Science Publishers B.V., 1992.

[4] HAWTHORNE, W.R. The early development of the dracone barge. In *Proceedings of the Institution of Mechanical Engineers*, vol. 175, pp. 52–83, 1961.

[5] HELLER, S.R. AND ABRAMSON, H.N. Hydroelasticity: a new naval science. *Journal of the American Society of Naval Engineers*, 71:205–209, 1959.

[6] KORSMEYER, F.T., LEE, C.-H., NEWMAN, J.N. AND SCLAVOUNOS, P.D. The analysis of wave interactions with tension leg platforms. In *Proc. Conference on Offshore Mechanics and Arctic Engineering*, Houston, 1988.

[7] LEE, C.-H., NEWMAN, J.N., KIM, M.-H. AND YUE, D.K.P. The computation of second-order wave loads. In *Proc. Offshore Mechanics and Arctic Engineering Conference*, Stavanger, 1991.

[8] NEWMAN, J.N. Wave effects on deformable bodies. *Applied Ocean Research,* (to be published).

[9] NEWMAN, J.N. Deformable floating bodies. In *8th International Workshop on Water Waves and Floating Bodies*, S. John's, Newfoundland, 1993.

[10] NEWMAN, J.N. AND LEE, C.-H. Sensitivity of the wave loads to the discretization of bodies. In *Proc. Int. Behaviour of Offshore Struct. (BOSS 92)*, vol. 1, pp. 50–64, London, England, 1992.

[11] NEWMAN, J.N. AND SCLAVOUNOS, P.D. The computation of wave loads on large offshore structures. In *Proc. Conference on the Behaviour of Offshore Structures*, Trondheim, 1988.

[12] WU, Y., XIA, J. AND DU, S. Two engineering approaches to hydroelastic analysis of slender ships. In *Dynamics of Marine Vehicles and Structures in Waves*, Elsevier Science Publishers B.V., 1991.

Skirt and bag systems

Hydroelasticity in Marine Technology, Faltinsen et al. (eds) © 1994 Balkema, Rotterdam, ISBN 90 5410 387 6

Hydroelastic analysis of surface effect ship (SES) seals

Paul Kaplan
Hydromechanics, Inc., Delray Beach, Fla., USA

ABSTRACT: A description is given of the analysis of motions and forces on seals used on high speed SES vessels. The coupled effects of hydrodynamic and hydrostatic forces; aerostatic and aerodynamic internal flow and deformations; and material structural elasticity are described and combined into appropriate dynamic analysis equations used to predict seal motions and forces (including effects on the vessel force system). Both bow and stern seals are considered, covering different design concepts used for such vessels.

1. INTRODUCTION.

A Surface Effect Ship (SES) is a twin hull configuration, similar to a catamaran, with a pressurized region between the separate hulls. This pressurized region, referred to as the air cushion, is sealed both at the bow and stern by flexible seals whose main purpose is to contain the air cushion. This feature is to be achieved by conforming to the encountered wave surface, while also providing minimum drag. In addition to the direct "sealing" action per se, the seals can also contribute to the vessel pitch stability, as well as providing some alleviation of slamming loads in large waves (primarily the bow seal, and depending on the nature of the design configuration).

Illustrations of representative seal designs are given in Fig. 1 and 2, which show a multilobe stern seal as well as a bow seal (bag and finger design). The use of fingers, either attached to an inflated bow bag or alone, is very common in many existing vessels (which are generally of relatively small size). However such bow seals tend to wear poorly (see Malakhoff and Davis, 1981), so that other concepts have been sought to improve the seal life and reliability. In general, seals should be light in order to minimize air leakage and weight, while still having adequate strength and durability in order to operate reliably at high speed in waves, and at the same time have low drag. An appropriate design will therefore involve a compromise between such conflicting requirements.

An extensive series of investigations was carried out as part of the U.S. Navy SES program, which was aimed at relatively large high speed vessels (3,000 tons or more, at speeds of 60 kt. and higher). With the trend in present day developments toward larger vessels, the same basic difficulties will manifest themselves. The present state-of-the art in SES structural design in regard to seals still recognizes this situation (see ISSC Technical Comm. V.4, Surface Effect Ships, Report, 1994). The present paper presents some of the analysis methods and results obtained concerning seal behavior, primarily in terms of hydroelastic phenomena, which were established as part of the U.S. development effort.

2. HYDROELASTIC CONSIDERATIONS

Any analysis of SES seal behavior will recognize the interaction between a number of different physical phenomena and/or features, viz. aerostatic and aerodynamic internal flow effects; hydrostatic and hydrodynamic forces; material deformation and structural elasticity properties. The aerostatic effects are represented by the different pressures affecting the seal, such as atmospheric pressure, cushion pressure, and internal seal pressure. Aerodynamic flow effect are present when considering dynamic variations in internal air flows between the cushion, seal and atmosphere. Hydrostatic forces arise due to immersion changes of enclosed portions of a seal in the water surface, with hydrodynamic forces also arising from orientation of seal elements relative to the water surface when moving at forward speed. Changes in shape of the seal occur due to deformations brought about by these different fluid forces, with such deformations also resisted by the elasticity of the seal structure stiffening members.

All of these separate factors interacting with each other represent a definite hydroelastic system, which must be properly analyzed in order to determine the seal orientation, shape, position, etc. The problem is further complicated by the presence of a time-varying surface wave elevation, which is encountered at high forward speed.

SES seals in general use are classified as either flexible or semi-flexible, with the use of rigid seals having being discarded as a result of their wave response performance limits. The nature of the hydroelastic features of each of these types of seal differs, as well as the resulting analysis and consequences of their use. Similarly different types of seals are used for either bow or stern seal applications, and separate discussions and analytical modeling procedures are applied in each case. The following sections provide descriptions of representative analyses for some of the seal concepts.

3. FLEXIBLE SEALS

The bag and finger bow seal, illustrated in Fig. 2, is a typical flexible seal. In some cases the bag is absent and the fingers are attached directly to the vessel wet deck. The bow seal is oriented at some angle with respect to the craft hull centerbody, which is also the angle between the undisturbed water surface and the seal at equilibrium, with this angle denoted as τ_{bs}. When the vessel moves vertically, the finger buckles at the water surface with a wetted length change of $z_r/\sin\tau_{bs}$, where z_r is the relative immersion change at the bow. This results in a force on the bow seal,, which is a pressure force that can be included as a cushion force acting on the vessel, given by

$$Z^{(1)}_{bow\ seal} = -(p_b - p_a)bz_r/\sin\tau_{bs} \qquad (1)$$

where b = cushion beam (seal lateral extent), p_b= cushion pressure, and p_a= atmospheric pressure.

There is also a small longitudinal force due to friction drag on the wetted seal surface. The resulting vessel pitch moment, primarily due to the vertical seal force, is interpreted as an effective shift in the craft cushion center of pressure due to the bow seal immersion changes.

In addition to the forces associated with the change in wetted length of the bow seal, another force mechanism is present if the seal elements are attached to a bow bag. The additional forces arise primarily due to the buoyancy of the bag when it is immersed in the water, and it is (initially) assumed that the bag does not undergo any deformation or change in shape due to its immersion. This statical analysis also does not consider any pressure changes in the bow bag. The only effect on the vessel in this case is due to changes of immersed volume when the bag is immersed in the water.

More detailed consideration of bag deformation and large bag pressure changes will be discussed when analyzing more complex bow seal configurations, including effects of slamming. The simple analysis covered above does not consider any dynamic effects, with all external forces acting on the seal entering into the total force system of the overall vessel.

4. SEMI-FLEXIBLE STERN SEAL

A semi-flexible seal consists of an upper flexible bag attached to a lower planing surface that may be rigid or semi-flexible. There are also restraint cables to support the lower element, which also provide a limit down-stop position. A representative analytical model is shown in Fig. 3, as applied to a particular type of stern seal. This seal is a two-lobe bag membrane system, with an effectively rigid planing surface (assumed) as its lower element in contact with the water. There are orifice openings connecting the two seal bag lobes, and also an orifice between the stern seal and the main cushion. A fan system is assumed to feed the seal in order to maintain the pressure there. The seal planing surface is close to a circular arc (in section), and in the analysis it is approximated by two straight line sections rigidly attached at a fixed angle (see Fig. 3).

The seal equation of motion, for the stern seal described above, is

$$\ddot{\theta}_{ss} = \frac{F_{ms}}{I_{ss} + I_{ss_{am}}} \qquad (2)$$

where θ_{ss} is the stern seal angle, measured positive counterclockwise from the craft deck, I_{ss} is the stern seal moment of inertia, and $I_{ss_{am}}$ is the added inertia due to fluid added mass effects. The total moment on the stern seal is given by

$$F_{ms} = F_{ms_{hydrodynamic}} + F_{ms_{pressure}} + F_{ms_{spring}} + F_{ms_{gr}}$$

$$(3)$$

The hydrodynamic forces acting on the stern seal, which contribute to the moment above, are determined in terms of the seal wetted length, which is calculated from the intersection of the actual water surface and the seal by accounting for both craft and seal motions as well as the water surface elevation and slope. This wetted length is denoted as ℓ_{ski}.

In accordance with the notation in Fig. 3, the hydrodynamic moment on the seal is given by

$$F_{ms_{hydro.}} = -(T_{ski} + N_{ski}\alpha_{geom.})\,\ell_1 - N_{ski}\ell_2 \quad (4)$$

where T_{ski} is the friction drag force on the ski, N_{ski} is the normal hydrodynamic force on the ski, ℓ_1 and ℓ_2 are vertical and horizontal distances from the seal hinge to the center of the average wetted length, respectively, and $\alpha_{geom.}$ is the geometric angle of the seal wetted surface relative to the calm water plane, defined by

$$\alpha_{geom.} = \beta - \theta_b \qquad (5)$$

where β and θ_b are defined in Fig. 4 and 3, respectively (β is the angle of the ski portion, measured relative to the horizontal).

The ski friction force is given by

$$T_{ski} = \frac{1}{2}\rho u_{ski}^2\, b\ell_{ski}\, C_{D_f} \qquad (6)$$

where C_{D_f} is the friction drag coefficient, defined in terms of the appropriate Reynolds Number, and u_{ski} is the total horizontal seal velocity given by

$$u_{ski} = u + \ell_1\dot{\theta}_{ss} + \ell_3\dot{\theta} \qquad (7)$$

where ℓ_3 is the vertical distance from the craft CG to the center of the seal ski wetted length (see Fig. 4).

The seal hydrodynamic lift force N_{ski} is given by

$$N_{ski} = \frac{1}{2}\rho u_{ski}^2\, b\ell_{ski}\, C_{L_\alpha}\left(\alpha_{ski} + \frac{w_0}{u_{ski}}\right) \quad (8)$$

where $C_{L_\alpha} = \frac{\pi}{2}$ = seal lift curve

slope (assumed for a high aspect ratio planing surface)
α_{ski} = seal angle of attack

w_0 = wave vertical velocity at seal location

The seal angle of attack is defined by

$$\alpha_{ski} = \alpha_{geom.} + \alpha_{vel.} \qquad (9)$$

which is the sum of the effective geometric angle and a part due to vertical velocity of the seal, given by

$$\alpha_{vel.} = \frac{w + \ell_2 \dot\theta_{ss} + \ell_4 \dot\theta}{u_{ski}} \qquad (10)$$

The other hydrodynamic quantity of interest in this seal dynamic model representation is the seal added moment of inertia, which is given by

$$I_{ss_{am}} = (\ell_1^2 + \ell_2^2) m_{am} \qquad (11)$$

where

$$m_{am} = \frac{1}{8} \rho \pi b \ell_{ski}^2 \qquad (12)$$

is the effective added mass of the wetted seal portion.

The stern seal moment due to pressure is made up of two distinct terms. The first term is due to the difference between the stern seal internal pressure p_{ss} and the cushion pressure p_b, which acts over the seal planing surface. The second pressure term arises from the difference between stern seal pressure and pressure external to the bag, which is assumed to be atmospheric pressure p_a, and results in a net horizontal force that acts as a reaction on the seal planing surface. This reaction force is calculated on the assumption that the attachment of the seal bag is a pin-type joint, and that there is a zero net vertical reaction force on the bag rear face. the net resultant moment on the seal due to pressure forces is then.

$$F_{ms_{pressure}} = \frac{1}{2} b (\ell_6^2 + \ell_5^2)(p_{ss} - p_b)$$
$$\qquad (13)$$
$$- \frac{1}{2} b \ell_5^2 (p_{ss} - p_a)$$

according to the sketch in Fig. 5.
A more refined analysis recognizes that the pressure acting in the stern seal planing surface varies from the cushion pressure at the forward hinge to atmospheric pressure at the aft tip (since that pressure must match the ambient pressure at the exit). This variation has been represented in an approximate model, which is not illustrated here for purposes of brevity.

The moment due to the seal effective mechanical spring (arising from the elasticity of the stiffening members) is represented by

$$F_{ms_{spring}} = -k_{ss} \left(\theta_{ss} - \theta_{ss_{preload}} \right) \qquad (14)$$

where k_{ss} is the spring constant and $\theta_{ss_{preload}}$ is the seal angle at which the spring is uncompressed. The preload angle in the various computations that were made when using this model was taken to be the same as the seal downstop angle. There is also a moment due to gravitational forces (i.e. weight of seal component terms) which is left out here for purposes of convenience and simplicity.

4.1 Aerodynamic Flow Relations

The pressure in the stern seal is calculated on the basis of an assumed isentropic process from atmospheric pressure to its own value, according to the relation

$$\frac{p_{ss}}{p_a} = \left(\frac{\rho_{ss}}{\rho_a} \right)^\gamma \qquad (15)$$

where ρ_{ss} is the mass density of the air in the stern seal and $\gamma = 1.4$ is the ratio of specific heats for air. This equation is rewritten in the form

$$p_{ss} = p_a \left(\frac{m_{ss}}{Vol._{ss} \, \rho_a} \right)^\gamma \qquad (16)$$

where m_{ss} and $Vol._{ss}$ are the mass of air and the total internal volume of the stern seal, respectively.

The remaining equations associated with the seal pressure are

$$\dot{m}_{ss} = \rho_a\left(Q_{in_{ss}} - Q_{out_{ss}}\right) \qquad (17)$$

with

$$Q_{in_{ss}} = f(p_{ss} - p_a) \qquad (18)$$

where the function $f(p_{ss} - p_a)$ is obtained from the stern seal fan maps, with the particular fan, flow rate, etc. properties established for the case being considered. The leakage flow $Q_{out_{ss}}$ is given by

$$Q_{out_{ss}} = A_{eff}.\sqrt{\frac{2\,(p_{ss} - p_b)}{\rho_a}} \qquad (19)$$

where $A_{eff.}$ is the "effective" leakage area between the stern seal and the main cushion plenum, which includes the orifice coefficient for the particular orifice between these regions. The seal volume for a multi-lobe bag stern seal is given by

$$Vol._{ss} = \frac{n_{lobe}b}{2}\left[\frac{c_{ss}^2\tan\left(\dfrac{\theta_{ss}}{2n_{lobe}}\right)}{2+\left(\pi+\dfrac{\theta_{ss}}{n_{lobe}}\right)\tan\left(\dfrac{\theta_{ss}}{2n_{lobe}}\right)}\right]$$

$$(20)$$

where
n_{lobe} = number of bag lobes (n = 2 in present study)
c_{ss} = circumferential length of one lobe

Equations (15)-(20) are used in determining the seal pressure via a differential equation in terms of various seal parameters. This dynamic model of stern seal pressure is thereby available for the present type of stern seal considered here. For other types of seals, such as the bag type used in many other SES craft designs, which do not have any dynamic representation of the seal motion, the general equations given above for determining the stern seal pressure in terms of inflow and outflow, etc. are directly applicable, with the appropriate geometry, volume, etc. characteristics employed.

5. SEMI-FLEXIBLE PLANING BOW SEAL

One of the concepts proposed for the 3,000 ton Rohr Marine SES design was a semi-flexible planing bow seal, as illustrated in Fig. 6. This seal had a uniform straight across bag, to which a set of individual planers was attached. The planers were joined together by flexible joints, and supported by flexible stays in front, a geometry strap in the middle, and retraction straps near the lower end. The lower end of each planer element had a "feather" made up of a tapered fiberglass panel, which were in contact with the water surface and had a high wear resistance. These feathers were designed to deflect easily in waves, so as to ride easily on the wave surface.

In order to analyze this complex configuration, a number of simplifying assumptions were made, using the representation of the bow seal in Fig. 7. The procedure that was established allowed the loads due to the planer and the loads due to the seal bag to be separately modeled. It was assumed that the load acting on the bag due to the planer is equivalent to an increased external pressure on the bag forward arc $\overline{AD}$ in Fig. 7), with this increased pressure equal to the difference between cushion pressure and atmospheric pressure. This would make the pressure difference $(p_{bs} - p_b)$ the same across the forward (A – D) and aft (C – A) arcs of the bag, so that in equilibrium they each had the same radius of curvature.

The planer feather was assumed to fold at the intersection with the water surface, and to lie along the water surface. Thus the planer element in contact with the water behaves like a membrane, and the hydrodynamic load on the planer is assumed to be equal and opposite to that due to the pressure difference across the wetted

315

portion. This is given by

$$F_{z_p} = -(p_b - p_a) \, b d_p \cotan(\theta_{bs}) \quad (21)$$

where d_p = immersion of planer tip below the water surface if no planer deformation occurred, and where the negative sign is due to vertical forces being positive downward in this analysis. The drag force on the planer due to skin friction drag is neglected here. The angle θ_{bs} is limited by the seal downstop angle.

In the analysis of the bag, it is assumed that the water surface is an inclined straight line with inclination calculated from the water elevation differences at the x-location of the forward and aft bag attachment points (D and C in Fig. 7). As part of this procedure it is also assumed that the bow bag does not deform the water surface during bag impact and immersion.

The bow bag loads are found from the deformed bag geometry, using the notation in Fig. 8 to define the geometry. There are five unknown variables; R, β_A, β_F, x_2 and x_3, with the remaining values known from craft geometry, the immersion at the reference station (d_{bow}) and the water slope relative to the craft baseline (α). To solve for these variables there are five equations. With the total perimeter of the bag cross-section denoted as s_p, then

$$s_p = R(\beta_A + \beta_F) + x_2 - x_3 \quad (22)$$

and from geometry

$$x_3 = x_4 + R\sin(\pi - \beta_A)$$
$$x_2 = x_1 - R\sin(\pi - \beta_F)$$
$$z_1 = R + R\cos(\pi - \beta_F) \quad (23)$$
$$z_4 = R + R\cos(\pi - \beta_A)$$

Combining these equations gives :

$$s_p = R(\beta_A - \sin\beta_A + \beta_F - \sin\beta_F) + x_1 - x_4$$

$$(24)$$

$$\beta_A = \cos^{-1}\left(\frac{R - z_4}{R}\right) \quad (25)$$

$$\beta_F = \cos^{-1}\left(\frac{R - z_4}{R}\right) \quad (26)$$

which are sufficient to iteratively determine R, β_A and β_F. From these, x_2 and x_3 are then found, and then the wetted length of the bag and the bag cross-sectional area are given by

$$\ell_w = x_2 - x_3 \quad (27)$$

$$A_{bag} = \frac{1}{2}[R^2(\beta_F - \sin\beta_F + \beta_A - \sin\beta_A)$$
$$+ z_k(x_1 - x_4) + z_4(x_k - x_3) + z_1(x_2 - x_1)$$

$$(28)$$

All of the expressions above apply to conditions where the bag is not sufficiently deformed that hard structure water contact will occur. In such cases the aft attachment point is under water, and the bag is assumed to be crushed along the craft structure. Appropriate expressions for the bag wetted length and cross-sectional area are also found for those cases. In all cases, once the bag cross-sectional area and wetted length are known, the vertical force due to the bow seal bag is given by

$$F_{z_{bag}} = -(p_{bs} - p_a) \int_{-b/2}^{b/2} \ell_w \, dy \quad (29)$$

5.1 Bow Seal Bag Pressure

The bow seal bag pressure is calculated using the instantaneous bag volume and assuming the air in the bag is compressed isentropically, i.e.:

$$p_{bs} = pa\left(\frac{m_{bs}}{\rho_a V_{bs}}\right)^\gamma \quad (30)$$

where V_{bs} is the seal volume given by

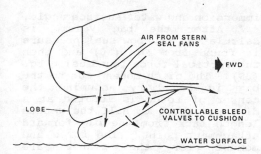

Fig.1 Stern seal schematic

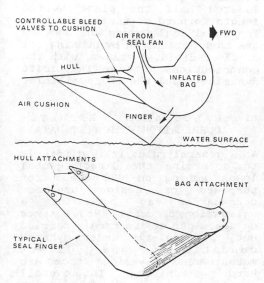

Fig.2 Bow seal schematic

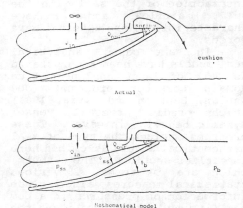

Fig.3 Representation of semi-flexible stern
seal and equivalent mathematical model

$$V_{bs} = \int_{-b/2}^{b/2} A_{bag} \, dy \qquad (31)$$

The quantity m_{bs} is the mass of air in the bow seal bag which is found from solution of the differential equation of mass conservation, i.e.

$$\frac{dm_{bs}}{dt} = \rho_a \left[Q_{in} - A_{bs} \sqrt{\frac{2\,(p_{bs} - p_b)}{\rho_a}} \right] \qquad (32)$$

where ρ_a = mass density of atmosphere air

A_{bs} = effective leakage area between seal and cushion including orifice corrections

Q_{in} = total volumetric bow seal inflow from fans

For this bow seal design, as shown in Fig. 7, the primary leakage path is provided by a series of slits on the aft arc of the seal bag. As the bag compresses, the lower portion of these slits are closed off thereby giving a time varying leakage area. In order to model this, the leakage area for a single slit was given as:

$$A_{bs_{slit}} = r \beta_A b_{slit} \qquad (33)$$

with a maximum slit length (and therefore area) determined by the seal construction.

5.2 Computer Implementation of the Bow Seal Model

In order to implement this model for use in the digital computer program for SES craft motions and loads (Kaplan, Bentson and Davis, 1981) the solution for the bag wetted length, cross-section area and slit length were found for a range of drafts and water slope angles. These calculations are performed off-line using the forward bag attachment point (x_1) as the reference longitudinal station at which the draft is measured. Values for the wetted length and cross-sectional area are then stored in two-dimensional data arrays in the bow seal subroutine and a table interpolation is performed to get the actual values at any instant of time given the instantaneous

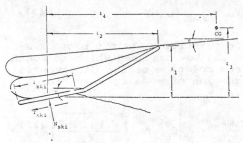

Fig.4 Forces and geometry of stern seal representation

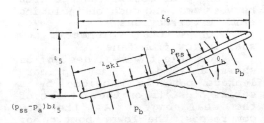

Fig.5 Stern seal pressure forces

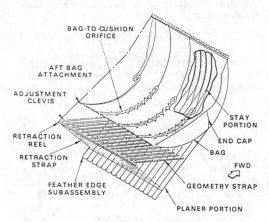

Fig.6 Planning bow seal

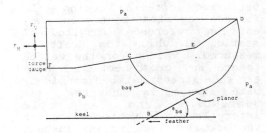

Fig.7 Bow seal elements

immersion and water surface angle.

After the bag volume is calculated, the bow seal pressure is found from Eq. (30) and then the vertical force found using Eq. (29). The pitch moment due to the bag is found by assuming the center of force is located at a distance equal to half the wetted length back from the bag forward attachment point. The force and moment for the planer are calculated in a similar manner, with the planer longitudinal center of pressure assumed to be located half the plane wetted length forward of the planer tip. The total seal forces and moments are then determined by summing all of the individual values, with the contributions to the overall craft dynamics model thereby available.

6. CALCULATED RESULTS-CORRELATION WITH TEST DATA

As a general rule, it is difficult to determine the direct global loads acting on SES seals for purposes of design use or correlation as part of a validation process. Model tests on seals per se are rarely carried out, particularly under dynamic conditions, and direct measurements of various forces are hard to achieve. The overall problem area of seal behavior, as described above, involves consideration of the dynamic nature of the seal itself, the interaction of the seal with the hull, airflow and leakage characteristics, and external wave forces.

In same cases, analytical seal models have been incorporated as part of large nonlinear time domain computer programs (see Kaplan, Bentson and Davis, 1981) which predict overall vessel dynamic behavior in waves. These programs include the type of seal mathematical models described here in earlier sections. The outputs, which are in terms of various statistical measures, also include information on pressures in a bow bag (if used), the cushion and the stern seal. Examples of such results, as given by (Kaplan, Bentson and Davis, 1981), are shown below in Tables 1 and 2.

TABLE 1   Comparison of model test and computer prediction.
Full-scale conditions: sea-state 5,50 knots (measured/prediction)

Variable	Mean	Rms	Peaks $^{1}/_{10}$ Amplitude	Peaks $^{1}/_{3}$ Amplitude	Troughs $^{1}/_{10}$ Amplitude	Troughs $^{1}/_{3}$ Amplitude
Wave, in.	0.0/0.0	1.15/1.12	3.10/2.70	2.39/2.23	2.78/2.67	2.19/2.22
Pitch, deg	1.27/1.06	1.91/1.92	5.29/4.99	4.02/4.10	4.57/4.06	3.85/3.43
Heave, in.	1.61/1.80	1.06/1.13	2.68/2.62	2.07/2.02	2.72/2.84	2.20/2.36
Bow acceleration, g	0.0/0.0	0.53/0.62	1.93/1.68	1.12/1.28	1.30/1.48	0.820/1.30
Knuckle acceleration, g	0.0/0.0	0.38/0.44	1.45/1.32	0.880/1.13	0.886/1.01	0.656/0.890
CG acceleration, g	0.0/0.0	0.26/0.23	1.07/0.905	0.602/0.641	0.594/0.525	0.422/0.436
Bow seal pressure, lb/ft^2	8.28/8.13	2.99/2.84	12.54/10.46	7.53/7.30	6.03/6.31	4.62/5.15
Cushion pressure, lb/ft^2	7.66/7.57	3.00/2.96	12.44/10.21	7.28/7.32	6.78/6.92	5.06/5.79
Stern seal pressure, lb/ft^2	12.59/10.63	3.60/3.31	13.08/8.38	8.38/6.38	8.15/8.11	5.49/6.38

TABLE 2   Comparison of model test and computer prediction.
Full-scale conditions: sea-state 6,50 knots (measured/prediction)

Variable	Mean	Rms	Peaks $^{1}/_{10}$ Amplitude	Peaks $^{1}/_{3}$ Amplitude	Troughs $^{1}/_{10}$ Amplitude	Troughs $^{1}/_{3}$ Amplitude
Wave, in.	0.0/0.0	1.51/1.46	4.15/3.51	3.14/2.92	3.76/3.47	2.93/2.90
Pitch, deg	1.34/1.37	2.61/2.70	7.23/6.73	5.50/5.62	6.26/5.42	5.08/4.70
Heave, in.	1.86/2.38	1.32/1.46	3.20/3.07	2.53/2.50	3.59/3.84	2.73/3.22
Bow acceleration, g	0.0/0.0	0.65/0.76	2.59/2.19	1.55/1.66	1.55/1.68	1.02/1.42
Knuckle acceleration, g	0.0/0.0	0.46/0.53	1.84/1.64	1.13/1.25	0.993/1.11	0.742/0.968
CG acceleration, g	0.0/0.0	0.29/0.27	1.29/1.08	0.719/0.728	0.705/0.633	0.504/0.501
Bow seal pressure, lb/ft^2	8.10/7.66	3.57/3.31	16.05/12.89	9.76/8.86	6.38/6.33	5.06/5.42
Cushion pressure, lb/ft^2	7.21/6.97	3.44/3.29	15.82/12.19	8.87/8.22	7.29/6.97	5.36/5.88
Stern seal pressure, lb/ft^2	12.67/10.54	4.25/3.96	15.84/13.22	9.83/8.39	9.36/9.28	6.60/7.57

The degree of agreement shown there provides adequate support for the validity of the model representation, from which details related to seal behavior can be extracted from within the overall computer simulation model.

There are also simpler models of SES motion responses in waves (e.g. McHenry et al, 1991) wherein much simpler seal models are also included. That type of analysis, via frequency response methods, can provide useful estimates of overall vessel response. However there is insufficient detail regarding all of the various aspects of seal forces, deflections, internal flows, etc. in such models to provide assistance for seal design. The main influence of the seals in such models is to provide their basic sealing behavior, without any concern as to their detailed properties.

6.1 Specialized Experimental Studies

In order to obtain some insight into the slam alleviation properties of bow seal bags, some special model tests were conducted under controlled conditions. Such tests were conducted on SES models with bow seal bags, using a vertical oscillator to impose known vertical velocities at the bag location while the craft is moving at forward speed. A

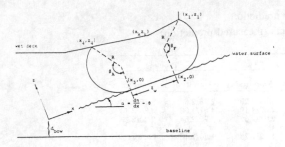

Fig.8 Geometry of deformed bow seal bag

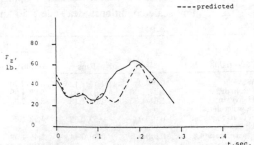

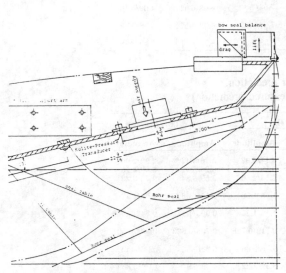

Fig.9 Bow seal, including measurement
instruments

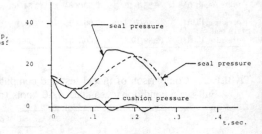

Fig.10 Correlation results for u=13.6ft/sec,
θ=3.13°, n=63.6 rpm, d=3 in.; Run36,
Point 105

description of the procedure, and some early representative results from such tests, is given in (Kaplan, Davis and Malakhoff, 1976). More extensive model tests were also carried out on a model equipped with the semi-flexible planing bow seal described above in Section 5.

These latter tests (Kaplan and Bentson, 1978) were carried out over a range of forward speeds, with a vertical oscillator applying forced pitch motion over a range of angular amplitudes and frequencies. A pictorial illustration of the model bow seal, including the measuring instruments, is given in Fig. 9 (a scale of 1:30 was used here). The quantities measured were the total

vertical and horizontal forces acting on the bow segment, the bag pressure, and the vertical and horizontal force components acting on the seal forward hinge point.

The observed test data showed that the maximum value of vertical force on the bow segment is essentially independent of forward speed, for non-zero forward speed conditions. Values obtained at zero speed differ from those at forward speed due to a different bag deformation, arising from the lack of hydrodynamic loading of the planer. The maximum value of this vertical force occurs at the same time as the maximum bow seal pressure. For lower values of pitch amplitude (where hard structure impact did not occur), the maximum pressure at non-zero forward speed was essentially independent of forward speed. These experimental findings support the assumption of the dependence of bag pressure, bag geometry, forces etc. on only "pneumatic" effects due to air pressure and related effects, with no direct dependence on

TABLE 3   Comparision of Predicted and Measured Bow Seal Hinge Loads

Case	C.G. Immersion (in.)	Indicated Trim at Max. Load (deg.)	Oscillation Frequency (rad./sec.)	Speed (fps)	$P_{bag_{exp.}}$ (psf)	$F_{V_{exp.}}$ (lb.)	$F_{V_{predict.}}$ (lb.)	$F_{H_{exp.}}$ (lb.)	$F_{H_{predict.}}$ (lb.)
1	3	-2.88	60	17.0	45	44	42.7	-43.1	-42.3
2	4	-2.65	40	13.6	31	25.1	29.4	-29.9	-29.3
3	4	-2.61	40	17.0	40	34.5	37.9	-36.7	-37.8
4	4	-2.18	70	10.2	38	33.6	36.5	-33.2	-36.6
5	3	-2.83	40	13.6	31	29.4	29.5	-32.1	-31.7

hydrodynamic effects for this type of bow seal.

A representative comparison between theory and experiment for this forced oscillation bag impact testing is shown in Fig. 10, which illustrates the good agreement between theory and experiment for the maximum force and maximum pressure. The comparisons between predicted and measured bag pressure time histories show the measured time history exhibited an earlier pressure rise than the predicted results (see Fig. 10). This is probably due to the model of the planer chosen here. An examination of the results of the computer simulation indicated that the experimental pressure rise starts at a point prior to the first indication by the computer program of bag-water impact. This is probably due to the action of the rigid planer on the bag, either through direct compression of the bag or through the reduction of the bag leakage slit length as the planer moves upward. With the membrane model of the planer used here neither of these effects is accounted for, and the pressure will not rise until the bag impacts the water (provided the cushion pressure is not increasing).

For the cases involving large amplitude motion, the predicted time histories showed good agreement only for the early portion of the time history. This is ascribed to two different effects, viz. the presence of

hard structure impact (whose force effect is not included within the present theoretical model) as well as the influence of the non-rigidity of the oscillator test rig, with an indication of unknown values of the resulting vertical motion during the tests which thereby preclude proper simulation and comparisons.

6.2 Bow Seal Hinge Loads

As a consequence of this problem, and other related measurement problems, the correlation of the seal hinge forces was performed using the *measured* pressures in conjunction with the predicted bag geometry to obtain predicted seal hinge loads. With the cushion pressure approximately equal to atmospheric pressure during most of the bag impact portion of the pitch cycle, the maximum seal hinge load is due primarily to the membrane tension in the bag. For a membrane, the tension is related to the pressure differential across the membrane by the relation

$$T = (p_{bs} - p_a)Rb \qquad (34)$$

where R is the radius of curvature of the forward bag segment. With the tension computed, it is then decomposed into vertical and horizontal components by using the bag geometric solution to find the bag angle at the hinge. The comparison of measured and predicted values for the maximum hinge load, when comparing both components at the maximum load point during bag impact, is shown in Table 3. This comparison, which involves predicted values of the bag radius of curvature and the bag angle relative to the body shows good agreement of the predicted forces compared to the measured values for these maximum hinge forces.

7. CONCLUSIONS

The mathematical models described herein are representative of methods used for analysis and prediction of SES seal loads, as part of the detailed computer simulation procedures that were established as part of the U.S. Navy SES development program. Results obtained from the overall study of vessel loads and motions were supportive of the general complete simulation methodology. On the basis of that successful use of the theoretical modeling, in addition to the effects on craft overall loads and motions, the mathematical models for seals illustrated here can be applied to determine design values of seal forces. The predictions of such forces can be made for standard operation in waves, as well as during impacts, so that appropriate design values of seal forces can be established on the basis of a rational theory.

REFERENCES

International Ship Structures Congress (ISSC), Tech. Comm. V.4, (Surface Effect Ships) Report, 1994.

Kaplan, P., Davis, S. & Malakhoff, A. 1976. The use of oscillators to obtain hydrodynamic and structural load data for SES craft. *Proc. Eleventh Symp. on Naval Hydrodynamics, ONR, Univ. College London*.

Kaplan, P. and Bentson, J. 1978. Further studies of the effect of a flexible bow seal on SES slam pressures and loads. *Oceanics, Inc. Rpt. 78-147E*.

Kaplan, P., Bentson, J. and Davis, S. 1981. Dynamics and hydrodynamics of surface effect ships. *Trans. SNAME*.

Malakhoff, A. and Davis, S. 1981. Dynamics of SES bow seal fingers. *AIAA 6th Marine Systems Conf., Seattle, Wash*.

McHenry, G., Kaplan, P., Korbijn, F. & Nestegard, A. 1991. Hydrodynamic analysis of surface effect ships; experiences with a quasi-linear model. *Proc. FAST'91, Trondheim*.

Hydroelasticity in Marine Technology, Faltinsen et al. (eds) © 1994 Balkema, Rotterdam, ISBN 90 5410 387 6

Hydroelastic behavior of the flexible bag stern seal of a SES

Tore Ulstein & Odd Faltinsen
The Norwegian Institute of Technology, Trondheim, Norway

ABSTRACT: A numerical and theoretical study of the interaction between a flexible bag structure and the free water surface is presented. The flexible bag is the stern seal of a SES. A two-dimensional analysis of the problem is used. The hydrodynamic part of the problem has similarities with the linearized unsteady foil problem. The unsteady deformation of the bag is found by numerical time integration. High numerical accuracy is needed. This has been achieved by use of "dry" normal modes for the flexible behavior of the bag in combination with extensive use of analytical expressions for the excitation and reaction forces. It has been found that it is necessary to account for the coupling between the elastic longitudinal and transverse oscillations as well as the curvature of the bag. Results indicate the importance of taking into account the interaction between the bag structure and the free water surface.

1 INTRODUCTION

During the later years an increasing interest for high speed marine vessels is observed. One concept out of many, is the surface effect ship (SES). The idea with this concept is to partly lift the vessel out of the water by trapping an air cushion between two catamaran hulls, a bow skirt and a stern seal (see Figure 1). A consequence is that the resistance of a typical SES is lower than the resistance of a similar sized catamaran in most sea states of practical interest.

A problem with the SES has been and still is, the cobblestone effect. This is a resonance phenomenon that occurs in low sea states due to the compressibility of air in the air cushion. The cobblestone effect is excited because the water waves dynamically changes the air cushion volume. This resonance phenomenon occurs at high frequencies compared to the resonance frequencies for the rigid body motions of conventional ships. The two lowest resonance frequencies in the air cushion of a 30-35 m long SES are approximately 2 Hz and 5 Hz, respectively. Due to the frequency of encounter effect there are waves with sufficient en-

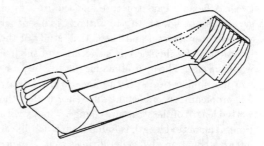

Figure 1: Sketch of an SES air cushion with a bow finger seal and a 3-loop flexible bag seal aft. Toyama, Ono and Nishihara (1992).

ergy in small sea states that excite these resonance oscillations. The eigenfunction for the dynamic air cushion pressure is constant in space for the lowest eigenfrequency and represent acoustic resonances for the higher eigenfrequencies. The rigid ship motions in this frequency region are small, but the vertical acceleration level is high. The hydrodynamic damping due to the rigid ship motions is negligible in this frequency range. Other damping mechanisms have to be considered such as the air flow into the air cushion through the fans and the air leakage underneath the seals and through louvers that are part of a ride control system. Steen (1993) studied the cobblestone effect, and found that the dynamics of the stern seal bag was important for the global acceleration level in low sea states. He considered the effect of a dynamic varying leakage area underneath the seal together with the deformation of the bag due to change in the air cushion pressure at the stern. The deformation of the bag was analyzed quasi-statically. The dynamic varying leakage area and the deformation of the bag will have a similar effect on the air-cushion as a moving piston at the end of a long tube. Acoustic waves have been shown to be excited by this mechanism, which in turn generate high vertical accelerations. Steen neglected the hydroelastic interaction between the bag stern seal and the free water surface. By hydroelastic interaction we mean that the hydrodynamic loading is a function of the structural deformations resulting from the hydrodynamic loading. This interaction is focused on in this paper and is believed to have an important effect on the cobblestone oscillations.

An analysis that accounts for the interaction between a flexible bag structure and the free water surface is not known to the authors. This paper presents a numerical and theoretical study of the interaction of a flexible bag structure and the free water surface. The hydrodynamic part of the problem has similarities with the linearized unsteady foil problem. An important difference is that the wetted length of the structure changes rapidly with time. The wetted length is found from a non-linear integral equation, by generalizing what Wagner (1932) did in the case of slamming.

The bag structure is pressurized with air, and is deformed due to the hydrodynamic pressure distribution on the wetted surface of the bag and the compressibility of air in the bag. The unsteady deformation of the bag is found by a numerical time integration. High numerical accuracy is needed. This has been achieved by use of "dry" normal modes for the flexible behavior of the bag in combination with extensive use of analytical expressions for the excitation and reaction forces. By "dry" normal modes it is meant that the eigenvalue problem is solved without taking into account the hydrodynamic reaction forces and the pressure forces due to the compressibility of air in the bag. It has been found that it is necessary to

account for the coupling between the elastic longitudinal and transverse oscillations as well as the finite radius of curvature of the bag.

Results presented in Figure 8 for the time derivative of the volume change of the air cushion due to the bag, indicate that the interaction between the free water surface and the bag structure is important for the excitation of the cobblestone oscillations.

2 STRUCTURAL MODELLING OF THE BAG

It is necessary to make simplifications in the analysis of the bag structure . The bag structure is modelled as a two-dimensional cable. The effect of gravity is disregarded, but not the inertia due to dynamic accelerations. This can be done due to the excess pressure inside the bag. In the static case, the pressure force acting on the bag is an order of magnitude larger than the gravity force. In the dynamic case it can also be shown that the contributions from gravity will appear as restoring terms and will be negligible compared to the tension terms. With these simplifications the bag structure can now be modelled as a weightless but not massless cable.

2.1 *The static solution*

The bag geometry is shown in Figure 2. In order to simplify the problem, it is assumed that the bag has only one loop. The bag geometry consist of two circle segments with different radii, because of different pressure differences over the segments. Force equilibrium considering only pressure and tension, results in two equations. The length L, and the height H, are known geometric parameters. These parameters are used to set up two additional equations. It is assumed that the tangent to the loop is continuous and horizontal at the connecting point of the two circle segments. The problem now consists of 4 equations and 4 unknowns. The unknowns are T_0, R_1, R_2 and θ_1, where T_0 is the constant tension in the cable. R_1 and R_2 are the radii of curvature for the two cable segments and θ_1 is the angle as shown in Figure 2.

A solution is now obtained by eliminating all the unknowns, except for R_1. This results in a nonlinear equation in R_1. This nonlinear equation is solved for R_1, by an accellerated iteration proce-

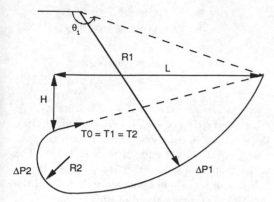

Figure 2: Static geometry of the one lobe bag. It is assumed to consist of two circle segments with constant curvature. The tension in the two segments are constant and equal.

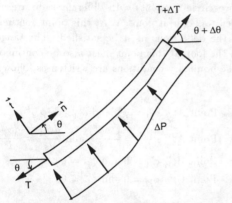

Figure 3: Infinitesimal element where tension and pressure forces are shown together with the definition of the coordinate system.

dure. When R_1 is found the other variables are found from back-substitution.

2.2 Linearized dynamics of the bag structure

In this section it is focused on the linear dynamics of the simplified bag structure. It is assumed small perturbations on the static solution. First in this section the governing linearized equations of motions with coupled transverse and axial motions of the cable are derived. Then in the following sections it is focused on the eigenvalue problem using "dry modes". In the last section linear response is studied.

2.2.1 Derivation of equations of motions

To derive the equations of motion for a cable, force equilibrium of an infinitesimal element must be satisfied. (See Figure 3.)

The orthogonal coordinate system that is used is fixed to the static geometry, where the longitudinal coordinate is pointing in the tangential direction of the cable. The force equilibrium is decomposed in the longitudinal and the transverse directions.

The equations of motions are now derived by making a perturbation on the known static solution. This is done by expanding the variables in a time independent static and a time dependent dynamic part. All second order terms like products of perturbation state variables are neglected. Compatibility gives a relationship between tension and the motion in the transverse and the longitudinal directions. It is further assumed that the radius of curvature is constant and equal to R.

The equations of motions can then finally be written as (See Bliek (1984)),

$$\rho\ddot{\eta}_n = T_0\frac{\partial^2 \eta_n}{\partial s^2} + \frac{(T_0 + EA)}{R}\frac{\partial \eta_t}{\partial s} - \frac{EA}{R^2}\eta_n + \Delta P \quad (1)$$

and

$$\rho\ddot{\eta}_t = EA\frac{\partial^2 \eta_t}{\partial s^2} - \frac{EA}{R}\frac{\partial \eta_n}{\partial s} \quad (2)$$

where η_n and η_t are respectively the motion in the transverse and longitudinal directions. T_0 is the static tension in the cable, E the elasticity modulus of the material and A the cross dimensional area of unit width. s is the longitudinal coordinate and ΔP is the dynamic pressure acting on the cable. ΔP includes the effect of internal and external pressure. Dot stands for time derivative. We will use equations (1) and (2) on the two circle segments of different radius.

2.2.2 The eigenvalue problem

In this section it is focused on how the eigenvalues and the corresponding mode shapes are found. We consider a solution of the form,

$$\eta_n(s,t) = e^{i\omega_j t}\phi_j^n(s)$$

$$\eta_t(s,t) = e^{i\omega_j t}\phi_j^t(s) \quad (3)$$

where i is the complex unit, t is the time variable, ω_j is the circular eigenfrequency and $\phi_j^n(s)$ and $\phi_j^t(s)$ are respectively the mode shapes in the transverse and longitudinal directions of vibration mode j. The equations in (3) are substituted into equations (1) and (2) where ΔP is set equal to zero.

This results in a set of two coupled homogeneous differential equations with constant coefficients that can be solved analytically. $\phi_j^n(s)$ and $\phi_j^t(s)$ are assumed to be proportional to $e^{D_j s}$. Using this assumption, the determinant of the equation system is set equal to zero. The roots of the resulting equation give the relation between the eigenfrequencies and the wavenumbers, D_j.

It is assumed that the $\frac{T_0}{EA} \ll 1$. Given this assumption, $D_j = \pm k_j$ and $D_j = \pm \mu_j$ are four possible roots and can be written as follows,

$$k_j^2 = \frac{-\rho\omega_j^2(1 + \frac{T_0}{EA} + K_j^1 + K_j^2)}{2T_0} \quad (4)$$

and

$$\mu_j^2 = \frac{-\rho\omega_j^2(1 + \frac{T_0}{EA} + K_j^1 - K_j^2)}{2T_0}. \quad (5)$$

Here $K_j^1 = \dfrac{T_0}{\rho\omega_j^2 R^2}$ and

$$K_j^2 = \sqrt{1 + 6K_j^1 + (K_j^1)^2}$$
$$+ \frac{(K_j^1 - 1)}{\sqrt{1 + 6K_j^1 + (K_j^1)^2}} \frac{T_0}{EA} + O((\frac{T_0}{EA})^2).$$

Since R is finite we cannot neglect K_j^1 for all values of ω_j. By studying equations (4) and (5), it can be seen that μ_j will be real for low frequencies while for higher frequencies it will become imaginary. The k_j-value is always imaginary. The solution can therefore be written as follows;
μ_j is real,

$$\phi_j^n(s) = \quad c_{j1}\cos k_j s + c_{j2}\sin k_j s$$
$$+ \quad c_{j3}\cosh \mu_j s + c_{j4}\sinh \mu_j s$$

$$\phi_j^t(s) = \quad c_{j5}\cos k_j s + c_{j6}\sin k_j s$$
$$+ \quad c_{j7}\cosh \mu_j s + c_{j8}\sinh \mu_j s \quad (6)$$

and μ_j is imaginary,

$$\phi_j^n(s) = \quad c_{j1}\cos k_j s + c_{j2}\sin k_j s$$
$$+ \quad c_{j3}\cos \mu_j s + c_{j4}\sin \mu_j s$$

$$\phi_j^t(s) = \quad c_{j5}\cos k_j s + c_{j6}\sin k_j s$$
$$+ \quad c_{j7}\cos \mu_j s + c_{j8}\sin \mu_j s. \quad (7)$$

The relationship between the coefficients c_{j1}, c_{j2}, c_{j3}, c_{j4} and c_{j5}, c_{j6}, c_{j7}, c_{j8} is found by substituting the solutions in equation (6) or (7) into the governing equations (1) and (2) with $\Delta P = 0$. The terms proportional to either sine, cosine, hyperbolicsine or hyperboliccosine are collected and the resulting coefficients ahead of each of these terms are set equal to zero.

In order to find the eigenfrequencies and the four remaining coefficients, the boundary conditions have to be used. The geometry is as defined in Figure 4.

Two circle segments with different radii are coupled together at point A. At this point continuity of the deflections must be satisfied. The tangent to the loop at this point must also be continuous. The boundary conditions are written as follows,

Point A: $\quad \phi_{j1}^n = \phi_{j2}^n$ and $\phi_{j1}^t = -\phi_{j2}^t$

Point A: $\quad \dfrac{d\phi_{j1}^n}{ds} = -\dfrac{d\phi_{j2}^n}{ds}$ and $\dfrac{d\phi_{j1}^t}{ds} = \dfrac{d\phi_{j2}^t}{ds}$

Point B: $\quad \phi_{j1}^n = 0$ and $\phi_{j1}^t = 0$

Point C: $\quad \phi_{j2}^n = 0$ and $\phi_{j2}^t = 0$. $\quad (8)$

where the second subscript 1 and 2 denotes respectively circle segment 1 and 2.

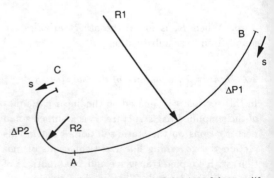

Figure 4: Static bag geometry with definition of the two different circle segments. The longitudinal coordinates used, are also shown for the two segments.

The solution described by equation (6) or (7) is now substituted into the boundary conditions defined in equation (8) for the points A, B and C. An equation system of eight equations with eight unknown coefficients are obtained (four unknowns respectively both for circle segment 1 and 2). In order to find the eigenfrequencies the determinant of this equation system must be equal to zero. The roots of this polynomial is found numerically by an accelerated iteration scheme.

When the eigenfrequencies are found, mode shapes and the unknown coefficients are found by setting $c_{j1}^1 = 1$. Here the superscript 1 denotes circle segment 1. Further the coefficients are all normalized with respect to the largest coefficient.

2.2.3 The linear response

Now the eigenfrequencies and their mode shapes are known. They are used to find the response by modal superposition. It is assumed that the eigenfunctions represent a complete set of solutions and that the motions in the transverse and longitudinal directions therefore can be written as follows,

$$\eta_n(s,t) = \sum_{i=1}^{\infty} a_i(t)\phi_i^n(s) \qquad (9)$$

and

$$\eta_t(s,t) = \sum_{i=1}^{\infty} a_i(t)\phi_i^t(s). \qquad (10)$$

Here $a_i(t)$ is the principal coordinate of vibration mode number i.

Equation (9) and (10) are substituted into the governing equations (1) and (2). The two equations are multiplied with ϕ_j^n and ϕ_j^t, respectively and integrated over the length L, of the cable. The two coupled equations of motions are added and the following result is obtained;

$$[M_{ji}^n + M_{ji}^t]\ddot{a}_i + [C_{ji}^n + C_{ji}^t]a_i = \int_0^L \Delta P \phi_j^n ds \qquad (11)$$

where

$$M_{ji}^n = \rho \int_0^L \phi_i^n \phi_j^n ds$$

$$C_{ji}^n = -T_0 \int_0^L \frac{d^2\phi_i^n}{ds^2}\phi_j^n ds$$

$$-(T_0 + EA)\int_0^L \frac{1}{R}\frac{d\phi_i^t}{ds}\phi_j^n ds$$

$$+EA \int_0^L \frac{1}{R^2}\phi_i^n \phi_j^n ds$$

$$M_{ji}^t = \rho \frac{(T_0 + EA)}{EA} \int_0^L \phi_i^t \phi_j^t ds$$

$$C_{ji}^t = -(T_0 + EA)\int_0^L \frac{d^2\phi_i^t}{ds^2}\phi_j^t ds$$

$$+(T_0 + EA)\int_0^L \frac{1}{R}\frac{d\phi_i^n}{ds}\phi_j^t ds.$$

Equation (11) is used for the numerical time integration.

In general the eigenfunctions do not satisfy the orthogonality condition,

$$\int_0^L [M_{ji}^n + M_{ji}^t]ds = 0, \qquad (12)$$

but other orthogonality conditions may be found. Care has to be taken when integrating over the cable length due to the fact that two different coordinate systems have been used. The integrals over the length are divided into integrals extending over the two different circle segments. This is done because of the two different radii of the two segments.

The forcing term consisting of the generalized pressure on the right hand side in equation (11), can be decomposed into a term dealing with volume change of the bag-volume and a term dealing with the hydrodynamic impact pressure. Given this decomposition the following can be written,

$$\int_0^L \Delta P \phi_j^n ds$$

$$= K_{bag} \sum_{i=1}^{\infty} \int_0^L \int_0^L a_i(t)\phi_i^n(\xi)d\xi \phi_j^n(s)ds$$

$$+ \int_0^L p(s,\eta_n,t)\phi_j^n(s)ds \qquad (13)$$

where K_{bag} is defined as $\Delta pressure = K_{bag} \cdot \Delta volume$. $\Delta volume$ is the dynamic volume change that causes a dynamic pressure change ($\Delta pressure$), due to the compressibility of the air in the bag. Based on results by Steen (1993), K_{bag} is approximated by zero in the present calculations. These results show small dynamic variations in the pressure difference over the bag-segment bounding the air cushion relative to the dynamic air cushion pressure acting on the bag. This is partly the reason for approximating K_{bag} by zero. The bag-segment bounding open air will be exposed to stronger dynamic variations in the pressure difference, but is neglected in the present paper. One reason for neglecting it is due to the coupling between the bag pressure and the air cushion pressure occurring in typical bag designs. This coupling of the bag pressure and the air cushion pressure will probably reduce the effect relative to the model described in equation (13). This effect will be studied in the near future.

A fourth order Runge Kutta method is used for the numerical time integration.

2.2.4 Verification of the structural model

In order to check the structural model some tests have been carried out. The radius of curvature has been increased to a very large number in order to check if the solution both for eigenfrequencies and eigenmodes in the asymptote goes to the solution of a string.

The equations of motions that are used for the numerical time integration are also checked. These equations can be written in matrix notation as shown on the left hand side of equation (11). The eigenvalue problem for this equation system is solved in order to compare the resulting eigenvalues with the eigenvalues obtained as described in the subsection "The eigenvalue problem".

3 THE BAG AS AN UNSTEADY PLANING SURFACE

In order to study the dynamics of a given bag design, one must be able to analyze the hydrodynamic impact forces on the bag when it hits the free water surface.

The following hydrodynamic analysis will assume a two dimensional geometry in the direction of the free stream, an incompressible medium and a high Froude number F_n, of the order of magnitude ten. Here $F_n = \frac{U}{\sqrt{gl}}$ where U is the forward speed of the vessel. Relative to the bag structure this velocity appears as a free stream velocity. l is a characteristic wetted length. In the following text it will be shown that the unsteady planing problem is mathematically equivalent to the unsteady foil problem when the effect of gravity is neglected and the body boundary conditions are linearized. It is assumed that neglecting the gravity is a good approximation due to the high Froude number. The effect of gravity for the steady planing problem is discussed by Ogilvie (1970) and Wagner (1948).

Given that one disregard the gravity and linearize the body boundary conditions, one will be able to predict the transient behavior of the planing bag bouncing up and down on the free water surface. The problem is transient in its behavior, and a steady state solution is of less interest. This is due to the fact that an impact with the free water surface will excite the bag. This can also be seen from the forces acting on the bag. When it bounces out of the water, there is only the excess pressure from the air cushion acting on the bag. When it bounces into the water again, there is a large planing force acting on the bag.

3.1 The theoretical approach

The hydrodynamic loading will be much larger then the aerodynamic loading due to the relative low pressure in the air cushion and inside the bag. This indicates that the immersion of the bag will be low. The boundary conditions can therefore as a good approximation be linearized and transferred to the undisturbed free water surface. The linearization of the body boundary conditions leads to a square root singularity in the hydrodynamic pressure at the spray root in the planing problem. This singularity has been shown to model the spray root at a global scale reasonable well as long as the effective angle of attack is kept small. (See Kochin, Kibel and Roze (1955)) By this we mean that the integrated pressure over the wetted length of the bag is not destroyed by the linearization. One may therefore conclude that the linearized approach is reasonable for small perturbations of the water flow. For the case of steady planing of a flat plate this means angles of attack less then approximately ten degrees.

The wetted length of the bag will vary strongly through the impact on the free water surface. Constant wetted length would then be a much too crude approximation. One therefore has to consider a time varying wetted length in the solution of this hydrodynamic problem.

The analysis that will be carried out here, is based on a right-handed xy-coordinate system. The origin is fixed at the lowest point of the bag. The x-axis is positive towards the upstream direction and the y-axis is positive pointing upwards. The free stream velocity U, is in the negative x-direction. Since a potential flow is considered, the separation point must be given a priori. Point A in Figure 4 is chosen. The body boundary conditions are linearized and transferred to a straight line that is the undisturbed free water surface ($y = 0$).

The boundary condition on the free water surface upstream the wetted surface of the bag (from here on, only wetted surface) is,

$$\phi = 0 \qquad \text{at} \quad x > 2c(t), \qquad (14)$$

where c(t) is an approximation of the half wetted length and ϕ is the velocity potential for the water flow caused by the bag. This is a conventional type of free surface condition used both in impact and high speed planing problems.

On the wetted surface, the boundary condition is no normal velocity through the surface. The mathematical condition is,

$$\frac{\partial \phi}{\partial y} = v_0(x, t) \qquad \text{at} \quad 0 < x < 2c(t), \qquad (15)$$

where $v_0(x, t) = V(t) + \frac{\partial \eta_{bag}(x,t)}{\partial t} - U \frac{\partial \eta_{bag}(x,t)}{\partial x}$. V(t) is the relative vertical rigid body velocity of the bag relative to the free water surface, that will appear as a vertical motion of the free water surface. U is the free stream velocity of the water and $\eta_{bag}(x, t)$ describes the geometry of the bag.

It is assumed that the flow separates and leaves the wetted surface smoothly (Kutta condition) at the separation point (the origin). Another aspect of the Kutta condition is that the hydrodynamic pressure p, has to be continuous and to be equal to the atmospheric pressure at the trailing edge. On the line downstream the wetted surface ($y = 0$, $x < 0$), atmospheric pressure is therefore pre-

scribed instead of the boundary condition $\phi = 0$. The mathematical condition is,

$$(p - p_a) = 0 \qquad \text{at} \quad x < 0, \qquad (16)$$

where p is the hydrodynamic pressure and p_a is the atmospheric pressure. The flow region is now confined to the region below $y = 0$.

We will now solve this boundary value problem by using the solution for an unsteady foil in infinite fluid. In this solution a vortex sheet is shed into the wake downstream of the foil. The boundary condition in the wake is that there shall be no pressure jump across this sheet. This boundary condition can be written as,

$$-\rho(\frac{\partial \phi}{\partial t} - U\frac{\partial \phi}{\partial x})^+ + \rho(\frac{\partial \phi}{\partial t} - U\frac{\partial \phi}{\partial x})^- = 0 \qquad (17)$$

for $x < 0$. Here the superscript $-$ denotes just below $y = 0$ and the superscript $+$ denotes just above $y = 0$. By using that

$$\phi(x, 0^+, t) = -\phi(x, 0^-, t), \qquad (18)$$

we obtain

$$(p - p_a)^- = -\rho(\frac{\partial \phi}{\partial t} - U\frac{\partial \phi}{\partial x})^- = 0 \qquad (19)$$

,for $x < 0$, which is equal to the boundary condition given in equation (16). In addition it should be noted that $\phi = 0$ on $y = 0$ and $x > 0$ for the linear unsteady foil problem in infinite fluid.

It has now been shown that the artificial unsteady foil problem satisfies the boundary condition for the unsteady planing problem in the lower region below $y = 0$. The solution of the unsteady foil problem is therefore used.

Given this hydrodynamic boundary value problem (HBVP) the solution in terms of vorticity distribution on the wetted surface is given by Newman (1977), i.e.

$$\gamma(x',t) = \frac{2}{\pi[c(t)^2 - x'^2]^{1/2}} \times$$

$$\left(PV \int_{-c(t)}^{c(t)} \frac{[c(t)^2 - \xi^2]^{1/2}}{\xi - x'} v_0(\xi,t)d\xi \right.$$

$$\left. + \frac{1}{2} \int_{-c(t)-Ut}^{-c(t)} \frac{[\xi^2 - c(t)^2)]^{1/2}}{\xi - x'} \gamma(\xi + Ut)d\xi \right)(20)$$

where $x' = x - c(t)$ and PV means the principal value of the integral.

In equation (20) the lower limit $-\infty$ in the integral over the wake, is substituted with $-c(t) - Ut$, since the problem is solved in the time domain and started from rest.

It is assumed that the velocity $v_0(x,t)$ at the wetted surface can be written as follows,

$$v_0(x,t) = v_0(t) + v_1(t)\cos\theta \qquad (21)$$

where $x - c(t) = c(t)\cos\theta$. $v_0(t)$ and $v_1(t)$ are defined as,

$$v_0(t) = \frac{1}{\pi} \int_0^\pi v_0(\theta,t)d\theta \qquad (22)$$

and

$$v_1(t) = \frac{2}{\pi} \int_0^\pi v_0(\theta,t)\cos\theta d\theta. \qquad (23)$$

The vorticity in the wake is found by using the Kutta condition. The following equation is obtained,

$$\int_{-c(t)-Ut}^{0} \sqrt{\frac{\xi - 2c}{\xi}} \gamma(\xi + Ut)d\xi$$

$$= \pi c(t)(2v_0(t) - v_1(t)). \qquad (24)$$

3.1.1 Approximation of the wetted length

A time domain analysis is carried out, where the wetted length will vary from zero to a finite value. The point where the fluid separates from the bag (trailing edge) is assumed to be fixed in space at the lowest point of the static bag configuration (point A in Figure 4).

An integral equation can be set up, based on keeping track of a particle on the free water surface that

hits a point on the bag at one time instant t. This particle is convected in the free stream direction, with the free stream velocity U. This follows from the linearization of the body boundary conditions. From a time step to another, we know on beforehand the horizontal displacement of the particle, but we do not know the vertical displacement. The vertical displacement can be found from the induced vertical velocity due to the wetted surface of the bag. This vertical displacement of the free water surface relative to the undisturbed free water surface, is denoted pile-up of water.

The non-linear integral equation in the unknown wetted length $(2\,c(t))$, can be stated as follows for $\eta_{gap}(x,t) = 0$,

$$\eta_{gap}(x,t) = \eta_0 + \eta_{bag}(x,t) + \int_0^t V(\tau)d\tau$$

$$- \int_0^t v(x = 2c(t) + U(t-\tau); c(\tau))d\tau \qquad (25)$$

where η_0 is the gap at the lowest point of the bag at $t = 0$. It is assumed that the bag is flexible, and has a shape described by $\eta_{bag}(x,t)$ in the impact region,

$$\eta_{bag}(x,t) = \bar{\eta}_{bag}(x) + \hat{\eta}_{bag}(x,t) \qquad (26)$$

where $\bar{\eta}_{bag}(x)$ describes the static configuration of the bag (measured from the lowest point of the bag), and $\hat{\eta}_{bag}(x,t)$ describes the dynamic perturbation of the bag $(\hat{\eta}_{bag}(x,t) = \eta_n(s,t))$. Due to the geometry, $L_1 - s \approx x$ in the impact region. L_1 is the length of segment 1. $v(x = 2c(t) + U(t-\tau); c(\tau))$ is the vertical velocity upstream the leading edge of the wetted surface. This velocity can be written as follows,

$$v(x',t) = -\frac{1}{2\pi} \int_{-c(t)-Ut}^{c(t)} \frac{\gamma(\xi,t)}{\xi - x'}d\xi \qquad (27)$$

where $x' > c(t)$.

The vortex distribution on the wetted surface (See equation (20)) is substituted, and the order of integration is interchanged.

This is a generalized procedure of what Wagner (1932) did in the case of slamming.

3.1.2 Modal hydrodynamic force

When the wetted length and the vorticity on the wetted surface is known, the modal hydrodynamic force can be found.

The hydrodynamic pressure acting on the bag can be expressed as follows, based on Bernoulli's equation,

$$p = -\rho(\frac{\partial\phi}{\partial t} - U\frac{\partial\phi}{\partial x})^-. \tag{28}$$

Here the superscript $-$ indicates the pressure side of the "foil". The modal hydrodynamic force may now be written as,

$$\int_0^L p(s,\eta_n,t)\phi_j^n(s)ds$$

$$= -\rho \int_0^{2c(t)} (\frac{\partial\phi}{\partial t})^-(x,t)\phi_j^n(L_1-x)dx$$

$$+\frac{1}{2}\rho U \int_0^{2c(t)} \gamma(x,t)\phi_j^n(L_1-x)dx$$

$$= F_{exc,j}(t) - \sum_{i=1}^{\infty} C_{ji}a_i(t)$$

$$-\sum_{i=1}^{\infty} B_{ji}\dot{a}_i(t) - \sum_{i=1}^{\infty} A_{ji}\ddot{a}_i(t). \tag{29}$$

The terms on the right hand side of equation (29) are obtained by collecting terms that explicitly depend on $a_i(t)$, $\dot{a}_i(t)$ and $\ddot{a}_i(t)$, respectively. The remaining terms are collected in $F_{exc,j}(t)$. One should note that $F_{exc,j}(t)$ and the coefficients A_{ji}, B_{ji} and C_{ji} are implicitly functions of the coefficients $a_i(t)$ through their dependence on $c(t)$ and $\frac{dc(t)}{dt}$. This decomposition is done in order to move as much as possible of the total modal hydrodynamic force over to the left hand side of equation (11). The terms proportional to $a_i(t)$, $\dot{a}_i(t)$ and $\ddot{a}_i(t)$ are moved over to the left hand side. This improves the numerical stability and accuracy in the numerical time integration. Equation (29) indicates that there are interaction effects between all vibration modes. Since $F_{exc,j}(t)$, C_{ji}, B_{ji} and A_{ji} depend on $a_i(t)$, an iteration for the solution of $a_i(t)$ has to be carried out at each time step. It is seen from equation (29) that the velocity potential is needed in order to find the modal force.

To find the velocity potential on the wetted surface, the following relation is used.

$$\left(\frac{\partial\phi}{\partial x}\right)^-(x,t) = \frac{1}{2}\gamma(x,t) \tag{30}$$

This leads to,

$$\phi(x,t) = -\frac{1}{2} \int_x^{2c(t)} \gamma(x,t)dx. \tag{31}$$

The fact that the potential goes to zero at the leading edge, has been used. The expression for the vortex distribution on the wetted surface given in equation (20) is substituted into equation (31), and the order of integration is interchanged. The resulting expression for the velocity potential is used in the first term of the hydrodynamic pressure defined by equation (28). In the last term of equation (28), equation (30) is used.

3.1.3 Numerical implementation

To implement the solution defined by equations (20) and (24), numerical approximations are necessary since no analytical solution has been found. In the present case it is assumed that the vortex strength in the wake can be approximated by piecewise constant segments of vorticity. This is a reasonable approximation as long as the time step is kept small in comparison with the time scale of the problem. The length of each segment is set equal to $U\Delta t$. Δt is the constant time step used during the impact. Analytical integration over each segment is performed. This integration has been found to be important only on the segments near the trailing edge of the bag (point A in Figure 4).

One should also note that the induced vertical velocity defined by equation (27) has a square root singularity at the leading edge. When this velocity is integrated over one time step, the square root singularity is integrated analytically. The remaining non-singular part of this velocity is evaluated at the middle of the time step. This is done in order to obtain a solution that converges and has a reasonable accuracy as the time step is reduced. The discretized version of equation (25) is used to find the wetted length at each time step. An accelerated iteration procedure is used to find an approximation of the wetted length.

In order to verify the implemented theory, lift on a foil with constant chord length is calculated for two different cases where analytical solutions exist. That is the Wagner problem and the problem of a steady oscillating foil. The Wagner problem is the problem where a flat plate of zero angle of attack, suddenly is given a constant angle of attack. Comparisons show good agreement. Comparisons with the steady oscillating foil show also good agreement. The present hydrodynamic analysis is a time domain approach, so transient response in the lift force must die out first, in order to obtain a steady state solution. This is normally achieved in a couple of cycles, depending on both the free stream velocity U, and the frequency of oscillation.

To check the procedure of finding the wetted length, the procedure has been checked against a case where an analytical solution exist. That is the problem of a wedge hitting the free water surface with a constant vertical velocity V, and a constant horizontal free stream velocity U. V is assumed to be small compared to U. The steady state foil solution for the vortex distribution on the wetted surface has been used. The effect of the free stream velocity U has not been accounted for in the calculation of the wetted length. The procedure is stable and converges quickly. The reason is that the square root singularity in the induced vertical velocity at the leading edge is integrated analytically. Before this analytic integration was used, large numerical problems occurred.

We have also compared the present solution with the linearized steady state planing problem. Our numerical solution seems to converge to the linearized steady state solution, but quite slowly. The linearized steady state planing solution is discussed by Squire (1957).

Convergence in terms of decreasing the time step has been found to be satisfying, as long as the time step is chosen to be small relative to the duration of the impact ($\Delta t \approx \frac{t_{impact}}{100}$). We should note that the length Δx of each vortex element is reduced when the timestep is reduced. The reason is that $\Delta x = U \Delta t$.

Simulation results for a bag with one lobe are presented. The data set used in the simulations are as follows. The static pressure differences are $\Delta P_1 = 500[N/m]$ and $\Delta P_2 = 5500[N/m]$. ΔP_1 is the excess pressure in the bag relative to the pressure in the air-cushion. ΔP_2 is the excess pressure in the bag relative to the atmospheric pressure. The height H is equal to $0.57[m]$ and the length L is equal to $3.17[m]$ (see Figure 2). Given these data the static analysis gives the static geometry and the static tension T_0. The calculated radii of curvature are respectively $R_1 = 4.26[m]$ and $R_2 = 0.39[m]$ for circle segment one and two. The arc length of the two segments are respectively $L_1 = 3.47[m]$ and $L_2 = 1.15[m]$. Further $T_0 = 2130[N]$ and the axial stiffness is set equal to $EA = 0.7 \cdot 10^6[N]$. The vertical velocity $V(t)$ is described by $V(t) = -2\sin(\frac{2\pi}{0.2}t)$ and η_0 is set equal to $\eta_0 = 0.99\frac{2 \cdot 0.2}{\pi}$.

The three lowest "dry" eigenmode shapes are shown in Figure 5 . The corresponding natural frequencies are 4.36 [Hz], 6.81 [Hz] and 12.48 [Hz]. The effect of curvature is very important for the lowest mode shapes. This effect can be explained as follows. If the effect of curvature is neglected, the classical string equation is obtained for the transverse oscillations. There would be no coupling between the transverse and the longitudinal oscillations. The transverse mode shapes would then become sinusoidal. In Figure 5 it is seen that the mode shapes differ from being sinusoidal. For higher mode shapes the effect of finite radius of curvature will be less. One may note that the longitudinal modal deflections are considerably less than the transverse modal deflections. In spite of this fact, it can be shown that the coupling is important for the lower eigenmodes. This can be illustrated by using the coupled equations of motions. By studying equations (1) and (2), it can be seen that the coefficient ahead of the coupling term in equation (1) for the transverse motion, is approximately proportional to the axial stiffness EA. The axial stiffness is much larger than the static tension T_0, so the coupling term will be important compared to the tension term $T_0\frac{\partial^2 \eta_n}{\partial s^2}$. It should also be pointed out that the transverse deflection of the lowest point on the bag (point A in Figure 4) is small compared to the transverse

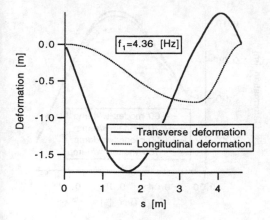

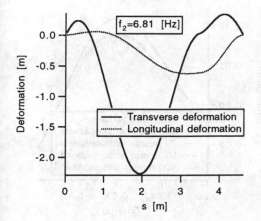

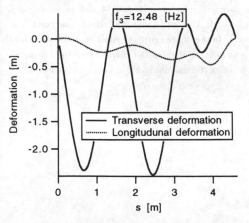

Figure 5: The eigenmode shapes for the three lowest eigenfrequencies. Deflection both in the transverse and the longitudinal directions are shown. s is the longitudinal coordinate. s is equal to zero at point B and L_1 at point A in Figure 4.

deflection of the point midways between point A and B in Figure 4. Since the bag hits the water near point A, a consequence is that the generalized modal hydrodynamic impact force will be small for the lowest eigenmode shapes. We will illustrate this by using a simplified estimate of the impact force. The results for the time independent generalized modal impact force are presented in Figure 6 as a function of the mode number. The impact pressure was assumed to be constant in space and to have a sinusoidal time variation. This means that the generalized modal impact force for mode j was written as,

$$\int_0^L p(s, \eta_n, t)\phi_j^n(s)ds$$

$$= \sin(\frac{\pi}{t_{imp}}t)\, p \int_0^{\Delta s} \phi_j^n(L_1 - x)dx. \quad (32)$$

Here $p\Delta s = 1000[N]$. Δs is the length of the area where the pressure acts. The term $p \int_0^{\Delta s} \phi_j^n(L_1 - x)dx$ is denoted the time independent generalized modal impact force. t_{imp} is the time duration of the impact. When t is greater than t_{imp}, the impact force is set equal to zero. To show the influence of varying Δs, three values has been used i.e. $\Delta s = 0.01, 0.05$ and 0.1 [m]. Δs is extending from point A towards point B in Figure 4. The effect of varying this length is seen to have a great influence on the time independent generalized modal force. It is illustrated that the modal force is small for the lower mode shapes. This can be explained by the fact mentioned above, that the modal deflections of the lower mode shapes are small in the impact region. One should also note that the time independent modal force for higher mode shapes are reduced for increasing length of the pressure region. For a small length Δs, it is seen that the magnitude of the time independent modal force is large for high mode numbers. This indicates that a high number of mode shapes is needed in the analysis.

The transverse motions of the point mid between point A and B are shown in Figure 7. $\Delta s = 0.015[m]$ and $t_{imp} = 0.03[s]$ are used in these calculations. The linear structural damping coefficient is set equal to $2\omega_i(M_{ii}^n + M_{ii}^t)\xi$ and included only on the diagonal of the structural modal damping ma-

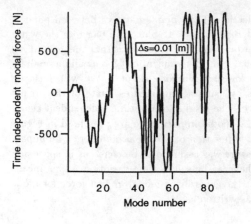

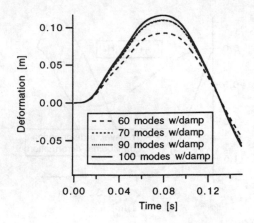

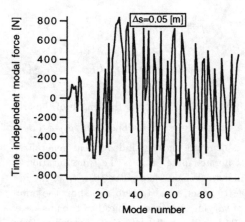

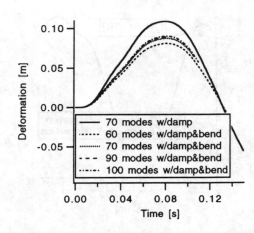

Figure 7: Dynamic transverse deflection of the mid point between point A and B in Figure 4, due to an idealized impulse load. This load acts in the region near point A in Figure 4. In the upper figure the effect of structural damping is included while both the effect of structural damping and bending stiffness are included in the lowest figure.

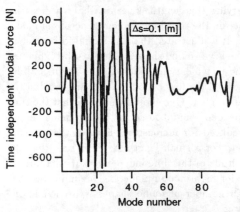

Figure 6: Time independent modal impact force is shown as a function of mode number. Δs is varied and is the length of the impact region measured from point A in Figure 4 towards point B. A line is drawn between the discrete points.

trix. Here ω_i is the eigenfrequency for mode number i. M_{ii}^n and M_{ii}^t are defined in equation (11). ξ is the relative critical damping ratio. $\xi = 0.02$ is used for the results presented in this paper. The results show a slow convergence as a function of number of mode shapes. The mechanism may be described as follows. The higher modes are excited during the impact. (See Fig. 6) The lower modes are then in turn triggered through the strong coupling between the mode shapes. The slow convergence can be explained by the fact that the bag will interpret the impulse load as very concentrated both in time and space. When this high number of mode shapes is needed to obtain convergence, the effect of the bending stiffness will be important for the higher mode shapes. In the results shown in the lower figure the effect of bending stiffness is included through the term $-EI\frac{\partial^4 \eta_n}{\partial s^4}$ in the equations of transverse motions. When the contributions to the equations of motions due to bending stiffness are derived, terms in addition to the term mentioned above will occur both in the equations of transverse and longitudinal motions. If the order of magnitudes are considered for the different terms, one will find that the term $-EI\frac{\partial^4 \eta_n}{\partial s^4}$ is an order of magnitude larger then the other terms. The other terms are therefore neglected in this analysis. The bending stiffness is set equal to $EI = 1.83[Nm]$. It is probably lower for the material of typical air bags on SES. A more realistic value may be in the order of $0.3[Nm]$. However this leads to a small effect of the bending stiffness. Larger values for the bending stiffness EI, is used to avoid problems with the hydrodynamic load model. These problems are indicated in Figure 9 and discussed in relation to this figure.

If the term $T_0\frac{\partial^2 \eta_n}{\partial s^2}$ is compared to the term $-EI\frac{\partial^4 \eta_n}{\partial s^4}$, it is found that these two terms in the equation give a contribution of the same order of magnitude for the mode 30. The higher the mode is, the larger the relative importance of the bending stiffness is. The results in Fig 7 show the expected trend that the transverse deflection response level is reduced when bending stiffness is introduced and that a lower number of mode shapes is needed to get convergence.

Figure 8 shows results where the complete hydrodynamic planing theory has been used. The bending stiffness has been set equal to $EI = 6.00[Nm^2]$. The points B and C (see Figure 4) are fixed and

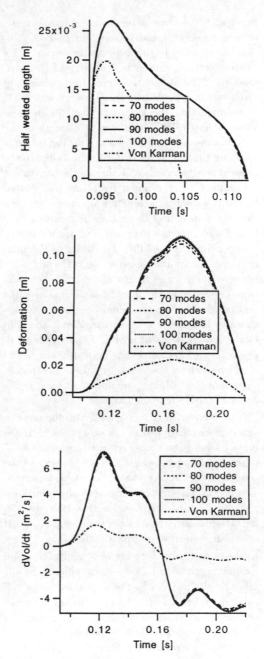

Figure 8: Dynamic response due to a hydrodynamic impulse load. Transverse deflection of the mid point between point A and B in Figure 4 is shown in the second figure from the top. The volume change of the air cushion due to the bag deformation is shown in the lowest figure. Structural damping and the effect of bending stiffness is included. Complete hydrodynamic impact planing is used in combination with a simplified Von Karman type of approach.

the undisturbed free water surface is moved vertically to model the relative vertical rigid body motion between the bag and free water surface. This relative vertical motion is consisting of the local motions of the SES at the stern, due to heave and pitch motions, and the small incident waves. It is assumed that the wave length of the incident waves is long relative to the wetted length of the bag. In the presented calculations the bag is barely touching the free water surface. This is due to the large hydrodynamic load acting on the bag during the impact. In the upper figure the half wetted length is shown as a function of time. We note that the duration of the impact is approximately $0.02[s]$ and that the maximal total wetted length is approximately $0.05[m]$. The figure in the middle shows the transverse deflection of the mid point between point A and B. The lower figure shows the time derivative of the integrated volume over the segment facing the air cushion (segment 1). This quantity is important when considering the cobblestone oscillations of a SES. The effect is analogous to the effect of a moving piston at the end of a long tube. The work by Steen (1993) indicates that this volume pumping effect triggers the acoustic cobblestone oscillations.

The problem with convergence is still the same in this case as mentioned in the case of the idealized impact load. The convergence is slow due to the concentrated loads both in time and space and the coupling effect between the different mode shapes. The maximum difference between the results for 100 and 70 mode shapes, for the transverse deformation of a point mid between point A and B shown in Figure 4, is approximately 5%.

In the same figure there are also results for a Von Karman type of solution, which differ from the other results in the way the wetted surface is calculated. In the Von Karman approach the wetted surface is found without taking into account the pile-up of water ahead of the wetted surface. We note that the results are strongly dependent on how the wetted surface is calculated. This is not unexpected when we note that the geometry in the impact region is close to being horizontal. The effect of pile-up of water becomes important in this case.

Results for the hydrodynamic impact on the bag is shown for two cases in Figure 9. One case is including bending stiffness in the same way as in the

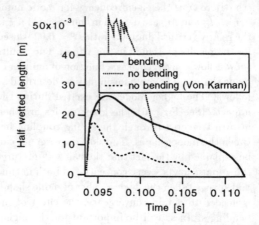

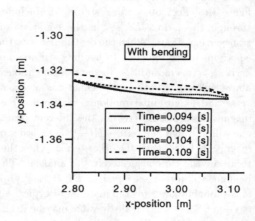

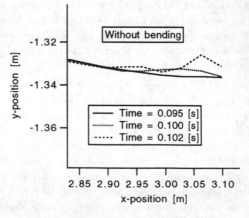

Figure 9: Results for a hydrodynamic impact load, where the effects of bending stiffness and the Von Karman approach is shown. Point A and B in Figure 4 have coordinates $x = 3.099[m]$, $y = -1.336[m]$ and $x = 0[m]$, $y = 0[m]$ respectively.

results presented in Figure 8, and the other is not including the bending stiffness. 70 mode shapes are used in both cases. The difference between no bending stiffness and a realistic value in the order of $0.3[Nm]$ is negligible for this number of mode shapes. These results are presented in order to illustrate a problem with the impact of the bag on the free water surface when the bending stiffness is not included. In the results where the bending stiffness are included it is seen that the bending stiffness is stabilizing the process of finding the wetted surface. In the case without including the bending stiffness it is found that the local deformation of the bag in the impact region leads to a negative pile-up of water ahead of the wetted surface of the bag for some time steps. By negative pile-up of water we mean a negative vertical deflection of the free water surface ahead of the wetted surface. This we interpret as unphysical, since one in this case will obtain negative pressures. Negative pressures will in turn lead to ventilation of air on the wetted surface of the bag. When this is happening, we are setting the total hydrodynamic force and the pile-up of water equal to zero. The rapid variations of the wetted surface when the effect of bending stiffness is not included, may be explained by this phenomenon. In the upper figure it is also shown results for the Von Karman approach for the case of not including the effect of bending stiffness. It is shown that this approach is more stable then the approach accounting for the pile-up of water ahead of the wetted surface of the bag. In the case of not including the bending stiffness, convergence both with respect to the number of mode shapes and the time step was not obtained.

We will now try to give one possible explanation of the slow convergence regarding the number of mode shapes, in the calculations. We focus on a simplified case where the radius of curvature is set equal to infinity in the equations of motions expressed by equations (1) and (2). The classical string equation is then obtained for the transverse oscillations. If the impact length is small compared to the total length of the cable and the duration of the impact is small compared to the first eigenfrequency, the solution will not depend on the boundary conditions during the impact. This is the case of an infinite string. The solution of this simplified case can be found by a Fourier transform in

the longitudinal coordinate. The impact load can be approximated by a delta function both in time and space. The solution written in terms of an inverse Fourier transform is an integral over the wavenumbers extending from zero to infinity. It is observed that this integral when truncated for high wavenumbers, is oscillating and converging very slowly.

In our modal solution we have an analogous discrete sum that is truncated after a certain number of modes assuming the sum has converged. The slow convergence in our modal solution can probably be explained by the behavior of the infinite string solution. If the effect of bending stiffness is included, the convergence of the modal solution is improved. This can also be seen in the simplified case of the infinite string. For high wavenumbers bending stiffness should be included in order to describe the local deformations in the impact region more realistic. The level of response is also reduced when including bending stiffness.

In this simplified model of the interaction between the bag structure and the free water surface there may be several error sources. An inviscid fluid has been assumed. This implies that the separation point must be determined on beforehand. The lowest point on the static bag configuration has been chosen (point A in Figure 4). Due to the high free stream velocity U, this may be a reasonable approximation. However in reality the fluid may be able to follow the curve of the bag further than the lowest point.

A problem discussed in flat bottom impact problems is the trapping of an air pocket between the body and the free water surface. The free water surface deflection plays an important role in this connection. In our case of high free stream velocity U, relative to vertical impact velocity V, it is believed that the free water surface deflection will be asymmetric and hinder the creation of an air pocket.

It has been found that the results dealing with the bag motion are depending on the structural damping. In our model, damping has been included as a percentage of critical modal damping. There are uncertainties both in the structural damping level and the structural damping model.

In order to understand the effect of the interaction between the bag structure and the air-cushion, the computations should be carried out for more then one impact. This has been tried, but large problems with finding a reasonable wetted surface has been obtained. In the case of more impacts the lowest point of the bag structure need not be the same as in the static configuration. This makes it necessary to have a trailing edge that is moving in space. This effect will be included in the near future.

5 CONCLUSIONS AND FURTHER WORK

A numerical and theoretical study of the interaction between the free water surface and the bag stern seal of a surface effect ship has been presented. The bag is pressurized with air, and is deformed due to the hydrodynamic pressure distribution on the wetted surface of the bag and the compressibility of air in the bag. High numerical accuracy is needed. This has been achieved by use of "dry" normal modes for the flexible behavior of the bag, in combination with extensive use of analytical expressions for the excitation and reaction forces.

It has been found important to account for the coupling between the elastic longitudinal and transverse oscillations as well as the curvature of the bag.

Slow convergence with respect to the number of mode shapes has been experienced. It has also been found that bending stiffness improves the convergence with respect to the number of mode shapes, and generates a more stable solution regarding the approximation of the wetted length. Bending stiffness does only have an influence in the impact region where the bag is wetted, and is needed in order to obtain convergence both with respect to the number of mode shapes and the time step, in the case of a hydrodynamic impact load.

It is also indicated that the effect of the interaction between the free water surface and the bag structure is important for the cobblestone effect, through the volume pumping caused by the bag.

Further work will concentrate on improving the hydrodynamic model to simulate several impacts and to relate the dynamic bag structure response between the bag and the water to a global response of surface effect ships.

6 ACKNOWLEDGEMENTS

This work is part of a dr.ing. study of one of the authors (T. Ulstein) and has received financial support by the The Research Council of Norway (NFR). The computer time is supported by the Norwegian Supercomputing Committee (TRU).

REFERENCES

Bliek, A. (1984) *Dynamic Analysis of Single Span Cables*, PhD Thesis, Dept. of Ocean Engineering, MIT.

Kochin, N.E., Kibel, I.A., Roze, N.V. (1964) *Theoretical hydromechanics.* New York:Interscience publishers.

Newman, J.N. (1977) *Marine hydrodynamics.* Cambridge:The MIT Press.

Ogilvie, T.F. (1970) *Singular perturbation problems in ship hydrodynamics.* Dept. of Naval Architecture and Marine Engineering, University of Michigan.

Steen, S. (1993) *Cobblestone Effect on SES.* Dr.ing thesis. Norwegian Institute of Technology.

Squire, H.B. (1957) The motion of a simple wedge along the water surface. *Proc. R. Soc.* A 243:48-64

Wagner, H. (1932) Über Stoß- und Gleitvorgänge an der Oberfläche von Flüßigkeiten. *Zeitschrift für Angewandte Mathematik und Mechanik* Vol.12 No. 4.:193-215

Wagner, H. (1948) *Planing of watercraft.* Washington:NACA, Technical memorandum no.1139.

Hydroelasticity in Marine Technology, Faltinsen et al. (eds) © 1994 Balkema, Rotterdam, ISBN 90 5410 387 6

Effects of combination motions on hydrodynamic forces induced on bodies in the sea

J.M.R.Graham, Y.D.Zhao & C.Y.Zhou
Department of Aeronautics, Imperial College, London, UK

M.J.Downie
Department of Marine Technology, University of Newcastle, UK

ABSTRACT: The paper examines two types of flow around long bodies with a range of cross-sections from circular to square. The first situation consists of oscillatory flows having two components at right angles to one another with a range of frequency and amplitude ratios. This type of flow occurs for example if a submerged pontoon hull moves in heave and sway simultaneously with different frequencies and amplitudes. Experimental measurements have been carried out for this case on two pontoon sections undergoing forced motion in two directions provided by a general planar motion mechanism, in a wave tank. The second type of flow consists of planar oscillatory flow around a circular cross-section having cycles of different period and amplitude following in sequence. This has been studied as a way of examining history effects which appear to occur for fixed cylinders in random waves or compliant cylinders such as risers moving back and forth through in-line waves. Both types of flow have also been studied using numerical simulation of the Navier-Stokes equations based on a vortex-in-cell method. The computations are at lower Reynolds numbers than the experiments but generally show very similar effects on the force coefficients.

1. INTRODUCTION

In contrast to the theoretically based computation of wave induced flows round bodies in the diffraction regime, forces induced on bodies in the drag-inertia regime are normally calculated by the empirical Morison's equation. For long fixed bodies such as circular cylinders this may be expressed as:

$$F(t) = 1/2 \, \rho \, U|U|D \, C_d + \rho \, dU/dt \, A \, C_m$$

F is the force / unit length of the body, $U(t)$ the relative cross-flow velocity and D the diameter and A the area of the body cross-section. It is usual to evaluate the force coefficients C_m and C_d for this equation from data obtained for the same section tested in simple planar sinusoidal flow. Extensive data for this, (e.g. Sarpkaya and Isaacson, 1981) has been published for common body cross-sections. Morison's equation with constant coefficients, depending only on the Keulegan-Carpenter number KC (= UT/D, where T is the period) and Reynolds number, does not give by any means a perfect fit to the time history of the force $F(t)$ but it does give a reasonable fit when the incident flow is sinusoidal.

However it is common practice to apply Morison's equation to the prediction of quite general time varying flows on long cylindrical bodies. Examples of such cases are the prediction of forces induced on vertical circular cylinders (such as riser pipes) subject to random incident waves and on bodies which due to their orientation and motion in waves are subject to components of velocity from different directions. In many of these cases it is found that Morison's equation gives a much poorer fit to the data than for regular sinusoidal flow. This is even true when the coefficients themselves are evaluated from measurements taken in the same incident flow field. Indicative of this is the very large amount of scatter found in successive wave by wave evaluation of coefficients at the same value of KC on vertical cylinders in random unidirectional waves in a flume (Davies, 1992). Similarly the measurements of Martigny et al. (1994) for pontoon sections undergoing different combined motions and of Chaplin (1993) for horizontal cylinders in orbital flow show the significant differences between coefficients evaluated in planar flows and those evaluated in combined perpendicular motions.

In the present work force measurements carried out on a body of square cross-section undergoing various combined forced motions at different depths in a wave tank, are described. The results of computations at much lower Reynolds number for similar combined motions are compared with these experiments. Further computations are also presented for a body undergoing periodic in-line motions composed of sequences of alternating large and small waves. This motion is used to study the effects of different size waves in a sequence of random waves. The computations were carried out using a two-dimensional numerical simulation of the unsteady Navier-Stokes equations. A laminar (constant) viscosity coefficient was assumed and the solution method was based on the viscous vortex-in-cell method described by Graham and Djahansouzi (1991).

The experiments were carried out at three different depths at which the flow approximately simulates flow around a barge, a semi-submersible pontoon and a TLP pontoon respectively. The present paper is concerned with the deepest submerged case only. The model was tested in pure heave and in pure sway. It was also tested with heave and sway combined. In the combined case the model was forced to undergo large amplitude sway with a relatively small higher frequency heave motion superposed, similar to the motions experienced by a TLP. Measurements were also made as the model decribed circular and elliptical orbits. The combined motion tests were intended to investigate the effect of mixed frequency motion on the force coefficients.

2. DETAILS OF EXPERIMENTS

The experiments were carried out in the wave basin of the Fluid Mechanics Laboratory of the Ecole Centrale, Nantes. The basin is 18 m in length by 9.5 m in width and has a mean water depth of 2 m. At one side of the tank there is a wave generator, and at the other a beach. The tank is spanned by a bridge that is supported on rails running along its length. A large digitally controlled motor driven mechanism, named the Generateur de Mouvements Plans (GMP), is supported by the bridge. There are two arms, linked to hydraulic rams, projecting down from the GMP and ending in a fixture to which a model may be attached. The arms are streamlined so as to minimise any disturbances as they pass through the water in the basin. When the motors are operating, the model is forced to describe planar motion, that is to say sway, heave and roll, or surge, heave and pitch, depending on the model orientation. Subject to limits on amplitude and acceleration, the GMP can be programmed to generate any required planar motion. The maximum surge/sway and heave half amplitudes are 0.3 m and roll/pitch amplitudes are 15deg. The maximum acceleration is 0.3 g for a mass of 500 kg. The frequency range lies between 0.05 Hz and 3 Hz.

The fixture by which the model may be attached to the GMP comprises a horizontal bar spanning the lower ends to the two arms and supporting the GMP connection plate and load cell. The fixture is wholly enclosed by the model itself. The load cell is made up of four piezoelectric multi-force transducers sandwiched between two plates, one of which is attached to the GMP connection plate, and the other fastened to the model connection plate. The load cell measures two orthogonal force components and the moment on the model in the same plane as the motion.

The model is rectangular in section, 0.38m square, and 1.6m long, and takes the form of a hollow watertight box. It has been designed so that its sharp square edges can be removed and replaced with rounded edges of 0.066m. With its rounded edges, it is approximately a 1:30 scale model of one of the pontoons of the Snorre TLP (Almeland, Gaul, Pettersen and Vogel, 1991). The arms of the GMP pass through openings in the lid which are sealed with latex gaskets. The gaskets are made of material 0.0003m thick with linear elastic properties, and do not induce any coupling effects. The model was constructed with a rigid internal aluminium frame clad with a 0.003 m thick PVC skin glued to the frame and welded along all external joints to make it watertight. The frame also supports the model connection plate which is aligned with the horizontal plane through the centre line of the model. The GMP is attached to the model only at the interface of the two connection plates, via the load cell, and by the gaskets sealing the openings for the arms of the GMP. In addition to the load cells, the model is fitted with an array of pressure transducers located on the surface in a line along the girth at the mid-section of the model and encompassing one of its edges. Transducers were also fitted at one end of the model to monitor the two dimensionality of the flow. The wires connecting the load cell and the pressure transducers to the data acquisition system were passed through the hollow arms of the GMP. The model was mounted on the GMP entailed a lengthy procedure involving passing the GMP arms through the lid, attaching the GMP connection plate to the arms, and then connecting the main body of the model to the GMP. The instrumentation was then linked to the data acquisition system and the desired edges fitted. Finally the lid was dropped back down the arms, secured to the model, and the gaskets around the arms sealed. A similar procedure was required to change the edges on the model since one of them was instrumented, although it was unnecessary to disconnect the main body of the model from the GMP. The model was mounted on the GMP so that its longitudinal axis was parallel to the beach.

The experiments included variation of the model edge geometry and variation of the motion prescribed by the GMP. The model geometry was varied by testing with sharp edges and with rounded edges of 0.066m radius. In addition three sets of experiments were carried out at different depths of submergence. In the first set, the set considered in this paper, the centreline of the model was 1m below the mean water level in a configuration corresponding to flow about a TLP pontoon. In the others smaller depths of the submergence were tested.

The motion prescribed by the GMP consisted of combined sway and heave motion and orbital motion and a number of experiments carried out for pure sway and for pure heave. In these the sway period ranged from 3.3 to 10s and the sway Keulegan Carpenter number based on sway velocity and period, KC_s, from 1 to 5. The heave period ranged from 2.8 to 5s and the heave Keulegan Carpenter number, KC_h, from 1 to 5.

Experiments with combined sway and heave were carried out with relatively large amplitude sway in comparison to the heave motions. The sway periods again ranged from 3.3 to 10s, with KC_s ranging from 1 to 5, whilst the heave periods ranged from 1.8 to 3.6s, with KC_h ranging from 1 to 3.

The sway and heave periods were designed not to be simple multiples in order to avoid short period repetition of the flow cycle. In the case of orbital motion, the experiments were conducted with circular orbits with a period of 3.3s and a Keulegan Carpenter number ranging from 1 to 3. An experiment was also performed with the model undergoing elliptical motion with the heave amplitude set to half the sway amplitude, which was kept the same as for the circular orbits.

3. RESULTS

The experimental data are presented in the form of force coefficients C_d and C_a for the particular force components and motions. Thus $C_{d(s)}$ implies the drag coefficient of force in the sway direction due to sway motion alone. $C_{d(s-h)}$ is the drag coefficient of the force in the sway direction in the presence of combined sway and heave motions. Thus comparison of $C_{d(s-h)}$ with $C_{d(s)}$ shows the effect of superposed heave motion on the sway drag force due to sway motion. F_{sh} is the sway force induced by combined sway and heave and all the force coefficients are non-dimensionalised by the component of relative velocity in the same direction as the force. For the forced motion tests the inertia coefficient is presented as an added mass coefficient C_a.

The force coefficients obtained with pure sway and pure heave motions at 1m depth of submergence are plotted in Figures 1 and 2 for the sharp edged model, and 3 and 4 for the round edged model. Also plotted on the figures are curves representing the results of experiments carried out at lower Reynolds numbers in an oscillating U-tube by Bearman et al. (1984).

The combined motion results are plotted in Figures 5 to 8. The sway-heave force coefficients are plotted against the heave Keulegan Carpenter number, KC_h, for the sharp edged and the round edged bodies respectively. Chaplin's (1993) theoretical curves for C_a in orbital flow are also shown in figures 6 and 8.

NUMERICAL RESULTS

The numerical method uses the discrete vortex-in-cell approach to calculate the forces on transverse body sections. It is a two dimensional method which solves the Navier Stokes equations applied to flow

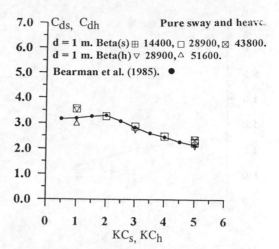

Figure 1. Sharp Edged Pontoon undergoing Sway or Heave (Drag Coefficients)

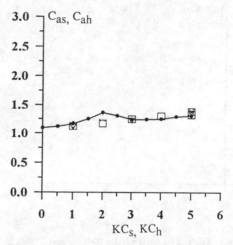

Figure 2. As (1) - (Added Mass Coefficients)

about an oscillating body without any consideration of free surface wave effects.

The vortex in cell method is used for computing incompressible time-dependent separated flows about two-dimensional bluff bodies. The particular approach used here has been reported by Graham and Djanhansouzi (1991). The flow is modelled using the vorticity/stream function form of the Navier-Stokes equations. The vorticity in the flow is represented by arrays of point vortices. At each time step the vorticity convection equation is solved by distributing the vorticity of the discrete moving vortices onto a grid, solving the Poisson equation and then updating the positions of the vortices by reverse interpolation. The diffusion

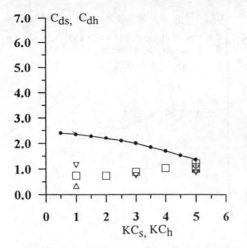

Figure 3. Round Edged Pontoon undergoing Sway or Heave (Drag Coefficients)

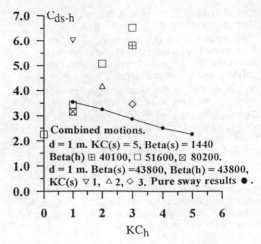

Figure 5. Sharp Edged Pontoon undergoing Combined Sway and Heave (Drag Coefficients)

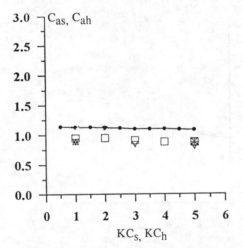

Figure 4. As (3) - (Added Mass Coefficients)

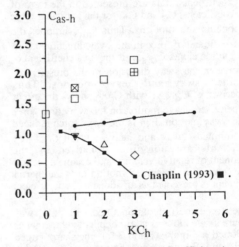

Figure 6. As (5) - (Added Mass Coefficients)

equation is solved using a finite difference scheme on the grid and the vorticity change is projected onto the point vortices. The force on the cylinder is calculated from a momentum balance from the vortex positions (Cozens, 1987). momentum balance using the vortex positions.

The circular cylinder can be most easily modelled by using a regular polar mesh for the computation. In the case of the pontoon section the physical plane was transformed into a circular geometry using a two stage conformal mapping procedure. A Schwartz Christoffel transformation was used for the sharp edged section, and a Lewis form transformation was used for the round edged section.

The vortex in cell method was used to calculate the flow about the sharp edged pontoon section for three different cases. In the first case pure sway

only was considered and the drag and added mass coefficents were computed for Keulegan Carpenter numbers ranging from 1 to 5 at a frequency parameter, β, of 213. β is the Stokes parameter $D^2/\nu T$ where ν is the kinematic viscosity. It is equal to the ratio of the Keulegan Carpenter number to the Reynolds number. Although the numerical model could include a turbulence model, the present calculations were performed at sufficiently low values of β for laminar flow to be assumed. The results with β = 213 are shown in Figures 9 and 10 Shown also in these figures are the experimental results with β = 28900. The data has been scaled here in the same way as the added mass, hence the factor $2KC/3\pi^2$ multiplies C_d.

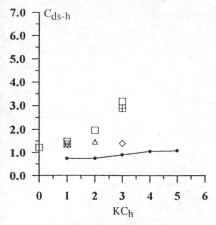

Figure 7. Rounded Edged Pontoon undergoing Combined Sway and Heave (Drag Coefficients)

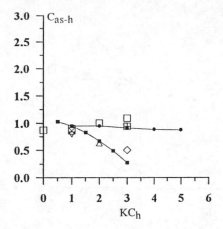

Figure 8. As (7) - (Added Mass Coefficients)

In the second case the drag and added mass coefficients have been computed for combined motion with KC_s and β_s for sway fixed at 5 and 400 respectively for two different heave frequency parameters giving β equal to 1429 and 2222 respectively. The results are plotted against KC_h in Figures 11 and 12. Also shown in these figures are experimental results for KC_s=5 and β_s=14400 and β_h=516004 and 80200.

In the third case, force coefficients were computed for orbital motion with $\beta_s=\beta_h=400$. The results are shown plotted against Keulegan Carpenter number in Figures 13 and 14, which also show the experimental results for $\beta_s=\beta_h=43800$.

Computations were also carried out for the second class of flows, in-line oscillatory flows for which succeeding wave cycles had different amplitudes and periods. In the cases shown here two different wave cycles alternated in a continuous sequence, the velocity, being in-line throughout, given by:

$$U(t) = U_1 \text{Sin } 2\pi t/T_1, \quad n(T_1+T_2) < t < n(T_1+T_2)+T_1$$

$$= U_2 \text{Sin } 2\pi t/T_2, \quad n(T_1+T_2)-T_2 < t < n(T_1+T_2)$$

for any integer n.

In-line forces were computed for a circular cylinder in this flow under a range of β parameters from 200 upwards based on the larger wave of the pair. In each case the wave designated 1 had either the larger velocity amplitude or the longer period or both. The force coefficients C_{d1}, C_{d2}, C_{m1} and C_{m2} were evaluated on a wave by wave basis after an approximately asymptotic state had been reached

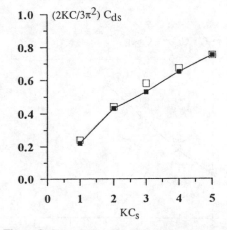

Figure 9. Sharp Edged Pontoon in Sway. Comparison of measured and computed C_d. ────

and similarly two Keulegan Carpenter numbers KC_1 and KC_2 were defined for the alternate cycles.

Figure 15 shows the results for the drag coefficient C_{d1} during the larger wave cycles plotted against KC_1 for wave sequences with amplitude ratios U_1/U_2 and period ratios T_1/T_2 both held constant at 2.0 giving a fixed ratio $KC_1/KC_2 = 4.0$. Also shown in this figure is the computed drag coefficient for purely sinusoidal flow.

Figures 16 and 17 show C_{d2} and C_{m2} for the smaller wave in the same sequences. Figure 18 shows C_{d2} and C_{m2} for the smaller wave for two sets of tests in which first the amplitude ratio was held constant at unity and the period ratio T_1/T_2 varied and then the period ratio was held constant at unity and the amplitude ratio U_1/U_2 varied.

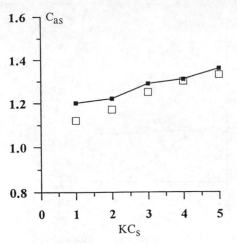

Figure 10. Sharp Edged Pontoon in Sway.
Comparison of measured and computed C_a.———

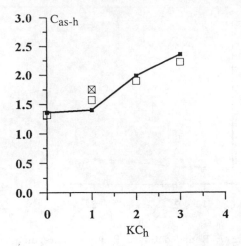

Figure 12. Sharp Edged Pontoon in Sway/Heave.
Comparison of measured and computed C_a. ———

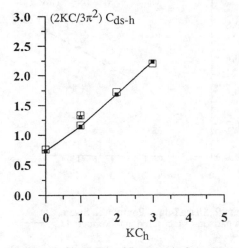

Figure 11. Sharp Edged Pontoon in Sway/Heave.
Comparison of measured and computed C_d. ———

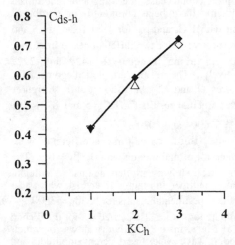

Figure 13. Sharp Edged Pontoon in Orbital Flow.
Comparison of measured and computed C_d. ———

4. DISCUSSION

The force coefficients for heave and sway alone are, to all intents and purposes, the same, as can be seen from Figures 1 to 4. This shows that the effect of the free surface is negligible at this depth. In the case of the sharp edged pontoon the experimental results measured at a β of the order of 10^4 generally agree very closely with those of Bearman et al. (1984) measured in a U-tube at a β of the order of 10^3. The results also agree well with the predictions of the vortex in cell approach shown in Figures 9 and 10. In contrast the drag coefficients, particularly, for the round edged pontoon are clearly influenced by the β

number, particularly at the lowest KC numbers where the GMP coefficients are half those measured in the U-tube. The two sets converge as KC increases and the edge radius becomes smaller in relation to the amplitude of the motion.

The pure sway coefficients for the sharp edged pontoon in pure sway (Figure 9) are well predicted by the vortex in cell method even though the β number in the computation is much smaller than that of the experiment.

Figures 5 to 8, show results for which a high frequency heave was superposed on the sway motion. This significantly increases the magnitude of the force coefficients in the sway direction. The sharp and round edged cylinders have a different

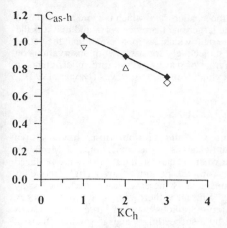

Figure 14. Sharp Edged Pontoon in Orbital Flow. Comparison of measured and computed C_a. ———

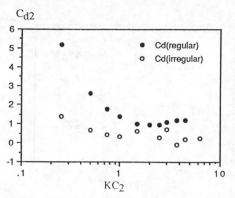

Figure 16. Circular Cylinder in waves of alternating size, $KC_1/KC_2 = 4$. Drag Coefficient in smaller wave.

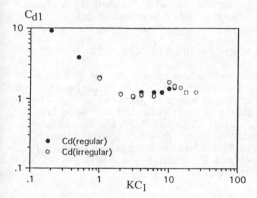

Figure 15. Circular Cylinder in waves of alternating size, $KC_1/KC_2 = 4$. Drag Coefficient in larger wave.

behaviour in this respect. If KC_s, β_s and β_h are constant (in this instance at 5, 14400 and 51600 respectively) the sharp edge drag coefficients increase linearly with the heave amplitude. At $KC_h=3$ the sway coefficient, C_{dsh} has increased from 2.3 at $KC_h=0$ to 6.5. The curve also shown in the figure represents the results C_{ds} vs KC_s for pure sway and C_{dh} vs KC_h for pure heave at $\beta=28900$. The difference between the points and the curve is an indication of the influence of the combined motion effect, although differences in β should also be borne in mind. It should also be noted that the combined motion increases significantly the velocity around the shedding edges but that in the results presented the forces are non-dimensionalised with respect to only the sway velocity component. The curves of C_{as} and C_{ah} are also shown on the added mass plots for comparison.

For the orbital motion, Figures 5 to 8, the highest drag coefficient for the sharp edged body is obtained with the smallest orbits. Its value falls rapidly tending towards the value obtained in pure sway at the higher Keulegan Carpenter numbers. The added mass coefficients agree well with Chaplin's results at low KC, tending to diverge as KC increases. For the sharp edged body, the combined motions are well predicted by the vortex in cell method, (Figures 11 to 14).

For the round edged pontoon, at constant KC_s, β_s and β_h, the drag coefficient C_{dsh} also increases with heave amplitude, but in a shallow curve relative to the sharp edged case. The added mass coefficient follows similar trends to those of the sharp edged body but are generally smaller in magnitude. The drag coefficient for orbital motion, by contrast, does not attain the high value at low KC_h shown for the sharp edged body. The added mass coefficient does however also follow Chaplin's theoretical curve.

The computed results for the circular cylinder in alternating waves of different size are shown in figures 15 to 19. The first set of tests in which the amplitude and period ratio were kept constant so that both the Keulegan Carpenter numbers varied in a fixed ratio show clearly that the drag coefficient measured in the larger wave (Figure 15) was very little affected by the presence of a smaller wave preceding it. Measurements of the inertia coefficient, not shown, similarly exhibited hardly any effect. On the other hand both drag and inertia coefficients (figures 16 and 17) for the smaller wave were strongly affected by the presence of larger waves in the sequence. Further tests which varied first the period ratio (figure 18) and then the amplitude ratio (figure 19) while keeping the other ratio constant showed that the effect of velocity amplitude differences in the wave sequence was much greater than the effect of changes in period ratio and that C_d was much more affected than C_m. These two results,

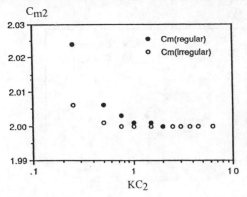

Figure 17. Circular Cylinder in waves of alternating size, $KC_1/KC_2 = 4$. Inertia Coefficient in smaller wave.

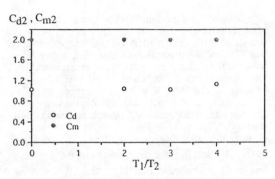

Figure 18. Circular Cylinder in waves of alternating size, T_1/T_2 varying. Drag and Inertia Coefficients in smaller wave.

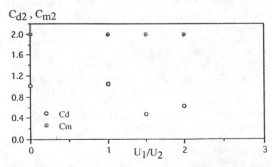

Figure 19. Circular Cylinder in waves of alternating size, U_1/U_2 varying. Drag and Inertia Coefficients in smaller wave.

that it is the smaller wave which is more affected by a preceding larger one than vice versa and that it is the amplitude ratio which has more effect than the period ratio are subjectively consistent with observations on the behaviour of wave by wave force coefficients in tests on cylinders in random waves (Davies, 1992).

5. CONCLUSIONS

Experiments and computations have shown significant changes in the coefficients of Morison's equation when planar oscillatory flows are replaced by more complex periodic flows. High frequency heave considerably increases lower frequency sway damping. Similarly there can be a strong history effect from a previous wave, if large, on the forces induced by a succeeding wave, if small. Effects of Reynolds number (or β parameter) have been shown to be small for sharp edged bodies where comparisons have been made but were found to be very significant for the drag or damping coefficients of pontoon sections with rounded bilges.

6. ACKNOWLEDGEMENTS

This work was supported by the Behaviour of Fixed and Compliant Offshore Structures Managed Programme, which is promoted by the Marine Technology Directorate Ltd and sponsored by SERC, HSE, Amoco, BP, Elf, Statoil, ARE, Aker Engineering and Brown and Root Marine. The valuable collaboration in the experiments of J. P. Le Goff and D. Martigny of SIREHNA is also acknowledged.

7. REFERENCES

Almeland, S, Gaul, TR, Pettersen, DJ and Vogel, H 1991. Snorre TLP Configuration & Analysis Technology, *Proc. 23rd OTC, Houston, Texas, USA, OTC 6622*, pp. 577-586.

Bearman, PW, Downie, MJ, Graham, JMR and Obasaju, ED 1985. Forces on Cylinders in Viscous Oscillatory Flow at Low Keulegan Carpenter numbers, *J Fluid Mech.*, vol 154, pp 337-356.

Chaplin, JR 1993. Orbital Flow around a Circular Cylinder : Pt II, Attached Flow at Larger Amplitudes, *J. Fluid Mech.*, 246, p. 397.

Cozens, PD 1987. Numerical modelling of the roll damping of ships due to vortex shedding. *Ph.D thesis. University of London.*

Davies, MJS 1992. Wave loading data from fixed vertical cylinders. *OTI.* 92, p558, *HMSO.*

Graham, JMR and Djanhansouzi, B 1991. Computation of Vortex Shedding from Rigid and Compliant Cylinders in Waves., *Proc. 1st ISOPE Conf., Edinburgh*, p 504.

Martigny, D, Molin, B and Scolan Y-M 1994. Slow Drift Viscous Damping of TLP Pontoons. To be presented at *13th OMAE Conf. Houston, TX*.

Sarpkaya T and Isaacson, M 1981. Mechanics of Wave Forces on Offshore Structures. *Van Nostrand*.

Floating airports

Hydroelasticity in Marine Technology, Faltinsen et al. (eds) © 1994 Balkema, Rotterdam, ISBN 90 5410 387 6

A preliminary to the design of a hydroelastic model of a floating airport

Seok-Won Lee & William C.Webster
Department of Naval Architecture & Offshore Engineering, University of California, Berkeley, Calif., USA

ABSTRACT: The problems posed in the design a hydroelastic model of a floating airport are explored. It is found that a scale ratio of near 500 is required. It was found that very few common materials can be used to model typical strucuture at this scale ratio. A review of available materials led to a preference for very flexible plastics for construction of a built-up model. Finally, a scheme is proposed to aid in the measurement of the complex, two-dimensional motion of the floating airport model in a wave tank.

1. INTRODUCTION

The twentieth century has been marked by a dramatic increase in air travel, resulting in congestion at many of the international airports around the world. Attempts to construct new airports or even to extend existing runways are hampered by the proximity of residential areas and by restrictive environmental regulations in some nations. The noise of jet aircraft, and the extensive amount of land required for aircraft operations make any airport expansion a difficult enterprise. In order to increase airport access for major coastal metropolitan areas around the world, without the difficulties associated with onshore expansion, several studies have investigated the feasibility of an offshore floating airport. In the United States, feasibility studies of such airports have been made in various localities, from the coast of California on the Pacific Ocean, to the shores of the Great Lakes. In other parts of the world where there is an acute shortage of land, such as Japan, floating airports have been seriously considered. In fact, a floating airport was evaluated as an early design option for Japan's Kansei International Airport, currently being built as an artificial island. However, most of the investigations have centered around a semisubmersible design. Little research has been carried out on the potential of a barge-like design.

Instead of linking together a series of semisubmersibles by flexible joints, an alternative concept is to have a structure that, in its final form, would be an enormous, continuous, barge-like structure. This floating airport, or commonly referred to as a very large floating structure(VLFS), built to satisfy the functions of an international airport, would be the largest water-borne structure ever built. Engineering

this structure would present serious challenges that would require the development of new technologies and innovations in a variety of areas from hydroelastic modeling to large scale construction methods.

Before any slab of concrete is pretensioned or a sheet of steel is welded, prudence would dictate that the physical characteristics of such a structure should be investigated and verified through, if possible, physical model testing. Due to the immense sizes involved, any such physical model will require scale ratios far outside those usually attempted for hydroelastic studies. Researchers who have had the experience of attempting to accurately model the hydroelastic characteristics of a ship can attest to the numerous difficulties involved.

This research resulted from a preliminary study we conducted in preparation for the construction of a suitable hydroelastic model of a floating airport. We intend to construct such a model in 1995 and test it in 1996. In the paper below we will address some of the problems we have had to deal with associated with the physical modeling of a barge-like VLFS and the potential solutions we are currently considering.

2. OVERVIEW OF RECENT RESEARCH

Regardless of design, there are still numerous engineering questions that must be answered in designing, constructing and operating a floating airport. How many new technologies will have to be developed? Is the structure constructable? Will the flight deck be flat enough for flight operations during normal sea conditions? Will the structure have a useful service life? Will the VLFS be economically viable? Current and past research thrusts can be divided into two categories, the semisubmersible design and the barge-like design.

2.1 Semisubmersible design

The semisubmersible design utilizes modules that are essentially variations of the mobile offshore drilling platform widely used in the oil and natural gas industry. Twin hull-like pontoons, that are normally submerged, provide the base on top of which are columns that support a flat deck, where machinery, accommodation areas, etc. are located. In the case of a floating airport, each platform, or module, would be connected in series with flexible joints so that the total structure would be long enough to accommodate a runway. There are some inherent advantages of this design. Since the waterplane area of the semisubmersible modules is only that of the columns, the effects of waves are lessened. An adaptation of the existing construction methods for offshore platforms can be used. Paulling & Tyagi [1] and Ertekin, et. al. [2], have carried out some extensive investigations into the hydroelastic responses of such a multi-module structure, using extensive computer modeling.

Even though this design has the benefit of basing some of its engineering on existing technologies, there are serious maintenance and economic hurdles to overcome. A semisubmersible structure would most likely be constructed out of steel, or a combination of concrete and steel. Such a structure will likely have relatively higher construction and maintenance costs than a barge-like VLFS that is constructed entirely out of prestressed concrete. Morgan's paper on constructing concrete barges highlights this economics[1]. However, it is important to note that regardless of design or construction material, any VLFS will be extremely expensive to construct. In addition, there are doubts as to the feasibility of actually linking together a large number of modules, even in protected waters. Even slight sea motions can create enormous difficulties in connecting these joints[2]. Indeed, the problem of linking joints in random sea motions merits further research.

---

[1] R. G. Morgan states in his paper in the Proceedings, Concrete Ships and Vessels 1975, that the "only recorded evidence for prestressed concrete comes from operators of Yee's barges where after many years, service average annual maintenance of 2000 tonne [concrete] barges was $2830 as against $8200 for 1750 tonne steel barges."

[2] Mr. Phillip Chow of T.Y. Lin, in a private communication, indicated that his company was involved in a project which involved joining two large platforms. Due to the random nature of waves, Mr. Chow has stated that even small motions of either one of the structures prevented the linking of these joints. Basically, they had to wait until the random motions of the water lined up the joints perfectly so that a pin could be immediately set in place. It is clear that a multimodule VLFS

2.2 Barge-like design

An alternative concept is the barge-like structure. Only a few papers have investigated the possibilities of this design for a floating airport. Currently, the only existing comparison to this concept would be floating bridges, such as those in Washington State in the USA. The Lake Washington Floating Bridge has five lanes of traffic over a distance of 2400m and more than 32m wide. However, this structure has the benefit of operating in the near calm waters of a lake. The VLFS structures we are investigating would have to withstand the rigors of open-sea waves.

Chow, Lin, Riggs and Takahashi [3] present engineering concepts from design to construction for a continuously-constructed, 10,000 ft long VLFS of the barge-like design. In completed form, the VLFS is essentially a mat-like prestressed concrete structure that is segmentally constructed. Like the semisubmersible design, this concept will be constructed in modules, but in far fewer numbers. In contrast, the modules will be rigidly connected together in protected waters. This rigid connection will allow the stresses arising from hydrostatic and hydrodynamic loads to be gradually applied over a large area, instead of being concentrated at joints.

3. PROTOTYPE VLFS

A prototype VLFS would be 4000m long, 300m wide and 30m deep (see Figures 1 and 2). The width dimension is primarily constrained by the space required for flight operations, passenger and cargo loading/unloading, and aircraft hangars. Aircraft performance specifications would determine the length while the depth dimensions is primarily constrained by the wave heights and volume requirements. A total of four decks, the flight, accommodation, machinery and bottom decks, would fulfill all other requisite functions of a fully functioning international airport. To support all these decks, there is a regular array of transverse and longitudinal bulkheads. With these minimum structural details, certain physical parameters to be observed in the model can be estimated for the VLFS.

The minimum required runway length of a Boeing 747-400 passenger jet during normal conditions at sea level is 3500m. An additional 305m at either end of the runway is recommended by the US Federal Aviation Administration. With the possibility of future passenger aircraft that require longer runway lengths, a 4000m long runway should be adequate.

---

will either have to be linked together in well protected waters large enough to accommodate the VLFS, or some sort of technique have to be developed to facilitate the linking of these modules in the open ocean. This consideration of these aspects is beyond the scope of this paper.

Figure 1 shows the general arrangement of such a floating airport with anticipated dimensions for the structural elements. The width of the flight deck, or the uppermost deck, would have a 60m wide runway with a 40m wide safety apron on either side of the runway. Beyond the safety apron, there is a 30m wide taxi way with a 10m shoulder, while on the other side of the apron there is a 130m wide area for passenger terminals, aircraft hangars, refueling and maintenance. These dimensions, and particularly the length, approximates those of planned onshore airport expansion and construction at many international airports around the world[3].

Interestingly, one of the conclusions reached in a recent research conducted by Mamidipudi and Webster [11], is that in response to waves, a VLFS mat-like structure experiences minimal motion along the centerline, in head seas. Hence, vertical runway motion can be minimized by placing the runway along the fore-aft centerline of the VLFS. This would minimize any sort of grade or vertical excursions that the aircraft would have to negotiate for takeoff and landing.

Ten meters below the flight deck is an accommodation deck that can serve to hold airport restaurants, stores, customs areas and VLFS crew accommodation areas. Machinery and vehicles can be stored on the next deck 10m below the accommodation deck. Lastly, there is a ballast compartment underneath the storage deck. All the decks below the flight deck would have thickness of 0.5m.

It should be noted that all these specifications are the result of a cursory look into airport design. Further research needs to be done in optimizing the internal and external VLFS structural layout to meet airport planning requirements. However, for our purposes, only rough specifications will be used for a preliminary hydroelastic modeling. In addition, one will see that further simplifications of this VLFS design will have to be made.

The preliminary dimensions of the VLFS platform are:

length	4000m
width	300m
depth	30m
draft	20m
longitudinal bulkheads	14
transverse bulkheads	99
displacement	2,472 Ktonnes

---

[3] Proposed or planned expansion lengths:

Denver International Airport (U.S.A):
4900m (runway & taxiway)
Hong Kong New International Airport (U.K.):
3,800m (runway)
Munich New International Airport (Germany):
4000m (runway)

In other words, this platform would have a displacement about 5 times larger than the largest supertanker.

4. MODELING TECHNIQUES

Testing of physical models in wave tanks has been a standard practice for determining the dynamics of floating rigid bodies such as ships and offshore platforms. There has been considerable progress in the theoretical foundations and in the use of computers for simulation of the motions of ships and offshore platforms. However, the complex nonlinear interactions of the mechanical loads, the wave forces and the viscous effects make physical model tests an important compliment to the theoretical studies, even for typical structures where there has been a wealth of experience. It seems reasonable that if any floating airports were to be built, model tests, particularly scaled dynamic tests, would be a central feature of the development program. Like the platforms themselves, there are a number of alternative approaches to the development of these models.

4.1 *Flexible Spine Model*

A traditional approach to the development of dynamic models is to concentrate the structural properties in a structural element, a spine, located near the center of the body. In the case of a floating airport this spine could be a relatively thin plate mounted near the half-depth of the platform. The plate could be constructed with transverse (or longitudinal elements so that it behaves as an orthotropic plate with properties which would be equivalent to the whole platform. The rest of the geometric configuration of the platform would be built up with purposely non-structural elements, such as foam segments which are connected only to the spine and not to each other. Lewis [10] and Nawwar et al. [14] among many others, used this approach to determine the dynamics of ships in a seaway.

This concept has several drawbacks in our view. First, the spine only approximates the primary structural bending responses of the platform. It is difficult in this approach to duplicate the significant torsional stiffness of a floating airport. Second, it is very difficult to assure that the gaps, which are necessary between the segments, are divorced from the rest of the structure and will not interfere with the platform dynamics or with the flow around the platform. If a membrane is used to keep water out of the gaps, then it can induce significant structural stiffness to the overall model. The difference in pressure between hydrostatic pressure on the outside of the membrane and air pressure on the inside of the membrane causes tensions in the membrane that, in turn, cause un-

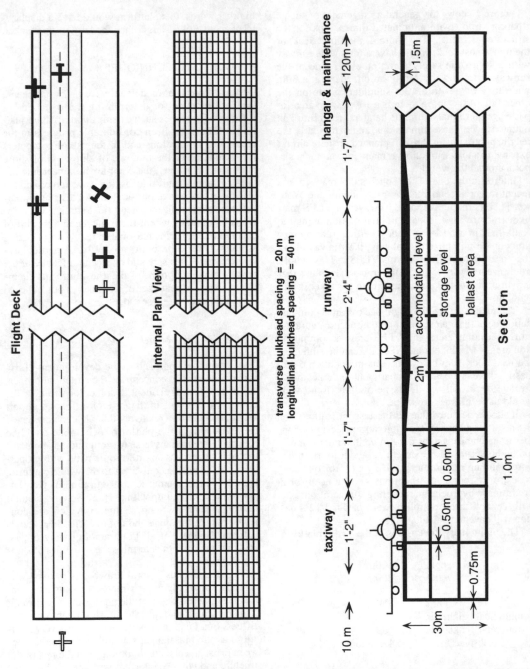

Flight Deck

Internal Plan View

transverse bulkhead spacing = 20 m
longitudinal bulkhead spacing = 40 m

hangar & maintenance

runway

taxiway

accomodation level

storage level

ballast area

Section

120m

1'-7"

2'-4"

1'-7"

1'-2"

10 m

2m

1.5m

0.50m

0.50m

1.0m

0.75m

30m

Figure 1. Arrangement of a potential Floating Airport

wanted loadings in the model. If the gaps are not covered, then the water trapped in the gaps is alternately squeezed out and ingested as the model deforms. This action can cause significant (and unrealistic) damping of the structural modes.

4.2 Built-up Models

A different approach is to try to duplicate the hydroelastic response of the platform by modeling more of the individual structure than that which would be done in the spine model. In this approach the primary structural elements such as decks, sides and bottom are modeled as separate plates and joined to form a scaled, box-like structure. It is unreasonable to attempt to model every detail of the structure of a floating airport, but it is feasible to model enough of the structure to incorporate some of the secondary structural elements, particularly the longitudinal and transverse bulkheads.

The built-up model approach has it own set of drawbacks. First, the material from which the model is to be built must have special properties (see the next section). Second, the construction of a built-up model is considerably more difficult than the spine model. There are many more elements which need to be assembled and assuring that these elements are connected effectively as a structure is difficult. Finally, assuring that a built-up model is watertight is difficult.

4.3 model selection

Based on the above considerations, the focus of the remainder of this paper is on the development of a built-up model for dynamic modeling of a floating airport. It was felt that this modeling would be more appropriate for investigating the motions which are likely to occur (see Mamidipudi & Webster), even though it is realized at the outset that the development is likely to be difficult.

5. SCALING THEORY

The built-up model under consideration here involves modeling some of the secondary structural elements but none of the tertiary detail. That is, the decks, sides, bulkheads, etc. will be represented locally as equivalent orthotropic plates, with the contributions of the frames, stiffeners, brackets, etc. incorporated into the properties of the plate. Procedures for estimating the properties of a equivalent orthotropic plate given the structural details are well known and will not be repeated here. If the plate coordinates are taken

locally to be x and y, then there are four parameters which describe the structural response of the plate in the prototype (subscripted by p):

$E_p A_{px}$, the axial stiffness in the x direction,

$E_p A_{py}$, the axial stiffness in the y direction,

$$D_{px} = \frac{E_p i_{px}}{1-\nu_p^2}, \text{ the x bending rigidity,}$$

$$D_{py} = \frac{E_p i_{py}}{1-\nu_p^2}, \text{ the y bending rigidity,}$$

(5.1)

where

E_p is the elastic modulus,

A_{px} and A_{py} are the x and y areas of the continuous structural components modeled by the orthotropic plate,

i_{px} and i_{py} are the x and y bending moments of inertia (per unit width) of the continuous structural components modeled by the orthotropic plate,

ν_p is Poisson s ratio.

For the sake of simplicity in the modeling of the local structure, torsional rigidity of the orthotropic plate elements will be neglected. Since Poisson's ratio has a small effect on the bending and axial rigidity values and does not vary much for typical materials, it, too, will be neglected. For scaling, the following relationships must be preserved (where the subscript m refers to the model):

$$
\begin{aligned}
E_p A_{px} &= \lambda^3 E_m A_{mx}, \\
E_p A_{py} &= \lambda^3 E_m A_{my}, \\
E_p i_{px} &= \lambda^5 E_m i_{mx}, \\
E_p i_{py} &= \lambda^5 E_m i_{my},
\end{aligned}
$$

(5.2)

where λ is the scale ratio. For dynamic scaling of the local plate motions the mass of the plate should be scaled as well. This requires that:

$$m_p = \lambda\, m_m,$$

$$\omega_p = \frac{1}{\sqrt{\lambda}}\, \omega_m$$

(5.3)

where m_p and m_m are the masses per unit area of the orthotropic plate element for the prototype and model, respectively, and ω_p and ω_m are the corresponding frequencies. In the proposed modeling of the VLFS no attempt will be made to model local (secondary) hydroelastic motions and, as a result, we will not attempt to meet this condition.

6. SELECTION OF SCALE

In selecting an appropriate scale, two important points must be taken into consideration; the size of research wave tanks and limits in generating miniature waves.

The size of the wave tank sets a lower limit to the scale ratio. For a jumbo passenger jet to land on a floating airport, the VLFS must have a length of at least 4000m (as discussed above). A scale of 500 would result in a model length 8m long and this is already near the upper limit of most wave tanks. Although it may be possible to consider models larger than this, the handling of these very flexible models is a severe problem (such a model will break if it is not supported continuously along its length while it is being put in the wave tank). As a result, we have chosen a minimum scale ratio in the neighborhood of 500 as the target.

On the other hand, the ability to generate and measure small waves sets the maximum scale ratio. Wave less than 10 cm long begin to exhibit surface tension effects which do not exist for the prototype. At a scale ratio of 500, this boundary corresponds to prototype waves that are 50m long, that is, waves of a period of about 5.5 sec (i.e., common waves in small to moderate sea states). Further increases in the scale ratio raises this boundary to include even more important portions of typical wave spectra. In addition, at a scale ratio of 500, a 2m high wave for the prototype becomes a 4mm high wave for the model. Most wave tanks are unable either to reliably generate waves of heights less than about 5mm or to measure these waves accurately. This is not to say that equipment can not be developed to do this. Rather, these limits are typical for equipment designed to do ordinary Offshore Engineering work that generally involves much different scale ratios. Thus, it appears that the limitations on the physics of modeling imposed by surface tension effects and that imposed by typical wave basin equipment limit the maximum scale ratio to about 500.

Evidently, 500 is a good scaling factor to use, and this has been selected for our further considerations.

7. MATERIAL SELECTION

Next, the model material must be selected. The properties which are of interest are:

a) Elastic modulus
b) Water absorption
c) Creep and internal damping.
d) Constructability (forming, bonding, welding, etc.)
e) Availability
f) Affordability

Essentially, the model's cross-sectional structural details will dictate the range of the elastic modulus. Materials that are too flexible will require the model to be constructed out of thicker members in order to maintain a particular stiffness. Material of large thickness and flexibility may be difficult to form and, if the material is exotic, expensive. On the other hand, a relatively stiff modeling material will require very thin sheets. It may be difficult with thin sheets to provide adequate bonding surfaces for assembly of the model. In particular, making the model watertight is extremely difficult when the sides and bottom are very thin sheets. That is, any extreme values of elastic modulus cause difficulties. For a steel or concrete structure dynamically modeled at a scale ratio 500, the elastic modulus should be in the range of 5 - 200 MPa.

Many of the materials available within the desired range of modulus of elasticity are plastic materials which are physically unlike the full-scale structural materials. Some candidate materials absorb water and, when they do so, change both in weight and in other physical properties. Some materials creep. That is, the elastic modulus depends on the rate of deformation of the material. A material with large creep will take a long time to achieve its final deformation under a load and will also take a long time to recover when the load is removed. It is typical for materials which exhibit a large creep to also have a large energy absorption (i.e., internal damping). Since both steel and high-performance concrete structures have very low internal damping, this property is of concern as well.

For the sake of practicality, the materials should be easy to form and to join. Many of the very flexible materials can not be easily cut and formed with ordinary tools. For joining it is necessary to have clean, square and accurate corners so that good contact is achieved. Some plastic materials can be welded together using solvents, some require expensive heat welding equipment, and yet others require adhesives. If the material can be joined together using adhesives or solvents, it is imperative that no softening or hardening occurs at the joined edges. In the case of adhesives, the bead of new material should be small and very flexible when dry or else the adhesive itself will make an important contribution to the overall stiffness of the model.

The last two requirements, that the material be available and affordable, may perhaps be the most important requirements to satisfy.

7.1 Plastic as the material of choice

In previous hydroelastic models of marine structures, metals, plastics and even rubbers have been used. But most, if not all, of these experiments have involved scaling factors far less than 500. The limited range of practical scales will have serious limitations on the choice of modeling material. Metals cannot be used because of their high values of elastic modulus, 45-200 GPa, almost 3 orders of magnitude too large.

Unfortunately, there seems to be a large gap in the elastic modulus of the plastics that are available, especially in the range which we are most interested in. Nonetheless, plastic appears as an feasible material, since there are some plastics with an elastic modulus within the desired range of elastic modulus, see Table 1. In addition, elastomers, or more commonly known as rubbers, also appear to be a viable modeling material candidate. The large number of plastic suppliers and the extensive amount of technical information available on plastics, in comparison to rubbers, led to a tentative selection of plastics for detailed investigations for model material selection.

Even though several plastics have the appropriate elastic modulus values, many of these candidates have prohibitive physical limitations. For some plastics the only method to join together pieces is to use ultrasonic, hot gas or friction welding tools. Equipment necessary to carry out these welding are expensive and, in some cases, more appropriate for use in mass

Table 1
Typical Properties of Unfilled Resin Plastics

Material[4]	Tensile Modulus (MPa)
Rubbery styrenic block polymers (Kraton_)	5
Polyurethane (thermoplastic)	6
Polyvinyl chloride (flexible)	7
Alkyd Polyesters cast rigid	33
Polyethylene (linear low density)	162
Polyethylene (low density)	224
Polyphenylene ether	255
Ionomers (Surlyn_)	345
Polybutylene (pipe grade)	380
Fluorocarbon polymers (PTFE, Teflon_)	410
Polyurethane (thermoset)	410
Polyvinylidine chloride (Saran_)	520
Fluorocarbon polymers (FEP, PFA)	550
Polymethyl pentene (TPX)	800
Fluorocarbon polmers (ETFE)	830
Cellulose butyrate	860
Polypropylene	1,380
Acrylonitrile-butadiene-styrene (ABS)	1,960

production factories. However, if adequate funds and manually skilled labor are available, hot gas welding may be possible.

---

[4] Significant variations exist in each material. Fillers and reinforcements may have a major effect. Material manufacturer should be consulted for further information.

Polyethylene is a widely available material commonly used in bottles, drums, trays and bins, ducting and as tank linings. In addition to being a tough yet flexible material, it is resistant to many chemicals, which makes it difficult to join using commonly available adhesives or solvents. Specialized equipment such as hot gas welding is required to join.

For the same reasons, no fluorocarbon polymers can be used. Fluorocarbons such as *polytetrafluoroethylene* (Teflon_), require the use of two component adhesives based on epoxide resins or cyanoacrylate adhesives on pretreated surfaces with etching reagents such as sodium in liquid ammonia or sodium napththyl. In addition to using these toxic adhesives, sufficient funds must be allotted for these expensive fluorocarbon polymers. Industry and government efforts to reduce the use of ozone layer depleting chemicals, namely fluorocarbons, CFCs, and the expense of processing CFC has resulted in prices of at least $13.00 per pound.

Many plastics such as *ionomers, polyamides, polybutylene, polymethylpentenes, polyphenylene ether, alkyd polyesters,* and *polyvinylidene chlorides* all have the appropriate elastic moduli and some are commonly available. Unfortunately, most are not suitable due to either its unavailability in sheet form, or in the thicknesses required, or impractical bonding and welding techniques required, or exorbitant material purchase costs.

7.2 *Flexible PVC*

Amongst all of the available plastics, flexible PVC appears to be one of the more attractive ones. This material has an elastic modulus in the right range and is the one of the most widely used and widely available plastics. This family of flexible PVC plastics are relatively inexpensive.

Appealing as flexible PVC may appear to the hydroelastic modeler, there are still many problems associated with the use of this or any other flexible material. Because the plastic is flexible, careful consideration must be given to the adhesive used. Certain adhesives such as *cyanoacrylate* based adhesives will harden the joined surfaces. This can drastically alter the overall stiffness of the model. There are adhesives available that will preserve the flexibility of the material being used. A potential area of further research is to investigate the physical effects of different adhesives on the flexibility of low modulus of elasticity materials.

Sensitivity to heat is an another physical feature of flexible PVC or most other plastics, that will require particular attention. The coefficient of linear expansion, $[(\Delta L/L)/\Delta T]$, of flexible PVC range from $10*10^{-5}/$ C to $25*10^{-5}/$ C. Since PVC is not a good heat conductor, deformations can occur in the model if that part exposed to air (and perhaps bright lights) attains a temperature much higher than that part in contact with the water. It is also important that the model is constructed at or near the same temperature in which it will be tested. On a related point, care must taken in limiting the exposure of the material and the model to the sun. Any long term exposure to ultraviolet (UV) radiation can have a deteriorating effect on the model and on its bonded surfaces, depending on the plastic used.

As stated earlier, creep is also a problem to be dealt with. Previous hydroelastic research with plastics seems to have yielded good results. However, most of these research involved scales around 100 or much less, using plastics that are much stiffer than flexible PVC. Flexible PVC has a definite creep which depends on its exact composition. One potential solution to this problem is to experimentally determine the creep rates of a miniature I-beam under various loadings. After each loading, the load is taken off and the beam is allowed to return to its normal undeflected shape. Then the deflections in the model that occur, say, 1 second immediately after the beam is loaded can be used to determine the modulus of elasticity. Since waves lead to dynamic loading, one will not have to worry about the deflections that occur in the model due to constant loading. This 1-second elastic modulus is probably more reflective of the calculated deflections that occur in the model under the influence of waves.

Finally, perhaps the most difficult problem of using this, or any other material, is availability and cost. In tests of a clear flexible PVC plastic, the modulus of elasticity was determined to be 21 MPa. With this value, model transverse bulkhead thicknesses of at least 1 cm spaced every 6 cm are required. Unfortunately, flexible PVC is more commonly available much smaller thicknesses. A 1 cm thickness is often only available as a special order, and expensive, item. A possible solution is to obtain sheets of flexible PVC in intermediate thicknesses so that two or three sheets can be bonded together to form the necessary thickness. Again, as stated previously, the use of adhesives on such large areas of the model pose potential problems.

7.3 Alternative stiffer plastics

Instead of laminating flexible PVC sheets together to obtain the desired thicknesses, one possibility is to obtain a slightly stiffer plastic. With a higher modulus of elasticity, the model would not need as thick bulkheads or walls. For example, if the maximum thickness that are to be used in the model is to be 0.5 cm, then a material with a modulus of elasticity of 137 MPa is required. Unfortunately, there are no easily obtainable material that have an elastic modulus in this range. The material with an elastic modulus closest to this that is commonly available is linear low density polyethylene which has and elastic modulus of 166 MPa. Unfortunately, this material has poor adhesion properties. A related and economically available plastic is polypropylene, but its modulus of elasticity of 1379 MPa makes it far too stiff for the model concerned, making the thicknesses required (about 1mm) far too thin. This thickness is not only extremely difficult to obtain but also questionable in providing an adequate bonding surface.

7.4 Foam materials and composites

As described in the previous section, the use of homogeneous plastic products is possible but not without substantial difficulties. In an effort to find more suitable alternatives, foam-like plastics have been investigated although, at the writing of this paper, this investigation is not yet complete. A large number of foam products are available and these are based on elastomers or plastic materials such as polyurethane or polystyrene.

The modulus of elasticity of these foams depends not only on the base material, but on the density of the foam and on the type of cell in the foam (open or closed cell). Closed cell foams appear the most attractive since they are generally resistant to water absorption. Open cell foams, on the other hand, require an additional barrier to make the model water tight. Use of open cell foams would require making a laminate sheet, with the foam sandwiched between a thin layer of plastic or rubber. The core would impart the required flexibility, while the outer membrane would provide the water impermeability. Developing such a composite is a difficult undertaking since not only does the membrane need to be bonded to the foam, but techniques for forming and bonding the composite would also need to be developed.

Of the various alternatives, closed cell polystyrene foam seems to be the most attractive of the foam options. Its modulus of elasticity is in the range of 20 - 100 MPa, placing it in the middle of the desirable range. It has some attractive properties. It is easy to form by cutting and sawing and there exist many bonding agents. Its application does have several serious problems stemming from the formation of the material. Polystyrene foam sheets are formed by the compressing and bonding (by slight melting) of Styrofoam beads. The resulting sheet still retains some of the character of the beads. Because the bonding may not be complete, there are interstices between the bead remnants in which water can intrude (the beads themselves absorb little water). Thus, the sheet may absorb water and even leak. Second, again as a result of the manufacturing process, polystyrene foam is brittle and lacks a high breaking strength in tension.

Fractures occur at the partially bonded bead surfaces and quickly spread when loaded.

In order to use *polystyrene* foam for a floating airport model, additional research will be required. However, amongst the small number of options available it appears to be one of the most attractive ones.

8. DESIGN OF THE MODEL

Scaling requires that the linear dimensions of the model are simply the scaled dimensions of the VLFS. At a scale ratio of 500 the overall dimensions of the model platform become:

length	8000mm
beam	600mm
depth	60mm
draft	40mm

The material to be selected will be either flexible PVC or polystyrene foam. With either of these materials, the modulus of elasticity (and other properties) will be known, but the thicknesses which the material is available will be limited.

Fashioning the structure of the model is a multi-stepped process. Because of the complexity of the prototype, it is not remotely feasible to duplicate much of the structural detail. This prototype includes sides, bottoms, decks and bulkheads composed of a plate (either steel or prestressed concrete) with an array of stiffeners in both directions form the same material. Simple plate elements equivalent in response to these orthotropic structures are then developed. These elements generally have a thickness somewhat smaller than the overall thickness of the stiffened plate and an elastic modulus smaller than the actual material. These equivalent plates are next scaled down to the model size.

With the model at this scale, it is not possible to duplicate installation of all three intermediate decks, all 14 longitudinal and all 99 transverse bulkheads. If this were to be attempted, not only would the model bulkheads have thicknesses which will not correspond to obtainable thicknesses, but the time required to construct the model will be enormous. Simplicity and structural feasibility while satisfying stiffness requirements is a necessity.

In order to accomplish the goal of developing a useful hydroelastic model within the constraints of available materials, the concept of a ∍homogeneous design has been adopted. The idea is to construct all elements of the model out of a material of a single or at most a few thicknesses of readily obtainable material. In this concept, the exact scale of the model is determined from the requirements for the thickness of the deck bottom and sides. The interior details of the VLFS are not be exactly duplicated in the model. The bending stiffnesses and axial stiffnesses of the built up model are matched to those of the prototype by adjusting the *spacing* of the model s longitudinal and transverse bulkheads, using one of the standard material thicknesses. The resulting model does not yield an exact representation of the secondary structural response of the VLFS. It is different from flexible spine modeling (see 4.1 above) in that the resultant model is a continuous structure with proper bending and with approximately correct torsional properties. Because the plating system is orthotropic, the plastic representing them may have to be formed using additional longitudinal or transverse strips joined to the sheet to represent the different properties in the x and y directions.

8.1 *A proposed model*

In order to examine the implications of the modeling, a preliminary design of a prototype was performed. For this design, it is assumed that the intermediate decks of the prototype do not play a central role in the overall structural response. These decks are near the platform neutral axis and may be discontinuous. At any rate, since the purpose of this model is to determine if it is possible to model accurately such a large floating structure, it is not so important that all the working details of a floating airport are included in the model at this time. At a later date, inclusion of the intermediate decks represents complexity and cost in the model construction, and is not a feasibility question. What is important is to make certain that model is structurally sound and matches the scaled stiffness values.

If flexible PVC is used with an elastic modulus of 20.7 MPa then the thicknesses of the various elements are:

	thickness (mm)
bow and stern	4
sides	4
top deck	8
bottom deck	4
5 longitudinal bhds.	10
39 transverse bhds.	10

Since the unweighted model will have a lighter displacement than what is required, sand ballast will be used to increase the weight of the model. pBecause the model itself will be quite flexible and sensitive, especially when loaded with the sand ballast, care must be taken when it is moved. The model should be constructed on a cradle which can be used to support it when it is lowered and raised from the water tank.

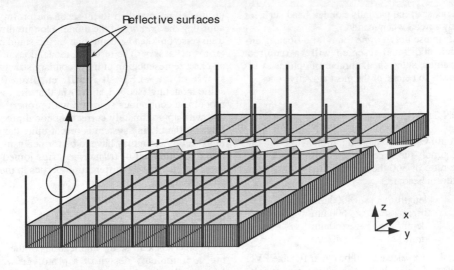

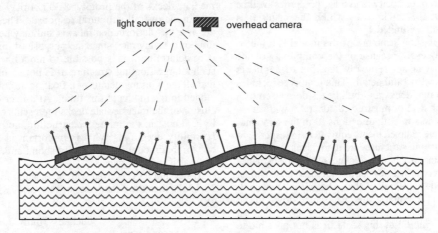

Figure 2. Scheme for measuring slopes distribution on the model

9. EXPERIMENTAL APPROACH

Experiments on a model floating airport in a wave tank will be primarily aimed at determining the motions of the platform in response to waves. Of particular interest will be the magnitude of the motions and, especially, the locations where the horizontal slopes are too severe for ordinary landing and takeoff operations (for airplanes, slopes are more important than the platform motions themselves because of the high speeds attained during these operations). As mentioned previously, the waves which the model must be immersed in are, by ordinary wave tank standards, relatively short in length and small in amplitude. The corresponding motions anticipated by the platform are equal to or smaller than the wave ampli-

tudes. In addition, one can expect the critical vertical motion pattern to have a complex distribution over the deck of the platform.

Ordinary measurement approaches such as the use of accelerometers, LVDT motion sensors, and the like would require an enormous number of sensors and may only have marginal accuracy for this application. As a result, an alternative measurement scheme is proposed which will allow the inexpensive measurement of a large number of slopes. Slender balsa wood sticks will be mounted vertically at regular intervals over the deck surface, as shown in Figure 2. It is intended that these sticks penetrate the deck and be glued to a bulkhead, either transverse or longitudinal. The sticks will extend vertically above the deck a considerable distance (say 20cm) and on their top reflective tape will be installed (see Figure xxx). Because of their length, the sticks act as amplifiers of the slope. By taking a picture of the platform during the tests with a single source light at the camera and with the lens open long enough to capture a complete wave period, the motion of the top of each individual stick will be captured (see Figure 2). The pattern of motion (really a Lissajou figure) shows not only the amplitudes of the x and y slopes, but the phase between these two as well. A separate picture, this time with a high-speed flash, can capture the phases between adjacent sticks. These two sets of photographs will quickly isolate, albeit more qualitatively than quantitatively, the ɔhot spots or regions of large slope amplitudes. It is intended to use this scheme to isolate the most important locations on the airport to install more precision equipment for careful measurements.

With the model deformations measured, the experimental data can be compared with theoretical calculations. If both the experimental data and calculated data can be brought into good correspondence, the modeling technique is viable and the appropriate physics will be understood. Using the same technique, a model of a far more accurately detailed VLFS can be constructed. This model can be used to measure deflections that will occur in the actual VLFS structure due to waves. If deflections are too large for aircraft operations, then appropriate design changes in the VLFS can be effected. Through experiments on physical models and theoretical computer models, one will know if aircraft operations on a potential VLFS design is possible.

10. Conclusion

This paper presents the preliminary considerations necessary for making a useful hydroelastic model of a floating airport. The focus of this study was on the construction of a built-up hydroelastic model where the decks, bottom, sides and bulkheads are all structural elements. An alternate approach using an internal spine was considered, but not selected because of the limitations concomitant with that approach.

The study indicates that this modeling will be particularly difficult. Important obstacles remain to be overcome and, even though we were able to make progress in a number of the areas. We did, however come to several important conclusions:

a. The scale ratio for a hydroelastic model of a floating airport will have to be in the neighborhood of 500 if it is to fit into typical wave tanks and be exposed to waves which are not affected by surface tension.

b. There are very few materials are suitable for constructing a built-up model. The required modulus of elasticity is about 5 - 200 MPa and this corresponds to very flexible material.

After a careful review of the literature, it appears that two materials are most attractive: flexible PVC and *polystyrene* foam sheets.

c. Constructability of a model using plastic materials may be very difficult because of cutting and bonding problems.

d. Measurement of the motions necessary to assess the suitability of the floating airport for air operations requires development of new measurement tools. A qualitative approach to obtaining the deck slope distribution using mechanical slope amplifiers was presented.

REFERENCES

[1] Paulling, J.R., and Tyagi, Sushil, 'Multi-Module Floating Ocean Structures,' *Proceedings*, First International Workshop on Very Large Floating Structures 1991, Honolulu, Hawaii, pp. 39-48.

[2] Ertekin, R.C., Riggs, H.R., Wang, Dayun, and Wu, Yousheng, 'Composite Singularity Distribution Method with Application to Hydroelasticity,' *Proceedings*, First International Workshop on Very Large Floating Structures 1991, Honolulu, Hawaii, pp. 59-80.

[3] Chow, Phillip Y., Lin, T. Y., Riggs, H. R., and Takahashi, Patrick K., 'Engineering Concepts for Design and Construction of Very Large Floating Structures,' *Proceedings*, First International Workshop on Very Large Floating Structures 1991, Honolulu, Hawaii, pp. 97-106.

[4] Dailey, J.E., Hickey, E.I., Gaul, R.D., Nolan, C.E., 'Mobile Offshore Bases,' *Proceedings*, First International Workshop on Very Large Floating Structures 1991, Honolulu, Hawaii, pp. 133-148.

[5] Li, Q., Lin, J., Qiu, Q., Wu, Y., 'Experiment of an Elastic Ship Model and Theoretical Predictions of its Hydroelastic Behavior,' *Proceedings*, First International Workshop on Very Large Floating Structures 1991, Honolulu, Hawaii, pp. 265-276.

[6] Evaluation of Construction Methods for Offshore Airports, Ralph M. Parsons Company and Tetratech for the Federal Aviation Administration, Washington D. C., August 1969, p.66, p.20.

[7] Lemeke, Eberhard, 'Floating Airports,' Concrete International, May 1997, p37 -41.

[8] ≅Kansai Airport , Civil Engineering in Japan, v29, December 1990, pp. 28 - 40.

[9] Lewin, M. M., 'Design of the Third Lake Washington Floating Bridge,' Concrete International, February 1989, pp.50-53.

[10] Lewis, E. V., 'Ship Model Tests to Determine Bending Moments in Waves,' *Transactions*, Society of Naval Architects and Marine Engineers, vol 62, 1954, pp. 426-490.

[11] Mamidipudi, P. and Webster, W.C., *Proceedings*, International Conference on Hydroelasticity in Marine Technology, May 1994, Trondheim, Norway.

[12] Ashford, N. J., Wright, P. H., Transportation Engineering: Planning and Design, John Wiley & Sons, Inc., 1989, pp. 653-655.

[13] Morgan, R. G., *Proceedings*, Concrete Ships and Vessels 1975, Berkeley, California, p.8, p.10.

[14] Nawwar, A. M., Godon, A., Roots, T., Howard, D., Bayly, I.M., 'Development of a Measuring System for Segmented Ship Models,' Experimental Mechanics, June 1989, pp. 101-108.

[15 Domininghaus, H., Plastics for Engineers, Hanser Publishers, Munich, 1993.

[16] Evans, Verney, Plastics as Corrosion-Resistant Materials, Oxford, Pergamon Press, 1966.

[17] Modern Plastics Encyclopedia, McGraw-Hill, New York, 1993.

[18] Rubin, I., I., Handbook of Plastic Materials and Technology, John Wiley & Sons, Inc., 1990, p.1723.

[19] Shackelford, CRC Materials Science and Engineering Handbook, CRC Press, 1992.

[20] Titow, W. V., PVC Plastics: Properties, Processing, and Applications, Elsevier Applied Science, New York, NY, USA, 1990

Hydroelasticity in Marine Technology, Faltinsen et al. (eds) © 1994 Balkema, Rotterdam, ISBN 90 5410 387 6

The motions performance of a mat-like floating airport

P. Mamidipudi & W. C. Webster
College of Engineering, University of California, Berkeley, Calif., USA

ABSTRACT: A linearized analysis of the motions of a very large floating mat-like structure for the use as an airport is presented. The analysis follows from the assumption of small amplitude incident waves and, hence, small resulting deformations. A panel method based on an extension of Frank's method is used to analyze the hydrodynamic aspects and orthotropic plate theory is used to deal with the structural aspects. The consequences of these motions based on the results obtained for typical airport dimensions and varying structural characteristics are presented.

1 INTRODUCTION

The concept of hydroelasticity gained prominence after the work of Bishop and Price [4]. Very large floating structures (VLFS) have been considered by many researchers in the past for uses like manufacturing and storage facilities, airports and even habitation. A great deal of study has gone into the analysis of such structures for the uses such as cited above, references can be found in [1] [2]. This paper focuses on the potential use of such a structure as an airport.

It is supposed that the airport is to be used for commercial purposes, in which case it will have to satisfy several regulations which limit the principal dimensions (see [3]). This structure will be much larger than any floating structures which now exist. Based on the requirements for the size of the runway, taxiway, administrative and storage structures, navigational requirements etc., the airport will have to be approximately 2500m long and 300m wide for a typical Boeing 747 traffic. Due to the small draft of this type of structure compared to its other two dimensions, it is inappropriate to use rigid body dynamics to evaluate its motions response. This structure will be very flexible compared to any offshore structure in existence and, thus, elastic deformations may be more important that the rigid body motions.

In this paper, a mat-like floating airport is considered. Several configurations have been proposed for a floating airport, the merits and demerits of which are discussed in [7]. Here, mat-like structures are considered primarily because of their simple geometry and possibly easier construction. Here, motions of mat-like structure are predicted, taking the flexibility of the structure into account, all within the scope of linear theory.

2 STATEMENT OF THE PROBLEM

A very large mat-like floating airport is considered, moored in deep waters. Although the construction and assembly of such a large structure is an important problem in itself, we only address the motion response of such a moored structure. Further, we focus on the operational aspects of the structure; the survivability of such a structure, which will presumably involve large deformations of the structure under inclimate conditions, requires proper treatment of steep waves and large amplitude deformations of the structure, both of which are excluded from this discussion. In this context, it is appropriate to assume the validity of linearized hydrodynamics to describe the wave environment and orthotropic plate theory to analyze the platform motions.

It is assumed that the mooring system provides the structure with the ability to align itself in the direction of the significant wave. Although this assumption facilitates the flight operations, it presents us with a more difficult hydrodynamic problem because the wavelengths of interest are small compared to the characteristic length of the structure. Figure 1 is a schematic sketch of the problem. The platform is assumed to have a length L, breadth B, and draft T. It is also assumed to remain horizontal in the free-surface, in the absence of any waves. For a wave incident along the positive x-direction, the vertical deformations of the platform are to be determined.

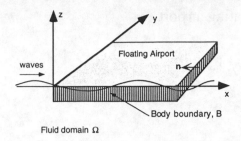

Figure 1: Schematic sketch of the coordinate system

2.1 Hydrodynamic model

In the coordinate system chosen to describe the problem (see Figure 1), x and y are the horizontal coordinates coincident with the undisturbed free-surface, and z is positive pointing upwards. For further analysis, it will be beneficial to use n for the outward unit normal; V to represent the velocity at any point on the surface of the mat; g for the gravitational acceleration; and Ω, the fluid domain.

The fluid is assumed to be incompressible and inviscid and the flow irrotational. The flow can therefore be represented by a velocity potential $\phi(x, y, z; t)$, which satisfies Laplace's equation. Following the small waves and small motions assumptions, we can linearize the free-surface conditions and the kinematic condition on the body. We decompose the velocity potential into three parts, namely, the incident potential ϕ_I, the diffraction potential ϕ_D, and the radiation potential ϕ_R. All three potentials, which are assumed harmonic in time, must individually satisfy the linearized free-surface conditions and collectively satisfy the kinematic condition on the body. The resulting problem can then be solved by superposing solutions. With the incident waves (and hence their potential) being specified, the diffraction and radiation potentials satisfy (in addition to the free-surface and body boundary conditions) a radiation condition, that specifies that the waves are outgoing. Thus, the problem reduces to field equation everywhere in Ω:

$$\Delta\phi = 0 \quad \text{where} \quad \phi = \phi_I + \phi_D + \phi_R \qquad (1)$$

The linearized combined kinematic and dynamic free-surface condition (FSBC) on F(z = 0):

$$\frac{\partial^2\phi}{\partial t^2} + g \frac{\partial\phi}{\partial z}\bigg|_F = 0 \qquad (2)$$

The kinematic body boundary condition (KBBC) or the "no leak" condition on the body B:

$$\frac{\partial\phi_D}{\partial n} = -\frac{\partial\phi_I}{\partial n} \qquad (3)$$

and

$$\frac{\partial\phi_R}{\partial n} = \vec{V}.\vec{n} \qquad (4)$$

given the incident wave whose elevation η, and its corresponding potential ϕ_I, respectively as:

$$\eta = A \cos(k(x \cos\theta + y \sin\theta) - \omega t)$$

and $\qquad (5)$

$$\phi_I = \frac{Ag}{\omega} e^{kz} \sin[k(x \cos\theta + y \sin\theta) - \omega t]$$

2.2 Structural model

Pressures will be exerted on the sides and the bottom of the mat due to hydrostatics, incident waves and the motions of the mat. It is assumed that the mooring system is such that it allows significant motions only in the vertical direction. We thus ignore the motions resulting from the forces acting on the sides of the mat and focus on the vertical deformations due to the pressure on the bottom of the mat. If z = W(x,y; t) - T is used to represent the vertical deformations of the mat, then the pressure acting on the mat is calculated using the linearized Euler's integral,

$$P_{bottom} = - \rho \phi_t - \rho g (W - T) \qquad (6)$$

It is further assumed that the structure is so constructed that the weight and the buoyancy distributions cancel out each other exactly in calm seas, thus, rendering the structure undeformed and perfectly horizontal in the absence of waves. The motions of the mat are therefore only as a result of the change in the bottom pressure due to the deformations and waves.

The motions of the mat are assumed to be governed by orthotropic plate theory which, for the above mentioned forcing function on the bottom of the mat yields,

$$D_x \frac{\partial^4 W}{\partial x^4} + 2H \frac{\partial^4 W}{\partial x^2 \partial y^2} + D_y \frac{\partial^4 W}{\partial y^4} +$$

$$m \frac{\partial^2 W}{\partial t^2} = -\rho \frac{\partial \phi}{\partial t} - \rho g W = P_{\text{bottom}} \qquad (7)$$

where, m is the mass per unit area of the mat, and all other terms have the usual meaning. It should be noted here that the rigidity terms represent the internal forces due to the motions W, the time derivative terms represent the acceleration per unit area of the mat and the equation essentially is a balance of these with the external forces on the bottom. Since the mat is freely floating, boundary conditions of "no shear" and "no moment" are imposed along the edges of this idealized plate.

3 SOLUTION PROCEDURE

The general method of the computation of vibrations of a body presented by Webster [8] [9] is taken as a template for the solution of the above problem. This procedure is an extension of Frank's method for computation of flow about two dimensional bodies with free-surface. The body is divided into panels and the kinematic boundary conditions are satisfied at discrete locations (called collocation points). The details of the procedure can be found in the original papers.

The solution process can be divided into three stages. First, the solution of the diffraction problem. Second, general solution of the radiation problem, which essentially is the recognition that the radiation potential can be written in terms of an arbitrary, yet unknown displacement of the mat. And finally, the dynamic boundary condition (pressure determination) is used to determine the forcing function and thus to obtain the deformation of the mat by solving the orthotropic plate equation subject to the aforementioned boundary conditions. It is however, easier for further discussion, to use time complex representation for the incident wave elevation and the resulting deformations of the mat, i.e.,

$$\eta(\vec{x},t) = \text{Rej}\left\{ \tilde{\eta}(\vec{x}) \, e^{j\omega t} \right\}$$

and $\qquad (8)$

$$W(\vec{x},t) = \text{Rej}\left\{ \tilde{W}(\vec{x}) \, e^{j\omega t} \right\}$$

3.1 Diffraction problem

Using the linearity of the problem and the aforementioned complex representation of the incident wave, the potential problem reduces to solving for the diffraction and radiation potentials separately. In both cases Green functions are used to find the required

potentials. The idea is to consider the submerged portion as a sum of panels fitted with pulsating sources. The resulting potential then describes the fluid motion. The corresponding unit strength sources are in fact the Green functions $G(x, \xi; t)$, which are distributed on the submerged portion B. For a given source and field point combination, this is evaluated using the MIT FINGREEN subroutine [10]. The details of this development follows very closely the procedure used by Webster [7]. With some manipulation, the complex amplitude of the diffraction potential can be represented as follows (see Appendix for details):

$$\tilde{\phi}_{D_m} = H_{mk} \left\{ \left(\vec{n} \cdot \vec{\nabla}_z \right) \tilde{\phi}_I \right\} \Big|_{(\overline{x}_k, \overline{y}_k, \overline{z}_k)} \qquad (9)$$

The complex matrix H_{mk}, which is referred to as the "hydrodynamic influence matrix", captures the hydrodynamics of the problem and the repeated indices indicate summation. This procedure for the solution of the diffraction problem is essentially finding the strength of the sources of the Green functions necessary to satisfy the kinematic boundary condition on the body (see eqn 3).

3.2 Radiation problem

By a similar procedure, the radiation potential on the body can be given in terms of the hydrodynamic influence matrix as (see Appendix for details):

$$\tilde{\phi}_{R_m} = j \, \omega \, H_{mk} \, \tilde{W}_k \qquad (10)$$

Further, the complex pressure p_{R_k} on the kth panel due to the motion W_m of the mth panel can be determined from Euler's integral, yielding,

$$\tilde{p}_{R_m} = -\rho \, \omega^2 \, H_{mk} \, \tilde{W}_k \qquad (11)$$

It is thus clear that if we know the complex amplitude W_k, of the normal deformation patterns on all the panels, we can compute the radiation potential ϕ_{R_m}, and the resulting pressure therefrom. However, the complex motion amplitudes are yet unknown.

3.3 Combined hydroelastic problem

In solving the combined hydroelastic problem, the following assumptions are made:

1) The bow, stern and the sides of the undeformed mat are vertical while, the bottom is flat and horizontal.

365

2) Only loading on the bottom of the mat is considered for the determination of the motions.

3) The kinematic conditions on the bow, stern and the sides must be satisfied.

The dynamic pressure acting on the bottom of the mat can be assembled using the results from the previous section and the Euler's integral for all three potentials. These can then be used in conjunction with Eqn. 7 to yield,

$$D_x \frac{\partial^4 \tilde{W}}{\partial x^4} + 2H \frac{\partial^4 \tilde{W}}{\partial x^2 \partial y^2} + D_y \frac{\partial^4 \tilde{W}}{\partial y^4} + \qquad (12)$$

$$\tilde{W} \{\rho g - m \omega^2\} = -\rho \frac{\partial}{\partial t} \{\tilde{\phi}_I + \tilde{\phi}_D + \tilde{\phi}_R\}$$

It was shown in the previous section that in order to know the pressure due to the radiation potential, the complex amplitude of the deformations must be known. From the above equation it is clear that the complex amplitude of the deformations can be obtained if the pressure due to the radiation potential along with the other pressures is known. Hence, these two problems need to be solved simultaneously. We use the representation of ϕ_R in terms of the hydrodynamic influence matrix (Eqn. 10) to simplify the above equation. The hydroelastic equation for the determination of motions at the panel m then reduces to

$$D_x \frac{\partial^4 \tilde{W}}{\partial x^4}\bigg|_m + 2H \frac{\partial^4 \tilde{W}}{\partial x^2 \partial y^2}\bigg|_m + D_y \frac{\partial^4 \tilde{W}}{\partial y^4}\bigg|_m +$$

$$\{\rho g - t \rho_s \omega^2\} \tilde{W}\bigg|_m + j \omega^2 H_{mk} \tilde{W}_k$$

$$\qquad (13)$$

$$= -\rho \frac{\partial}{\partial t} \{\tilde{\phi}_I + \tilde{\phi}_D\}\bigg|_m$$

The above equation is solved subject to "free edge" conditions that were stated earlier. Mathematical representation of these boundary conditions are:

$$M_x = -\left(D_x \frac{\partial^2 W}{\partial x^2} + D_1 \frac{\partial^2 W}{\partial y^2}\right)$$

$$M_y = -\left(D_y \frac{\partial^2 W}{\partial y^2} + D_1 \frac{\partial^2 W}{\partial x^2}\right)$$

$$\qquad (14)$$

$$V_x = -\left(D_x \frac{\partial^3 W}{\partial x^3} + H \frac{\partial^3 W}{\partial y^2 \partial x}\right)$$

$$V_y = -\left(D_y \frac{\partial^3 W}{\partial y^3} + H \frac{\partial^3 W}{\partial x^2 \partial y}\right)$$

4 NUMERICAL SOLUTION

The numerical solution of the problem follows a two step procedure. First, the diffraction problem is solved. The solution of the diffraction problem is the determination of the hydrodynamic influence matrix. For this purpose, two matrices B_{kl} and U_{mi} are evaluated (see Appendix) by using the MIT FINGREEN subroutine. The mat is discretized into panels in a manner similar to that used in strip theory computations, so as to yield a better "conditioned" matrix. The mathematics of this computation is in the Appendix. It is, however, to be noted that the hydrodynamic influence matrix, H_{mk}, is calculated only once for each frequency; it is used in the representation of the pressure due to the radiation potential, therefore the procedure doesn't require the explicit solution of the radiation potential. This part of the program was tested by comparison with the test cases used by Faltinsen [6]. With the diffraction problem solved, and the hydrodynamic influence matrix evaluated, the next step is the solution of the combined hydroelastic problem by using centered finite difference approximation of the combined hydroelastic equation (Eqn. 13). The boundary conditions, discretized using one-sided differences, are suitably applied at the panels close to the edges. The following linear system results from the discretizations.

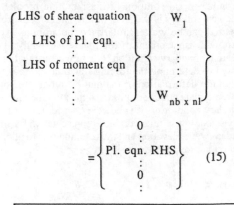

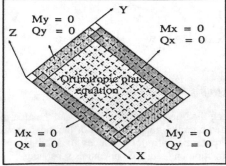

Figure 2: Schematic sketch showing the application of boundary conditions at the edges of the mat.

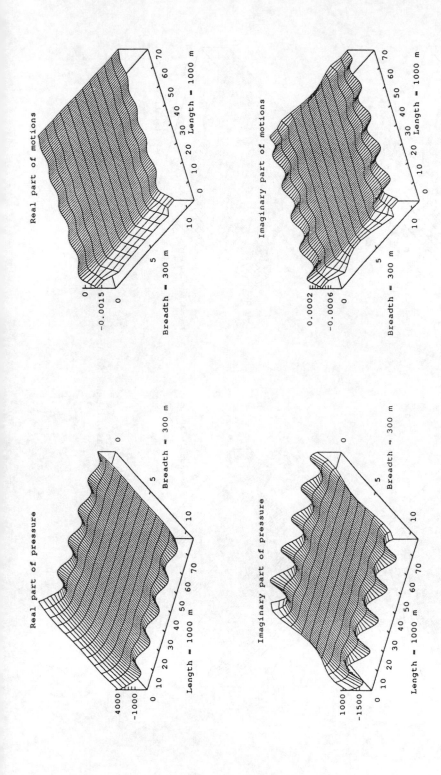

Figure 3: Pressure and motions of the mat for $\lambda = 200$m and $D = 4.04 \times 10^2$ per meter.

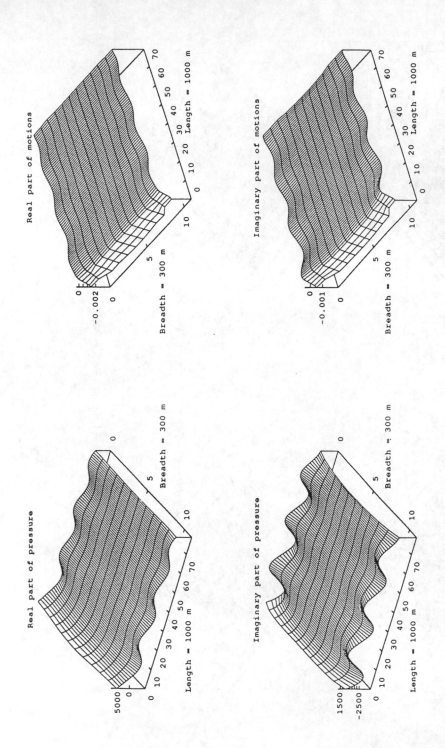

Figure 4: Pressure and motions of the mat for $\lambda = 300\text{m}$ and $D = 4.04 \times 10^2$ per meter.

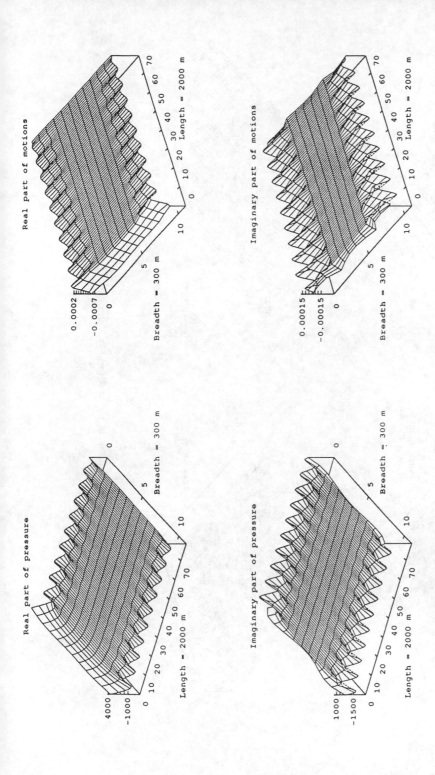

Figure 5: Pressure and motions of the mat for $\lambda = 200$m and $D = 4.04 \times 10^2$ per meter.

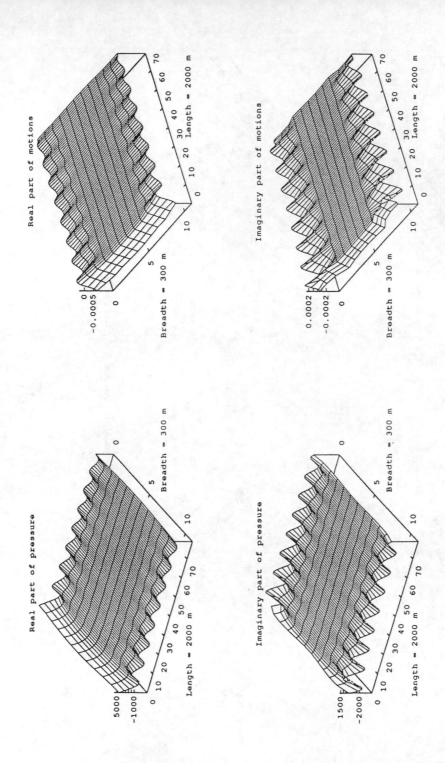

Figure 6: Pressure and motions of the mat for $\lambda = 250$m and $D = 4.04 \times 10^2$ per meter.

370

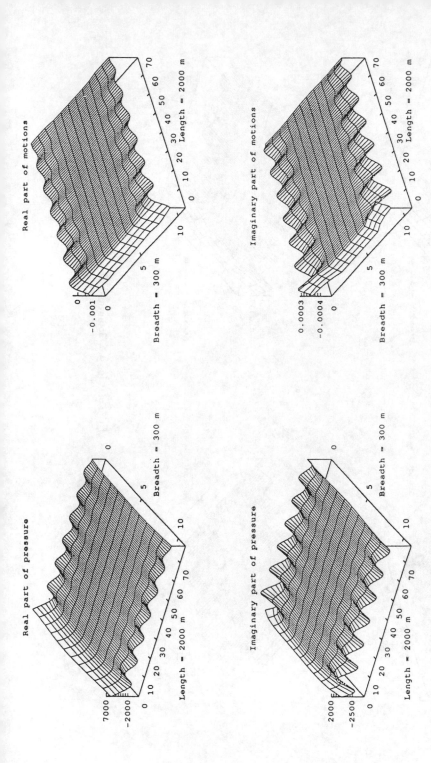

Figure 7: Pressure and motions of the mat for $\lambda = 300$m and $D = 4.04 \times 10^2$ per meter.

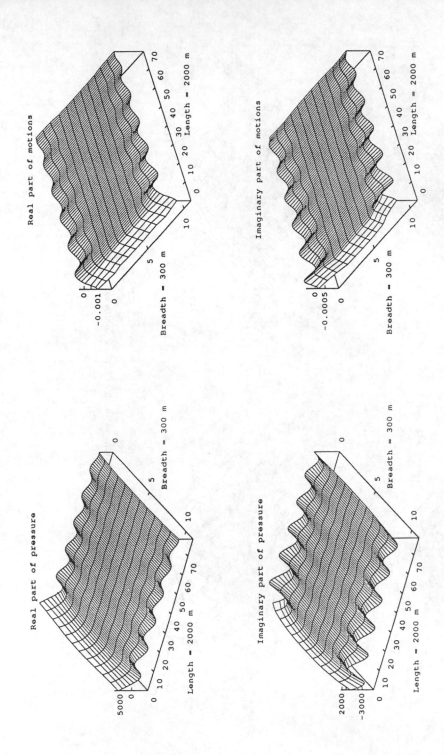

Figure 8: Pressure and motions of the mat for $\lambda = 350$m and $D = 4.04 \times 10^2$ per meter.

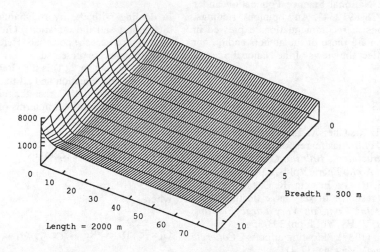

Amplitude of pressure

8000
1000
0
10
20
30
40
50
60
70
Length = 2000 m

0
5
10
Breadth = 300 m

Figure 9: Amplitude of pressure on the bottom of the mat for $\lambda = 350m$ and $D = 4.04 \times 10^2$ per meter

The above linear system is solved by gaussian elimination. The minimum number of panels is chosen so that the incoming wave and the resulting motions are resolved. Figure 2 shows the application of the boundary conditions in forming the above linear system.

5 RESULTS

The figures that follow are the results obtained from the computer program developed to solve the problem. Due to restrictions on memory, the results are provided for structures that are 1000m and 2000m long, instead of the required length of 2500m. The results can be used to understand the behavior of the structure in waves. Figures 3 and 4 are for structure that is 1000m long and has been provided for better resolution of the motions. The panel discretization used is 80 x 12 along the length and breadth respectively. Figure 5 to Figure 8 are for 2000m long structures of the same structural characteristics and discretization as the 1000m structure.

5.1 Discussion

For the cases presented here, the waves were assumed to be incident along the length of the mat. The following observations were made:

1) For the rigidity ranges in both the steel and concrete regimes the deformations of the structure seemed to be diffraction driven.

This can be seen from the plot of the pressure and the corresponding motions in the various figures.

2) Most significant motions were found to be at the forward end of the structure. This can be understood as a consequence of the very large pressure gradients at the bow of the structure. The figures that follow show the real and imaginary parts of the pressures on the bottom and the resulting motions of the mat. A plot of the complex amplitude of the pressure on the bottom of the mat is shown to give an estimate of the pressure gradients along the length of the mat (see Figure 9).

3) It was observed that the motions not only die out from the port (or starboard) end inwards (towards the centerplane), but also diminish along the length of the mat.

ACKNOWLEDGMENTS

This paper is funded in part by a grant from the National Sea Grant College Program, National Oceanic and Atmospheric Administration, U.S. Department of Commerce, under grant number NA89AA-D-SG138, project number R/OE-23, through the California Sea Grant College. The views expressed herein are those of the authors and do not necessarily reflect the views of NOAA or any of its sub-agencies. The U.S. government is authorized to reproduce and distribute for governmental purposes.

373

This material is based on the work partially supported by the National Science Foundation under Award No. BCS-8912341. Any opinions, findings, and conclusions or recommendations expressed in this publication are those of the authors and do not necessarily reflect the views of the National Science Foundation.

REFERENCES

[1] Riggs, H. R., Che, X. L. and Ertekin, R. C. (1991). "Hydroelastic response of very large floating structures", *10th Int. Conf. on Offshore Mech. and Arctic Engr.*, Vol I-A, pp. 291-300.

[2] Riggs, H. R. and Ertekin, R. C. (1991). "Hydroelastic response of very large floating structures", *1st Int. Conf. on Very Large Floating Structures*, VLFS, Vol I, pp. 333-354.

[3] Lemke, E. (1987). "Floating airports", *Concrete International*, May, pp. 37-41.

[4] Bishop, R. E. D. and Price, W. G. (1979). *Hydroelasticity of ships*, Cambridge University Press, Cambridge, U. K.

[5] Faltinsen, O. M. (1990). *Sea loads on ships and offshore structures*, Cambridge University Press, Cambridge, U. K.

[6] Faltinsen, O. M. and Michelsen, F. (1974). "Motions of large structures in waves at zero Froude numbers", Proc. Int. Symp. on Dynamics of Marine Vehicles and Structures (eds. Bishop, R. E. D. and Price, W. G.), Mechanical Engineering Publications Ltd., London, pp 91-106.

[7] Webster, William C. (1991). "Considerations for the design of floating airports", *U. S. Marine Facilities Panel, UJNR Meeting*, May.

[8] Webster, William C. (1975). "The flow about arbitrary, three dimensional smooth bodies", *Journal of Ship Research*, Vol. 19, No. 4, pp 206-218.

[9] Webster, William C. (1979). "Computation of hydrodynamic forces induced by general vibration of cylinders", *Journal of Ship Research*, Vol. 23, No. 1, pp. 9-19.

[10] Newman, J. N. and Sclavounos, P. D. (1986). *User manual for FINGREEN*, Department of Ocean Engineering, MIT.

APPENDIX

In dealing with the hydrodynamic problem (both diffraction and the radiation), Green functions are used to represent the potentials. Here the details of the developments are presented.

If $G(x, y, z, \xi, \eta, \zeta; t)$ is the Green function satisfying the field equation and all the boundary conditions except the KBBC, then the required diffraction potential can be written in terms of the Green functions as follows:

$$\tilde{\phi}_D = \int_B Q(s) \; \tilde{G}(\vec{x},\vec{\xi};\omega) \, ds \qquad (16)$$

where,

$$G(\vec{x},\vec{\xi};t) = \mathrm{Re}_i \{\tilde{G}(\vec{x},\vec{\xi};\omega)\} \qquad (17)$$

and $Q(s)$ is the source distribution which is to be determined from the KBBC. The KBBC can then be written as:

$$\left\{ \left(\vec{n}.\vec{\nabla}_z\right) \int_B Q(s) \; \tilde{G}(\vec{x},\vec{\xi};\omega) \, ds \right\} \Bigg|_B$$

$$= -\left(\vec{n}.\vec{\nabla}_z\right) \tilde{\phi}_I \Big|_B \qquad (18)$$

Further simplification can be made by discretizing the structure into panels and thereby writing the above equation in matrix form. It is assumed at this point that the source strength remains a constant over each panel but however, could vary from one panel to the other. The boundary conditions are to be satisfied at the center of each panel. Therefore, the surface integral which appears in the exact representation has to be evaluated approximately using a suitable quadrature rule. Upon discretization, the matrix representation of the above equation would be:

$$B_{kl} Q_l = \left\{ \left(\vec{n}.\vec{\nabla}_z\right) \tilde{\phi}_I \right\} \Bigg|_{(\bar{x}_k,\bar{y}_k,\bar{z}_k)} \qquad (19)$$

where B_{kl} is the matrix of coefficients evaluated from the following integral over panel l, the collocation point being at the center of the panel k :

$$B_{kl} = \left\{ \left(\vec{n}.\vec{\nabla}_z\right) \int_l \tilde{G}(\vec{x},\vec{\xi};\omega) \, ds \right\} \Bigg|_k \qquad (20)$$

374

Since the integration is with respect to the source point, the expression can be further simplified to

$$B_{kl} = \int_l \left. \frac{\partial \tilde{G}(\vec{x}, \vec{\xi}; \omega)}{\partial n} \, ds \right|_{(\overline{x}_k, \overline{y}_k, \overline{z}_k)} \tag{21}$$

where the differentiation within the integral sign is with respect to the field point. Therefore, Q_l can be obtained by:

$$Q_l = - B_{kl}^{-1} \left. \left\{ \left(\vec{n} . \vec{\nabla}_z \right) \tilde{\phi}_I \right\} \right|_{(\overline{x}_k, \overline{y}_k, \overline{z}_k)} \tag{22}$$

Once Q_l is known, the diffraction potential can be written using Eqn. 16 as:

$$\tilde{\phi}_{Dm} = U_{mi} \, Q_i \tag{23}$$

where, U_{mi} represents:

$$U_{mi} = \left. \int_i \tilde{G} \, ds \right|_{(\overline{x}_m, \overline{y}_m, \overline{z}_m)} \tag{24}$$

Using a similar representation as above, the radiation potential can be written as follows:

$$\tilde{\phi}_{Rm} = U_{mi} \, Q_i \tag{25}$$

In the above equation,

$$Q_l = B_{kl}^{-1} \left. \{ \vec{n} \cdot \vec{V} \} \right|_{(\overline{x}_k, \overline{y}_k, \overline{z}_k)} \tag{26}$$

where $\vec{V}$ is the induced velocity at each panel collocation point. For the case of harmonic motion of the panels, this expression can be simplified to:

$$Q_l = j \omega B_{kl}^{-1} \left. \{ \vec{n} \cdot \vec{W} \} \right|_{(\overline{x}_k, \overline{y}_k, \overline{z}_k)} \tag{27}$$

where, $\vec{W}$ is the complex displacement of the collocation point on the panel at the at the specified location. The hydrodynamic influence matrix referred to in the paper represents

$$H_{mk} = B_{ml}^{-1} \, U_{lk} \tag{28}$$

Hydroelasticity in Marine Technology, Faltinsen et al. (eds) © 1994 Balkema, Rotterdam, ISBN 90 5410 387 6

Influence of flexibility on the motions and deflections of an airport-oriented floating long offshore structure

T. Hirayama, N. Ma & S. Ueno

Naval Architecture and Ocean Engineering, Yokohama National University, Japan

ABSTRACT: The hydroelastic responses of an airport-oriented floating long offshore structure in short-crested irregular waves are dealt with in this paper using both experimental and theoretical approaches. Two types of models are used for studying the influence of flexibility. The transfer functions of the displacements and structural deformations of the structure in regular waves are calculated based on a practical method which combining hydrodynamic force calculation and structural modal analysis using wet mode, and the results are verified by experiments. A systematic model tests are conducted to investigate the responses of flexible structure in short-crested irregular waves moored by TLP type buoys. The influences of flexibility of structure and directional distributions of irregular waves on motions and deflections are clarified. Furthermore, the responses in short-crested irregular waves are estimated and comparisons of predicted spectra and statistical values of displacements and bending moments with experimental ones showed good agreements.

1 INTRODUCTION

Recently, a large number of plans for offshore development and ocean-space utilization have been proposed such as large offshore plants, floating airports, artificial islands and so on. In the cases of these supra-structures, generally the scale of structure has become larger and natural frequencies of structural vibrations become lower. If a large structure becomes long in its size, its structural flexibility, such as stiffness of bending, is consequently reduced, and elastic vibration due to waves increasing relatively. The relation between frequency of elastic vibration and that of ocean waves, slowly varying external forces is therefore very important. When the structure has extremely lower structural flexibility, the mode shape of vibration will correspond with the wave profile and a large deflection of structure will be possibly caused by waves.

When a floating airport was once proposed in Japan, Ando et al. (1983) have made a feasibility study concerning to the local behavior of the structure, Aoki et al. (1985) calculated the wave-induced vibration and bending moment of the structure based on beam theory and compared with their experiments. However, these studies did not consider the effect of the shape of directional distribution of irregular waves. When a large flexible structure is subjected to short-crested waves the wave induced responses including the structural deformations are possibly considered as more complicated than the case of long-crested waves. For these reasons, Takezawa et al. (1992a, 1992b, 1993) carried out a series of experiments in short-crested irregular waves generated at the towing tank of Yokohama National University. The motion responses and structural deformations of the body under various conditions of mooring with tensioned leg buoys has being reported.

In this paper, emphasis is placed on the influence of flexibility of structure on the response, such as local displacement and bending moments at deck part. Theoretical calculations for these responses in regular waves are performed by combining the wave force calculation and structural modal analysis. This method is considered to be useful for the complicated structure. The calculated results are compared with experiments to verify the applicability of applied method. Two models with different structural flexibilities are considered, and the influence of flexibility on structural deformation and local displacement are discussed. The experiments were conducted with different sorts of short-crested irregular wave. Since the mooring is important for realization of this type of structure, the effects of mooring upon motions and structural responses are also taken into account.

Finally, the response in short-crested irregular waves are estimated and compared the results with measured one. The effects of the short crestedness on response are clarified and from the viewpoint of design of a floating airport some comments are given.

2 EXPERIMENTS

2.1 Outline of experiments

The experiments were carried out at towing tank of Yokohama National University which is 100m long, 8m wide and 3.5m deep. The generation of the directional (short-crested) wave having a mean direction parallel to the longitudinal wall of long tank is one of the important points of the experiments. This was made possible through the research of Takezawa et al.(1988) and others. A system using laser beams which measures wave height

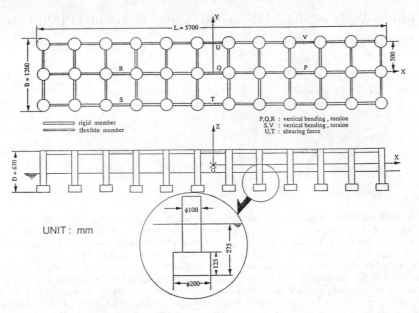

Fig.1 **Configuration of model, definition of coordinate system and arrangement of strain measurements**

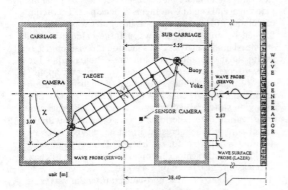

Fig.2 Scheme of the set-up for tests in longitudinal waves with mooring of two TL-type buoys

and wave slope developed by Takezawa et al.(1989) has been established for multi-direction wave measurement. The two dimensional wave spectrum, frequency based power spectrum and directional distribution function, can be obtained by means of MLM(Maximum Likelihood Method) analysis.

Two types of directional distribution of short-crested irregular waves, one with wide directional distribution (referred as wide) and the other with narrow directional distribution (referred as narrow) in addition of long-crested irregular waves (referred as long) are were applied for experiments. The short-crested irregular waves have JONSWAP type spectrum and were generated between the

periods 0.8 to 1.2 sec, having significant wave height from 2 to 8 cm. In order to obtain the wave transfer functions of responses tests in transient water waves and regular waves were performed. The frequency range of transient water wave is between 2.0 to 8.0 rad/sec.

2.2 The applied models

An airport-oriented floating offshore structure, which has quite large length-to-breadth ratio is applied in this study. The model used in the experiment is composed of 36 column with underwater footings which are connected to each other by aluminum pipe beams (60mm height, 30mm wide and 3mm thickness) from above structure and its configurations together with a structure-fixed coordinate system are shown in Fig.1.

Two models having different structural flexibilities were applied in experiments, one is called as rigid structure and the other is called as flexible structure. The only difference of these two models is that the connections has different flexibilities. The model is divided into four parts 1/4L, midship and 3/4L (R,Q,P points as illustrated in Fig.1), each of them having 9 columns. For the flexible structure, instead of using 60mm pipe beam, 4mm thick aluminum plates with same breadth were used for connecting at these connections (See Fig.1). In this case the stiffness of vertical bending is 0.001 times that of rigid structure. The principal dimensions of each model is shown in table 1.

The flexible model dealt with in this paper is link-connected type structure with locally reduced EI and it has very small flexural rigidity comparing with a conventional one. From this reason the occurrence of resonant vibra-

Table 1. Principal dimensions of model

Length(m)	5.70
Breadth(m)	1.20
Depth(m)	0.676
Draft(m)	0.275
Displacement(Kg)	183.78
KG(m)	0.327
GMT(m)	0.025

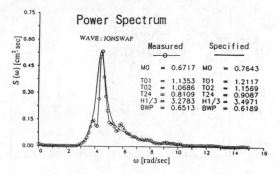

Power Spectrum

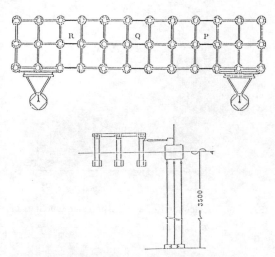

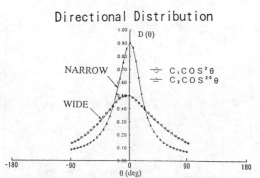

Fig.4   Measured power spectrum and directional
distributions of short-crested waves

Fig.3   Arrangement of mooring system with two
TL-type buoys for tests in beam seas

Table 2. Principal dimensions of buoy

Diameter(m)	0.38
Depth(m)	0.30
Draft(m)	0.16
Displacement(Kg)	18.0

tions among waves will be expected and magnitudes due
to such vibrations may be of dominant order.

2.3 Mooring system

The mooring for such large structure is a one of most important items because of the small motions are requested
especially for an airport. In this study, the tensioned leg
buoys mooring is adopted to prevent the drift of the structure. For the sake of simplicity of experiments, two
tensioned leg (TL) buoys, each of them having four tethers
that were hooked at the bottom of the tank, were used. The
buoys are connected to the structure by a yoke. The structure and the yoke are connected to each other with single
degree of freedom joint which can freely rotate around y-
axis. Between the buoy and the yoke, a single degree of
freedom (pitch) joint and a bearing which can freely rolls
are used. These connection are necessary so that the motions (heave,roll,pitch and yaw) of the buoys will not affect the motions of the structure. The buoy is a cylinder
with circular section, the main particulars are shown in
table 2.
The initial tension of tether and arrangement of buoys
were designed for the survival condition and initial tension

for each tether was adjusted to 4.0 kgf. In the case of head
sea condition (the waves propagate in direction of -x axis,
referred as 0 deg), the buoys were connected to the structure, one at fore (weather side) and the other at the aft (lee
side) as shown in Fig.2. For beam sea condition (the waves
come from starboard, referred as 90 deg), two buoys were
installed at weather side and connected to the structure at
the position against fore and aft as shown in Fig.3.

2.4 Measurements

The three dimensional rigid motions and displacement of
the structures were measured by a optical measuring system with a special TV camera and a number of LED(light
emitting diode) targets. Displacements of buoys in x and
y directions are recorded by the same system.
Waves are measured at three points as illustrated in Fig.2,

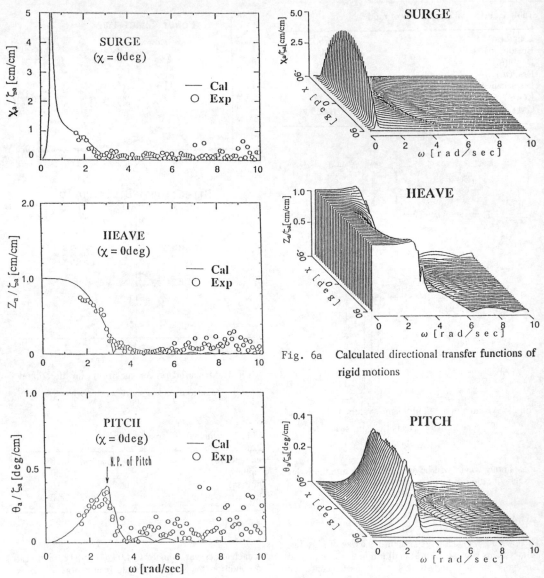

Fig.5 Transfer functions of rigid motions in head seas

Fig. 6a Calculated directional transfer functions of rigid motions

Fig. 6b Calculated directional transfer functions of rigid motions

two of them are servo needle type wave probes for the wave height and one is the laser beam wave surface probe as mentioned earlier for 3-dimensional wave measurement (wave height and wave slope in x, y directions) respectively. An example of measured power spectrum and directional distribution functions of short-crested irregular waves are shown in Fig.4. It was found that the short-crested irregular waves are simulated precisely.

The structural deformations at beam element are also measured simultaneously using strain gauge, the positions

of strain measurements are shown in Fig.1. As for mooring tension in each tether, ring gauge type tensionmeter was used. Measured data were recorded on a personal computer with a length of 10 and 25 minutes and 10 Hz sampling time.

Before the tests in waves, free oscillation tests in still water under various modes of motions and structural vibrations were carried out. These tests enable us to experimentally determine the hydrodynamic coefficients which are indispensable for developing the theoretical model of

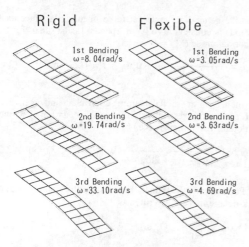

Fig.7 Mode shapes of rigid and flexible structure

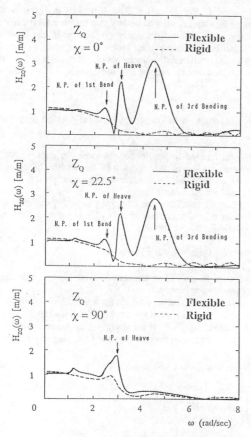

Fig.8 Calculated transfer functions of Z-displacement at point Q of rigid and flexible structure

calculation. Through the free Oscillation tests the added masses, damping coefficients, natural periods of motions and structural vibrations were obtained from measured timehistory.

3 THEORETICAL CALCULATIONS

In this chapter, the theoretical background of calculation is discussed. When estimating the responses of motion and deflection in waves it is important to calculate the transfer functions. In the calculation , incident wave is taken as a regular sinusoidal wave, the fluid is ideal and the amplitudes of motion and deflection are both of small order. It is assumed that there is no interactions between columns and the wave is traveling without damping in its height. The water depth is assumed to be infinitely deep.

3.1 Transfer functions in regular waves

The calculation of hydroelastic responses in regular waves are composed from two parts; hydrodynamic analysis and structural modal analysis. By combining of these two, the displacements and bending moments can be calculated. The method is considerably benefit for a complicated structure because it decouples the structural analysis from hydrodynamics.

The first order wave exciting forces on each column are evaluated from the incident wave potential. For each individual column the phase difference of encountered wave are taken into account according to the position of column. The whole wave forces and the rigid motions are calculated based on a well established 'Hooft' method, see Hooft (1971). The detail formulation is omitted here. By this hydrodynamic analysis, the rigid motion and the wave exciting forces can be obtained. The wave forces on each column are provided for calculation of deflections of structure as the nodal external forces as explained later.

With regard to the deflections of structure in waves, an approximated method of combining wave force calculation and structural modal analysis is employed. The equation of displacement of a mass-damping vibration system with multi-degrees of freedom can be written as

$$[M]\,\ddot{u} + [C]\,\dot{u} + [K]\,u = f\,(\,\ddot{u},\dot{u},u,t\,) \qquad (3.1)$$

where $u = u(x,y,z,t)$ is the displacement vector, $[M],[C],[K]$ are the structural mass-, damping- and stiffness matrices, respectively; f is the vector of external force. Overdot denote differentiation with respect to time t. For the global deformation, u may be approximated adequately by an aggregate of the m lowest eigenmodes $u_r(x,y,z)$,

$$u\,(\,x,y,z,t\,) = \sum_{r=1}^{m} u_r(\,x,y,z\,)\,p_r(t) \qquad (3.2)$$

in which $p_r(t)$, r=1,2,..,m are the principal coordinates. The first 6 modes corresponding to the rigid motions, and the remaining are flexible modes. From Eq.(3.1) and

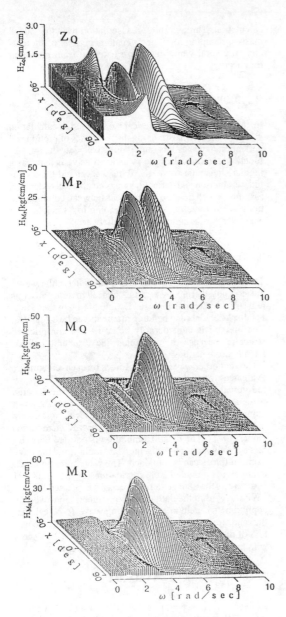

Fig.9 Calculated directional transfer functions of Z-displacement and vertical bending moments of flexible structure

Eq.(3.2) the following equation can be obtained.

$$\sum_{r=1}^{m} \left[M_{kr} \ddot{p}_r(t) + C_{kr} \dot{p}_r(t) + K_{kr} p(t) \right] = F_k \qquad k=1,2,\cdots,m \qquad (3.3)$$

where $M_{kr} = \mathbf{u}_k^T[M]\mathbf{u}_r$, $C_{kr} = \mathbf{u}_k^T[C]\mathbf{u}_r$, $K_{kr} = \mathbf{u}_k^T[K]\mathbf{u}_r$; $F_k = \mathbf{u}_k^T\mathbf{f}$

is the generalized external force in mode k. Since each mode is considered to be independent each other, for the normal mode ($M_{kr} = 1$) we have

$$p_r = \frac{1}{\omega_r^2 - \omega^2 + 2i\zeta_r \omega \omega_r} \mathbf{u}_r F_r, \quad r = 1,2,\cdots,m \qquad (3.4)$$

where ζ_r, ω_r are modal damping coefficient and eigenvalue for r-th mode respectively. Thus the steady transfer function for r-th mode is given by

$$G_r(\omega) = \frac{1}{\omega_r^2 - \omega^2 + 2i\zeta_r \omega \omega_r} \mathbf{u}_r F_r, \quad r = 1,2,\cdots,m \qquad (3.5)$$

Furthermore, the transfer function for the displacement of k-th column can be written as

$$\mathbf{H}_k(\omega) = \sum_{r=1}^{m} \mathbf{u}_{kr} G_r(\omega) \qquad (3.6)$$

where $\mathbf{H}_k(\omega)$ includes the translatory and rotational displacements of k-th node, m is the total number of modes including rigid modes.
$\mathbf{u}_r$ in Eq. (3.5) is calculated from structural analysis based on finite element method and by that the structural deformations and natural frequency of each mode can be obtained. As for the bending moment, it can be calculated from the rotational displacements of structural member. In our calculation the highest flexible mode up to 8 is considered.

3.2 Prediction of responses in short-crested irregular waves

It is attempted to predict the responses in short-crested waves by a stochastic approach in this study. The spectra and statistical values of maxima of irregularly varying response are estimated.

The spectrum of linear responses in a long-crested irregular waves can be calculated by linear superposition as following

$$S_x(\omega) = |H_x(\omega)|^2 S_\zeta(\omega) \qquad (3.7)$$

where $H_x(\omega)$, $S_\zeta(\omega)$ are the transfer function and spectrum of wave respectively.

On the other hand the formula for estimation of responses in short-crested irregular waves can be introduced by extension of Eq. (3.7). If we assuming the short-crested irregular waves is represented by the following expressions

$$S_\zeta(\omega,\theta) = D(\omega,\theta) S_\zeta(\omega) \qquad (3.8)$$

$$\int_{-\pi}^{\pi} D(\omega,\theta) \, d\theta = 1 \qquad (3.9)$$

where $D(\omega,\theta)$, $S_\zeta(\omega)$ are the directional distribution function and power spectrum respectively. Accordingly, the response spectrum is given by

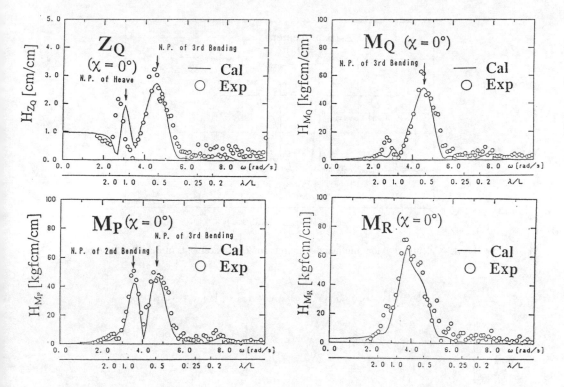

Fig.10 Transfer functions of Z-displacement and bending moments of flexible structure in head seas

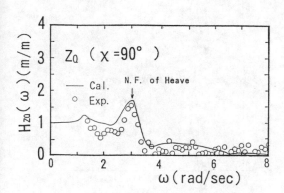

Fig.11 Transfer functions of Z-displacement of flexible structure in beam seas

$$S_x(\omega) = \int_{-\pi}^{\pi} |H_x(\omega)|^2 D(\omega,\theta) S_\zeta(\omega) \, d\theta \quad (3.10)$$

The maxima of response is one of most interesting items when evaluating the performances of the structure. According to Cartwright and Longuet-Higgins (1956), the general probabilistic density function of a steady random process x(t) can be given by the following expression

$$P(\xi,\varepsilon) = \frac{\varepsilon}{\sqrt{2\pi}} e^{-\frac{\xi^2}{2\varepsilon^2}} + \sqrt{\frac{1-\varepsilon^2}{2\pi}} \cdot \xi \, \varepsilon^{-\frac{\xi^2}{2}} \cdot \int_{-\infty}^{\xi\sqrt{\frac{1-\varepsilon^2}{\varepsilon^2}}} e^{-\frac{u^2}{2}} du \quad (3.11)$$

where $\xi=h/\sigma_x$, h is maxima of x(t), σ_x is standard deviation of x(t). ε is the band width parameter of spectrum and is able to be calculated from spectrum directly.

From above descriptions, the procedure for calculation of hydroelastic responses in short-crested wave is summarized as following.

(1) calculating wave exciting forces etc. in regular waves
(2) performing the structural modal analysis
(3) combining above two to obtain the transfer functions
(4) estimating the response spectrum
(5) predicting the statistical values of maxima

4 RESULTS AND DISCUSSIONS

The results of calculations and comparisons with measured results are shown and discussed in this chapter. The main concern is placed on displacement and bending moment.

4.1 Natural frequencies

The measured and calculated natural frequencies of rigid

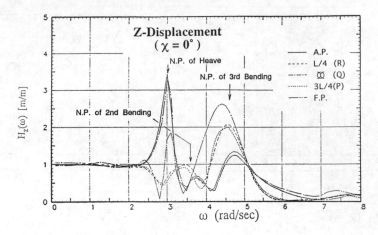

Fig.12 Calculated longitudinal distributions of Z-displacements of flexible structure

WAVE : JONSWAP T_{02}= 1.1s , $H_{1/3}$ = 4cm

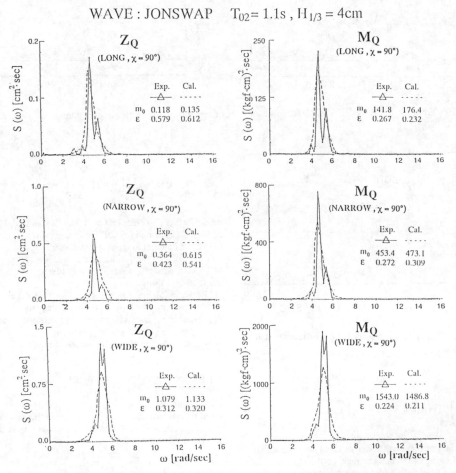

Fig.13 Response spectra of Z-displacement and vertical bending moment in long and short-crested irregular beam seas

Table. 3 Rigid and elastic natural frequencies (rad/sec)

	no mooring	moored**
Surge	---	0.45
Sway	---	0.49
Heave	3.02(3.07)*	3.08
Roll	0.72(0.65)*	0.94
Pitch	2.86(2.74)*	2.78
Yaw	---	0.74
1st bending	2.67(3.05)*	2.71
2nd bending	3.67(3.63)*	3.34
3rd bending	5.51(4.69)*	5.41
1st torsion	14.61	14.61

* denotes the calculated ones
**moored by TL buoys

motions and structural vibrations of flexible structure are shown in table.3 in model scale. However the calculation which neglects the effects of mooring shows overall coincidence with measured one. It also can be recognized that the natural frequencies of heave, pitch, roll and structural vibrations are not affected so much by the effects of mooring. Each natural frequencies are of smaller values except the first torsion mode.

4.2 Transfer functions

The transfer functions (amplitude) of rigid motions in head sea condition are shown in Fig.5. The calculation shows good agreement with experimental data except in very high frequency region for heave and pitch. In Fig.6, calculated directional frequency transfer functions of motions are shown. It can be seen that at all wave incident angles, surge , pitch show large amplitude at their natural frequencies but very small amplitude at other frequencies.

The calculated mode shapes of structures are shown in Fig.7. It can be seen that there is a apparent difference in rigid structure and flexible structure and the frequency of mode of vibration are of different order.

The calculated transfer functions of vertical displacement at center of structure (point Q as shown in Fig.1) for three angles of wave propagation are shown in Fig.8 with comparing with the rigid and flexible structure. With regard to wave frequency lower than 2 rad/sec there is no apparent difference between two structures. For the region of wave frequency higher than 2 rad/sec, the flexible structure shows drastical differences to rigid one and resonant phenomena with first and third bending mode give large response amplitudes. 3-D views of calculated directional transfer function of vertical displacement at point Q and bending moment at point P,Q,R are shown in Fig.9. It can be clearly observed that the 2nd and 3rd bending modes affect these responses strongly especially in head seas. Also the bending moment at point P shows different tendency with point R and this is considered to be caused by the slight difference in phase lag between each flexible mode. The calculation is verified by experiments and the results for displacement and bending moment in head sea and displacement in beam sea are shown in Fig.10, Fig.11

respectively. Both in head sea and in beam sea, a good correlation between calculation and experiment for displacement and bending moment is obtained.

Distribution of Z-displacement along centerline of flexible structure is shown in Fig.12. There is a apparent difference between the longitudinal positions especially in frequency from 2 up to 6 rad/sec. The response of the edge (end) of structure shows lager amplitude at natural frequency of heave, while the response of the center of structure has larger values at the frequency corresponding to natural frequency of 3rd bending mode than at elsewhere. This implies that it is important to consider the longitudinal distribution when evaluating the displacement of such type of structure.

4.3 Response spectra and maxima

The response spectra is therefore estimated using method described in section 3.2. The Z-displacement and vertical bending moment in short-crested irregular beam waves are estimated for 3 types of directional distribution and the comparison with experiments are shown in Fig.13. It can be observed that the measured spectra of displacement and bending moment has two peaks occurring. One where the wave has considerable power (left peak), and the other where the wave frequency is equal to the 3rd bending natural frequency (right peak). However these peaks are mainly in wave frequency region. The calculated spectra have slight difference from the measured ones, but generally the agreements is good for each directional distribution. The response tends increasing when the directional distribution becomes wide. Understandably, these response are easily affected by the mean direction of waves.

From the estimated response spectra, the significant values of responses (denoted as $\sqrt{m_0/m_{0\zeta}}$ in figure), which are divided by the significant wave heights, are calculated consequently. The comparisons of calculated significant values of Z-displacement and vertical bending moment with measured ones in short-crested irregular beam waves for 3 types of directional distribution are shown in Fig.14. Also in this case, a fairly good agreement between calculation and experiment is obtained and effects of short-crestedness is similar to the case of spectra.

The prediction of statistical values of maxima in irregular waves is then performed. As an example of results, Z-displacement along the longitudinal axis of structure is shown in Fig.15 with respect to maximum and 1/3 highest mean values. The experiment value shown in figure is analyzed directly from the measured timehistory of length about 7 minutes (249 peaks). It can be recognized that the displacement at midship is of larger order than others, and this can be considered as the contributions of 3rd bending is dominant. This is a remarkable feature of flexible structure of the type dealt with in this study.

5 CONCLUSIONS

The hydroelastic responses of floating long structure, composed with four relatively rigid body and elastic connection, moored by TL buoys in long and short-crested waves are investigated focusing attention on the displace-

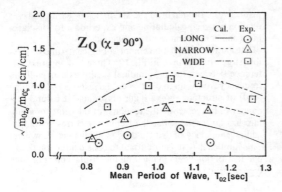

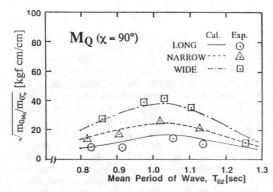

Fig.14 Significant values of Z-displacement and

vertical bending moment in long and short-

crested irregular beam seas

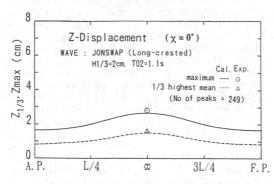

Fig.15 Statistical values of maxima of Z-displacement

in long-crested irregular head seas

ment and bending moment. Through extensive experiments and estimation works, it is proved that the proposed estimation method is available for assessment of an airport-oriented floating long structure , and following concluding remarks are derived.

(1) For large and long offshore structures as dealt with

in this study , structural deformations affect both the first and second order motions of structure, so that this effect should be taken into account carefully in calculations.
(2) The results from experiments indicates that the effects of directional spreading of waves on motions and deflections may be important, it is necessary to include this effect on the response estimation to ensure the safety of structure.
(3) The response transfer functions are calculated with good accuracy by approximated method combining hydrodynamic and hydroelastic structural analysis.
(4) The characteristic features of spectra and significant values of displacement and bending moment in short-crested waves are properly described, and the estimated values show rather good agreements with the results of model test.

Furthermore, the required flexural rigidity including deck part for an airport and the optimum arrangement of mooring buoys from the viewpoint of not only motion characteristics but also the deflections in waves should be investigated in detail as a future research.

6 ACKNOWLEDGMENTS

The authors appreciate the kindly advise from professor emeritus S.Takezawa of Yokohama National University and his developments of advanced experimental apparatuses. The assistance from Mr.K.Miyakawa and Mr. T.Takayama in execution of experiments and editing of manuscripts and the help received from graduated students of Seakeeping Laboratory of Y.N.U. who supported the experiments and data analyses are gratefully acknowledged.

REFERENCES

Ando,S., Ohkawa,Y. & Ueno,I. 1983. Feasibility study of a floating airport, *Report of Ship Research Institute, Japan, Supplement No.4*

Aoki,M.,Yago,k.,Hoshino,Endo,H., et al. 1985. Study on the structural strength and vibration of floating structure, *Report of Ship Research Institute, Japan, Supplement No.6*: 71-105

Cartwright,D.E., Longuet-Higgins,M.S. 1956. The statistical distribution of the maxima of a random function, *Proc. Royal Society, A 239*: 212-232

Hooft,J.P. 1971. A mathematical method of determining hydrodynamically induced forces on a semisubmersible, *SNAME, vol.79*: 28-70

Takezawa,S., Hirayama,T., Ueno,S., Kajiwara, H. 1992a. Experiments on responses of very large floating offshore structures in directional spectrum waves (first report), *Journal of Society of Naval Architects of Japan,Vol.171*:51- 63, (in Japanese)

Takezawa,S., Hirayama,T., Ueno,S., Akin,S.T. & Kajiwara, H. 1992b. Experiments on responses of very large floating offshore structures in directional spectrum waves(second report), *Journal of Society of Naval Architects of Japan, Vol.172*: 57- 68, (in Japanese)

Takezawa,S., Hirayama,T., Ueno,S., Akin,S.T. &
Kajiwara, H. 1993. Experiments on responses of very
large floating offshore structures in directional spectrum waves(third report), *Journal of Society of Naval
Architects of Japan,Vol.173*: 147-159

Takezawa,S., Kobayashi,K., Kasahara, A. 1988. Directional irregular waves generated in a long tank, *Journal
of Society of Naval Architects of Japan,Vol.163*: 222-
232, (in Japanese)

Takezawa,S., Miyakawa,K., Takayama, T., Itabashi, M.
1989. On the measurement of directional wave spectra
by the new wave measuring system using laser beams,
*Journal of Society of Naval Architects of
Japan,Vol.166*: 173- 185, (in Japanese)

Hydroelasticity in Marine Technology, Faltinsen et al. (eds) © 1994 Balkema, Rotterdam, ISBN 90 5410 387 6

Hydroelastic response of a floating runway

R.C.Ertekin & S.Q.Wang
Department of Ocean Engineering, University of Hawaii at Manoa, Honolulu, Hawaii, USA

H.R.Riggs
Department of Civil Engineering, University of Hawaii at Manoa, Honolulu, Hawaii, USA

ABSTRACT: The hydroelastic response of a floating runway under the action of waves is studied by using the Elastic Module Flexible Connector (EMFC) method with frequency-dependent hydrodynamic coefficients. A three-dimensional finite element method is used to model the structural loads and Morison's equation is used to model the hydrodynamic loads. Present predictions are compared with the available results based on the panel method and EMFC method with constant hydrodynamic coefficients.

1 INTRODUCTION

There will be a need in the future for floating runways and airports mainly due to the lack of adequate land space near coast lines where there are densely populated metropolitan areas. In addition, there already are environmental concerns such as pollution and noise associated with having an airport especially near residential areas.

A possible solution to those problems can be seen in a floating structure that can accommodate landing and take-off of large aircraft. Such a structure is envisioned as many modules connected by connectors, each module being a semi-submersible or a mat-like structure, such that the construction, transportation and deployment of each module can be achieved without much technical and economic difficulties.

A very large floating structure (VLFS) requires special response analysis techniques because of the presence of flexible modes of motion in addition to the usual rigid-body modes. There are several major approaches to the solution of the problem of determining the flexible modes of motion that eventually lead to the calculation of stresses and deformations either in the modules themselves and/or in the connectors.

One of these methods is the Rigid Module flexible Connector (RMFC) method (Wang et al. 1991) in which the rigid modules are connected by flexible connectors and the full three-dimensional hydrodynamic interaction between the modules is taken into account by using the single-symmetry, composite source distribution method (Wu 1984). This method is later extended to double-symmetry and applied to a five-module VLFS (Wu et al. 1993) and a 16-module VLFS (Ertekin et al. 1993).

The three-dimensional hydroelastic analysis of a VLFS requires significant computational resources in comparison with any other available method. This difficulty still exists despite the exploitation of any symmetries present in a VLFS. Therefore, an approximate method such as the RMFC method is necessary to obtain response predictions. On the other hand, in cases where the flexibility of each module becomes important and/or the nodal forces/moments are needed for detailed structural analysis, one must resort to alternate methods. One of these methods is the Elastic Module Flexible Connector (EMFC) method in which the three-dimensional finite element method is used for structural loading and Morison's equation with constant inertia and drag coefficients is used for hydrodynamic loading, and then the equations of motion are solved to directly obtain the nodal forces, displacements, and so forth. This direct method was devised recently and applied to a 16-module VLFS (Ertekin et al. 1993; Che 1993).

The first applications of hydroelasticity theories were on rather slender structures such as ships (Bishop and Price 1979). In such applications, the structure is typically modeled as a single beam and strip theory is used to model the fluid flow. For a semi-submersible, the assumption that each module can be modeled as a non-uniform beam is a dramatic departure from the usual slenderness assumptions valid for ships for example. However, such an assumption made in Riggs and Ertekin (1993) showed that with the proper handling of the forces (surge), that can not be supplied by the strip theory, such an approximation may be accurate enough to be used in the preliminary design of a VLFS.

In the classical use of the strip theory via Frank's close-fit method, it is assumed that the cross-section of the submerged structure does not deform. This allows the use of the two-dimensional hydroelasticity theory where the local distortions are assumed negligible compared with the longitudinal ones. However, such an assumption is not necessary in general, for it is possible to represent the three-dimensional finite element deformational response by the 'basic modes' of the structure cross-section (Wang et al. 1991; Che et al. 1992) or by directly calculating the generalized radiation forces and then the diffraction forces by using the Haskind-Hanaoka relationship in two dimensions (Yu 1980; Che et al. 1994).

All the approximate methods of hydroelasticity analysis are alternatives to the three-dimensional hydroelasticity theory which is based on the mode superposition approach where the fluid loading is supplied by the usual linear source distribution method. In this work, the EMFC method is extended by determining the frequency dependent hydrodynamic coefficients via the potential theory, and then using them in Morison's equation for the fluid loading. This way, it becomes possible to analyze both the rigid- and flexible-mode dynamics of a VLFS as large as a floating runway. Furthermore, in comparison with the prediction of fluid loading based on constant hydrodynamic coefficients, it is expected that the present refinement will lead to more accurate results at high frequencies.

We will first present the EMFC method of hydroelasticity analysis and then discuss the structural and hydrodynamical models we use in the response calculations of semi-submersible modules which are flexibly connected. For some details left out here the reader is referred to Ertekin et al. (1993), Che (1993) and Wang et al. (1993).

2 ELASTIC MODULE FLEXIBLE CONNECTOR (EMFC) METHOD

In this hydroelasticity method, the entire VLFS which is made up of many modules is assumed elastic. The modules are flexibly connected at the deck lever or at both the deck and pontoon levels. The finite-element method will be used to model the flexibility of modules. Assuming that each module is discretized into many elements, each being connected by nodes, we can write the equations of motion of the body as

$$[M_s]\{\ddot{D}(t)\}+[C_s]\{\dot{D}(t)\}+[K_s]\{D(t)\}$$
$$=\{F_f(t)\}+\{F_r(t)\} \qquad (2.1)$$

where $[M_s]$, $[C_s]$, $[K_s]$ are the NxN structural mass, damping and stiffness matrices, respectively, N is the number of degrees-of-freedom of the structure, $\{D(t)\}$ is the vector of nodal displacement and superposed dot refers to time derivatives, $\{F_f(t)\}$ is an Nx1 vector of hydrodynamic forces, and $\{F_r(t)\}$ is an Nx1 vector of hydrostatic restoring forces. $\{F_f(t)\}$ includes the added-mass, damping and exciting forces. We will assume that the nodal displacements are periodic with a frequency ω, and therefore, t in Eq.(2.1) is understood to represent this periodicity which is originated by the regular, long-crested waves impinging on the structure.

We will next briefly present the structural model used and then discuss to a greater extent the hydrodynamic model.

2.1 Structural EMFC model

The displacement-based finite element method is used to analyze the dynamic response of the

VLFS. It is convenient to employ one-dimensional slender structural elements to model the submerged members of the VLFS since the hydrodynamic loading will be based on Morison's equation, i.e. the pontoons and columns are modeled by three-dimensional frame elements. The deck of the semi-submersible module is modeled by a frame element grid.

A frame element is defined by two end nodes, each of which has three translational and three rotational displacement degrees-of-freedom. Hence, each node has one axial force, two shear forces, one torque and two bending moments. The displacements of the element at any point can be obtained from the displacements of the nodes by using interpolation functions:

$$\{\overline{u}(\overline{x}_1)\} = [N]\{\overline{d}\} \tag{2.2}$$

where $\{\overline{u}\}$ is a 6x1 vector of translational and rotational displacements at the element local coordinate $\overline{x}_1$, [N] is a 6x12 matrix of interpolation functions, and $\{d\}$ is the 12x1 vector of nodal displacements at the two end nodes of the element, [N] is a function of spatial variables, not time.

In addition to the strain-displacement equation, $\{\varepsilon\} = [B]\{\overline{d}\}$ and stress-strain equation, $\{\sigma\} = [E]\{\varepsilon\}$, where [B] and [E] are the strain-displacement and material stiffness matrices, respectively, one needs the equilibrium equations to obtain the equation of the motion in terms of nodal displacements. These equations can be obtained by using the principal of virtual displacements as it is well known in finite element analysis.

The mass density matrix, $[\overline{m}_s]$, of an element is a diagonal matrix if the local coordinate system is defined such that its origin is at the center of mass of the cross section and the axes correspond to the principal axes of the cross section (with respect to the mass distribution), both of which we assume here. The local mass matrix is given by

$$[\overline{m}] = \int_{L_e} [N]^T [\overline{m}_s][N] d\overline{x}_1 \tag{2.3}$$

Once the principle of virtual displacements is written for a single element, one can use the strain-displacement relationship and the constitutive equations to obtain the nodal

displacement equations in terms of the consistent element mass, damping and stiffness matrices and the external and internal load vectors. The equations for the nodal displacement vector must now be transformed to global coordinates, and the mass, damping and stiffness matrices in Eq. (2.1) are assembled.

2.1.1 Hydrostatic forces

The hydrostatic forces $\{F_r\}$ in Eq. (2.1) requires spacial attention within the frame work of the three-dimensional structural frame model since sectional restoring forces must be considered rather than the restoring forces for the whole structure as is the case in rigid-body motion analysis. The major difference between the hydrostatic restoring force of a unit section and that of the whole body is that the displacement of a unit section beam may not be equal to its weight.

It is convenient to write the restoring forces in terms of the hydrostatic stiffness matrix, i.e.

$$\{F_r(t)\} = -[K_r]\{D(t)\} \tag{2.4}$$

This stiffness matrix has both the stabilizing component, due to the waterplane area of the surface-piercing structural members, and destabilizing component, due to the shift of the application point of the net buoyancy forces of columns and pontoons. The hydrostatic restoring coefficient matrix in the local coordinates is a diagonal matrix because of the symmetry of the beam element.

The stabilizing component of the hydrostatic stiffness is modeled by locating a node at each intersection of the still-water plane and column and attaching discrete vertical and rotational springs to these nodes. The only non-zero components of the hydrostatic stiffness are given by

$$k_{x_3 x_3} = \rho g A_c \ , \ k_{x_1 x_1} = \rho g I_{x_1 x_1} \ , \ k_{x_2 x_2} = \rho g I_{x_2 x_2} \tag{2.5}$$

where A_c is the water-plane area of the column, $I_{x_1 x_1}$ and $I_{x_2 x_2}$ are the second moments of area of the column with respect to the principal axes of the water plane area, x_1 and x_2, respectively.

A shift in the center of buoyancy such that it is no longer vertically aligned with the center of gravity causes an overturning moment on the deck. This destabilizing effect of buoyancy is caused by the pitching/rolling motion. This overturning moment is transmitted to the deck through the columns. To incorporate this into the structural model, the geometric stiffness of the column frame elements is used. This is equivalent to subjecting the column to an axial force. This additional term in the hydrostatic stiffness matrix in the local coordinates can be written as

$$[\overline{k}_g] = \int_{L_e} f_a [N']^T [N'] d\overline{x}_1 \tag{2.6}$$

where f_a is the axial force and primes indicate differentiation of interpolation functions.

The consistent element hydrostatic stiffness matrix is then assembled by adding the stabilizing and destabilizing components and then transforming them into the global coordinates.

2.2 Hydrodynamic EMFC model

The distributed fluid forces and moments must be replaced by equivalent nodal forces to calculate $\{F_f(t)\}$ in Eq. (2.1). This can be achieved by calculating

$$\{\overline{F}_f\} = \int_{L_e} [N]^T \{\overline{p}\} d\overline{x}_1 \tag{2.7}$$

where $\{\overline{p}\}$ is the 6x1 vector of distributed forces and moments and $\{\overline{F}_f\}$ is the 12x1 vector of equivalent nodal fluid forces and moments.

As in Ertekin et al. (1993), we employ Morison's equation to determine the fluid loading. The main reason for this choice is the high efficiency of this method compared with the three-dimensional hydroelasticity theory based on potential flow. On the other hand, it is well known that the use of constant coefficients in Morison's equation leads to less accurate results especially at high frequencies for which wave scattering is important. In an attempt to improve the prediction for fluid loading at high frequencies, we will use potential theory to calculate the added mass, damping and exciting forces/moments for the columns and pontoons. It is noted however that these calculations are based on isolated structural members to be consistent with the use of Morison's equation, and therefore, they can not be accurate in cases where there is significant hydrodynamic interaction between neighboring members of the semi-submersible.

The distributed fluid loading due to waves in Eq. (2.7) includes the structural displacement forces, namely the added-mass forces proportional to acceleration and the damping forces proportional to velocities, as well as the exciting forces due to the incoming and scattered waves.

In terms of Morison's equation, we can write these distributed forces as

$$\{\overline{p}\}_s = \frac{1}{2} \rho C_D D \, |\{\dot{u}_{fn}\} - \{\dot{u}_n\}| (\{\dot{u}_{fn}\} - \{\dot{u}_n\}) $$
$$ + \frac{1}{4} \rho \pi D^2 (C_M \{\ddot{u}_{fn}\} - (C_M - 1)\{\ddot{u}_n\}) \tag{2.8}$$

where C_D is the form-drag coefficient; C_M the inertia coefficient; $\{\dot{u}_{fn}\}$ the water particle displacement vector normal to the element; $\{\dot{u}_n\}$ the displacement vector of a point on the element in the normal direction; D the diameter of the tubular member of the structure; and $\{\overline{p}\}_s$ the sub-vector of $\{\overline{p}\}$ in Eq. (2.7) representing the forces. The nonlinear drag force can be written in terms of the equivalent linear drag coefficient:

$$\{\overline{p}_D\}_s = \frac{1}{2} \rho C_{DL} D (\{\dot{u}_{fn}\} - \{\ddot{u}_n\}) \tag{2.9}$$

As mentioned before, we will consider modules, each being a semi-submersible, linked together to form a floating runway. We will further consider a semi-submersible which has columns and pontoons only. To obtain the approximate hydrodynamic coefficients and exciting forces for columns, we will use the extended MacCamy & Fuchs' (1954) method proposed by Garrison (1984), and for pontoons we will use Frank's (1967) close-fit method based on the two-dimensional source-distribution technique. Obviously, the use of such methods imply that any hydrodynamic interference effects between members are negligible. Moreover, such approximations can not predict the three-dimensional flow structure at the end of a column or pontoon. With these shortcomings indicated we now proceed to describe briefly the theories behind such approximations. The details can be found in Wang et al. (1993).

2.2.1 *Columns*

It is assumed that columns of a module can be isolated hydrodynamically. If this assumption proves to lead to great inaccuracies in certain cases, it is possible to consider the mutual interaction between neighboring columns by using the interaction theory for a matrix of columns (see for example, Kagemoto and Yue 1993).

In the extended MacCamy-Fuchs approach, the total potential is decomposed into incoming, diffraction and radiation potentials, i.e.

$$\Phi_T = \Phi_I + \Phi_D + \Phi_R \qquad (2.10)$$

where $\Phi_R = \psi(x_3)\phi_r$ and $\psi(x_3)$ denotes the shape function for each mode of motion. The incident wave potential is known. As is well known, the incident potential is expressed in cylindrical polar coordinates and then the application of the no-flux condition on the cylinder provides the solution for the diffraction (or scattering) potential.

The pressure due to the incoming and scattered waves can then be calculated by Euler's integral to give the force per unit length of the cylinder. Making this force equal to the force obtained by Morison's equation leads to the following coefficients:

$$C_M = C\cos\delta, \quad C_{DL} = \pi\omega a C \sin\delta \qquad (2.11)$$

where

$$C = \left(\frac{2\lambda}{\pi D}\right)^2 / \pi \sqrt{J_1'(ka)^2 + y_1'(ka)^2}$$
$$\delta = \tan^{-1}[J_1'(ka)/Y_1'(ka)] \qquad (2.12)$$

where k is the wave number, ω the wave frequency, a the cylinder radius (=D/2), λ the wave length, J_1' the derivative of the Bessel function of the first kind and of order one, and Y_1' the derivative of the Bessel function of the second kind and of order one. These coefficients then provide approximations to the exciting forces that could have been obtained by potential theory when used in Morison's equation. Note that these coefficients are depth independent. Once the sectional forces are calculated, they are used only from the still-water plane down to the bottom of a column. This is equivalent to using a shape function for the exciting forces also, for which $\psi(x_3)=0$ if $x_3 < -d$ (d=draft of the column).

To determine the radiation potential, the Laplace equation for ϕ_r can be solved through the use of the separation of variables technique. The radiation potential satisfies the same conditions as the diffraction potential except the body-boundary condition which is given by

$$\frac{\partial\phi_r}{\partial r} = -i\psi(x_3) \quad \text{on} \quad r=a \qquad (2.13)$$

By imposing this condition, the radiation potential can explicitly be obtained (see Garrison, 1984). Then the pressure and the force can be calculated. The shape functions for surge are

$$\psi(x_3)=1 \quad 0<x_3<-d, \quad \psi(x_3)=0 \quad x_3<-d \qquad (2.14)$$

and for pitch they are

$$\psi(x_3)=x_3/d \quad 0<x_3<-d, \quad \psi(x_3)=0 \quad x_3<-d \qquad (2.15)$$

The sectional radiation forces arranged as proportional to the velocities and accelerations lead to equivalent damping and added-mass coefficients:

$$C_{dl} = \text{Re}(\phi_r')/a\psi(x_3)$$
$$C_m = \text{Im}(\phi_r')/a\psi(x_3) \qquad (2.16)$$

where primes indicate differentiation with respect to r. If needed, the overall added-mass and damping coefficients can be obtained by integration over the length of the column to determine A_{ij} and B_{ij}, i,j=1 (surge), 5 (pitch), as the added-mass and damping coefficients, respectively.

In the next section, these coefficients will be compared with their counterparts obtained by using the three-dimensional Green function method as well as Morison's equation with constant coefficients.

2.2.2 *Pontoons*

The pontoons of the semi-submersible are relatively slender and therefore it appears suitable to use Frank's close-fit method to calculate the hydrodynamic coefficients. To do this, we solve the two-dimensional problem of a submerged cylinder to obtain the radiation potentials for surge and heave and then used the Haskind-Hanaoka relationship to calculate the exciting forces (see also Ogilvie, 1963). Our calculations are checked with Ogilvie's results and we found excellent agreement.

3 RESULTS

3.1 *Single column*

To determine the accuracy of the approximate theory based on the extended MacCamy-Fuchs approach, we calculated the hydrodynamic coefficients and exciting forces for a single column and compared the results with calculations based on the panel method. For this, we used a vertical cylinder of diameter 15.8m and draft 14.7m. This cylinder has the draft and volume equivalent to the draft and volume of a column (which has a rectangular cross section) of the semi-submersible used as a VLFS module.

Fig.1 shows the dimensionless surge exciting force calculated by Morison's equation with constant coefficients (C_M=2.0, C_D=0 for consistency) where A refers to wave amplitude. As expected, Morison's equation predicts inaccurate results at high frequencies. The present calculations for the extended MacCamy & Fuchs theory are quite satisfactory in comparison with the Green function method results. Note that A_{11} for Morison equation is given by $A_{11} = \rho \pi D^2 (C_m - 1) d / 4$. The dimensionless pitch

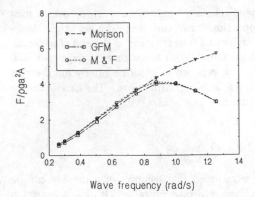

Fig.1 Normalized surge exciting force for a vertical cylinder

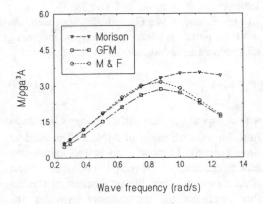

Fig.2 Normalized pitch exciting moment for a vertical cylinder

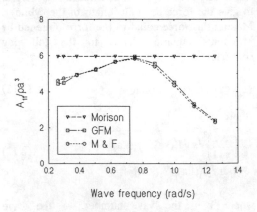

Fig.3 Normalized surge added mass coefficients for a vertical cylinder

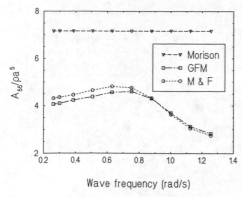

Fig.4 Normalized pitch added moment coefficients for a vertical cylinder

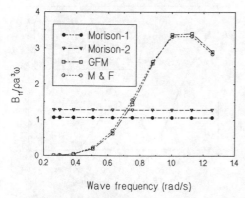

Fig.5 Normalized surge damping coefficients for a vertical cylinder

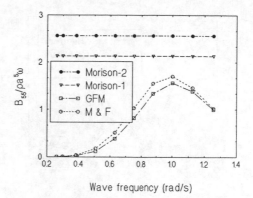

Fig.6 Normalized pitch damping coefficients for a vertical cylinder

(GFM) when Morison's equation predicts unsatisfactory results.

3.2 *Floating runway*

We consider the same runway used in Ertekin et al. (1993) which is made of 16 modules connected either at the deck level or both at the deck and pontoon levels. In Ertekin et al., only the calculations for the former case were done.

The main particulars of a single module are given in Table 1. More details of the module

Table 1 Main particulars of a single module

Length x width x height, m	100x100x59
Column (width x depth x height), m	12x17x35
Pontoon (width x height x length), m	18x10x96
Operating draft, m	25
Displacement (mass) ,kg	46440×10^3
I_{x1}, kg–m^2	7.91×10^{10}
I_{x2}, kg–m^2	6.49×10^{10}
I_{x3}, kg–m^2	9.35×10^{10}
KG, KB, m	30.67, 8.25
GM_L, GM_T, m	5.01, 4.13

exciting moment shown in Fig. 2 also shows the same trend.

The surge added mass and damping are shown in Figs 3 and 5, and the pitch added moment and damping are shown in Figs. 4 and 6. Note that in Figs. 5 and 6, two constant damping lines are shown since the damping in Morison's equation depends on the amplitude of motion due to its nonlinearity. Morison-1 refers to displacement over cylinder radius ratio of 0.033 and Morison-2 refers to 0.05. The damping B_{11} for Morison's equation is given by $B_{11} = \rho C_{DL} D d / 2$ where d is the submerged length (or draft) of the column.

In all the results presented, it is seen that the present results (M&F) are very close to the results obtained by the 3-D source distribution method

geometry can be found in Riggs et al. (1991). The section structural properties are given in Ertekin et al. (1991). For simplicity, we only used the four columns and two pontoons to model the semi-submersible. Some results for head,

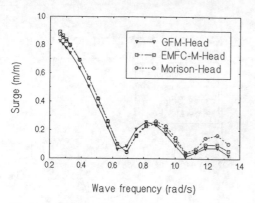

Fig.7 Surge transfer functions of a single
       rigid module in head seas

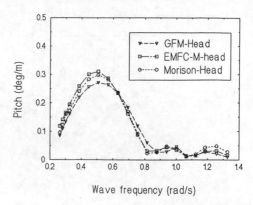

Fig.9 Pitch transfer function of a single rigid
       module in head seas

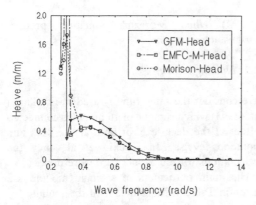

Fig.8 Heave transfer functions of a single
       rigid module in head seas

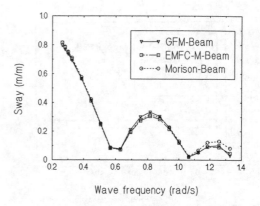

Fig.10 Sway transfer functions of a single rigid
        module in beam seas

quartering and beam seas will be shown here. In the Morison's equation results we use $C_M = 2.0$ and $C_D = 1.0$ as the constant inertia and drag coefficients.

The entire structure is modeled by frame elements; a total of 171 elements per module is used. There are 2464 nodes for the entire VLFS. We denote Module 1 as the bow and Module 16 as the stern. Module 8 is near the origin. In the following results presented, the motions of the center of gravity are shown. This is done by extending a massless element whose end node coincides with the center of gravity of each module.

Figures 7 through 9 show the motions of a single module, in the absence of all others, to check the accuracy of the present predictions. In these

figures, GFM refers to the source-distribution method and Morison refers to the EMFC method with constant coefficients. Both of these results were presented in Ertekin et al. 1993. The new results are indicated by EMFC-M (extended MacCamy & Fuchs method or the EMFC method with frequency-dependent coefficients). It is seen in these figures, which are for head seas, that there is some improvement in the high frequency range. Note that in the EMFC method, the module was made rigid by substantially increasing the moduli of elasticity and shear.

Figures 10 through 12 show the single module results for beam seas. It is clear that in the results where EMFC method does not predict results close to the GFM results, the three-dimensional flow structure must be important.

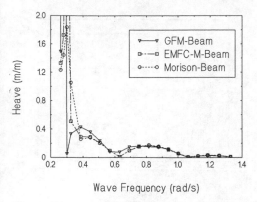

Fig.11 Heave transfer functions of a single rigid
module in beam seas

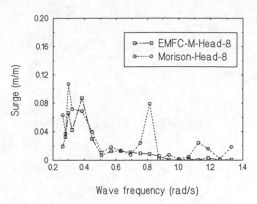

Fig.14 Surge transfer functions of Module 8
in head seas

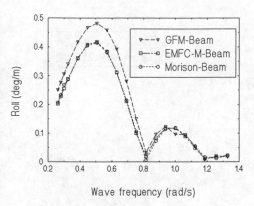

Fig.12 Roll transfer functions of a single rigid
module in beam seas

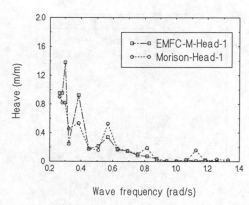

Fig.15 Heave transfer functions of Module 1
in head seas

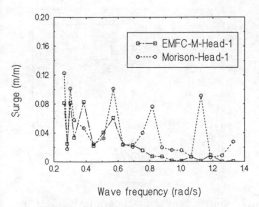

Fig.13 Surge transfer functions of Module 1
in head seas

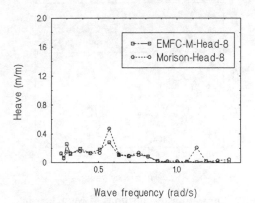

Fig.16 Heave transfer functions of Module 8 in
head seas

397

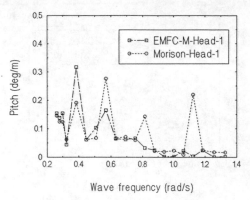

Fig.17 Pitch transfer functions of Module 1 in head seas

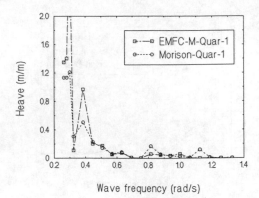

Fig.20 Heave transfer functions of Module 1 in quartering seas

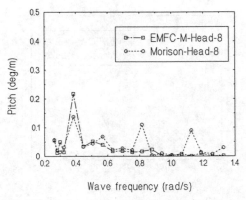

Fig.18 Pitch transfer functions of Module 8 in head seas

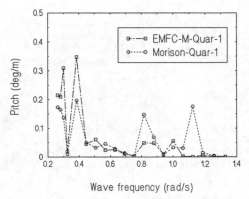

Fig.21 Pitch transfer functions of Module 1 in quartering seas

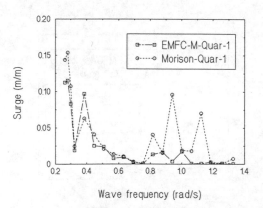

Fig.19 Surge transfer functions of Module 1 in quartering seas

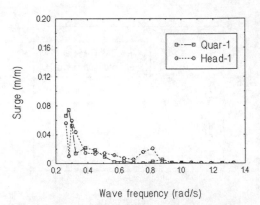

Fig.22 Surge transfer functions of Module 1 for the deck and pontoon connector case

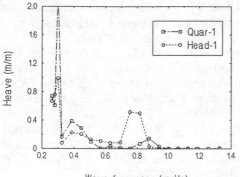

Fig.23 Heave transfer functions of Module 1 for the deck and pontoon connector case

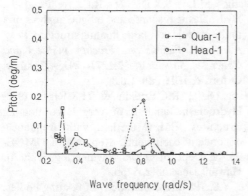

Fig.24 Pitch transfer functions of Module 1 for the deck and pontoon connector case

We now present the motions of the center of gravity of Modules 1 and 8, predicted by the EMFC method with constant coefficients (Morison) and by the frequency-dependent coefficients (EMFC-M) as developed here. Figures 13 through 18 show the surge, heave and pitch motion comparisons for the head-sea case. Note that Figs. 13 through 21 correspond to the deck connector (at x_3 = 26.5 m above SWL) only case.

These results consistently indicate that when wave scattering and radiation are included, the motions become less at most frequencies. Similar conclusion was reached in Ertekin et al. (1993) when Morison's equation results were compared with the Green function results. The peaks in constant-coefficient EMFC results can be shown

to correspond to flexible-mode natural frequencies.

Figures 19 through 21 show the quartering sea cases for Module 1, again for connectors located at the deck level only. Except at low frequencies, Morison's equation with constant coefficients predicts large motion amplitudes as before.

As mentioned before, we also calculated the case when there are connectors at the pontoon level in addition to the deck connectors. The connectors' structural properties were assumed to be the same as the deck connectors'. Figs. 22 through 24 show the motion responses for this case when EMFC method with frequency-dependent coefficients are used in quartering and head seas. Comparison of this case with earlier ones for deck-connector only shows that the motions are damped significantly.

4 SUMMARY AND CONCLUSIONS

The extended MacCamy-Fuchs approach is used to determine the frequency dependent added-mass, damping and force coefficients to be used in Morison's equation to supply the fluid loading in the EMFC method. This is done for the columns, and Frank's close-fit method is used to do the same for the pontoons.

When these frequency-dependent hydrodynamic coefficients are used in the EMFC method, it is predicted that the motions are significantly suppressed at most wave frequencies. This comparison leads to the same conclusion reached in earlier works, namely that the Morison's equation with constant coefficients over predicts the hydroelastic response.

On the other hand, the assumption that there is negligible interaction between neighboring members may prove to lead to rather inaccurate results that are seen when motions of a module of VLFS obtained by the extended MacCamy-Fuchs theory and the Green function method are compared. This then suggests that in a hydroelasticity problem where there are many modules, it is necessary to further extend the present approach to: (i) multiple vertical cylinders, and (ii) endplane forces.

As mentioned before, it is possible to determine the interference effects when there are neighboring columns, and we intend to pursue this. However, it is not clear how we can model the three-dimensional flow between two neighboring pontoon ends other than directly using the hydrodynamic coefficients obtained by the GFM method.

Acknowledgement : This work is sponsored by the U. S. National Science Foundation under Grant No. BCS-9200655. SOEST Contribution No. 3518.

REFERENCES

Bishop, R.E.D. & W.G.Price 1979. *Hydroelasticity of Ships*. Cambridge, UK: Cambridge University Press.

Che, X.L.; H.R.Riggs; R.C.Ertekin; Y.S.Wu & M.L.Wang 1992. Two-dimensional analysis of prying response of twin-hull floating structures. *Proc. 2nd Int. Offshore and Polar Engineering Conference*, 1:187-194.

Che, X.L. 1993. Techniques for hydroelastic analysis of very large floating structures. Ph.D dissertation, Dept. of Ocean Engineering, Univ. of Hawaii at Manoa, Honolulu, 226pp.

Che, X.L.; H.R.Riggs & R.C.Ertekin 1994. Composite 2D/3D hydroelastic analysis method for floating structures. in press, *J. Engineering Mechanics*, ASCE, July.

Ertekin, R.C.; H.R.Riggs; X.L.Che &S.X.Du 1993. Efficient methods for hydroelastic analysis of very large floating structures. *J. of Ship Research*, 37(1):58-76.

Frank, W. 1967. Oscillation of cylinders in or below the free surface of deep fluids, NSRDC Rep. No. 2375, Naval Ship Res. & Dev. Center, Bethesda, Maryland.

Garrison, C.J. 1984. In Specialty Conf. on Computer Methods in Offshore Engineering, *Proc. Canadian Society of Civil Eng.*, Halifax, Nova Scotia, Canada, May, 1-72.

Kagemoto, H. & D.K.P. Yue 1993. Hydrodynamic interaction Analysis of very large floating structures. *Marine Structures*, 6(2-3): 295-322.

Mac Camy, R.C. & R.A. Fuchs 1954. Wave forces on piles: a diffraction theory. Thech. Memo. No. 69, Beach Erosion Board, Coastal Eng. Res. Center, U.S. Army, Washington, D.C.

Ogilvie, T.F. 1963. First- and second-order forces on a cylinder submerged under a free surface. *J. Fluid Mechanics*, 16: 451-472.

Riggs, H.R.; X.L. Che & R.C. Ertekin 1991. Hydroelastic response of very large floating structures. *Proc. 10th Int. Conf. Offshore Mech. & Arctic Eng.*, ASME, Stavanger, Norway, I-A:291-300.

Riggs, H.R. & R.C.Ertekin 1993. Approximate methods for dynamic response of multi-module floating structures. *Marine Structures*. 6(2-3):117-141.

Wang, D.Y.; H.R.Riggs & R.C.Ertekin 1991. Three-dimensional hydroelastic response of a very large floating structure. *Int. J. Offshore and Polar Engineering*, 1 (4):307-316.

Wang, M.L.; S.X.Du & R.C.Ertekin 1991. Hydroelastic response and fatigue analysis of a multi-module very large floating structure. *Proc. Int. Symp. Fatigue and Fracture in Steel and Concrete Structures*, 2:1277-1291. Bombay: Oxford & IBH Pub. Co.

Wang, S.Q.; R.C.Ertekin & H.R.Riggs 1993. Hydroelastic analysis of very large floating structures: EMFC method with frequency-dependent coefficients. Rep. No. UHMOE-93112, Dept. of Ocean Engineering, Univ. of Hawaii, December, 57pp.

Wu, Y.S. 1984. Hydroelasticity of floating bodies. Ph.D dissertation, Brunel University, U.K.

Wu, Y.S.; D.Y.Wang; H.R.Riggs and R.C.Ertekin 1993. Composite singularity distribution method with application to hydroelasticity. *Marine Structures*, 6(2-3):143-163.

Yu, B.K. 1980. The calculation of hydrodynamic forces acting on a deformable cylinder in the free surface. Ph.D. dissertation, Dept. of Naval Architecture, Univ. of Calif., Berkeley, 80pp.

Large floating structures

Hydroelasticity in Marine Technology, Faltinsen et al. (eds) © 1994 Balkema, Rotterdam, ISBN 90 5410 387 6

Wave drifting forces on multiple connected floating structures

Mikio Takaki
Department of Naval Architecture and Ocean Engineering, Hiroshima University, Japan

Yoshihiko Tango
Ishikawajima-Harima Heavy Industries Co., Ltd, Kure, Japan

ABSTRACT: The wave drifting forces acting on a very large floating structure consisting of multiple barge type modules is studied theoretically and experimentally in this paper. We estimate the wave drifting forces on the multiple floating structure with two module connectors; a rotationally-hinge connector and a rotationally-rigid connector. The effects of connector conditions and the number of bodies on the wave drifting forces are discussed.

1. INTRODUCTION

Very large floating structures have been proposed for a number of applications such as airport, runways (Takarada, 1982), entire floating cities (Takarada,1989), oil storage (Uki, 1988). It is very important for estimating a wave drifting force on a very large floating structure from a view point of the structure safety, because it is impossible to catch up the drifting huge structure.

So far the strip theory has been used for estimating the seakeeping performance from a practical point of view (Maeda et al,1979). It is, however, seemed the hydrodynamic forces on the huge structures are affected by effects of three dimensional interactions, and many researchers have studied three dimensional hydrodynamic forces on the multiple structures. However, most of them are the studies on the first order force (Ertekin et al,1991, Goo et al,1989, Kagemoto, 1991, Riggs et al, 1991), and there are few studies on the wave drifting force on a large floating structure.

We have shown that the wave drifting force coefficient on the multiple connected floating structure with rationally-hinge connector is larger than 2.0 to 3.0 in the theoretical study (Takaki et al, 1994). Its maximum value is 1.0 at most in two dimensional condition. Moreover Kudo showed that the maximum wave of the wave drift force coefficient on three dimensional body exceeds 1.0 slightly (Kudo et al, 1978). However, the values of the wave drifting force coefficient larger than 2 to 3 are too large. This paper, therefore, deals experimentally and theoretically with wave drift-

ing forces acting on a very large floating structure consisting of multiple barge type modules to verify the above phenomena. We have considered two module connectors ; a rotationally hinge connector and a rotationally rigid connector. The three dimensional panel method is used to determine the hydrodynamic forces to take into account accurately the effect of hydrodynamic interactions among the modules, and the coupled equations of motions are solved directly. The number of modules affects amplitudes of motions in head sea condition. Furthermore we discuss the effect of connecting conditions on the wave drifting forces. It is verified experimentally that the coefficients of the wave drifting force take the values more than 2 to 3 in the case of rotationally-hinge connector. In addition, it is made clear that the wave drifting force on the multiple structure with hinge connectors becomes to zero in the range of a long wave length, as the number of the modules with the hinge connectors becomes larger.

2. EXPERIMENT

Three kinds of models are used in the experiments. They are consisted of 2-, 4- and 6-barge type modules respectively which have all the same dimensions. The principal dimensions of the module are shown in Table 1. These modules are connected with two kinds of connectors; a hinge connector and a rigid connector. Fig. 1 shows the connecting conditions. We performed free decay tests and measured a drifting force on the multiple connected floating structures in a head

Table 1   Principal particulars of barge type module

Length	L = 0.9 m
Breadth	B = 0.9 m
Draft	d = 0.09 m
Displacement	W = 72.9 Kg
Center of gravity	KG = 0.062 m
Metacenter height	GM = 0.733 m
Radius of gyration	RJ = 0.28 m

Rigid Connection

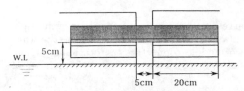

Hinge Connection

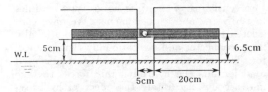

Fig. 1   Connecting conditions

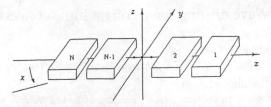

Fig. 2   Coordinate systems (for N-modules)

$$\Phi(x, y, z; t) = \mathrm{Re}[\phi(x, y, z)e^{i\omega t}]. \qquad (1)$$

$$\phi = \frac{g\zeta_a}{\omega}(\phi_0 + \phi_7) + i\omega \sum_{j=1}^{6} \xi_j \phi_j \qquad (2)$$

where, $\qquad \phi_0 = ie^{Kz - iK(x\cos\chi + y\sin\chi)} \qquad (3)$

ζ_a ; incident wave amplitude
ξ_j ; motion amplitude of j-mode
g ; gravitational acceleration

The velocity potential ϕ is governed by the three dimensional Laplace equation, the linearized free surface condition and bottom condition.

The body surface condition for the radiation and the diffraction problem can be represented respectively as follows.

For radiation potential;

$$\frac{\partial \phi_j}{\partial n} = n_j, \qquad (4)$$

where, $\quad \boldsymbol{n} = (n_1, n_2, n_3),$
$\qquad \boldsymbol{r} \times \boldsymbol{n} = (n_4, n_5, n_6).$

$\boldsymbol{n}$ is the unit normal vector directed into fluid.

For diffraction potential;

$$\frac{\partial \phi_{7c}}{\partial n} = -\frac{\partial \phi_{0c}}{\partial n}, \quad \frac{\partial \phi_{7s}}{\partial n} = -\frac{\partial \phi_{0s}}{\partial n}, \qquad \text{on } S_H \qquad (5)$$

where, the subscript c, s denotes the symmetry and asymmetry mode respectively.

Using the Green's theorem, the velocity potential is represented as follows.

$$\phi_j(P) = -\iint_{S_H} \left[\frac{\partial \phi_j(Q)}{\partial n} - \phi_j(Q)\frac{\partial}{\partial n} \right] G(P, Q) dS \qquad (6)$$

sea conditions under the assumption of a single point mooring. The experiments were carried out at the towing tank of Hiroshima University (Length = 100m, Breadth = 8m and Water depth = 3.5m). Incident waves are only regular waves. The range of wave length λ/L is 0.8 to 5.0, where the module length L is 0.9m. The standard wave height H is $\lambda/50$, while H =3.0 & 4.0cm are used only for $\lambda/L \leq 1.6$.

3. THEORETICAL CALCULATION

3.1 Integral equation

Let us consider the linear interaction of regular waves with three-dimensional bodies that consists of N floating bodies. Cartesian coordinate system (x,y,z) with z positive upwards is used and the origin is on the mean free surface as shown in Fig. 2. Assuming ideal fluid and small amplitude waves, a velocity potential around the body can be expressed as a sum of incident ϕ_0, diffraction ϕ_7 and radiation potentials ϕ_j :

or

$$\phi_j(P) = -\iint_{S_H} \sigma_j(Q)G(P,Q)dS \qquad (7)$$

where,

$$
\begin{aligned}
G(P,Q) &= G(x,y,z;x',y',z') \\
&= \frac{1}{4}\left(\frac{1}{r} + \frac{1}{r'}\right) \\
&\quad - \frac{K}{2\pi}e^{K(z+z')}\int_{-(z+z')}^{\infty}\frac{e^{K\xi}}{\sqrt{R^2+\xi^2}}d\xi \\
&\quad - i\frac{K}{2}e^{K(z+z')}H_0^{(2)}(KR),
\end{aligned}
\qquad (8)
$$

$$
\left.\begin{aligned}
r \\
r'
\end{aligned}\right\} = \sqrt{(x-x')^2 + (y-y')^2 + (z\mp z')^2}
$$

$$R = \sqrt{(x-x')^2 + (y-y')^2}$$

$H_0^{(2)}$; hankel function of the 2nd kind

As multiple-connected structure is generally huge, and the draft is smaller in comparison with the breadth and the length of it (Matsunaga et al,1979). Therefore we can neglect the effect of draft from hydrodynamic force ultimately. Yamashita (1979) and Bessho (1979) applied the pressure distribution method for solving a boundary value problem of a thin flat body. We have, however, solved the boundary problem by applying eq.(6) in this study. As a point P approaches a boundary point on the hull surface S_H, the integral equation to determine ϕ_j can be represented in the form of

$$
\begin{aligned}
\frac{\partial\phi_j(P)}{\partial n} &= \tfrac{1}{2}\phi_j(P) \\
&\quad - \iint_{S_H}\left[\frac{\partial\phi_j(Q)}{\partial n} - \phi_j(Q)\frac{\partial}{\partial n}\right]G(P,Q)dS,
\end{aligned}
\qquad (9)
$$

or

$$
\begin{aligned}
\frac{\partial\phi_j(P)}{\partial n} &= \tfrac{1}{2}\sigma_j(P) \\
&\quad - \iint_{S_H}\sigma_j(Q)\frac{\partial}{\partial n}G(P,Q)dS.
\end{aligned}
\qquad (10)
$$

For numerical calculation, we discretize the body surface S_H into N number of plane panels, and derive expression of eq.(9) assuming a constant potential distribution on each panel.

When the multiple-connected structure is oscillating in waves, each body is moving around the center of gravity of each body, and the effect of hydrodynamic interactions among floating bodies appears. Thereby the boundary conditions of body surface can be rewritten as follows;

$$
\begin{aligned}
\frac{\partial\phi_i^{(n)}}{\partial n} &= n_i^{(n)} \quad \text{on } S_n, \\
\frac{\partial\phi_i^{(k)}}{\partial n} &= 0 \quad \text{on } S_n \quad (k \neq n)
\end{aligned}
\qquad (11)
$$

where, a superscript (n) and a subscript j denote the N-th body of the multiple-connected structure and the mode of motion respectively.

On the other hand, the boundary conditions for diffraction problem are the same as eq.(5).

3.2 Hhydrodynamic forces of the first order

The linearized pressure is represented with fluid density ρ and the unsteady velocity potential ϕ as follows.

$$p(x,y,z) = -\rho i\omega\phi(x,y,z) \qquad (12)$$

Furthermore, the hydrodynamic force in the j-th direction for radiation problem can be expressed by using the second term of eq.(2) as follows.

$$
\begin{aligned}
F_j &= -\iint_{S_H} p(x,y,z)n_j dS \\
&= -(i\omega)^2\sum_{i=1}^{6}\xi_i\left[M_{ji} + \frac{N_{ji}}{i\omega}\right]
\end{aligned}
\qquad (13)
$$

The term M_{ji} and N_{ji} correspond to the added mass and damping force in the j-th direction due to unit amplitude of i-th mode respectively, which are represented in the forms.

$$M_{ji} = -\rho\mathrm{Re}\left[\iint_{S_H}\phi_i n_j dS\right] \qquad (14)$$

$$N_{ji} = \rho\omega\mathrm{Im}\left[\iint_{S_H}\phi_i n_j dS\right] \qquad (15)$$

As we can evaluate the velocity potential in the form of $\phi_i^{(m)}$ as shown in eq.(11), the radiation forces can be rewritten in the alternative form.

$$M_{ji}^{(mn)} = -\rho\mathrm{Re}\left[\iint_{S_H}\phi_i^{(n)}n_j^{(m)}dS\right] \qquad (16)$$

$$= -\rho\mathrm{Re}\left[m_{ji}^{mn}\right] \qquad (17)$$

$$N_{ji}^{(mn)} = \rho\omega\mathrm{Im}\left[\iint_{S_H}\phi_i^{(n)}n_j^{(m)}dS\right] \qquad (18)$$

These are added mass and damping force coefficient in the j-th direction of m-th module due to the motion with i-mode of n-th module respectively, which are a new term added in solving the motions of the multiple-connected structures.

405

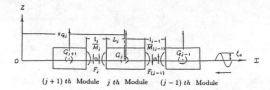

Fig. 3   Coordinate systems (for j-modules)

The wave exciting force of j-th direction on it can be represented in the form.

$$
E_j = -\iint_{S_H} p(x, y, z)n_j dS
$$
$$
= \rho g \zeta_a i \iint_{S_H} (\phi_0 + \phi_7)n_j dS \qquad (19)
$$

3.3 Equations of motions

We consider a floating structure consisted of N modules in a regular wave with a frequency ω as shown in Fig. 2. We assume that incident wave is propagating from the χ direction with respect to the negative x-axis. Then the floating body is always in a head sea condition under the assumption of a single point mooring. We can, therefore, consider only the motions of heave and pitch, and neglect surge motion because of very small draft in this problem.

Generally we cannot neglect the effect of hydrodynamic interaction among the multiple connected bodies and the coupling terms between heave and pitch. In particular it is seemed that their effects on the wave drifting forces are significant. Thereby we need the estimation of three dimensional hydrodynamic forces with the interaction each other and the motion equations including the coupling terms between heave and pitch. In Fig. 3 the motion equations with the coupling terms can be represented as follows.

For heave;

$$
(\frac{W_1}{g} + M_{33}^{11})\ddot{Z}_1 + N_{33}^{11}\dot{Z}_1 + \rho g A_{W_1} Z_1
$$
$$
+ M_{33}^{12}\ddot{Z}_2 + N_{33}^{12}\dot{Z}_2 + \cdots\cdots
$$
$$
\cdots + M_{33}^{1n}\ddot{Z}_n + N_{33}^{1n}\dot{Z}_n
$$
$$
+ M_{35}^{11}\ddot{\theta}_1 + N_{35}^{11}\dot{\theta}_1 + \cdots\cdots
$$
$$
\cdots + M_{35}^{1N}\ddot{\theta}_n + N_{35}^{1n}\dot{\theta}_n
$$
$$
= F_1 + E_{Z_1} \qquad (19)
$$

$$
(\frac{W_j}{g} + M_{33}^{jj})\ddot{Z}_j + N_{33}^{jj}\dot{Z}_j + \rho g A_{W_j} Z_j
$$
$$
+ M_{33}^{j1}\ddot{Z}_1 + N_{33}^{j1}\dot{Z}_1 + \cdots\cdots
$$
$$
\cdots + M_{33}^{jj-1}\ddot{Z}_{j-1} + N_{33}^{jj-1}\dot{Z}_{j-1}
$$
$$
+ M_{33}^{jj+1}\ddot{Z}_{j+1} + N_{33}^{jj+1}\dot{Z}_{j+1} + \cdots\cdots
$$
$$
\cdots + M_{33}^{jn}\ddot{Z}_n + N_{33}^{jn}\dot{Z}_n
$$
$$
+ M_{35}^{j1}\ddot{\theta}_1 + N_{35}^{j1}\dot{\theta}_1 + \cdots\cdots
$$
$$
\cdots + M_{33}^{jn}\ddot{\theta}_n + N_{53}^{jn}\dot{\theta}_n
$$
$$
= -F_{j-1} + F_j + E_{Zj} \qquad (20)
$$

$$
(\frac{W_n}{g} + M_{33}^{nn})\ddot{Z}_n + N_{33}^{nn}\dot{Z}_n + \rho g A_{W_n} Z_n
$$
$$
+ M_{33}^{n1}\ddot{Z}_1 + N_{33}^{n1}\dot{Z}_1 + \cdots\cdots
$$
$$
\cdots + M_{33}^{nn-1}\ddot{Z}_{n-1} + N_{33}^{nn-1}\dot{Z}_{n-1}
$$
$$
+ M_{35}^{n1}\ddot{\theta}_1 + N_{35}^{n1}\dot{\theta}_1 + \cdots\cdots
$$
$$
\cdots + M_{35}^{nn}\ddot{\theta}_n + N_{53}^{nn}\dot{\theta}_n
$$
$$
= -F_{N-1} + E_{ZN} \qquad (21)
$$

For pitch;

$$
(I_1 + M_{55}^{11})\ddot{\theta}_1 + N_{55}^{11}\dot{\theta}_1 + W_1\overline{GM_1}\theta_1
$$
$$
+ M_{55}^{12}\ddot{\theta}_2 + N_{55}^{12}\dot{\theta}_2 + \cdots\cdots
$$
$$
\cdots + M_{55}^{1n}\ddot{\theta}_n + N_{55}^{1n}\dot{\theta}_n
$$
$$
+ M_{53}^{11}\ddot{Z}_1 + N_{53}^{11}\dot{Z}_1 + \cdots\cdots
$$
$$
\cdots + M_{53}^{1N}\ddot{Z}_n + N_{53}^{1n}\dot{Z}_n
$$
$$
= \frac{L_1 + \ell_1}{2}F_1 + \overline{M}_1 + E_{\theta_1} \qquad (22)
$$

$$
(I_j + M_{55}^{jj})\ddot{\theta}_j + N_{55}^{jj}\dot{\theta}_j + W_j\overline{GM_j}\theta_j
$$
$$
+ M_{55}^{j1}\ddot{\theta}_1 + N_{55}^{j1}\dot{\theta}_1 + \cdots\cdots
$$
$$
\cdots + M_{55}^{jj-1}\ddot{\theta}_{j-1} + N_{55}^{jj-1}\dot{\theta}_{j-1}
$$
$$
+ M_{55}^{jj+1}\ddot{\theta}_{j+1} + N_{55}^{jj+1}\dot{\theta}_{j+1} + \cdots\cdots
$$
$$
\cdots + M_{55}^{jn}\ddot{\theta}_n + N_{55}^{jn}\dot{\theta}_n
$$
$$
+ M_{53}^{j1}\ddot{Z}_1 + N_{53}^{j1}\dot{Z}_1 + \cdots\cdots
$$
$$
\cdots + M_{53}^{jn}\ddot{Z}_n + N_{53}^{jn}\dot{Z}_n
$$
$$
= \frac{L_j + \ell_{j-1}}{2}F_{j-1} + \frac{L_j + \ell_j}{2}F_j
$$
$$
- \overline{M}_{j-1} + \overline{M}_j + E_{\theta_j} \qquad (23)
$$

$$(I_n + M_{55}^{nn})\ddot{\theta}_n + N_{55}^{nn}\dot{\theta}_n + W_n\overline{GM_n}\theta_n$$
$$+ M_{55}^{n1}\ddot{\theta}_1 + N_{55}^{n1}\dot{\theta}_1 + \cdots\cdots$$
$$\cdots + M_{55}^{nn-1}\ddot{\theta}_{n-1} + N_{55}^{nn-1}\dot{\theta}_{n-1}$$
$$+ M_{53}^{n1}\ddot{Z}_1 + N_{53}^{n1}\dot{Z}_1 + \cdots\cdots$$
$$\cdots + M_{53}^{nn}\ddot{Z}_n + N_{53}^{nn}\dot{Z}_n$$
$$= \frac{L_N + \ell_{N-1}}{2}F_{N-1} - \overline{M}_{N-1} + E_{\theta_N} \quad (24)$$

Compatibility conditions;

$$Z_j - Z_{j+1} = -\frac{L_j + \ell_j}{2}\theta_j - \frac{L_{j+1} + \ell_j}{2}\theta_{j+1} \quad (25)$$

For a rigid connector;

$$\theta_j = \theta_{j+1}, \qquad j = 1, 2, \ldots, n-1 \qquad (26)$$

For an elastic connector;

$$\overline{M}_j = k(\theta_j - \theta_{j+1}) + n(\dot{\theta}_j - \dot{\theta}_{j+1}) \qquad (27)$$

For a hinge connector;

$$\overline{M}_j = 0, \qquad j = 1, 2, \ldots, n-1 \qquad (28)$$

Where,

W_j	;	weight of j-th block
I_j	;	mass moment of inertia of j-th block
A_{Wj}	;	water plane area of j-th block
$\overline{GM}_j$	;	longitudinal metacentric height of j-th block
M_j	;	added mass of j-th block
N_{Zj}	;	damping force coeff. of $j$the block
E_{Zj}	;	wave exciting force for heave on j-th block
J_j	;	added mass moment of inertia of j-th block
$N_{\theta j}$	;	damping moment coeff. of j-th block
$E_{\theta j}$	;	wave exciting moment for pitch on j-th block
F_j	;	shering force beween j-th and $(j+1)$-th block
$\overline{M}_j$	;	bending moment between j-th and $(j+1)$-th block
$M_{ij}^{k\ell}$	;	added mass to the i-th direction of k-th block due to unit motion with j-th mode of ℓ-th block
$N_{ij}^{k\ell}$	;	damping coeff. to the i-th direction of k-th block due to unit motion with j-th mode of ℓ-th block

The terms of $M_{ij}^{k\ell}$, $N_{ij}^{k\ell}$ ($i \neq j$, $k \neq \ell$) are the coupling terms due to the hydrodynamic interactions among N bodies, which are added newly in these equations. Namely the above equations are the exact expressions which include all the effect of hydrodynamic interaction.

3.4 Wave drifting force

The wave drifting force in the x-th direction on a floating body can be represented using the Kochin function in the form.

$$\overline{F_x} = \frac{\rho K^2}{8\pi}\int_0^{2\pi}\left|H(K, \theta)\right|^2(\cos\chi - \cos\theta)d\theta \qquad (29)$$

where,

$$\left.\begin{array}{l} H(K, \theta) = \displaystyle\sum_{j=1}^{7}i\,\omega\xi_j H_j(K, \theta) \\[2mm] H_j(K, \theta) = \displaystyle\iint_{S_H}\left(\frac{\partial\phi_j}{\partial n} - \phi_j\frac{\partial}{\partial n}\right) \\[2mm] \hspace{3cm} f(K; x, y, z)ds \\[2mm] f(K; x, y, z) \\[1mm] = \exp\{K\,z + i\,K(x\cos\theta + y\sin\theta)\} \end{array}\right\} \qquad (30)$$

$$\xi_7 = -\frac{i\,\zeta_a}{K} = -\frac{i\,g\,\zeta_a}{\omega^2} \qquad (31)$$

$H_j(K, \theta)$ is the Kochin function which represents the amplitude and the phase due to the motion of j-th mode.

Next we consider a wave drifting force on the multiple-connected structure. We have to involve not only the radiation waves due to the motions of each body but also the diffraction waves due to the interaction among the bodies in this condition. The Kochin function of radiation problem due to heave and pitch of N floating bodies is represented in the form.

$$H_j(K, \theta)$$
$$= i\omega\xi_3^{(1)}\left\{H_3^{(11)} + H_3^{(21)} + \cdots + H_3^{(n1)}\right\}$$
$$+ i\omega\xi_3^{(2)}\left\{H_3^{(12)} + H_3^{(22)} + \cdots + H_3^{(n2)}\right\}$$
$$\cdots\cdots\cdots\cdots$$

$$+ i\omega \xi_3^{(n)} \left\{ H_3^{(1n)} + H_3^{(2n)} + \cdots + H_3^{(nn)} \right\}$$

$$= i\omega \xi_5^{(1)} \left\{ H_5^{(11)} + H_5^{(21)} + \cdots + H_5^{(n1)} \right\}$$

$$+ i\omega \xi_5^{(2)} \left\{ H_5^{(12)} + H_5^{(22)} + \cdots + H_5^{(n2)} \right\}$$

$$\cdots \cdots \cdots \cdots$$

$$+ i\omega \xi_5^{(n)} \left\{ H_5^{(1n)} + H_5^{(2n)} + \cdots + H_5^{(nn)} \right\} \tag{32}$$

Finally we get the general equation of the Kochin function for N floating bodies in a regular wave as follows.

$$H(K,\theta) = i\omega \sum_{j=1}^{7} \sum_{k=1}^{N} \xi_j^{(k)} \sum_{\ell=1}^{N} H_j^{(k,\ell)}(K,\theta) \tag{33}$$

For $k = \ell$;

$$H_j^{(k,l)}(K,\theta) = \iint_{S_\ell} \left(\frac{\partial \phi_j^{(\ell)}}{\partial n} - \phi_j^{(\ell)} \frac{\partial}{\partial n} \right)$$
$$f(K; x, y, z) ds,$$

For $k \neq \ell$;

$$H_j^{(k,l)}(K,\theta) = \iint_{S_\ell} -\phi_j^{(\ell)} \frac{\partial}{\partial n}$$
$$f(K; x, y, z) ds,$$

$$f(K; x, y, z) =$$
$$\exp \left\{ Kz + iK(x \cos \theta + y \sin \theta) \right\} \tag{34}$$

Substituting eq.(34) into eq.(29), We can get the wave drifting force on the multiple-connected structure in a regular wave.

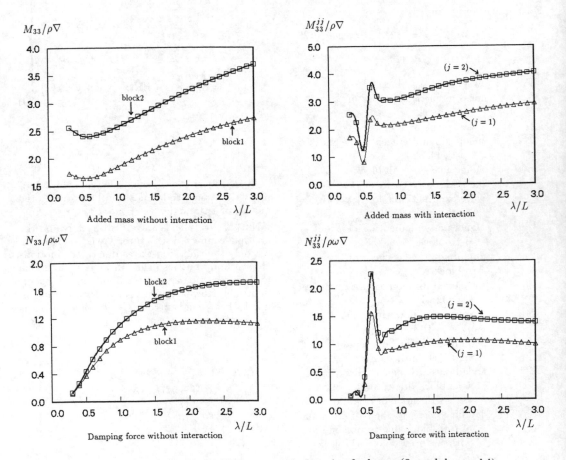

Fig. 4 Interaction effect on added mass and damping for heave (3-modules model)

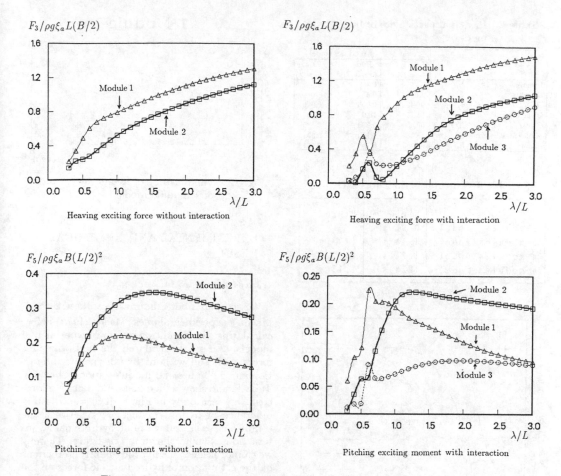

Fig. 5 Interaction effect on wave exciting forces (3-modules model)

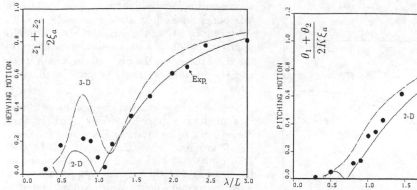

Fig. 6 Average heaving amplitude of 2-modules model

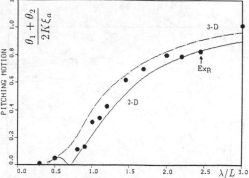

Fig. 7 Average pitching amplitude of 2-modules model

Table 2 Principal particulars of barge type model

| | | 2-modules | 3-modules model | |
| | | | End | Center |
		$(j = 1, 2)$	$(j = 1, 3)$	$(j = 2)$
Length	$L_j(m)$	0.75	0.3	0.6
Breadth	$B(m)$	1.0	0.6	0.6
Draft	$d(m)$	0.045	0.056	0.056
Displacement	$W_j(Kg)$	33.75	10.0	20.0
Center of gravity	$KG_j(m)$	0.046	0.053	0.036
Metacenter height	$GM_j(m)$	1.018	0.110	0.531
Radius of gyration	$RJ_j(m)$	0.0242	0.108	0.161

Table 3 Natural periods for pitch

	Rigid	Hinge
1-module model	0.91 s.	—
2-modules model	1.18 s.	1.07 s.
4-modules model	1.21 s.	1.12 s.
6-modules model	1.23 s.	1.13 s.

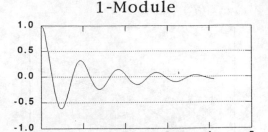

Fig. 10 Time history of free decay for pitch of 1-module model

4. EXPERIMENTAL AND NUMERICAL RESULTS

4.1 *Hydrodynamic Interaction*

We compared our numerical results with Maeda's experimental ones (Maeda, 1979) before calculating the conditions of our experimental models. First we calculated the hydrodynamic forces with and without hydrodynamic interaction among bodies of 3 modules' model shown in Table 2, and show added mass and damping force coeff. on each body in Fig. 4. It is seen that big difference of hydrodynamic forces between the conditions with and without hydrodynamic interaction exists as shown in this figure. Thereby we cannot neglect the hydrodynamic interactions on the estimation of hydrodynamic forces on the multiple-connected structure.

Fig. 5 shows the effects of hydrodynamic interaction on wave exciting forces and moments. The wave exciting force and moment estimated by the 3-D panel method on the body of the weather side are larger than those on the body of the lee side. On the other hand, the results estimated by the strip theory on the body of the weather side and the lee side are same because of the same body form.

Next we have calculated the motions of 2 modules' model shown in Table 2 by using the three dimensional hydrodynamic forces and moments. Fig. 6 and Fig. 7 show the comparisons between 3-D results and Maeda's 2-D results concerning the heaving and the pitching amplitudes at the hinge connection point, which are evaluated by averaging the amplitudes of No. 1 and No. 2 module. The results of 3-D theory are slightly larger than those of 2-D theory, and the differences be-

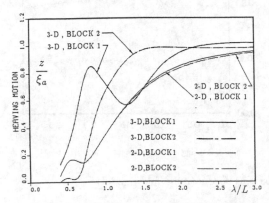

Fig. 8 Heaving amplitude of each module

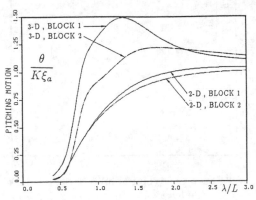

Fig. 9 pitching amplitude of each module

2-Modules

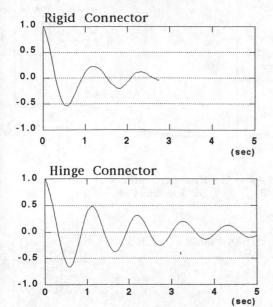

Rigid Connector

Hinge Connector

Fig. 11 Time history of free decay for pitch
of 2-modules model

4-Modules

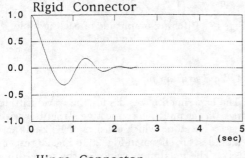

Rigid Connector

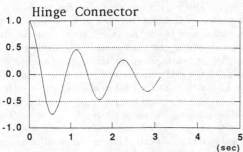

Hinge Connector

Fig. 12 Time history of free decay for pitch
of 4-modules model

6-Modules

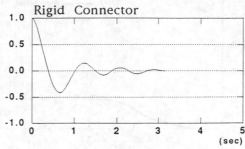

Rigid Connector

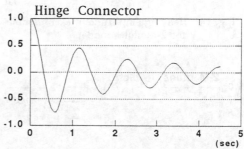

Hinge Connector

Fig. 13 Time history of free decay for pitch
of 6-modules model

tween them are small. Fig. 8 and Fig. 9 show the
heaving and pitching amplitudes of each body.
These amplitudes due to 3-D theory show the
large differences, which become larger in particular around $\lambda/L = 1$, while those due to 2-D theory are almost same. Thereby it can be said that
3-D theory is a useful tool for estimating the seakeeping performance of the multiple-connected
structure.

4.2 *Free decay test*

Fig. 10 to Fig. 13 show the time histories of free
pitching decay of the end module in the weather
side for various connecting conditions of 1-, 2-,
4- & 6-modules model. The natural periods for
pitching motion of the end module in the weather
side are evaluated from these free decay tests
and are shown in Table 3. The natural period
of the structure with rigid connectors is larger
than those of the multiple connected structure
with hinge connectors, because the inertia term
of the rigid body is much larger than that of the
module. These natural periods correspond to the
wave length ratio of 1.435 to 2.622, which are included in our experimental conditions measuring
the wave drifting forces.

411

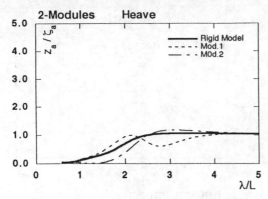

Fig. 14　Heaving amplitude of each module for 2-modules model

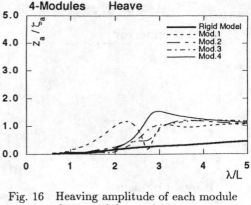

Fig. 16　Heaving amplitude of each module for 4-modules model

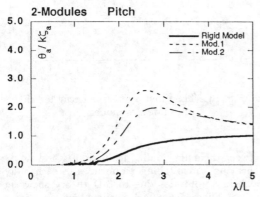

Fig. 15　Pitching amplitude of each module for 2-modules model

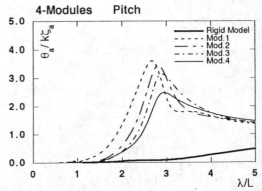

Fig. 17　Pitching amplitude of each module for 4-modules model

Furthermore the connecting conditions affect the damping ratios as shown in Fig. 10 to Fig. 13. In particular the damping moments on the structure with rigid connectors are much larger than those on the structure with hinge connector.

4.3 Motion amplitude

Fig. 15 to Fig. 18 show the comparison of motion amplitudes between each module of the multiple connected structure and the rigid structure. The heaving amplitudes of the rigid structure become smaller with the length of the structure. Similarly the pitching amplitudes of it have the same tendency as the heaving amplitudes. On the other hand, the motion amplitudes of each module of the multiple connected structure become much larger than those of the rigid structure with the number of the modules. In particular the pitching amplitude of the end module in the weather side becomes larger around the natural period.

4.4 Wave drifting force

The experimental results for the wave drifting forces on the multiple connecting structures with hinge connectors, which consists of 1-, 2-, 4- and 6-modules, are shown in Fig. 20 to Fig. 23 respectively. The maximum values of the wave drifting force coefficients increase with the number of modules, and takes the value larger than 3. The wave period of the maximum wave drifting force coefficient corresponds to the natural period for pitch of the end module in the weather side. We could watch that the end module in the weather side is pitching more largely than any other module is oscillating around these natural periods in our tank test. The numerical results are in a good agreement with the experimental ones.

Similarly the numerical estimation shown in Figs. 15, 17, 19 also show that the pitching amplitudes of the end module in the weather side become lager than those of any other modules

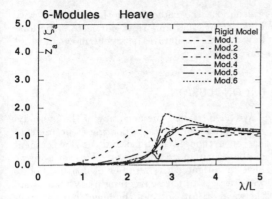

Fig. 18 Heaving amplitude of each module
for 6-modules model

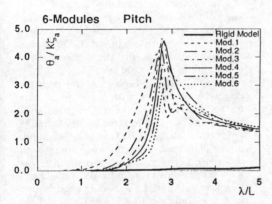

Fig. 19 Pitching amplitude of each module
for 6-modules model

1-Module

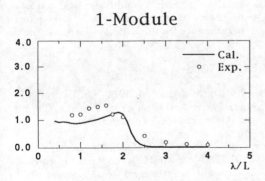

Fig. 20 Wave drifting force on 1-module model

2-Modules

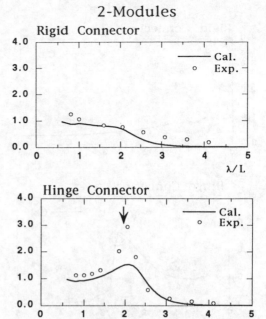

Fig. 21 Wave drifting force on 2-modules model

4-Modules

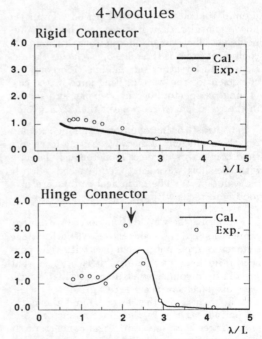

Fig. 22 Wave drifting force on 4-modules model

6-Modules

Rigid Connector

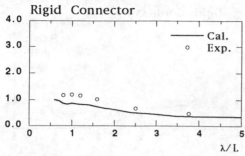

Hinge Connector

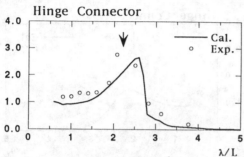

Fig. 23 Wave drifting force on 6-modules model

around the pitching natural period. From the above consideration, it is seemed that the phenomena, which the maximum wave drifting force coefficient exceeds the value 3, is relating the motions of the weather side module. Moreover the coefficient of the wave drifting force in the case of hinge connectors becomes smaller and smaller in the wave length range lager than $\lambda/L = 3.0$.

In contrast to the condition with hinge connector, the coefficient of the wave drifting force with rigid connectors becomes monotonously larger with the short wave length, and exceeds slightly the value of 1. The numerical results are in a good agreement with the experimental ones.

Comparing the wave drifting forces with hinge connectors and rigid connectors, ones with hinge connectors are smaller than ones with rigid connectors in long wave range. It is because that the multiple connected structure oscillates with wave displacement and scarcely generates the radiation waves. In the range around the natural period, ones with the hinge connectors are much larger than ones with rigid connectors. It is because that the end module in the weather side is more violently oscillating than the rigid body is and its radiation waves become larger. In the short wave range, the wave drifting forces

are almost the same for both structures with hinge connectors and rigid connectors, because any floating structures scarcely move in short wave range.

5. CONCLUSION

We have studied the accuracy of the hydrodynamic forces on the huge floating structure that has been thought to be useful for development of the ocean space utilization. Furthermore we have investigated the number of floating bodies with respect to the seakeeping performance and the wave drifting force of the huge floating structure in a regular wave. The main conclusions obtained in this study can be summarized as follows:

(1) It is verified experimentally that the coefficient of the wave drifting force on the multiple connected structure with hinge connectors can exceed the value of 3. This phenomenon relates deeply the motions of the end module in the weather side.

(2) The wave drifting forces on the multiple connected structure with hinge connectors are smaller than those of the structure with rigid connectors in long wave length range larger than $\lambda/L = 3.0$. On the other hand, in the short wave length range there is no difference between the wave drifting forces with hinge connector and rigid connector.

(3) We have estimated the 3-D hydrodynamic forces on individual bodies of the multiple-connected structure. It is made clear from the above results that the effect of the hydrodynamic interaction is strongly affected by the connecting conditions, and cannot be neglected for calculations of seakeeping performance for the huge floating structure.

REFERENCES

Bessho, M., Komatsu, M. 1974 "On Hydrodynamical Forces Acting on a Flat Plate Oscillating on Water Surface," JKSNAJ, No. 154, pp69-76.

Ertekin, R., Riggs, H.R., Che, X.L. and Du, S.X. 1991 "Efficient Methods for Hydro Elastic Analysis of Very Large Floating Structures," The Hawaii Section of The Society of Naval Architects and Marine Engineers.

Goo, J. and Yoshida, K. 1989 "Hydrodynamic interaction between multiple three-dimensional bodies of arbitrary shape in waves," JSNAJ, Vol. 165, pp.193-202.

Kagemoto, H. 1991 "Hydrodynamic Optimization of Floating-Bodies Arrays (Part 1. Minimization of Wave Forces)," JSNAJ, Vol. 170, pp.65-72.

Kudo, K. et al 1978 "The Drifting Force Acting on a Three-dimensional Body in Waves, SNAJ, Vol. 144, pp.155-162

Maeda, H. et al 1979 "On the Motions of a Floating Structure which Consists of Two or Tree Blocks with Rigid or Pin Joints," JSNAJ, Vol. 145, pp.71-78.

Matsunaga, K. et al 1979 "A Measurement of Bending and Torsional Moments on Floating Box Type Barges in Oblique Regular Waves," JKSNAJ, No. 174, pp35-43.

Ronald Riggs, H., Xiling Che, and Cengiz Eltekin, R. 1991 "Hydroelastic Response of Very Large Floating Structures," Proc. of OMAE, Vol. I-A, pp.291-300.

Takarada, N. 1982 "One Example of Technology and Accessment for Ultra Huge Off-Shore Structure (1st Report)," Bulletin of SNAJ, No. 638, pp.48-61.

Takaki,M. et al 1994 "Wave Drifting Force on Very Large Floating Structures", 12th Ocean Engineering Symposium, SNAJ, pp.129-136.

Takarada, N. et al 1989 "Model Test od a Huge Off-Shore Structure," 9th Ocean Engineering Symposium, SNAJ, pp. 347-354.

Uki, K. et al 1988 "The Initial Planning of the Floating Oil Storage System," Proc. of TECHNO-OCEAN'88, Vol.1, pp.416-420.

Yamashita, S. 1979 "Motions and Hydrodynamic Pressures of a Box-Shaped Floating Structure of Shallow draft in Regular Waves," JSNAJ, Vol. 146, pp.165-172.

Hydroelastic analyses of a structure supported on a large number of floating legs

Hiroshi Kagemoto
Naval Architecture and Ocean Engineering, The University of Tokyo, Japan

Dick K. P. Yue
Massachusetts Institute of Technology, Mass., USA

ABSTRACT:Hydroelasic analyses of a structure supported on a large number of legs are conducted. The structure considered is one of the prospective candidates for future very large floating structures. Theoretical methods for the analysis of dynamic behaviours of such structures in waves are presented. The hydroelastic interactions of structural deformations with the hydrodynamic forces acting on a structure are taken into account. The hydrodynamic interactions among supporting legs are also included. Numerical analyses are carried out for several example structures with various structural rigidity. The effects of structural rigidity on the dynamic behaviours are discussed.

1.Introduction

A number of proposals for the construction of a floating airport have been made both in the U.S. and in Japan. There exist, however, problems that should be cleared before we go ahead with the construction. Although some of the problems have little to do with technical matters but rather stem from sociological issues, the big question is that if it is really feasible from a pure technical point of view. A detailed feasibility study on a floating airport was once carried out in Japan more than ten years ago(Ando et al. 1983), when a then new international airport was decided to be constructed off Osaka city, which is the second largest city in Japan. The proposed structural concept of the airport was, as shown in Figure 1.1, composed of a flat deck structure for runways, which is several kilometers in length, and more than 30,000 equally-spaced legs that support the deck. Since no

such huge floating structures have ever been built, we should examine very carefully from every conceivable point of view so that we do not overlook unexpected phenomena that can cause serious trouble after the completion. One that is not commonly observed in conventional floating structures is the interactions between the elastic deformations of a structure and the hydrodynamic forces acting on the structure. Since, as shown in Figure 1.1, the horizontal dimension of a structure is expected to be far larger than its vertical dimension, the relative rigidity of the structure should be quite smaller than that of conventional structures and thus results in large structural deformations in waves. These large deformations of a structure may in turn affect the hydrodynamic forces acting on the structure. These hydroelastic interactions of a huge floating structure are one of the phenomena which are not fully understood and this is the subject of the present paper. Although no consensus has been ob-

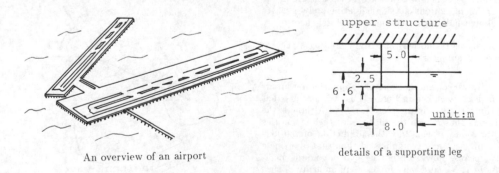

An overview of an airport

details of a supporting leg

Figure 1.1 A floating airport proposed in Japan.

tained yet as to what type of structure is appropriate for future very large floating structures, we restrict our attention to that supported on a large number of legs, which is, at least, one of the prospective candidates of such structures.

For hydroelastic analyses of a ship, the ship is usually replaced with a one dimensional beam. This technique was extended so that it can be applied to a floating structure supported on a large number of legs(Toki 1979). In this paper we present a more sophisticated method in which hydroelastic interactions of structural deformations with hydrodynamic forces as well as hydrodynamic interactions among the supporting legs are taken into account simultaneously. Numerical results on some typical structures are presented and the effects of hydrelasticity on dynamic behaviours of a floating structure supported on a large number of legs are discussed.

2. Theory

2.1 Analyses of motions of a single floating body

First we consider a single floating body in waves. The equations of motions of the body are written as:

$$M_k \ddot{x}_k + \sum_{j=1}^{6}(m_{jk}\ddot{x}_j + N_{jk}\dot{x}_j + C_{jk}x_j) = f_k \quad (2.1)$$

$$(k = 1, 2, \cdots, 6)$$

where $x_k, \dot{x}_k, \ddot{x}_k$ represent the displacement, velocity, acceleration of the corresponding body in the k-th direction respectively and:

M_k: mass or mass moment of inertia in the k-th mode.
m_{jk}: added mass or adde mass moment in the k-th direction due to the motion in the j-th direction.
N_{jk}: damping coefficient in the k-th direction due to the motion in the j-th direction.
C_{jk}: restoring force coefficient in the k-th direction due to the motion in the j-th direction
f_k: wave force in the k-th direction.

Within a linear potential theory, these coefficients m_{jk}, N_{jk}, C_{jk} and the wave force f_k can be calculated fairly easily for bodies of arbitrary geometry. Once the coefficients and the forces are known, the motions of a single floating body in waves can be obtained by solving eqn(2.1). Therefore the theoretical aspects in the prediction of motions of a single floating body in waves are almost finished in the linear potential theory.

2.2 Analyses of motions of a rigid floating structure supported on an array of legs

Secondly we consider a structure supported on an array of legs as shown in Figure 1.1. We assume that the the structure is completely rigid. Although the equations that govern the motions of such structures are the same as eqn(2.1), the prediction of motions are in practice more complicated than that of a single body because now we need the hydrodynamic coefficients m_{jk}, N_{jk} and wave forces f_k of an array of legs. (Ususally C_{jk} can be estimated without much difficulty.) When a body is not isolated but arrayed with

others in waves, it is affected hydrodynamically by the presence of the other bodies because waves scattered by or radiated from the body are bounced back from the other bodies. The waves produced due to the other bodies are also added to the waves incident to the body. When the number of bodies involved is not large, these hydrodynamic interactions among floating bodies can be accounted for by conventional numerical methods such as a boundary element method or a finite element method. However, if the number of bodies is large as is the case for structures being considered in this paper, these conventional methods become practically prohibitive because the required computational effort is enormous. In order to cope with these problems, the authors, using a matrix method(Simon 1982), have developed a theory that can estimate hydrodynamic interactions among a large number of floating bodies with reasonable computational efforts without loss of accuracy within a linear potential theory(Kagemoto and Yue 1986a). This method can be used for the prediction of the hydrodynamic coefficients m_{jk}, N_{jk} and the wave forces f_k that appear in eqn(2.1). Although the details of the method can be found in the reference, it is repeated here briefly for the sake of subsequent descriptions of the theory for hydroelastic analyses.

Let us first consider the plane-wave diffraction by an array of N floating three-dimensional bodies *fixed* in the waves as shown in Figure 2.1. Under the assumptions of usual linearized potential flow in the frequency domain, the complex velocity potential, $\Phi_i^s(\vec{x}, t)$ $= \Re\{\phi_i^s(\vec{x})e^{-i\omega t}\}$, representing the scattered wavefield around body-i can in general be expressed as a summation of cylindrical partial waves in the local coordinate system fixed in the center position O_i of body i:

$$\phi_i^s(r_i, \theta_i, z_i)$$
$$= \frac{\cosh k_0(z_i + h)}{\cosh k_0 h} \sum_{n=-\infty}^{\infty} A_{0n_i} H_n(k_0 r_i)e^{in\theta_i}$$
$$+ \sum_{m=1}^{\infty} \cos k_m(z_i + h) \sum_{n=-\infty}^{\infty} A_{mn_i} K_n(k_m r_i)e^{in\theta_i}$$

$$(2.2)$$

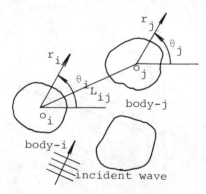

Figure 2.1 An array of N floating bodies

where H_n and K_n are respectively the n-th order Hankel function of the first kind and modified Bessel function of the second kind.

Assuming an appropriate truncation of the infinite summations over n, m, (2.2) can be expressed in a matrix form as:

$$\phi_i^s(r_i, \theta_i, z_i) = \tilde{A}_i^T \tilde{\psi}_i^s(r_i, \theta_i, z_i) , \qquad (2.3)$$

where $\tilde{A}_i$, $\tilde{\psi}_i^s$ are respectively the vectors of coefficients of A_{mn}, $m = 0, 1, 2, \ldots$, and the scattered partial waves in terms of H_n and K_n. The superscript T indicates the transpose. Using the addition theorem for Bessel functions, ϕ_i^s can be expressed in the near field of body $j(r_j < L_{ij})$ in terms of the local cylindrical coordinate system (r_j, θ_j, z_j) fixed at body j:

$$
\begin{aligned}
\phi_i^s \mid_j = {} & \frac{\cosh k_0(z_j + h)}{\cosh k_0 h} \sum_n C_{0n_j} J_n(k_0 r_j) e^{in\theta_j} \\
& + \sum_m \cos k_m(z_j + h) \sum_n C_{mn_j} I_n(k_m r_j) e^{in\theta_j}
\end{aligned}
$$

$$(2.4)$$

Here J_n, I_n are respectively the n-th order Bessel function of the first kind and modified Bessel function of the first kind. In matrix form, (2.4) can be written as:

$$\phi_i^s \mid_j = \tilde{A}_i^T \mathbf{T_{ij}} \tilde{\psi}_j^I , \qquad (2.5)$$

where $\tilde{\psi}_j^I$ is the vector of the incident cylindrical partial waves expressed in terms of J_n, I_n, $\mathbf{T_{ij}}^T \tilde{A}_i$ is the vector of coefficients C_{mn}, $m = 0, 1, 2, \ldots$, and $\mathbf{T_{ij}}$ is a coordinate transformation matrix from i to j representation for the set of partial waves.

Upon combining the scattering waves due to all the other bodies represented by $i = 1, 2, \ldots, N$, $i \neq j$, the total incident potential at body-j is obtained:

$$\phi_j^I = \phi_0 \mid_j + \sum_{\substack{i=1 \\ i \neq j}}^N \tilde{A}_i^T \mathbf{T_{ij}} \tilde{\psi}_j^I , \qquad j = 1, 2, \ldots, N .$$

$$(2.6)$$

where $\phi_0 \mid_j$ is the incident wave field expressed in the (r_j, θ_j, z_j) coordinate system which is also written in terms of partial waves:

$$\phi_0 \mid_j = \tilde{a}_j^T \tilde{\psi}_j^I . \qquad (2.7)$$

The incident and the scattered waves around any body-j are related by the diffraction characteristics of that body such that:

$$\tilde{A}_j = \mathbf{B_j}(\tilde{a}_j + \sum_{\substack{i=1 \\ i \neq j}}^N \mathbf{T_{ij}}^T \tilde{A}_i) , \qquad (2.8)$$

where $\mathbf{B_j}$ represents the diffraction characteristics of body-j. $\mathbf{B_j}$ is not affected by the presence of other bodies and can be obtained by solving the diffraction problem for body-j as a single body. Eqn(2.8) is a set of linear simultaneous equations for $\tilde{A}_j$. Once $\tilde{A}_j$, $j = 1, 2, \ldots, N$, are determined from (2.8), the velocity potential ϕ_j^s for the wave field around body-j, $j = 1, 2, \ldots, N$, are evaluated from (2.3) and the hydrodynamic pressure and thus the associated forces on individual bodies and on the array(f_k) are determined.

Radiation problems can be solved in a similar manner with the only difference that the velocity potential $\phi_0 \mid_j$ in (2.6) is now replaced by

$$\phi_0 \mid_j \equiv \sum_{\substack{i=1 \\ i \neq j}}^N \sum_{\ell=1}^6 x_\ell^i \mathbf{R}_\ell^{i^T} \mathbf{T_{ij}} \tilde{\psi}_j^I , \qquad (2.9)$$

where x_ℓ^i is the displacement of body-i in the ℓ-th direction and $\mathbf{R}_\ell^i$ represents the single-body radiation characteristics of body-i due to unit displacement amplitude in the ℓ-th direction, which, like $\mathbf{B_j}$, can be obtained by solving the radiation problems for body-j as a single body.

In this way after solving one diffraction problem and six radiation problems, f_k, m_{jk}, N_{jk} in eqn(2.1) can be determined. Once these forces and coefficients are obtained, the motions $x_k(k = 1, 2, \cdots, 6)$ of the corresponding structure can be predicted by solving eqn.(2.1).

2.3 Analyses of motions of an elastic structure supported on an array of legs

Next we consider a structure that is not completely rigid and thus can be deformed elastically. If the elastic deformations are small compared to the displacements due to motions as a rigid body, the analyses of the elastic behaviours can be carried out by making use of the results of rigid-body motion analyses. First the inertia forces due to motions and the hydrodynamic forces on a structure are calculated while treating the structure as a rigid body. This can be achieved by the method described in 2.2. After obtaining the inertia forces and the hydrodynamic forces, analyses of elastic behaviours of the corresponding structure are conducted while using the obtained forces as external forces acting on the structure. This technique is effective as far as elastic deformations are sufficiently small that they do not affect hydrodynamic forces acting on the structure, which is usually the case for conventional floating structures. However, as the horizontal scale of a structure becomes large compared to the vertical dimension, the relative rigidity of the structure decreases and elastic deformations become as large as rigid-body motions. When elastic deformations of a structure are comparable to rigid-body motions, their effects on hydrodynamic forces acting on the structure can no longer be neglected. In this case, although elastic deformations are induced by hydrodynamic forces, the hydrodynamic forces are in turn affected by the

419

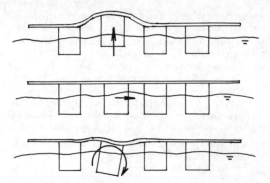

Figure 2.2 Some of the independent modes of motions of an elastic structure supported on legs.

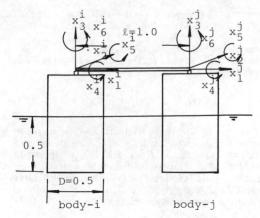

Figure 2.3 Two cylinders connected by a linear elastic beam

induced elastic deformations. Therefore the hydrodynamic forces and the elastic deformations should be determined simultaneously. When elastic deformations are taken into account, the number of independent modes of motions of a structure supported on N legs are no longer 6 but 6×N as schematically shown in Figure 2.2. Under these circumstances, the equations of motions should be designated for each of N legs:

$$M_k^i \ddot{x}_k^i + \sum_{j=1}^{N} \sum_{\ell=1}^{6} (m_{\ell k}^{ji} \ddot{x}_\ell^j + N_{\ell k}^{ji} \dot{x}_\ell^j + C_{\ell k}^{ji} x_\ell^j) = f_k^i$$

$$(k = 1, 2, \cdots, 6 \quad : i = 1, 2, \cdots, N) \qquad (2.10)$$

where M_k^i is the mass or moment of inertia of body-i in the k-th mode, $x_k^i, \dot{x}_k^i, \ddot{x}_k^i$ respectively the displacement, velocity, acceleration of body-i in the k-th direction, $m_{\ell k}^{ji}$, $N_{\ell k}^{ji}$ and $C_{\ell k}^{ji}$ respectively the added mass, damping and restoring force coefficients for body-i in the k-th mode due to the motion of body-j in the ℓ-th direction, and f_k^i the wave force on body-i acting in k-th direction.

The effects of the structural restraints on the motions of legs can be incorporated as additional restoring forces to $C_{\ell k}^{ji}$. For example, if two legs are connected by a linear elastic beam of length ℓ, section area A and rigidity EI as shown in Figure 2.3, the restoring force coefficients $\triangle C$ induced by the structural restraints that should be added to $C_{\ell k}^{ji}$ are written as:

$$\triangle C_{11}^{ii} = \frac{EA}{\ell}, \quad \triangle C_{11}^{ji} = -\frac{EA}{\ell}$$

$$\triangle C_{11}^{ij} = -\frac{EA}{\ell}, \quad \triangle C_{11}^{jj} = \frac{EA}{\ell} \qquad (2.11)$$

$$\triangle C_{33}^{ii} = \frac{12EI}{\ell^3}, \quad \triangle C_{33}^{ji} = -\frac{12EI}{\ell^3}$$

$$\triangle C_{33}^{ij} = -\frac{12EI}{\ell^3}, \quad \triangle C_{33}^{jj} = \frac{12EI}{\ell^3} \qquad (2.12)$$

Here only the restraints in surge(k=1) and heave(k=3) direction are shown. The restraints in all 6 degrees of

motions can be incorporated in the same way as the additional restoring forces and moments exerted on each leg that are given by:

$$\begin{bmatrix} \mathcal{E} & 0 & 0 & 0 & 0 & 0 & & & & \\ 0 & \alpha P_z & 0 & 0 & & & & & \\ 0 & 0 & \alpha P_y & \mathcal{G} & & & & sym & \\ 0 & 0 & -P_y & 0 & Q_y & & & & \\ 0 & P_z & 0 & 0 & 0 & Q_z & & & \\ -\mathcal{E} & 0 & 0 & 0 & 0 & 0 & \mathcal{E} & & \\ 0 & -\alpha P_z & 0 & 0 & 0 & -P_z & 0 & \alpha P_z & \\ 0 & 0 & -\alpha P_y & P_y & 0 & 0 & 0 & 0 & \alpha P_y \\ 0 & 0 & -\mathcal{G} & 0 & R_y & 0 & 0 & 0 & 0 & \mathcal{G} \\ 0 & 0 & -P_y & 0 & R_y & 0 & 0 & 0 & P_y & Q_y \\ 0 & P_z & 0 & 0 & 0 & R_z & 0 & -P_z & 0 & 0 & Q_z \end{bmatrix} \begin{bmatrix} X_1^i \\ X_2^i \\ X_3^i \\ X_4^i \\ X_5^i \\ X_6^i \\ X_1^j \\ X_2^j \\ X_3^j \\ X_4^j \\ X_5^j \\ X_6^j \end{bmatrix} = \begin{bmatrix} f_1^i \\ f_2^i \\ f_3^i \\ f_4^i \\ f_5^i \\ f_6^i \\ f_1^j \\ f_2^j \\ f_3^j \\ f_4^j \\ f_5^j \\ f_6^j \end{bmatrix}$$

where $\mathcal{E} \equiv EA/\ell$, $\mathcal{G} \equiv GJ/\ell$, $P_y \equiv 6EI_y/\ell^2(1 + \Phi_z)$, $P_z \equiv 6EI_z/\ell^2(1 + \Phi_y)$, $Q_y \equiv (4 + \Phi_z)EI_y/\ell(1 + \Phi_z)$, $Q_z \equiv (4 + \Phi_y)EI_z/\ell(1 + \Phi_y)$, $R_y \equiv (2 - \Phi_z)EI_y/\ell(1 + \Phi_z)$, $R_z \equiv (2 - \Phi_y)EI_z/\ell(1 + \Phi_y)$, and $\alpha \equiv 2/\ell$.

EA, GJ, and $EI_{x,y,z}$ are the axial, torsional and bending rigidities respectively, and $\Phi_{x,y,z}$ represent the effects of shearing deformations. Incorporating structural restraints in this way, motions x_k^i ($k=1,2,\cdots$, 6; $i=1,2,\cdots$, N) of each leg can be estimated from eqn(2.10). The coefficients $m_{\ell k}^{ji}, N_{\ell k}^{ji}, f_k^i$ in eqn(2.10) can, in principle, be obtained by the method described in 2.2. Ertekin et al.(1993) analysed elastic behaviours of a structure composed of a linear array of semisubmersibles. However, for N legs and $6N$ independent degrees-of-freedom, $6N$ radiation problems for the hydrodynamic coefficients $m_{\ell k}^{ji}$ and $N_{\ell k}^{ji}$ must be solved. This is quite tedious to conduct as the number of legs increases. The authors have developed an alternative method wherein the diffraction and radiation forces on each leg in the array as well as its motions can be solved *simultaneously* in the same computation(Kagemoto and Yue 1987).

Corresponding to (2.6), the total incident potentials to body-j, $j = 1, 2, \ldots, N$, is given by:

$$\phi_j^I = (\tilde{a}_j^T + \sum_{\substack{i=1 \\ i \neq j}}^{N} \tilde{A}_i^T \mathbf{T_{ij}}) \tilde{\psi}_j^I + \sum_{\substack{i=1 \\ i \neq j}}^{N} \sum_{\ell=1}^{6} (x_\ell^i \mathbf{R}_\ell^{i T} \mathbf{T_{ij}}) \tilde{\psi}_j^I ,$$

$$(2.13)$$

where the incident and the scattered waves around any body-j are related by:

$$\tilde{A}_j = \mathbf{B_j}\{(\tilde{a}_j + \sum_{\substack{i=1 \\ i \neq j}}^{N} \mathbf{T_{ij}}^T \tilde{A}_i) + \sum_{\substack{i=1 \\ i \neq j}}^{N} \sum_{\ell=1}^{6} x_\ell^i \mathbf{T_{ij}}^T \mathbf{R_\ell^i}\} ,$$

$$(2.14)$$

For a diffraction problem $x_\ell^i \equiv 0$ whereas for a radiation problem x_ℓ^i is given. Now x_ℓ^i is unknown. Therefore the unknowns are $\tilde{A}_i$, $i = 1, 2, \ldots, N$, and x_ℓ^i, $\ell = 1, 2, \ldots, 6; i = 1, 2, \ldots, N$.

Since (2.14) is a set of N equations, the number of equations is less than the number of unknowns by $6N$. This can be supplemented by equations of motions, which are written as eqn(2.10). Eqn(2.10) can also be written in the following form.

$$M_k^i \ddot{x}_k^i + \sum_{j=1}^{N} \sum_{\ell=1}^{6} C_{\ell k}^{ji} x_\ell^j = F_k^i \qquad (2.15)$$

where F_k^i represents the total hydrodynamic forces (including those associated with the hydrodynamic coefficients) acting on body-i, that is:

$$F_k^i \equiv f_k^i - \sum_{j=1}^{N} \sum_{\ell=1}^{6} (m_{\ell k}^{ji} \ddot{x}_\ell^j + N_{\ell k}^{ji} \dot{x}_\ell^j), \qquad (2.16)$$

F_k^i is expresses by the integration of hydrodynamic pressure over the submerged surface S_b^i of body-i:

$$F_k^i = \iint_{S_b^i} i\omega\rho \phi_i^s n_k dS . \qquad (2.17)$$

By making use of $\mathbf{B_i}$ from (2.8), ϕ_i^s in the above equation is given as:

$$\phi_j^s = \{(\tilde{a}_j^T + \sum_{\substack{i=1 \\ i \neq j}}^{N} \tilde{A}_i^T \mathbf{T_{ij}}) + \sum_{\substack{i=1 \\ i \neq j}}^{N} \sum_{\ell=1}^{6} x_\ell^i \mathbf{R_\ell^i}^T \mathbf{T_{ij}}\} \mathbf{B_j}^T \tilde{\psi}_j^s .$$

$$(2.18)$$

Therefore the equations of motions(eqn(2.15)) can be written in terms of a linear combianation of $\tilde{A}_i$, x_ℓ^i:

$$M_k^j \ddot{x}_k^j + \sum_{i=1}^{N} \sum_{\ell=1}^{6} C_{\ell k}^{ij} x_\ell^i = i\omega\rho\{(\tilde{a}_j^T + \sum_{\substack{i=1 \\ i \neq j}}^{N} \tilde{A}_i^T \mathbf{T_{ij}})$$

$$+ \sum_{\substack{i=1 \\ i \neq j}}^{N} \sum_{\ell=1}^{6} x_\ell^i \mathbf{R_\ell^i}^T \mathbf{T_{ij}}\} \mathbf{B_j}^T \iint_{S_b^j} \tilde{\psi}_j^s n_k dS$$

$$(k = 1, 2, \ldots, 6; \; j = 1, 2, \ldots, N \,) \; (2.19)$$

The feature of the above equation is that hydrodynamic forces are not divided into radiation and diffraction ones in the equations of motions but instead they are treated as the integration of the total hydrodynamic pressure. Eqn(2.19) is a set of $6N$ linear equations for $\tilde{A}_i$ and x_ℓ^i, $(\ell = 1, 2, \cdots, 6)$. Solving the N+6N set of linear equations (2.14),(2.19), $\tilde{A}_i, x_\ell^i (\ell = 1, 2, \cdots, 6; i = 1, 2, \cdots, N)$ are obtained simultaneously.

Figure 2.4(a),(b) compare the results estimated in this way with experimental data obtained in a model basin on surge and heave motions in head waves measured at the center of the upper structure supported on 3 by 11 composite cylinders. (The geometry of the cylinders are the same as that shown in Figure 1.1.) The rigidity of the model structure is quite large and elastic deformations are very small. The agreement is quite satisfactory except for the heave responses at the resonant frequency. This discrepancy is mainly due to the damping forces induced by votex sheddings, which are not the subject of the present study.

The effects of structural damping forces or mooring forces can be taken into account in the same way as

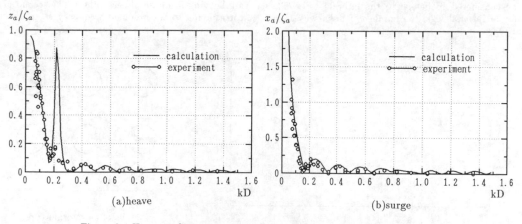

(a)heave

(b)surge

Figure 2.4 Heave and surge response charactereistics in waves of a structure supported on 3 × 11 composite cylinders

the effects of the structural restraints as far as the forces are linear functions of $\ddot{x}, \dot{x}, x$.

2.4 Extension of the theory to a structure supported on a extremely large number of legs

Although the theory described in the previous sections reduces the required computational effort considerably, it is still prohibitive to apply when the number of bodies involved is extremely large, as is the case for the proposed floating airport structure in Japan, which was to be supported on more than 30,000 legs.

For the hydroelastic analyses of such structures, we can exploit the fact that supporting legs are, as shown in Fig.2.5, usually identical and arrayed with equal space. Besides, the rigidity of the upper structure may be treated as uniform. From eqn(2.13), the velocity potential for the waves incident to body j is written:

$$\phi_j^I = (\tilde{a}_j^T + \sum_{\substack{i=1 \\ i \neq j}}^{N} \tilde{A}_i^T \mathbf{T_{ij}}) \tilde{\psi}_j^I + \sum_{\substack{i=1 \\ i \neq j}}^{N} \sum_{\ell=1}^{6} (x_\ell^i \mathbf{R}_\ell^{i^T} \mathbf{T_{ij}}) \tilde{\psi}_j^I ,$$

(2.20)

Since

$$\tilde{a}_j = e^{ik_0 L_{ij} \cos(\theta_{ij} - \theta_I)} \tilde{a}_i .$$

(2.21)

if the number of the legs is really infinite, $\tilde{A}_i, x_\ell^i$ respectively should also have the same interrelationship:

$$\tilde{A}_j = e^{ik_0 L_{ij}(\theta_{ij} - \theta_I)} \tilde{A}_i$$

(2.22)

$$x_\ell^j = e^{ik_0 L_{ij}(\theta_{ij} - \theta_I)} x_\ell^i$$

(2.23)

When legs are not infinitely arrayed but their number is still quite large, it will probably be justified to assume that the flow field in a certain interior core of the array is the same as that of an infinite array of legs and eqns(2.22),(2.23) hold for the legs that can be considered to belong to the internal core. Then for M intrenal legs, the number of unknowns among

$\tilde{A}_j, x_\ell^j (j = 1, 2, \cdots, M)$ is only one (say, $\tilde{A}_j, x_\ell^j$) because the others $(\tilde{A}_i, x_\ell^i (i = 1, 2, \cdots, j - 1, j, \cdots, M))$ are related to $\tilde{A}_j, x_\ell^j$ through eqns(2.22),(2.23). Thus, in the case of N legs, of which M legs belong to an internal core, for example, M of the unknown $\tilde{A}_j$s, x_ℓ^js are linearly interdependent and the total number of unknown vectors is reduced from N to $N - M + 1$. These approximations result in a tremendous reduction in the number of unknowns.

3. Results and discussion

In the following calculations, we assume that relative motions of legs in surge, sway directions are zero and only those in heave direction is allowed. Rotational motions of legs are also assumed to be completely restiricted. This is, firstly for the sake of simplicity of numerical computations and, secondly because these assumptions may be justified for a structure supported on a large number of legs considered in this paper.

3.1 A structure supported on two legs

We first consider a structure supported on two legs that was shown in Figure 2.3. The two legs are assumed to be connected by a linear elastic beam of length ℓ, rigidity EI. Figure 3.1, Figure 3.2 respectively show the response amplitude operator of heave motions and surge motions. Figure 3.3 shows the frequency characteristics of drift forces acting on the two-leg structure. All the results are those in head waves. In each figure, results for four different rigidity EI are compared in order to examine the effects of elastic restraints on motions. In heave motions slight differences between the two legs are observed. When the upper structure is completely rigid(Figure 3.1(a)), only one peak is observed whereas as the rigidity decreases second peak appears and the second peak frequency becomes lower as the rigidity decreases. The first peak corresponds to the resonant frequency ω_1 of rigid-body heave motions, in which the two cylinders are heaving in phase of each other. The second peak corresponds to the resonant frequency ω_2 of the 1st

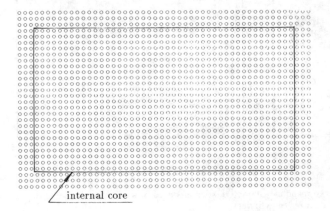

internal core

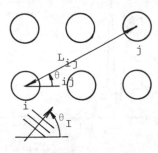

details of an internal core

Figure 2.5 An equally-spaced array of a large number of legs

elastic mode of motions, in which the two cylinders are heaving out of phase with each other. The resonant frequencies ω_1, ω_2 can be roughly estimated as follows.

Since the equations of vertical motions of each leg are written:

$$(M + m_{33}^{11})\ddot{x}_3^1 + kx_3^1 + A(x_3^1 - x_3^2) = f_3^1$$
$$(M + m_{33}^{22})\ddot{x}_3^2 + kx_3^2 + A(x_3^2 - x_3^1) = f_3^2$$

(3.1)

Here the restoring coefficients due to the hydrostatic pressure and those due to the structural restraints are written as k and A respectively, where k, A are:

$$k \equiv \rho g A_w, \quad A \equiv \frac{12EI}{\ell^3} \quad (3.2)$$

A_w represents the waterplane area of a leg. $m_{\ell k}^{ji}$ for $j \neq i$ or $\ell \neq k$ and $N_{\ell k}^{ji}$ for all j, i are neglected. Further assuming $m_{33}^{11} \sim m_{33}^{22}$, which we denote as $\triangle m$, the resonant frequency ω is obtained as the eigenvalues of the following matrix.

$$\begin{bmatrix} k+A & -A \\ -A & k+A \end{bmatrix}$$

The eigenvalues are obtained analytically as k, $k+A$ and ω_1, ω_2 are given as:

$$\omega_1^2(M + \triangle m) = k$$
$$\omega_2^2(M + \triangle m) = k + A \quad (3.3)$$

By virtue of the fact that an added mass does not vary much with oscillation frequencies and substituting the average value of the added mass of a single cylinder($\sim 0.3M$) into $\triangle m$, ω_1, ω_2 can be determined, which are indicated by arrows in Figure 3.1.

Although the surge motions of the two-leg structure are affected little by the structural rigidity, an interesting phenomenon is observed in Figure 3.2 in which the response characteristics of surge motions vary quite discontinuously at certain frequencies. If we examine the frequencies, it turns out that the frequencies correspond to the resonant frequncy ω_2 of the first elastic mode of motions. At this resonant frequency, the phase of the waves radiated due to the heave motions changes by π and this results in the abrupt change of the surge forces acting on the cylinders and result in the sharp change of surge responses. These changes at the resonant frequencies of heave motions can also be found in drift force characteristics as shown in Figure 3.3.

3.2 A structure supported on a 5×10 array of legs

We next consider a structure suported on a 5×10 array of legs in head waves. The legs are assumed to be connected with adjacent legs by linear elastic beams of rigidity EI (see Figure 3.4).

Figure 3.5(a),(b),(c),(d) show the response amplitude

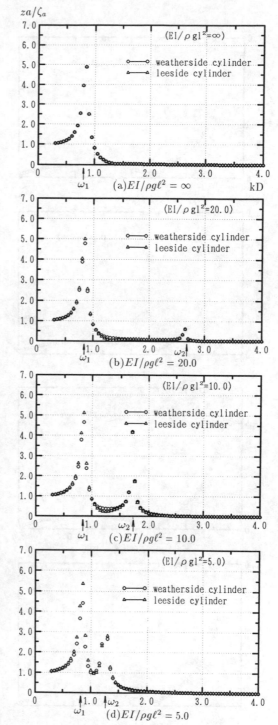

Figure 3.1 The response amplitude operator of heave motions of each of the two legs of a two-leg structure

423

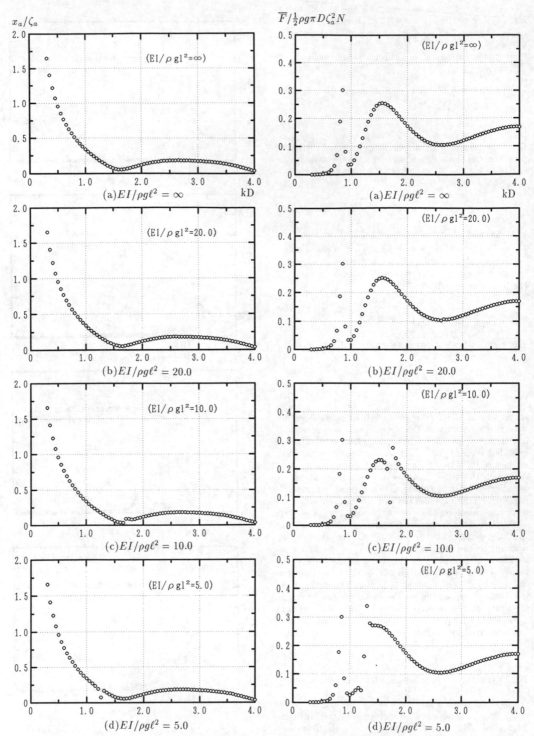

x_a/ζ_a

2.0

1.5

1.0

0.5

0

$(EI/\rho\, gl^2{=}\infty)$

0   1.0   2.0   3.0   4.0

(a)$EI/\rho g\ell^2 = \infty$ kD

2.0

1.5

1.0

0.5

0

$(EI/\rho\, gl^2{=}20.0)$

0   1.0   2.0   3.0   4.0

(b)$EI/\rho g\ell^2 = 20.0$

2.0

1.5

1.0

0.5

0

$(EI/\rho\, gl^2{=}10.0)$

0   1.0   2.0   3.0   4.0

(c)$EI/\rho g\ell^2 = 10.0$

2.0

1.5

1.0

0.5

0

$(EI/\rho\, gl^2{=}5.0)$

0   1.0   2.0   3.0   4.0

(d)$EI/\rho g\ell^2 = 5.0$

Figure 3.2 The response amplitude operator
of surge motions of a two-leg structure

$\overline{F}/\frac{1}{2}\rho g\pi D\zeta_a^2 N$

0.5

0.4

0.3

0.2

0.1

0

$(EI/\rho\, gl^2{=}\infty)$

0   1.0   2.0   3.0   4.0

(a)$EI/\rho g\ell^2 = \infty$ kD

0.5

0.4

0.3

0.2

0.1

0

$(EI/\rho\, gl^2{=}20.0)$

0   1.0   2.0   3.0   4.0

(b)$EI/\rho g\ell^2 = 20.0$

0.5

0.4

0.3

0.2

0.1

0

$(EI/\rho\, gl^2{=}10.0)$

0   1.0   2.0   3.0   4.0

(c)$EI/\rho g\ell^2 = 10.0$

0.5

0.4

0.3

0.2

0.1

0

$(EI/\rho\, gl^2{=}5.0)$

0   1.0   2.0   3.0   4.0

(d)$EI/\rho g\ell^2 = 5.0$

Figure 3.3 The frequency response characteristics
of drift forces acting on a two-leg structure

424

operators of vertical motions of the legs for four different rigidity of the connecting beams. Figure 3.5(a),(b) show the results of a structure supported on 50 legs which are connected by beams of infinite rigidity. Figure 3.5(c) shows those of the structure connected by an elastic beam of rigidity $EI/\rho g \ell^2 = 2.0 \times 10^2$ and Figure 3.5(d) shows those of the structure connected by an elastic beam of rigidity $EI/\rho g \ell^2 = 2.0 \times 10^1$. The diffrerence between the results shown in Figure 3.5(a) and Figure 3.5(b) is that in obtaining the results shown in Figure 3.5(a) hydrodynamic interaction effects among the legs are not taken into account whereas the hydrodynamic interactions are considered in the calculations to obtain the results shown in Figure 3.5(b). (These effects are also taken into account for the calculations to obtain the results shown in Figure 3.5(c),(d).) Since there are 50 legs, only the response characteristics of No.1,No.3,No.5,No.7,No.10 legs are shown in each figure. (Refer to Figure 3.4 to identify the leg No.) (Since the structure is completely rigid in obtaining the results shown in Figure 3.5(a),(b), the vertical responses are the same for all the legs in the two figures.) Comparing Figure 3.5(a) and (b), the hydrodynamic interaction effects on the vertical motions are not distinct. The only noticeable effects are the reduction of peak values at resonant frequencies. As the rigidity of the upper structure decreases, other peaks appear, which correspond to resonant modes of elastic motions. The shielding effects of waves by the weath-

erside legs are not clearly observed or even sometimes the leeside leg(No.10) moves with a larger amplitude than the weatherside leg(No.1). In general when the elastic motions become large, the legs at the both ends of the structure (No.1 leg and No.10 leg) experience, as expected, larger motions than the interior legs(No.3,No.5,No.7 legs). As mentioned before, the resonant frequencies can be roughly predicted as the eigenvalues of the following matrix.

$$
\begin{bmatrix}
k+A & -A & 0 & 0 & \cdots & 0 \\
-A & k+2A & -A & 0 & \cdots & 0 \\
0 & -A & k+2A & -A & \cdots & 0 \\
\vdots & & & & & \vdots \\
0 & 0 & \cdots & -A & k+2A & -A \\
0 & 0 & \cdots & 0 & -A & k+A
\end{bmatrix}
$$

where $k \equiv \rho g A_w$, $A \equiv \frac{12EI}{\ell^3}$. The eigenvalues $\lambda_i (i = 1, 2, \cdots, 50)$ correspond to $\omega_i^2 (M + \triangle m)$ as mentioned in 3.1, where M is the mass of each leg (including the associated mass of the upper structure) and $\triangle m$ is the added mass of a leg. The first frequency ω_1 corresponds to the natural frequency of rigid-body motions in which all the legs are heaving with almost the same phase. The subsequent frequencies are those of the elastic modes of motions. Figure 3.6(a),(b) show the amplitudes of the vertical motions of the legs that are arrayed along the center line (Leg No.1 ~ No.10) for $EI/\rho g \ell^2 = 2.0 \times 10^2$ (Figure 3.6(a)) and for $EI/\rho g \ell^2 = 2.0 \times 10^1$(Figure 3.6(b)). Five figures are included in Figure 3.6(a) and in Figure 3.6(b). These correspond to the vertical responses at the lowest 5 resonant frequencies, the first of which is the natural frequency of the rigid motions. When the rigidity is large(Figure 3.6(a)), all the legs are moving vertically with almost the same amplitude at the first resonant frequency ω_1. At the second resonant frequency ω_2, which is the first mode of elastic motions, the legs located near the center of the array are moving little while the legs located at the both ends of the structure are moving with large amplitudes. The higher modes of resonant motions at $\omega_n (n \geq 3)$ are not clear. As the rigidity decreased(Figure 3.6(b)), higher modes of resonant motions are enhanced and the nodes of the oscillations can be clearly identified for each mode of motions. Figure 3.7, Figure 3.8 show respectively the frequency response characteristics of the surge motions of and the horizontal drift forces acting on the 50-leg structure. For surge motions and drift forces the hydrodynamic interaction effects are quite distinct. For the surge motions, the clearly observed motionless points in the results obtained without the consideration of the hydrodynamic interaction effects disappear when the hydrodynamic effects are taken into account. For drift forces the value is almost halved as the hydrodynamic interaction effects are taken in the calculations.
Finally Figure 3.9(a),(b) show the examples of the deformation patterns of the upper deck of the 50-leg structure($EI/\rho g \ell^2 = 2.0 \times 10^1$) at the third($\omega_3$) and the fourth($\omega_4$) resonant frequencies respectively.

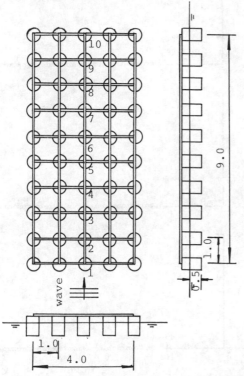

Figure 3.4 A structure supported on a 5×10 array of legs

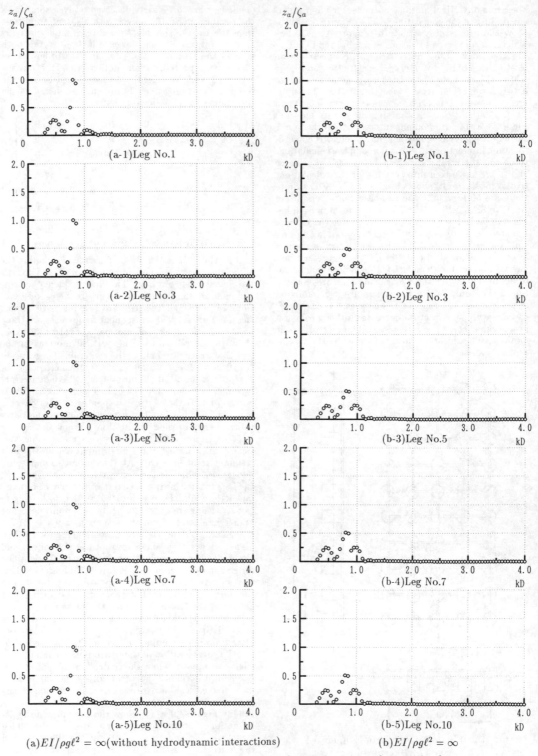

(a)$EI/\rho g\ell^2 = \infty$(without hydrodynamic interactions)   (b)$EI/\rho g\ell^2 = \infty$

Figure 3.5 The response amplitude operators of vertical motions of supporting legs

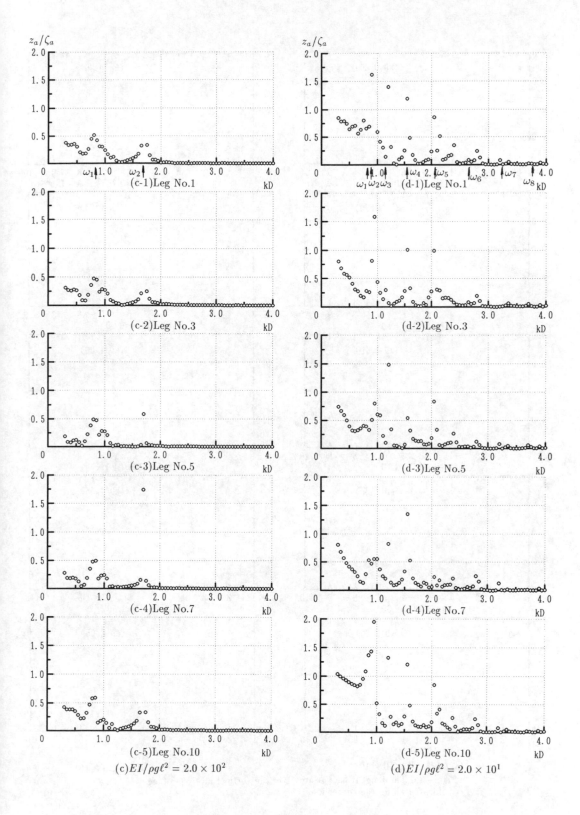

(c-1)Leg No.1

(c-2)Leg No.3

(c-3)Leg No.5

(c-4)Leg No.7

(c-5)Leg No.10

(c)$EI/\rho g\ell^2 = 2.0 \times 10^2$

(d-1)Leg No.1

(d-2)Leg No.3

(d-3)Leg No.5

(d-4)Leg No.7

(d-5)Leg No.10

(d)$EI/\rho g\ell^2 = 2.0 \times 10^1$

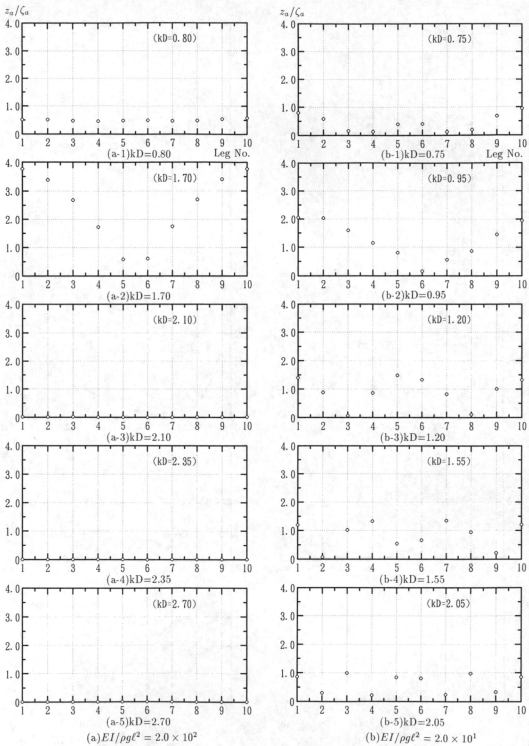

z_a/ζ_a

(a-1)kD=0.80
(kD=0.80)
Leg No.

(b-1)kD=0.75
(kD=0.75)
Leg No.

(a-2)kD=1.70
(kD=1.70)

(b-2)kD=0.95
(kD=0.95)

(a-3)kD=2.10
(kD=2.10)

(b-3)kD=1.20
(kD=1.20)

(a-4)kD=2.35
(kD=2.35)

(b-4)kD=1.55
(kD=1.55)

(a-5)kD=2.70
(kD=2.70)

(b-5)kD=2.05
(kD=2.05)

(a)$EI/\rho g\ell^2 = 2.0 \times 10^2$ (b)$EI/\rho g\ell^2 = 2.0 \times 10^1$

Figure 3.6 The lengthwise variations of the vertical motions of legs
located along the centerline

428

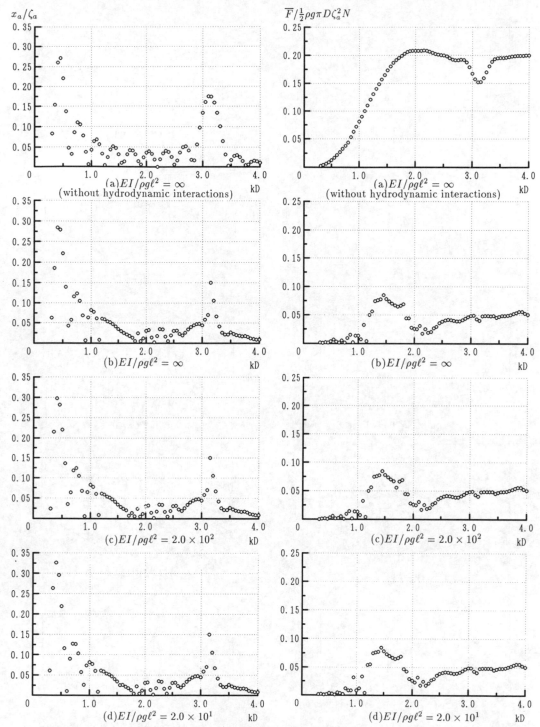

Figure 3.7 The response amplitude operators
of surge motions of a 50-leg structure

Figure 3.8 The frequency response characteristics
of drift forces acting on a 50-leg structure

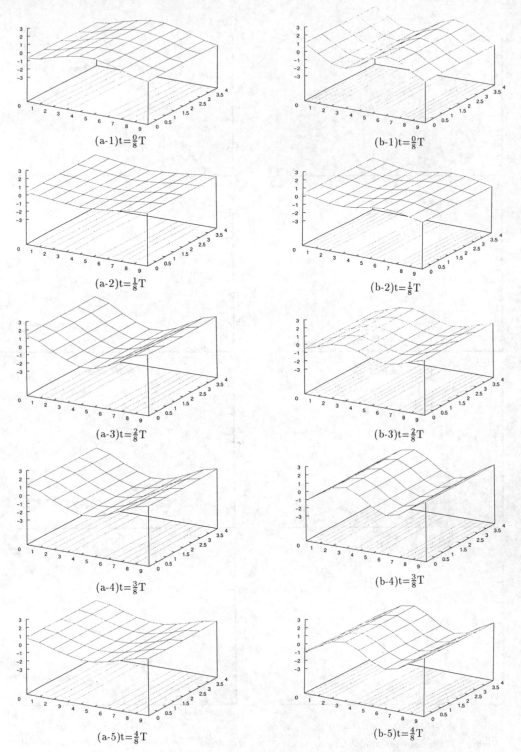

$(a\text{-}1)t=\frac{0}{8}T$ $(b\text{-}1)t=\frac{0}{8}T$

$(a\text{-}2)t=\frac{1}{8}T$ $(b\text{-}2)t=\frac{1}{8}T$

$(a\text{-}3)t=\frac{2}{8}T$ $(b\text{-}3)t=\frac{2}{8}T$

$(a\text{-}4)t=\frac{3}{8}T$ $(b\text{-}4)t=\frac{3}{8}T$

$(a\text{-}5)t=\frac{4}{8}T$ $(b\text{-}5)t=\frac{4}{8}T$

Figure 3.9 Examples of deformations patterns of the upper structure of a 50-leg structure

4. Conclusions

It has been confirmed that the presented methods for the analyses of hydroelastic behaviours of a floating structure supported on a large number of legs give physically reasonable results. Experimental data associated with hydroelastic behaviours should be obtained in order to compare the calculation results. Although the theoretical formulation for the analyses of hydroelastic behaviours of a structure supported on an extremely large number of legs are given, the example calculations were not conducted due to the lack of time. (It has been confirmed that the proposed method really work for diffraction problems(Kagemoto and Yue 1986b).) Extension of the existing computer code to enable the calculations of motions according to this formulation is now in progress. The results will be reported in the near future.

The experimental data shown in Figure 2.4(a),(b) were obtained by one of the authors(H.K.) when he was working for the Ship Research Institute of the Ministry of Transport, Japan(Kagemoto 1982).

References

Ando, S., Okawa, Y. & Ueno, I. 1983. Feasibility study of a floating offshore airport (in Japanese), Rep. Ship Res. Inst., Suppl. 4.

N. Toki 1979. A study on the behavior of huge floating structure in regular waves, J. Soc. Naval Arch. Japan), **155**, 185-194.

Simon, M.J. 1982. Multiple scattering in arrays of axisymmetric wave-energy devices. Part 1. A matrix method using a plane-wave approximation. J. Fluid Mech., **120**, 1-25.

H. Kagemoto & Dick K.P. Yue 1986a. Interactions among multiple three-dimensional bodies in water waves: an exact algebraic method. J. Fluid Mech., **166**, 189-209.

R.C.Ertekin, H.R.Riggs, X.L.Chen and S.X.Du 1993. Efficient methods for hydroelastic analysis of very large floating structures, J.S.R, **37**, No.1, 58-76.

H. Kagemoto & Dick K.P.Yue 1987. Wave-induced motions of multiple floating bodies, J. Soc. Naval Arch. Japan, **161**, 159-165,

H. Kagemoto & Dick K.P.Yue 1986b: Wave forces on a platform supported on a large number of floating legs, In Proc. 5th Intl. Offshore Mechanics & Arctic Engineering Symp. **1**, 206-211.

H. Kagemoto 1982. On hydroelastic responses of a very large structure in waves (Part1, Part2), Internal report of the Ship Research Institue (Ocean Engineering Division).

Hydroelasticity in Marine Technology, Faltinsen et al. (eds) © 1994 Balkema, Rotterdam, ISBN 90 5410 387 6

Dynamic response of flexible circular floating islands subjected to stochastic waves and seaquakes

T. Hamamoto
Musashi Institute of Technology, Tokyo, Japan

ABSTRACT: An analytical approach which predicts the dynamic response of a flexible circular floating island subjected to stochastic wind-waves and seaquakes is presented, taking into account structural flexibility and fluid-structure interaction. The mooring system is assumed to be composed of a series of tension-legs which are uniformly distributed on the lower surface of the island. Based on a linear potential flow theory, the hydrodynamic pressure generated on the lower surface of the island is obtained in closed form. The fluid-coupled free vibration and stochastic wave and seaquake responses are formulated by using the Lagrange's equation. The response quantities, such as displacements, accelerations and internal forces, are evaluated as standard deviations (root-mean-squared values) on the basis of a wet-mode superposition approach and stationary random vibration theory. Numerical examples are presented to discuss the effects of the hydrodynamic added mass and radiation damping due to island motion, the added stiffness due to mooring system, and the added excitation which is caused by the direct propagation of seismic waves through the mooring system on both wave and seaquake responses.

1 INTRODUCTION

With the recent trend toward the utilization of ocean space, a number of projects related to artificial large floating islands which serve as floating cities or airports have been proposed in Japan. The designs of these large floating structures demand a rational prediction of structural response in an ocean environment. With the increase of the dimensions of floating structures, it is expected that structural deformation becomes more significant than rigid body motion. As a result, the spatial distribution of structural response tends to be much more complicated than that of existing floating structures. Thus, the adequate evaluation of the dynamic response of large floating structures is inevitable to assure structural safety and keep human comfort.

The stochastic response of an existing tension-leg platform subjected to wind-induced waves has been evaluated (Soong and Prucz, 1984). Different from existing platforms, however, a problem arises that the spatial distribution of wave response statistics of large floating islands considerably varies according to load intensity because of structural flexibility. Moreover, it is pointed out that, especially in the seismically active regions, seaquake loading will be another threat to large floating islands in addition to wave loading (Ambraseys, 1985).

In the wave response analysis, the dynamic interaction between elastic deformation of a rectangular floating structure and sea water has been investigated (Wen, 1974). So far, however, only rigid body motion has been taken into account for circular floating structures (Garret, 1971; Ijima *et al.*, 1972). In the seaquake response analysis, on the other hand, Liou,

Penzien and Yeung (1988) have studied the rigid body motion of a circular floating structure with tension-legs, although structural flexibility has been introduced using two-dimensional finite element formulation (Babu and Reddy, 1986).

In this study, the spatial distribution of the response statistics of a flexible circular floating island are evaluated with the aids of short-term and long-term descriptions of wind-waves and seaquakes. The long-term description of environmental forces is represented by the occurrence rate of each load intensity at a specific offshore site. The short-term description of environmental forces is represented by the spectral density functions in terms of each load intensity. Numerical examples are presented to discuss the dynamic response of a large floating island with or without mooring system against both wind-waves and seaquakes. Emphasis is also placed on the hydrodynamic added mass and radiation damping in the overall response of large floating islands.

2 ANALYTICAL MODEL AND ASSUMPTIONS

The analytical model of a large floating island subjected to wind-waves and seaquakes is shown in Figure 1. The floating island is idealized as an elastic circular plate with or without mooring system. The mooring system is presumed to be composed of tension-legs which are uniformly distributed over the lower surface of the island.

The following assumptions are introduced in this study:
1 The floating island is guided vertically without friction around the circumference.

2 The floating island is isotropic, elastic and of constant thickness.
3 The sea water is inviscid and irrotational.
4 The sea is of constant depth and extends to infinity in the radial direction.
5 The tension-legs transfer only axial force and are elastically anchored to the sea-bed.
6 The mass effect of tension-legs can be disregarded.
7 The wind-waves propagate in one direction only.
8 The harmonic components of random wind-waves are governed by the Airy's liner wave theory.
9 The random wind-waves are stationary, ergodic and a zero-mean Gaussian process.
10 The second-order drift and springing forces are disregarded.
11 The seaquake is induced by vertical ground motion without time-lags at sea-bed.
12 The stationary part of vertical ground motion is ergodic and a zero-mean Gaussian process.

3 STOCHASTIC DESCRIPION OF ENVIRONMENTAL FORCES

Environmental forces, such as wind-waves and seaquakes, may be described from both short-term and long-term points of view. The long-term description is concerned with the frequencies of occurrence of multiple levels of load intensity. On the other hand, the short-term description is concerned with the details of time-series during each load intensity. In what follows, both short-term and long-term descriptions of environmental forces to be used in this study are illustrated.

3.1 Wave loading

3.1.1 Short-term description

It is clear that the sea surface is not a stationary process because of changing meteorological conditions. For relatively short time period, however, the assumption of stationarity is reasonable. The spectral density function of wave height in a fully developed sea is given by the Pierson-Moskowitz (1964) power spectrum,

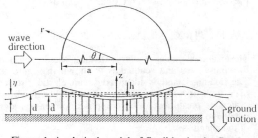

Figure 1 Analytical model of flexible circular floating island

$$S_{\eta\eta}(\omega) = \frac{\alpha g^2}{\omega^5} \exp\left[-\beta\left(\frac{g}{\omega \cdot V}\right)^4\right] \quad 0 \le \omega \le \infty, \quad (1)$$

in which V is the mean wind velocity at 19.5m above the sea surface, g is the acceleration due to gravity, and α and β are the spectral parameters. Common values $\alpha = 0.0081$ and $\beta = 0.74$ are adopted in this study.

3.1.2 Long-term description

In the study of a long-term safety and serviceability, an adequate description of sea-state is needed for time intervals corresponding to the lifetime of floating islands. In accordance with the short-term description of sea-state, the long-term description is described by the frequencies of occurrence of the maximum mean wind velocity above the sea surface at a site. The frequencies of occurrence of the mean wind velocity are often assumed to follow the Weibul distribution (Davenport, 1967). In this case, the maximum mean wind velocity, V_p, of independent samples taken from the parent population approaches a type I (Gumbel) asymptotic distribution of the largest extreme given by

$$p\left[V_p < v\right] = F_{V_p}(v) = \exp\left\{-\exp\left[-\frac{(v - \mu)}{\gamma}\right]\right\}, \quad (2)$$

in which μ and γ are the location and dispersion parameters of the distribution, respectively.

3.2 Seaquake loading

3.2.1 Short-term description

Seaquake loading is a hydrodynamic pressure due to shock waves consisting solely of compressional waves traveling through the water by earthquake, since water cannot transmit shear wave. Earthquake ground motion is generally nonstationary with respect to frequency content and intensity. In this study, however, motion during each loading event is modeled as a random process.

The short-term description of seaquake loading is described by the power spectral density function of vertical ground acceleration at sea-bed as follows (Clough and Penzien, 1975):

$$S_{\ddot{U}_s \ddot{U}_s}(\omega) = G_0 \cdot H_1(\omega) \cdot H_2(\omega) \quad 0 \le \omega \le \infty, \quad (3)$$

in which

$$H_1(\omega) = \frac{(\omega/\omega_k)^4}{\left\{1 - (\omega/\omega_k)^2\right\}^2 + 4\xi_k^2(\omega/\omega_k)^2}, \quad (4a)$$

$$H_2(\omega) = \frac{1 + 4\xi_g^2(\omega/\omega_g)^4}{\left\{1 - (\omega/\omega_g)^2\right\}^2 + 4\xi_g^2(\omega/\omega_g)^2}, \quad (4b)$$

G_0 is the spectral intensity, ω_k and ξ_k are the high-pass filter parameters, and ω_g and ξ_g are the low-pass filter parameters and may be interpreted as the predominant ground circular frequency and ground damping ratio, respectively. In this study, we assume $\omega_k = 1.0$ rad/sec, $\xi_k = 0.6$, $\omega_g = 10.0$ rad/sec, $\xi_g = 0.6$.

3.2.2 *Long-term description*

The long-term description of earthquake is usually represented by the seismic hazard curve at a site. The seismic hazard curve is a plot of annual exceedance probability versus the peak ground acceleration or velocity. Cornell (1968) showed that the peak ground acceleration follows a type II (Frechet) extreme value distribution. Ozaki *et al.* (1978) reported that the peak ground velocity at bed-rock, V_0, also follows the type II extreme value distribution,

$$P\left[V_0 > v\right] = 1 - F_{V_0}(v) = 1 - \exp\left[-\left(\frac{\mu}{v}\right)^k\right],$$
(5)

in which μ and k are the size and shape parameters of the distribution, respectively. The relationship between the peak ground velocity at ground surface, V_p, and that at bed-rock, V_0, is given by

$$V_p = 5\sqrt{T_g}V_0 = 5\sqrt{\frac{2\pi}{\omega_g}}V_0,$$
(6)

in which T_g is the predominant ground period at sea-bed.

The variance of ground velocity is calculated by

$$\sigma_V^2 = \int_0^\infty S_{\dot{U}_s\dot{U}_s}(\omega)\,d\omega = \int_0^\infty \frac{S_{U_sU_s}(\omega)}{\omega^2}\,d\omega.$$
(7)

Consequently, the relationship between the spectral intensity, G_0, and the peak velocity, V_p, may be obtained as

$$G_0 = \frac{4\omega_k}{\pi}\frac{f_1}{f_2}\left(\frac{V_p}{Z}\right)^2,$$
(8a)

in which

$$f_1 = \xi_k\xi_g\left(\omega_k^2 - \omega_g^2\right)^2 + 4\xi_k^2\xi_g^2\omega_k\omega_g + 4\xi_k\xi_g\omega_k^2\omega_g^2\left(\xi_k^2 + \xi_g^2\right),$$
(8b)

$$f_2 = 2\xi_k\omega_k\omega_g^3 + \xi_g\omega_g^2\left(\omega_k^2 + \omega_g^2\right),$$
(8c)

and Z is the peak factor, assumed to be 3.0 in this study.

4 HYDRODYNAMIC PRESSURE

Due to wind-waves or seaquakes, the hydrodynamic pressure is generated on the lower surface of the floating island. The hydrodynamic pressure may be evaluated as a linear combination of the pressure component, p_f, acting on the motionless island subjected to wind-waves or sea-quakes and the pressure component, p_m, due to the island motion which contains rigid body motion and elastic deformation, as shown in Figure 2. Moreover, as shown in Figure 3, p_f may be divided into two components, p_i and p_s, due to incident and scattered waves, respectively. Because of the geometric simplicity, each pressure component can be obtained in closed form based on a linear potential flow theory (Hamamoto and Tanaka, 1992b, 1993).

4.1 *Hydrodynamic pressure p_f*

4.1.1 *Hydrodynamic pressure p_i for wind-waves*

The hydrodynamic pressure acting on the lower

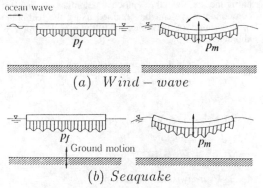

(a) *Wind − wave*

(b) *Seaquake*

Figure 2 Hydrodynamic pressure components p_f and p_m

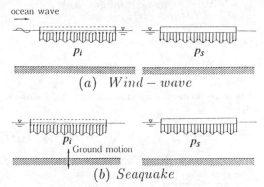

(a) *Wind − wave*

(b) *Seaquake*

Figure 3 Hydrodynamic pressure components p_i and p_s

surface of the motionless island due to incident wave may be evaluated in the free-field without floating island. The linearized governing equation for an incompressible and inviscid fluid is

$$\nabla^2 \phi_i = 0 , \tag{9}$$

in which ϕ_i is the velocity potential function of incident wave and ∇^2 is the Laplace operator.

The boundary conditions are as follows:

Free water surface condition,

$$\frac{\partial \phi_i}{\partial z} + \frac{1}{g} \frac{\partial^2 \phi_i}{\partial t^2} = 0 \qquad at \quad z = d , \tag{10a}$$

$$\eta = -\frac{1}{g} \frac{\partial \phi_i}{\partial z} \qquad at \quad z = d . \tag{10b}$$

Sea-bed condition,

$$\frac{\partial \phi_i}{\partial z} = 0 \qquad at \quad z = 0 , \tag{11}$$

in which η is the height of sea surface measured from the still water level and may be simulated by the superposition of harmonic wave components as follows:

$$\eta = \sum_{l=1}^{\infty} \sqrt{2 S_{\eta\eta}(\omega_l) \Delta\omega} \exp\left[i\left(\omega_l t - k_l r \cos\theta + \varphi_l\right)\right] , \tag{12}$$

in which ω_l, k_l and φ_l are the circular frequency, wave number and phase of the l-th component wave, respectively, and $S_{\eta\eta}(\omega)$ is the wave height spectrum given by equation(1). The potential function of incident component wave can be obtained as a closed form solution using the technique of separation of variables as follows (subscript l is abbreviated):

$$\phi_i = \sum_{n=0}^{\infty} A_{n0} \frac{J_n(kr)}{J_n(ka)} \cosh kz \cos n\theta \exp\left(i\omega t\right) , \tag{13}$$

in which the coefficient A_{n0} is given by

$$A_{n0} = (-i)^n \gamma_n J_n(ka) \eta_0 g / \omega \cosh kd , \tag{14}$$

The wave number k is satisfied with the following equation:

$$\omega^2 = kg \tanh kd , \tag{15}$$

and $J_n(kr)$ is the Bessel function of n-th order of the first kind, $\gamma_0 = 1$, $\gamma_n = 2 \ (n \geq 1)$ and

$$\eta_0 = \sqrt{2 S_{\eta\eta}(\omega) \Delta\omega}$$

4.1.2 Hydrodynamic pressure p_i for seaquakes

The hydrodynamic pressure during a seaquake is associated with the water-transmitted seismic vibration from sea-bed to the floating island. The incident wave consists of vertical pulses which induce a series of compression and tension waves in the fluid medium. The amplification of the seaquake force through the water must be evaluated by taking into account the compressibility of water. The linearized governing equation for a compressible and inviscid fluid is

$$\nabla^2 \phi_i = \frac{1}{c^2} \frac{\partial^2 \phi_i}{\partial t^2} , \tag{16}$$

in which c is the compressional wave velocity in sea water.

The boundary conditions are as follows:

Free water surface condition,

$$\frac{\partial \phi_i}{\partial z} + \frac{1}{g} \frac{\partial^2 \phi_i}{\partial t^2} = 0 \qquad at \quad z = d , \tag{17}$$

Sea-bed condition,

$$\frac{\partial \phi_i}{\partial z} = \frac{\partial U_g(t)}{\partial t} , \qquad at \quad z = 0 , \tag{18}$$

in which $U_g(t)$ is the vertical ground displacement at sea-bed and may be expressed in terms of the vertical acceleration power spectral density function. The vertical acceleration at sea-bed may be expressed as

$$\ddot{U}_g(t) = \sum_{l=1}^{\infty} \sqrt{2 S_{\ddot{U}_g \ddot{U}_g}(\omega_l) \Delta\omega} \exp\left[i(\omega_l t + \varphi_l)\right] , \tag{19}$$

in which $S_{\ddot{U}_g \ddot{U}_g}(\omega)$ is the power spectral density of vertical ground acceleration given by equation(3), ω_l and φ_l are the circular frequency and phase of the l-th component wave, respectively, and $\Delta\omega = (\omega_{l+1} - \omega_{l-1}) / 2$. When the sea-bed is vertically moving with harmonic motion, the velocity potential of incident component wave can be obtained using the technique of separation of variables as follows (subscript l is abbreviated):

$$\phi_i = \frac{V_0}{\mu} \frac{\mu g \cos\mu(d-z) - \omega^2 \sin\mu(d-z)}{\mu g \sin\mu d + \omega^2 \cos\mu d} \exp\left(i\omega t\right) , \tag{20}$$

in which $\mu = \omega / c$, $V_0 = A_0 / i\omega$ and

$$A_0 = \sqrt{2 S_{\ddot{U}_g \ddot{U}_g}(\omega_l) \Delta\omega}$$

4.1.3 Hydrodynamic pressure p_s

Due to the existence of the floating island, the incident wave of wind-waves and seaquakes is reflected and scattered at the interface between the island and sea water. For the frequency of interest, the sea water may be assumed to be incompressible for both wind-waves and seaquakes.

The governing equation for an incompressible and inviscid fluid is the Laplace's equation:

$$\nabla^2 \phi_s = 0,$$

(21)

in which ϕ_s is the velocity potential function of scattered wave.

The boundary conditions are as follows:
Free water surface condition,

$$\frac{\partial \phi_s}{\partial z} + \frac{1}{g}\frac{\partial^2 \phi_s}{\partial t^2} = 0 \qquad at \quad z = d,$$

(22a)

Sea-bed condition,

$$\frac{\partial \phi_s}{\partial z} = 0 \qquad at \quad z = 0,$$

(22b)

Structure-water interface condition,

$$\frac{\partial \phi_s}{\partial z} = -\frac{\partial \phi_i}{\partial z} \qquad at \quad z = \overline{d},$$

(22c)

Radiation condition,

$$\lim_{r \to \infty} \sqrt{r}\left(\frac{\partial \phi_s}{\partial r} + ikr\right) = 0 \qquad at \quad r \to \infty.$$

(22d)

To solve equation(21) and satisfy the boundary conditions given by equations(22a) to (22d), the whole fluid domain is divided into two regions: an exterior region $(r > a)$ and interior region $(r < a)$. In what follows, the exterior region is characterized by the potential function $\phi_s^{(e)}$, whereas the interior region is characterized by the potential function $\phi_s^{(i)}$. On the cylindrical interface between the exterior and interior regions, it is necessary to impose the two continuity conditions:
Kinematic continuity condition,

$$\frac{\partial \phi_s^{(e)}}{\partial r} = \frac{\partial \phi_s^{(i)}}{\partial r} \qquad 0 \le z \le \overline{d}, \quad r = a, \quad (23a)$$

$$\frac{\partial \phi_s^{(e)}}{\partial r} = 0 \qquad \overline{d} \le z \le d, \quad r = a. \quad (23b)$$

Pressure continuity condition,

$$\phi_s^{(e)} = \phi_s^{(i)} \qquad 0 \le z \le \overline{d}, \quad r = a. \quad (23c)$$

The velocity potential $\phi_s^{(i)}$ may be expressed as follows:

$$\phi_s^{(i)} = \phi_{sh}^{(i)} + \phi_{sp}^{(i)},$$

(24)

in which $\phi_{sh}^{(i)}$ and $\phi_{sp}^{(i)}$ are the homogeneous and particular solutions of the Laplace's equation, respectively. Both the solutions can be found by the technique of separation of variables as follows:

$$\phi_{sh}^{(i)} = \sum_{n=0}^{\infty}\left\{D_{n0}\left(\frac{r}{a}\right)^n + \sum_{s=1}^{\infty}D_{ns}\frac{I_n(l_s r)}{I_n(l_s a)}\cos l_s z\right\}$$
$$\cdot\cos n\theta \, \exp(i\omega t),$$

(25a)

and

$$\phi_{sp}^{(i)} = -\sum_{n=0}^{\infty}\sum_{s=1}^{\infty}\frac{2kA_{n0}}{a\lambda_{ns}}\frac{\sinh k\overline{d}}{J_n(ka)}\frac{J_n(\lambda_{ns}^* r)}{J_{n+1}^2(\lambda_{ns})}$$
$$\cdot\frac{\cosh \lambda_{ns}^* z}{\sinh \lambda_{ns}^* \overline{d}}\int_0^a J_n(kr)\,J_n(\lambda_{ns}^* r)\,rdr\,\cos n\theta\,\exp(i\omega t)$$

for wind-waves,   (25b)

$$\phi_{sp}^{(i)} = -2aV_0\frac{\mu g\sin\mu(d-\overline{d})+\omega^2\cos\mu(d-\overline{d})}{\mu g\sin\mu d+\omega^2\cos\mu d}$$
$$\cdot\sum_{s=1}^{\infty}\frac{J_0(\lambda_{0s}^* r)\cosh\lambda_{0s}^* z}{\lambda_{0s}^2 J_1(\lambda_{0s})\sinh\lambda_{0s}^*\overline{d}}\exp(i\omega t)$$

for seaquakes.   (25c)

In equations(25a) to (25c), $I_n(l_s r)$ is the modified Bessel function of n-th order of the first kind, $J_n(\lambda^*_{ns}r)$ and $J_{n+1}(\lambda_{ns})$ are the Bessel functions of n-th and $n+1$-th orders of the first kind, respectively, $l_s = s\pi/\overline{d}$, λ_{ns} is the s-th positive root of $J_n(\lambda_{ns}) = 0$, $\lambda^*_{ns} = \lambda_{ns}/a$, and D_{ns} $(s = 0,1,2,\cdots)$ are unknown coefficients.

The velocity potential $\phi_s^{(e)}$ can be also obtained using the technique of separation of variables as follows:

$$\phi_s^{(e)} = \sum_{n=0}^{\infty}\left\{B_{n0}\frac{H_n^{(2)}(kr)}{H_n^{(2)}(ka)}\cosh kz\right.$$
$$\left.+\sum_{j=1}^{\infty}C_{nj}\frac{K_n(k_j r)}{K_n(k_j a)}\cos k_j z\right\}\cos n\theta \, \exp(i\omega t),$$

(26)

in which k and k_j are the roots of the transcendental equation:

$$\omega^2 = kg\tanh kd = -k_j g\tan k_j d,$$

(27)

$H_n^{(2)}(kr)$ is the Hankel function of n-th order of the second kind, $K_n(k_j r)$ is the modified Bessel function of n-th order of the second kind, and B_{n0} and C_{nj} $(j = 1,2,\cdots)$ are unknown coefficients.

To determine the coefficients D_{ns}, B_{n0} and C_{nj}, we can enforce the continuity conditions given by equations(23a) to (23c). Making use of the orthogonality properties of $\cosh kz$ and $\cos k_j z$ in the range $0 \le z \le d$, sets of algebraic equation concerning the coefficients are derived. These equations are infinite in number and have an infinite number of unknowns. The number of terms required in the series depends on the accuracy desired.

Once the coefficients are obtained, the pressure component p_f may be obtained by the Bernoulli's equation,

$$p_f = -\rho_W \left(\frac{\partial \phi_i}{\partial t} + \frac{\partial \phi_s}{\partial t} \right) \Bigg|_{z=\bar{d}} , \tag{28}$$

in which ρ_W is the mass density of sea water.

4.2 Hydrodynamic pressure p_m

When subjected to wind-waves or seaquakes, the floating island responds with its rigid body motion and elastic deformation. Using a wet-mode superposition approach, the total displacement of the floating island can be expressed as

$$\zeta(r,\theta,t) = \sum_{n=0}^{\infty} \sum_{m=1}^{\infty} \zeta_{nm}(r) \, \cos n\theta \, q_{nm}(t) , \tag{29}$$

in which n is the circumferential Fourier wave number, m is the radial mode number, $\zeta_{nm}(r)$ is the nm-th wet-mode shape in the radial direction along $\theta = 0$, and $q_{nm}(t)$ is the nm-th generalized coordinate. The wet-mode shape of the floating island is obtained in the fluid-coupled free vibration analysis.

Due to the low frequency nature of the island response, the effect of the compressibility of water may be disregarded. When a floating island is moving and vibrating with harmonic motion, the velocity potential function, ϕ_m, for inviscid flow can be obtained by solving the Laplace's equation,

$$\nabla^2 \phi_m = 0 . \tag{30}$$

The boundary conditions are as follows:
Free water surface condition,

$$\frac{\partial \phi_m}{\partial z} + \frac{1}{g} \frac{\partial^2 \phi_m}{\partial t^2} = 0 \qquad at \quad z = d , \tag{31a}$$

Sea-bed condition,

$$\frac{\partial \phi_m}{\partial z} = 0 \qquad at \quad z = 0 , \tag{31b}$$

Structure-water interface condition,

$$\frac{\partial \phi_m}{\partial z} = \frac{\partial \zeta}{\partial t} \qquad at \quad z = \bar{d} , \tag{31c}$$

Radiation condition,

$$\lim_{r \to \infty} \sqrt{r} \left(\frac{\partial \phi_m}{\partial r} + ik\phi_m \right) = 0 \qquad at \quad r \to \infty . \tag{31d}$$

To solve equation(30) and satisfy the boundary conditions given by equations (31a) to (31d), the fluid domain is divided into exterior and interior regions as in the case with hydrodynamic pressure p_s. The velocity potential of interior region, $\phi_m^{(i)}$, can be expressed as

$$\phi_m^{(i)} = \phi_{mh}^{(i)} + \phi_{mp}^{(i)} , \tag{32}$$

in which $\phi_{mh}^{(i)}$ and $\phi_{mp}^{(i)}$ are the homogeneous and particular solutions of the Laplace's equation. Both the solutions can be found by the technique of separation of variables. The solution of $\phi_{mh}^{(i)}$ is the same form as that of previously given by equation(25a). The solution of $\phi_{mp}^{(i)}$ is obtained as

$$\phi_{mp}^{(i)} = \sum_{n=0}^{\infty} \sum_{m=1}^{\infty} \sum_{s=1}^{\infty} \frac{2i\omega}{a\lambda_{ns}} \frac{J_n(\lambda_{ns}^* r)}{J_{n+1}^2(\lambda_{ns})} \frac{\cosh \lambda_{ns}^* z}{\sinh \lambda_{ns}^* \bar{d}}$$

$$\cdot \int_0^a \zeta_{nm}(r) \, J_n(\lambda_{ns}^* r) \, r dr \, q_{nm}(i\omega) \, \cos n\theta \, \exp(i\omega t) , \tag{33}$$

in which $q_{nm}(i\omega)$ is defined as $q_{nm}(t) = q_{nm}(i\omega)$ $\exp(i\omega t)$. The potential function of the exterior region, $\phi_m^{(e)}$, can be also obtained in the same form as equation(26).

To determine the coefficients D_{ns} ($s = 0,1,2,\cdots$), B_{n0} and C_{nj} ($j = 1,2,\cdots$), both interior and exterior regions are combined by equations(23a) to (23c). This procedure leads to sets of algebraic equation concerning unknown coefficients. Once the coefficients are obtained, the pressure component p_m may be obtained by the Bernoulli's equation,

$$p_m = -\rho_W \frac{\partial \phi_m}{\partial t} \Bigg|_{z=\bar{d}} - \rho_W g\zeta . \tag{34}$$

5 TENDON FORCE OF MOORING SYSTEM

The floating island is maintained at operational draft by the mooring system. The tendon force of the mooring system acts on the floating island in addition to hydrodynamic pressure. When a series of tension-legs are uniformly distributed on the lower surface of the island, the stiffness per unit area of the lower surface is given by

$$k_d = \frac{E_t A_t}{\pi a^2 \bar{d}} , \tag{35}$$

in which E_t is the Young's modulus of tension-legs and A_t is the total area of tension-legs. The dynamic tendon force in the mooring system is given by

$$p_d(r,\theta,t) = -k_d \left\{ \zeta(r,\theta,t) - U_g(t) \right\} \tag{36}$$

6 FLUID-COUPLED FREE VIBRATION

The applications of the Rayleigh-Ritz procedure to the free vibration of elastic plates in air have been carried out using beam characteristic functions for rectangular plates by Warburton(1954) and using polynomical functions for circular plates by Timoshenko (1955). In this study, wet-mode shapes of the floating island are approximated by the superposition of the mode shapes in air and two-degrees-of-freedom of rigid body motion, heave and pitch, as follows:

$$\zeta_{nm} = \sum_{i=1}^{N} W_{nmi} \, R_n(r) \, \cos n\theta , \tag{37}$$

in which $R_{in}(r)$ ($i \geq 1$) is the in-th radial mode function of the island in air which satisfies the boundary conditions for the free edge, $R_{00}(r) = 1$, $R_{01}(r) = r / a$, and $R_{0n}(r) = 0$ ($n \geq 2$), W_{nmi} is the amplitude coefficient to be determined, and N is the number of superposition of $R_{in}(r)$. Using $W_{nmj}(t) = W_{nmj}$ $\exp(i\overline{\omega}_{nm}t)$, in which $\overline{\omega}_{nm}$ is the nm-th natural circular frequency of the island coupled with liquid, as the generalized coordinates, the motion of a freely vibrating island coupled with liquid is governed by the Lagrange's equation:

$$\frac{d}{dt} \left(\frac{\partial T}{\partial \dot{W}_{nmj}(t)} \right) - \frac{\partial T}{\partial W_{nmj}(t)} + \frac{\partial S}{\partial W_{nmj}(t)} = Q_{nmj}(t)$$
$$(j = 1, 2, \cdots N) \tag{38}$$

in which $\bullet$ denotes time derivative, T and S are kinetic energy and strain energy of the island, respectively, and given by

$$T = \frac{\rho_p h}{2} \int_0^a \int_0^{2\pi} \left(\frac{\partial \zeta_{nm}}{\partial t} \right)^2 r d\theta \, dr , \tag{39a}$$

$$S = \frac{D}{2} \int_0^a \int_0^{2\pi} \left[\left(\frac{\partial^2 \zeta_{nm}}{\partial r^2} + \frac{1}{r} \frac{\partial \zeta_{nm}}{\partial r} + \frac{1}{r^2} \frac{\partial^2 \zeta_{nm}}{\partial \theta^2} \right)^2 \right.$$
$$-2(1-\nu) \left\{ \frac{\partial^2 \zeta_{nm}}{\partial r^2} \left(\frac{1}{r} \frac{\partial \zeta_{nm}}{\partial r} + \frac{1}{r^2} \frac{\partial^2 \zeta_{nm}}{\partial \theta^2} \right) \right.$$
$$\left. \left. - \left(\frac{1}{r} \frac{\partial^2 \zeta_{nm}}{\partial r \partial \theta} + \frac{1}{r^2} \frac{\partial \zeta_{nm}}{\partial \theta} \right)^2 \right\} \right] r d\theta \, dr , \tag{39b}$$

in which ρ_p is the mass density of the island, and $D = Eh^3 / 12(1-\nu^2)$ is the flexural rigidity of the island (E and ν are Young's modulus and Poison's ratio, respectively). The generalized force, $Q_{nmj}(t)$, is given by

$$Q_{nmj}(t) = \int_0^a \int_0^{2\pi} (p_m + p_d) \cdot R_{jn}(r) \cos n\theta \, r d\theta dr . \tag{40}$$

Substituting equation(37) into equation (33) and making use of equations(23a) to (23c) yields the matrix form

$$[H]\{C\} = [B]\{W\} , \tag{41}$$

in which

$$\{C\}^T = \left\{ B_{n0}, C_{n1}, C_{n2}, \cdots, D_{n0}, D_{n1}, D_{n2}, \cdots \right\} , \tag{42a}$$

$$\{W\}^T = \left\{ W_{nm1}, W_{nm2}, \cdots, W_{nmN} \right\} , \tag{42b}$$

and the details of matrices $[H]$ and $[B]$ have been given elsewhere (Hamamoto and Tanaka, 1992a). Hence, the unknown coefficients may be expressed in terms of the amplitude coefficients as,

$$\{C\} = [H]^{-1}[B]\{W\} . \tag{43}$$

Substituting equation(43) into equation (25a) and then equation(40), equation(38) can be written in the matrix form as follows:

$$\left[([K_P] + [K_W] + [K_T]) - \Delta ([M_P] + [M_W]) \right] \{W\} = \{0\} , \tag{44}$$

in which $[K_P]$ is the stiffness matrix (N x N) of the island, $[K_W]$ is the added stiffness matrix (N x N) due to the cushioning effect of the supporting fluid, $[K_T]$ is the added stiffness matrix (N x N) associated with distributed tension-legs, $[M_P]$ is the mass matrix (N x N) of the island, $[M_W]$ is the added mass matrix (N x N) which causes the fluid in contact with the island to move with it, and Δ is the frequency parameter. Hence, the frequency equation is given by

$$\left| ([K_P] + [K_W] + [K_T]) - \Delta ([M_P] + [M_W]) \right| = 0 \tag{45}$$

The above equation is solved by an interaction procedure, since the added mass matrix $[M_W]$ is frequency dependent.

7 WET-MODE SUPERPOSITION

Having obtained the hydrodynamic pressure induced by wind-waves or seaquakes, the equation of motion of the floating island may be obtained in the frequency domain. The dynamic behavior of the floating island is governed by the Lagrange's equation:

$$\frac{d}{dt} \left(\frac{\partial T}{\partial \dot{q}_{nm}(t)} \right) - \frac{\partial T}{\partial q_{nm}(t)} + \frac{\partial S}{\partial q_{nm}(t)} = Q_{nm}^L(t) + Q_{nm}^D(t) , \tag{46}$$

in which $\bullet$ denotes time derivative, T and S are the kinematic and strain energy of the island, respectively,

439

$Q^L_{nm}(t)$ is the nm-th generalized loading force and $Q^D_{nm}(t)$ is the nm-th generalized damping force of the island.

Making use of the orthogonal properties of wet-mode shapes of the floating island, the nm-th uncoupled modal equations of motion can be obtained as follows:

$$\left(M^P_{nm} + M^W_{nm}\right)\ddot{q}_{nm}(t) + \left(C^P_{nm} + C^W_{nm}\right)\dot{q}_{nm}(t)$$
$$+ \left(K^P_{nm} + K^W_{nm} + K^T_{nm}\right)q_{nm}(t) = Q^F_{nm}(t) + Q^T_{nm}(t),$$

(47)

in which M^P_{nm}, C^P_{nm} and K^P_{nm} are the nm-th generalized mass, generalized damping and generalized stiffness of the island, respectively, given as

$$M^P_{nm} = \varepsilon_n \, \pi \, \rho_p h \int_0^a \zeta^2_{nm}(r) \, rdr,$$

(48a)

$$C^P_{nm} = 2\xi_{nm}\omega_{nm}M^P_{nm},$$

(48b)

$$K^P_{nm} = \omega^2_{nm}M^P_{nm},$$

(48c)

where $\varepsilon_0 = 2$, $\varepsilon_n = 1$ $(n \geq 1)$, ω_{nm} and ξ_{nm} are the nm-th natural circular frequency and material damping ratio of the island in air, respectively. M^W_{nm}, C^W_{nm} and K^W_{nm} are the nm-th generalized added mass, generalized added damping and generalized added stiffness, respectively, and given as

$$M^W_{nm} = \frac{\varepsilon_n \, \pi}{\omega^2 |q_{nm}(i\omega)|^2} \left\{ \mathrm{Re}\left[\int_0^a p_{mh}(r,\omega) \, \zeta_{nm}(r)rdr\right] \right.$$

$$\cdot \mathrm{Re}\left[q_{nm}(i\omega)\right] + \mathrm{Im}\left[\int_0^a p_{mh}(r,\omega) \, \zeta_{nm}(r)rdr\right]$$

$$\left. \cdot \mathrm{Im}\left[q_{nm}(i\omega)\right]\right\} + \int_0^a p_{mp}(r,\omega) \, \zeta_{nm}(r)rdr ,$$

(49a)

$$C^W_{nm} = \frac{\varepsilon_n \, \pi}{\omega \, |q_{nm}(i\omega)|^2} \left\{ \mathrm{Re}\left[\int_0^a p_{mh}(r,\omega) \, \zeta_{nm}(r)rdr\right] \right.$$

$$\cdot \mathrm{Im}\left[q_{nm}(i\omega)\right] - \mathrm{Im}\left[\int_0^a p_{mh}(r,\omega) \, \zeta_{nm}(r)rdr\right]$$

$$\left. \cdot \mathrm{Re}\left[q_{nm}(i\omega)\right]\right\} ,$$

(49b)

$$K^W_{nm} = \varepsilon_n \, \pi \, \rho_W g \int_0^a \zeta^2_{nm}(r)rdr ,$$

(49c)

in which $p_{mh}(r,\omega)$ and $p_{mp}(r,\omega)$ are satisfied with $p_{mh}(t) = -\rho_w \, \partial \phi_{mh} / \partial t = p_{mh}(r,\omega)\exp(i\omega t)$, and $p_{mp}(t) = -\rho_w \, \partial \phi_{mp} / \partial t = p_{mp}(r,\omega)\exp(i\omega t)$, respectively, $\mathrm{Re}[x]$ and $\mathrm{Im}[x]$ represent the real and

imaginary parts of x, respectively. The hydrodynamic radiation damping is directly proportional to the energy loss associated with the generation of the surface waves which radiate outwardly to infinity. K^T_{nm} is the nm-th added stiffness due to the attachment of distributed tension-legs and given by

$$K^T_{nm} = \varepsilon_n \, \pi \, k_d \int_0^a \zeta^2_{nm}(r)rdr ,$$

(50)

$Q^F_{nm}(t)$ and $Q^T_{nm}(t)$ are the nm-th generalized force associated with hydrodynamic pressure p_f, and tendon force, respectively, and given by

$$Q^F_{nm}(t) = \varepsilon_n \, \pi \int_0^a p_f(r,t) \, \zeta_{nm}(r)rdr,$$

(51a)

$$Q^T_{nm}(t) = \begin{cases} 2\pi \, k_d U_g(t) \int_0^a \zeta_{0m}(r)rdr & ; \quad n = 0, \\ 0 & ; \quad n \geq 1. \end{cases}$$

(51b)

Dividing both sides of equation(29) by $(M^P_{nm} + M^W_{nm})$ and making use of the relation:

$$K^P_{nm} + K^W_{nm} + K^T_{nm} = \overline{\omega}^2_{nm}\left(M^P_{nm} + M^W_{nm}\right),$$

(52)

the following equation may be obtained,

$$\ddot{q}_{nm}(t) + 2\left(\overline{\xi}_{nm} + \overline{\xi}^*_{nm}\right) \overline{\omega}_{nm}\dot{q}_{nm}(t) + \overline{\omega}^2_{nm}q_{nm}(t)$$
$$= \frac{Q^F_{nm}(t) + Q^T_{nm}(t)}{M^P_{nm} + M^W_{nm}} ,$$

(53)

in which $\overline{\omega}_{nm}$ is the nm-th wet-mode circular frequency, and $\overline{\xi}_{nm}$ and $\overline{\xi}^*_{nm}$ are the nm-th material damping ratio of the island and hydrodynamic radiation damping ratio, respectively.

8 STOCHASTIC RESPONSE

On the basis of a linear random vibration theory, the response power spectrum of displacement ζ may be obtained as

$$S_{\zeta\zeta}(r,\theta;\omega) = \sum_{n=0}^{\infty}\sum_{m=1}^{\infty} \zeta^2_{nm}(r)\cos^2 n\theta \, |H_{nm}(\omega)|^2 S_{Q_{nm}Q_{nm}}(\omega) ,$$

(54)

in which $S_{Q_{nm}Q_{nm}}(\omega)$ is the power spectral density function of the sum of $Q^F_{nm}(t)$ and $Q^T_{nm}(t)$, and $|H_{nm}(\omega)|^2$ is the nm-th transfer function given by

$$|H_{nm}(\omega)|^2 = \frac{1}{\left(M^P_{nm} + M^W_{nm}\right)^2 \left[\left(\overline{\omega}^2_{nm} - \omega\right)^2 + 4\left(\overline{\xi}_{nm} + \overline{\xi}^*_{nm}\right)\overline{\omega}^2_{nm}\omega^2\right]} .$$

(55)

440

In equation(54), the cross spectra related to the coupling between different modes of vibration are disregarded.

The variance of displacement response is obtained by white noise approximation as

$$\bar{\zeta}^2(r,\theta) = \int_0^\infty S_{\zeta\zeta}(r,\theta;\omega)\, d\omega$$

$$\approx \sum_{n=0}^\infty \sum_{m=1}^\infty \frac{\pi}{4} \frac{\zeta_{nm}^2(r)\cos^2 n\theta}{\left(M_{nm}^P + M_{nm}^W\right)^2 \left(\bar{\xi}_{nm} + \bar{\xi}_{nm}^\bullet\right)\bar{\omega}_{nm}^3} S_{Q_{nm}Q_{nm}}\left(\bar{\omega}_{nm}\right).$$

(56)

Other response statistics, such as accelerations and internal forces, can be obtained in the same manner.

9 NUMERICAL RESULTS AND DISCUSSION

For the numerical computations, the dimensions and material constants are assumed as follows: radius of island = 1000 m, thickness of island = 50 m, Young's modulus of island = 2.0×10^5 kg / cm^2, Poison's ratio of island = 0.15, mass density of island = 0.4×10^{-6} kg sec^2 / cm^4, material damping ratio in air = 0.05 for all modes of vibration, spring stiffness of mooring system = 5.0×10^{-2} kg / cm^3, water depth = 200 m, mass density of sea water = 1.046×10^{-6} kg sec^2 / cm^4 and compressional wave velocity in water = 1500 m / sec. The response statistics are evaluated at a presumed site in the sea off Japan. The distribution parameters of the long-term description of wind-waves and seaquakes are determined as; μ = 24.3m/sec and γ = 1.24 for wind-waves and μ = 0.23m/sec and k = 1.80 for seaquakes. The contributions of 18 and 10 wet-modes are considered to calculate the wave and seaquake responses, respectively.

Before going into the response analysis, vibrational characteristics which are obtained by the fluid-coupled free vibration analysis are presented. Figure 4 shows the variation in wet-mode circular frequencies due to island thickness when the island is freely floating without mooring system. The frequencies of the modes with m = 1 are almost constant irrespective of the island thickness because of the dominance of rigid body motion, whereas the frequencies of the modes with $m \geq 2$ increase rapidly with the island thickness. Figure 5 shows the variation in wet-mode circular frequencies due to water depth. As the water depth becomes shallower, the frequencies decrease gradually because of the existence of sea-bed. On the other hand, as the water depth becomes deeper, the frequencies converge to certain constant values. Figure 6 shows the variation in wet-mode circular frequencies due to anchor stiffness. The frequencies rapidly increase with anchor stiffness and diverge to infinity, although they remain constant within the region of small stiffness.

The stochastic responses are calculated for four different return periods, i.e., 10, 50, 100 and 200 year-return-periods, against wind-waves and seaquakes. Figure 7 shows the standard deviations of displacement response along θ = 0 for both cases

with and without mooring system. The elastic deformation is significant for the case without mooring system, while the rigid body motion becomes predominant for the case with mooring system against both wave and seaquake loadings. The attachment of mooring system effectively reduces the wave response, while adversely amplifies the seaquake response. This seems to be due to the direct propagation of seismic waves through the mooring system during seaquakes. In the seaquakes response, the distribution pattern of displacement response is independent of return period. In the wave response, on the other hand, the distribution pattern is dependent on load intensity. This is mainly due to the change in the contributions of excited modes of vibration.

Figure 8 shows the standard deviations of vertical acceleration response along θ = 0 for both cases with and without mooring system. The mooring system reduces the wave response effectively, although the seaquake response is not so changed for the cases with and without mooring system. The effect of higher modes of vibration can be clearly seen in the seaquake response, whereas only the effect of lower modes of vibration is observed in the wave response. Figure 9 shows the standard deviations of bending moment in the radial direction along θ = 0 for both cases with and without mooring system. In spite of the large difference in the acceleration response, the bending moment in the island is of same order for wind-waves and seaquakes.

Figure 10 shows the effect of hydrodynamic radiation damping on the displacement and acceleration responses for the case without mooring system. The displacement is considerably reduced by hydrodynamic radiation damping for both wind-waves and seaquakes. On the other hand, the effect of hydrodynamic radiation damping on the acceleration response is negligibly small for seaquakes, although it is still large for wind-waves.

10 CONCLUSIONS

A unified approach is presented to evaluate the stochastic responses of a flexible circular floating island subjected to wind-waves and seaquakes, taking into account structural flexibility and fluid-structure interaction. Hydrodynamic pressure acting on the lower surface of the vibrating island is obtained in closed form both for wind-waves and seaquakes. The modal equations of motion are derived by energy method. The response statistics are calculated using a wet-mode superposition approach and stationary random vibration theory.

Based on the numerical results, the dynamic behaviors of large floating islands are summarized as follows:

1   When the island is freely floating, the elastic deformation is predominant for both wind-waves and seaquake. However, once the mooring system is attached to the island, the rigid body motion becomes dominated.

2   The effect of added mass is varied due to the existence of sea-bed. If the water depth becomes shallower, the added mass effect increases

441

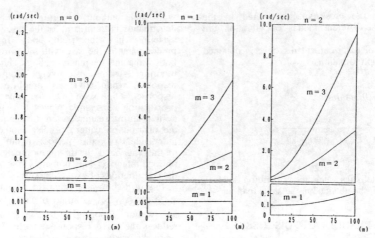

Figure 4 Variation in wet-mode frequencies due to island thickness

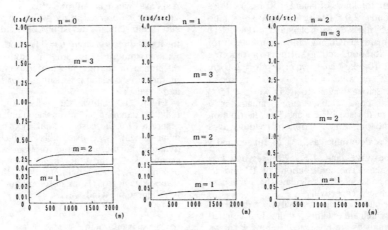

Figure 5 Variation in wet-mode frequencies due to water depth

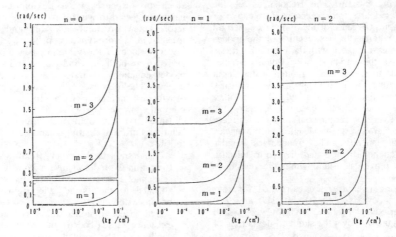

Figure 6 Variation in wet-mode frequencies due to anchor stiffness

442

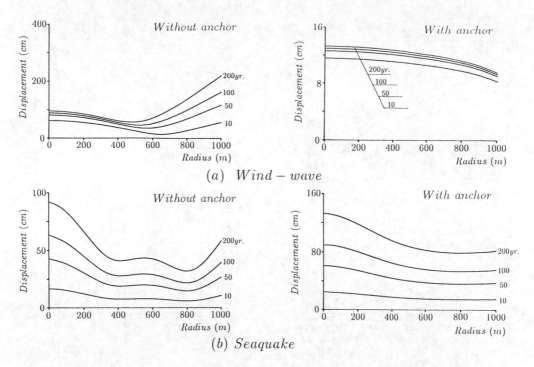

Figure 7 Standard deviations of displacement response for different return periods

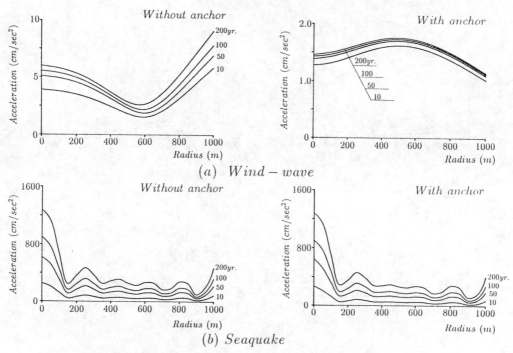

Figure 8 Standard deviations of acceleration response for different return periods

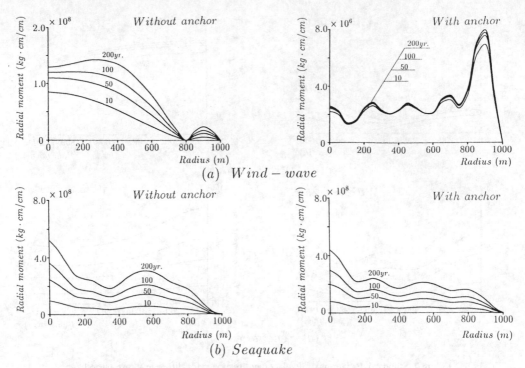

(a) *Wind − wave*

(b) *Seaquake*

Figure 9 Standard deviations of bending moment response for different return periods

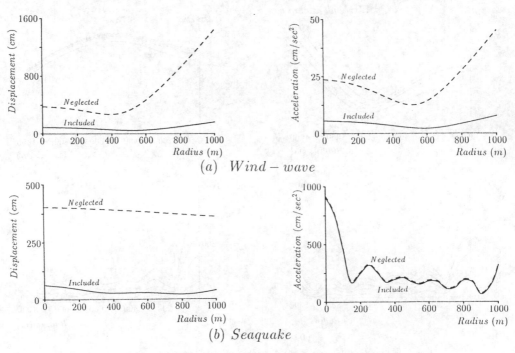

(a) *Wind − wave*

(b) *Seaquake*

Figure 10 Effects of hydrodynamic radiation damping on the response

especially for lower modes of vibration.

3   The effect of hydrodynamic radiation damping on the response is large for lower modes of vibration. It reduces both displacement and acceleration responses in the wave response, whereas it does not reduce the acceleration but the displacement in the seaquake response.

4   Distributed tension-legs are very effective to reduce the response statistics under wind-waves, whereas the seaquake responses are occationally amplified by the attachment of tension-legs.

5   In the acceleration response, only lower modes of vibration are excited for wind-waves, whereas higher-modes of vibration are significantly excited for seaquakes.

6   The internal forces during seaquakes does not become so large owing to the dominance of rigid body motion, although seaquakes induce much higher acceleration response than wind-waves.

REFERENCES

Ambraseys, N., 1985, "A Damaging Seaquake", Earthquake Engineering and Structural Dynamics, Vol.13, pp.421-424.

Babu, P.V.T., Reddy, D.V., 1986, "Dynamic Coupled Fluid-Structure Interaction Analysis of Flexible Floating Platforms", J. of Energy Resources Technology, ASME, Vol.108, pp.297-304.

Clough, R.W., Penzien, J., 1975, "Dynamics of Structures", McGraw-Hills.

Cornell, C.A., 1968, "Engineering Seismic Risk Analysis", BSSA, 58(5), pp.1583-1606.

Davenport, A.G., 1967, "The Dependence of Wind Loads on Meteorological Parameters", in 'Wind Effects on Buildings and Structures, Vol.1', pp.19-82.

Garret, C.R.J., 1971, "Wave Forces on a Circular Dock", J. of Fluid Mechanics, Vol.46, pp.129-139.

Hamamoto, T., Tanaka,Y., 1992a, "Coupled Free Vibrational Characteristics of Artificial Floating Islands", J. of Struct. Constr. Engng., AIJ, No.438, pp.165-177, (in Japanese).

Hamamoto, T., Tanaka,Y., 1992b, "Response Behavior of Artificial Floating Islands Subjected to Wind-Induced Waves", J. of Struct. Constr. Engng., AIJ, No.442, pp.157-167, (in Japanese).

Hamamoto, T., Tanaka,Y., 1993, "Response Behavior of Artificial Floating Islands Subjected to Seaquakes", J. of Struct. Constr. Engng., AIJ, No.448, pp.173-185, (in Japanese).

Ijima, T., Tabuchi, M., and Yumura, Y., 1972, "On the Motion of a Floating Circular Cylinder in Water of Finite Depth", Trans. JCE, No.206, pp.71-84.

Liou, G-S., Penzien, J., Yeung, R.W., 1988, "Response of Tension-Leg Platforms to Vertical Seismic Excitation", Earthquake Engineering and Structural Dynamics, Vol.16, pp.157-182.

Ozaki, M., Kitagawa, Y., Hattori, S., "Study on Regional Characteristics of Earthquake Motions in Japan (Part 1)", Trans. of AIJ, No.266, pp.31-40, (in Japanese).

Pierson, W.J., Moskowitz, Z., 1964, "A Proposed Spectral Form for Fully Developed Wind Seas Based on the Similarity Theory of S.A. Kitaigorodoskii", J. of Geophysical Research, Vol.69, pp. 5158-5190.

Timoshenko, S. P., 1955, "Vibration Problems in Engineering", 3rd. ed., D. Van Nostrand.

Warburton, G. B., 1954, "The Vibration of Rectangular Plates", Proc. Institute of Mechanical Engineers, Vol.168, pp.371-384.

Wen, Y-K., 1974, "Interaction of Ocean Waves with Floating Plate", ASCE, Vol.100, No.EM2, pp.375-395.

Hydroelasticity in Marine Technology, Faltinsen et al. (eds) © 1994 Balkema, Rotterdam, ISBN 90 5410 387 6

Author index

HYDROELASTICITY IN MARINE TECHNOLOGY

PROCEEDINGS OF THE INTERNATIONAL CONFERENCE ON HYDROELASTICITY
IN MARINE TECHNOLOGY / TRONDHEIM / NORWAY / 25-27 MAY 1994

Hydroelasticity in Marine Technology

Edited by

O.FALTINSEN, C.M.LARSEN & T.MOAN
The Norwegian Institute of Technology

K.HOLDEN
MARINTEK

N.SPIDSØE
SINTEF

A.A.BALKEMA / ROTTERDAM / BROOKFIELD / 1994

Published by
A.A. Balkema, P.O. Box 1675, 3000 BR Rotterdam, Netherlands
A.A. Balkema Publishers, Old Post Road, Brookfield, VT 05036, USA

ISBN 90 5410 387 6
© 1994 A.A. Balkema, Rotterdam
Printed in the Netherlands